Conjugate Heat and Mass Transfer in Heat Mass Exchanger Ducts

Li-Zhi Zhang

ELSEVIER

AMSTERDAM • BOSTON • HEIDELBERG • LONDON
NEW YORK • OXFORD • PARIS • SAN DIEGO
SAN FRANCISCO • SINGAPORE • SYDNEY • TOKYO
Academic Press is an imprint of Elsevier

AP

Academic Press is an imprint of Elsevier
225 Wyman Street, Waltham, MA 02451, USA
The Boulevard, Langford Lane, Kidlington, Oxford, OX5 1GB, UK

First edition 2014

Library of Congress Cataloging-in-Publication Data
A catalog record for this book is available from the Library of Congress

British Library Cataloguing in Publication Data
A catalogue record for this book is available from the British Library

ISBN: 978-0-12-407782-9

For information on all Academic Press publications
visit our website at http://store.elsevier.com

Printed and bound in United States of America
14 15 16 17 18 10 9 8 7 6 5 4 3 2 1

Contents

Preface

In modern society, new materials and new processes are required to solve the emerging new energy and environmental problems. Increasingly, novel heat and mass exchangers are being invented and used in various industries, for instance, in energy recovery, gas separations, air purification, air cleaning, waste water treatment, liquids separations, air dehumidification, and so on. Adsorbent beds, desiccant wheels, membrane contactors, membrane total heat exchangers, and air cleaners all belong to this type of heat and mass exchangers.

Ducts are the basic elements in these heat and mass exchangers. In this book, a systematic description of the conjugate heat and mass transfer in such ducts is presented. The basic physical phenomena in these ducts are described. The methods to treat the conjugate boundaries, either steady-state or transient, are introduced. The detailed transport data for ducts of various heat and mass exchangers are presented: adsorbent beds, desiccant wheels, membrane total heat exchangers, membrane contactors, either of parallel-plate type or of hollow fiber bundle type. The book also illustrates some examples of the applications of novel heat and mass exchangers, with theoretical analysis combined with experimental work: for instance, air dehumidification by honeycomb adsorbent beds, air dehumidification and energy recovery by desiccant wheels, heat and moisture recovery by membrane total heat exchangers, air humidification by hollow fiber membrane contactors, air dehumidification by salt solutions with parallel-plate, or hollow fiber membrane contactors. Both the working theories and the heat and mass transfer problems are addressed. The exchangers are constructed and their performances are optimized with the obtained heat and mass transfer data. This book provides a reference book for the design of these novel heat and mass exchangers.

The book combines theoretical analysis with engineering practices. It covers a wide range of knowledge from fundamental heat mass transfer to novel system design and performance analysis. It also gives deep insights and design guidelines for the heat and mass exchangers used in other industries such as air conditioning, energy engineering, mechanical engineering, chemical engineering, environmental engineering, food, and drug engineering, etc.

This is the first book systematically presenting conjugated heat and mass transfer in the ducts of various heat and mass exchangers. Various duct cross-sectional shapes are included. The book may also serve as a reference book for scientists, engineers, practitioners, and students in energy, chemical, production, and environmental industries for component design.

Acknowledgments

After 15 years accumulating knowledge and expertise, I began to write this book. After 8 months of preparation and hard work, I am relieved that it is now completed. It is like a new born son to me. I am full of joy, but more, a sense of responsibility. I hope readers can share with me the knowledge in this book, as well as the heart of curiosity and the feeling of accomplishment in finishing this work.

This book is a cross-disciplinary endeavor which relies heavily on numerical heat mass transfer. I hope peer engineers, scientists, and research students will benefit from the methodologies exhibited in this analysis and extend them to analyses of other energy and environmental systems. Your feedback of suggestions and comments, as well as criticisms is also valuable. Readers' support is critical to the success of my continuing writings.

I am deeply grateful to the Academician of the Chinese Academy of Sciences, Prof. Wenquan Tao at, Xi'an Jiaotong University, for the precious time that he squeezed from his tense schedule to write a Foreword for this book which has become an introduction to the content. I would also like to thank the Natural Science Foundation of China (NSFC) for their continuing financial support. I am indebted to my colleagues at South China University of Technology, and others all over China and throughout the world who have provided suggestions and ideas which, in no small way, have contributed to the fabric of this text. They all contributed a great deal to the results described in this book. I am also grateful to my colleagues at South China University of Technology and elsewhere, who have provided positive reinforcement for my efforts.

Li-Zhi Zhang
South China University of Technology
Guangzhou, China
Email: Lzzhang@scut.edu.cn

Foreword

Energy and environment are two critical issues today. Mankind is facing increasing energy and environmental problems. New materials and new processes are required to solve these problems. More frequently than before, novel heat and mass exchangers are being invented and applied in various industries; for instance, in energy recovery, gas separations, air purifications, air cleaners, waste water treatment, liquid separations, air dehumidification, toxic gas and liquid disposals, and so on. Depending on the application fields, these exchangers can be adsorbent beds, desiccant wheels, membrane contactors, membrane total heat exchangers, or cleaners. Regardless of the names, they all belong to these types of exchangers.

For these new exchangers, the ducts are the basic elements where the processes take place. Heat and mass transfer usually accompany each other. Simultaneous heat and mass transfer in the ducts are the fundamental and basic phenomena in these novel heat and mass exchangers. Therefore, the data on heat and mass transfer in these ducts are the fundamentals for component design, system optimization, and performance evaluation.

Compared to traditional metal-type heat exchangers, there are special features in these new heat and mass transfer ducts:

1. The solid walls have new functional materials. The heat and mass transfer in fluids are closely related to those in the solid walls. So, they are conjugate heat and mass transfer problems. The governing equations for the fluids and the solids should be solved together; simultaneously.
2. The Biot numbers of these ducts are high. They are large Biot number ducts. So, material properties have decisive impacts on the conjugate heat and mass transfer in ducts. The thermochemical and interfacial characteristics of wall materials, which are new and always nonlinear, should be considered, in addition to thermal and fluid-dynamics problems, in the fluids.
3. The duct geometries are always irregular. The simple analytical solutions of fluid flow for common round or rectangular tubes are no longer applicable.
4. There are multi-variables that need to be solved. Velocity, temperature, concentrations are usually coupled together. The complexities also include chemical reactions, sometimes in the solids.

These special features have led to many new phenomena:

1. The boundary conditions on duct walls have neither uniform values nor uniform fluxes. Rather, they are formed by the coupling of the fluids and materials.
2. It is often difficult to find analytical solutions. Therefore numerical schemes are highly depended on. Sometimes, new numerical schemes are required to solve a specific problem.
3. Transport data in the entry regions should be considered.
4. The final Nusselt and Sherwood numbers are highly dependent on materials, duct geometries, and operating conditions. The traditional way of estimating mass transfer coefficients from heat transfer coefficients with analogy correlations is not valid anymore.

Due to these new features and new difficulties, a comprehensive analysis and data collection of the conjugate heat and mass transfer in ducts are highly desired, both for industries and for academics.

Regretfully, such studies are still very limited now. It is ridiculous that nowadays, transport data in old textbooks that are proposed for traditional heat transfer-only in simple cases, are usually adopted in the design and optimization of emerging novel heat and mass exchangers. This is a dilemma facing us.

Over the past 15 years, the author of this book, Professor Li-Zhi Zhang, has conducted a systematic study of the conjugate heat and mass transfer in the ducts of heat and mass exchangers. The basic transport phenomena in the ducts were investigated and new data on Nusselt and Sherwood numbers for these ducts were disclosed. He published many papers in this area. However, the data on this topic is quite sporadic and a systematic introduction is still not available. From this background Professor Zhang wrote this book: *Conjugate Heat and Mass Transfer in Heat Mass Exchanger Ducts*.

This is the first book that systematically presents the conjugated heat and mass transfer in ducts of various heat and mass exchangers. The basic physical phenomena in these ducts are introduced. The governing equations for mass, momentum, and energy are presented and solved. Detailed numerical schemes to solve these conjugate problems are given. The detailed finite difference forms to solve the equations are specifically presented. The techniques of coordinate transformations from the physical domain to computational domain on various duct cross-sections are elucidated. The final algebraic forms of the partial differential equations are presented. The iterative procedures to solve these multi-elements and non-linear equations are also introduced. The methods to treat the conjugate boundaries, either steady state or transient, are introduced. Most of them are from the experiences of the author in solving them, besides the knowledge from classical textbooks. Experimental procedures and results are presented to validate the calculation results with various practical heat and mass exchangers. To some extent, this is also a reference book for numerical heat and mass transfer. Finally the detailed transport data for duct of various heat and mass exchangers is presented: adsorbent beds, desiccant wheels, membrane total heat exchangers, membrane contactors, either in parallel-plates type, plate-fin type, plate-fin and tube type, or in hollow fiber bundles type.

Based on the fundamental heat mass transfer data, this book illustrates some examples of the applications of novel heat and mass exchangers. For instance, air dehumidification by honeycomb adsorbent beds, air dehumidification and energy recovery by desiccant wheels, heat and moisture recovery by membrane total heat exchangers, air humidification by hollow fiber membrane contactors, air dehumidification by solutions with parallel-plates or hollow fiber membrane contactors. Both the working theories and the heat and mass transfer problems are discussed. The exchangers are constructed and their performances are optimized with the obtained heat and mass transfer data.

This book combines theoretical analysis with engineering practices. It covers a wide range of knowledge from fundamental heat mass transfer, to novel systems design and performance analysis, from materials introduction, characterization, to heat and mass exchanger thermodynamics and fluid dynamics. It can also provide insights and design guidelines for the other heat and mass exchangers, in various industries like air conditioning, energy engineering, mechanical engineering, chemical engineering, environmental engineering, food and drug engineering, etc.

This book is a cross-disciplinary endeavor which relies heavily on numerical heat mass transfer. I hope peer researchers, engineers, and research students could benefit from the methodologies exhibited in this analysis and extend them to the analyses of other energy and environmental systems.

At last, let me extend my congratulations to Professor Zhang on his new book. I hope he can continue his researches on this topic, which in no doubt, will benefit both industrial progress and academic developments.

Professor Wenquan Tao
Academician of the Chinese Academy of Science
Xi'an Jiaotong University
Xi'an, China
Email: wqtao@mail.xjtu.edu.cn

CHAPTER

1 An Introduction to Conjugate Heat and Mass Transfer in Ducts

1.1 Heat and mass transfer ducts

Energy and the environment are of increasing concern today, because mankind is facing greater energy and environmental problems than ever before. According to a report released recently [1], world energy consumption is projected to increase by 47% from 2010 through 2035. Most of the growth is projected for emerging economies outside the OECD (Organization for Economic Cooperation and Development), where robust economic growth is accompanied by increased demand for energy. Total non-OECD energy use grows by 72%, compared with an 18% increase in OECD energy use. Energy consumption in non-OECD Asia shows the most robust growth among the non-OECD regions, rising by 91% from 2010 to 2035. However, strong growth also occurs in many of the rest of the non-OECD regions: 69% in Central and South America, 65% in Africa, and 62% in the Middle East. Accompanying energy use, global carbon emissions from fossil fuels have significantly increased since 1900. Emissions increased by over 16 times between 1900 and 2008 and by about 1.5 times between 1990 and 2008. It is projected that the trend will continue for the next decades. These two factors have led to many energy and environmental problems such as fossil fuel depletion, global warming, ozone depletion, air pollution, new diseases, etc. The increased concerns with regard to energy and environmental problems have placed demands on new technologies.

To date, many new technologies are emerging to help to solve these energy and environmental problems. New types of equipment are invented to reclaim waste heat and to increase energy utilization efficiency; for example, total heat exchangers, desiccant wheels, etc. Other new equipment is constructed to get rid of air pollution from our environment, from either air or water, for example air cleaners and water cleaners. The basic processes in these types of equipment are heat and mass transfer. The nature of these processes is that substances and energy are transported from one place to another, and from one medium to another. Energy and pollutants are treated, transferred and processed in terms of heat and mass transfer. Without heat and mass transfer, energy and environmental problems cannot be controlled. All these novel forms of equipment can be called heat and mass exchangers.

Ducts are the basic units in these heat and mass exchangers. There is no site to accomplish heat and mass transfer without these ducts. They provide the interfaces for gases, solids and liquids to make contact and be transported. These ducts are usually very fine and are numerous, to provide sufficient area for gas-solid or gas-liquid contact. So the basic transport phenomenon in these small ducts, or channels, is the key for system design and performance optimization. Here are some examples.

Figure 1.1 shows a total heat exchanger unit for the reclamation of waste heat from buildings. The whole machine includes two fans, the ducting work for the two flows and the sealing walls separating the two flows, and the shell. The heart of the machine is the core. Figure 1.2 shows the core inserted in the total heat exchanger [2]. It is in either a parallel-plate ducts or a plate-fin ducts structure. Figure 1.3 shows a schematic of the heat and moisture exchange between the outside air and the indoor air through these numerous ducts. It is a cross-flow membrane based total heat exchanger. Because the heat and mass transfer between a solid surface and an air stream is usually very low, to achieve the required magnitude of treatment capability, large contact areas between the two flows are required. Therefore the cores are made of numerous fine channels.

With total heat exchangers, the efficiency of the existing HVAC systems can be improved substantially. With them, 20–40% of the energy use for air conditioning can be saved. The reason is that normally the fresh air is dehumidified by through condensation in a cooling coil followed by a reheating process, which is very energy intensive. This energy can be saved if total heat exchangers are installed to reduce the dehumidification load. Besides energy conservation, total heat exchangers have the additional benefits of ensuring a sufficient supply of fresh air, which is crucial for the prevention of epidemic respiratory diseases such as SARS and bird flu. As can be seen from Figure 1.2, the exchanger is composed of numerous ducts. Fresh air and exhaust air exchange heat and moisture simultaneously in these ducts. In this way, fresh air is dehumidified and cooled by the exhaust air in summer and heated and humidified by the exhaust air in winter. The hot and humid incoming fresh air becomes cool and dry in summer, thus decreasing the cooling load in summer. The cold and dry incoming fresh air is heated and humidified in winter, thus decreasing the heating load in winter. The ducts are the basic units for heat and moisture exchange between fresh air and exhaust air in a total heat exchanger. Two

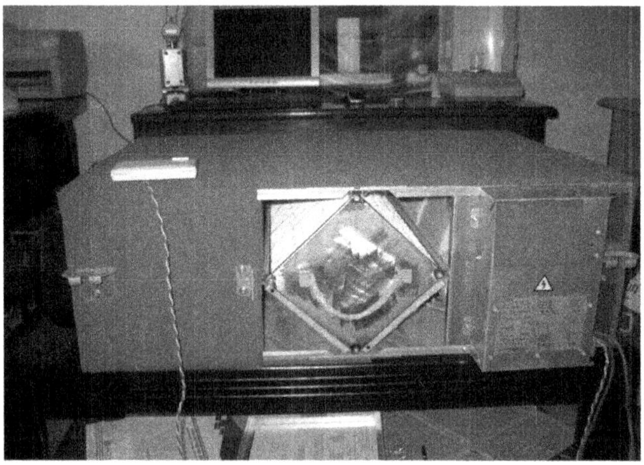

FIGURE 1.1

A total heat exchanger for heat and moisture recovery.

FIGURE 1.2

The core inserted in the total heat exchanger.

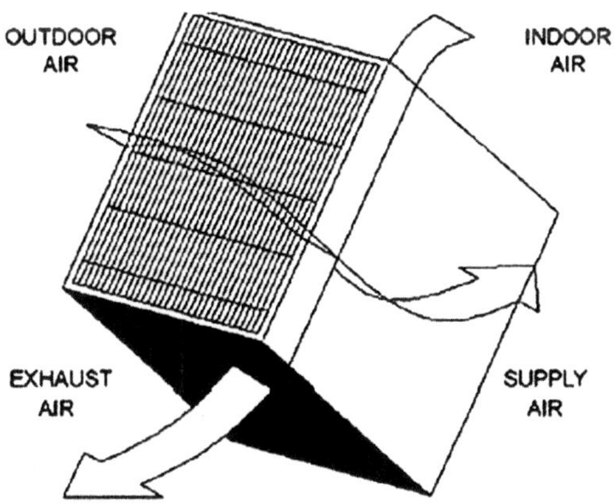

FIGURE 1.3

Schematic of heat and mass transfer through ducts of a total heat exchanger.

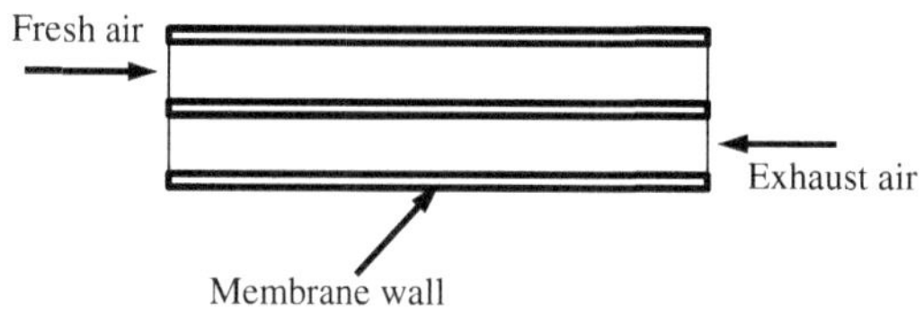

FIGURE 1.4

Two neighboring representative membrane-formed ducts in a membrane total heat exchanger.

neighboring membrane-formed ducts in the core are shown in Figure 1.4. They are of the counter-flow parallel-plate type. The ducts are composed of air flows and membrane walls. The membranes, though very thin, are closely related to the heat and mass transfer of air flows. Fluid dynamics and transport data in these membrane-formed ducts are the key issues. System design is possible only when the fundamentals of heat and mass transfer in the ducts are known.

Adsorbent beds have been extensively used in many applications, such as air purification, waste water treatment, gas separations, adsorption cooling, etc. They can be classified into two categories: fixed beds, and desiccant wheels or moveable beds [3]. Figure 1.5 shows a honeycomb type adsorbent bed for a multi-bed dehumidification system and Figure 1.6 shows a desiccant wheel. Both have very high packing densities and large contact areas.

The fixed beds and the desiccant wheels are both made of adsorptive materials in the form of honeycomb channels. There are numerous small ducts in the beds or wheels. For instance for the adsorbent bed in Figure 1.5, the parameters are: dimensions, $555 \times 415 \times 35\,mm^3$; total weight, 3528 g; effective adsorbent weight, 1141 g; contact area, $7.24\,m^2$; packing density, $898\,m^2/m^3$; duct geometry, $2 \times 1.5 \times 30\,mm^3$. A system with 10 such beds can handle and dehumidify $500\,m^3/h$ of fresh air. Five beds are cascaded together to dehumidify air, while another five are cascaded to be regenerated by the regeneration air simultaneously. The process air stream flows through these mini-channels. They

FIGURE 1.5

A fixed honeycomb-type adsorbent bed for multi-bed dehumidification.

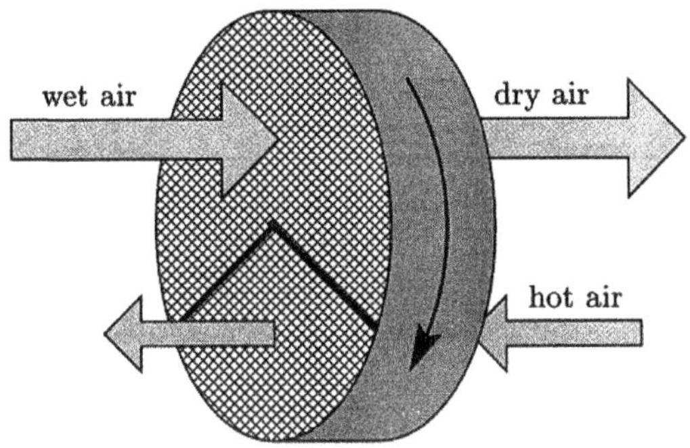

FIGURE 1.6

A desiccant wheel.

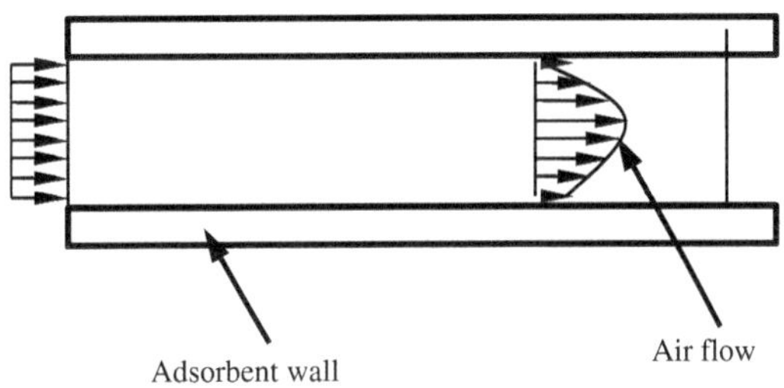

FIGURE 1.7

A representative duct in an adsorbent bed or desiccant wheel.

react with the solid walls, which are highly adsorptive. This is called the adsorption mode. The whole operation is dynamic and the materials work in transient mode. Their temperatures, humidity, and other properties change with time if necessary. When the solid walls become saturated, a regeneration air stream is passed through the ducts, and the working mode is switched to regeneration mode. The fixed bed relies on the alternate switching between process air and regeneration air to achieve continuous working. In comparison to a fixed bed, the desiccant wheel relies on the wheel revolving between the process air and regeneration air to achieve continuous working. A representative adsorptive duct in an adsorbent bed or desiccant wheel is shown in Figure 1.7.

Due to the adsorptive nature, there is strong simultaneous heat and mass transfer in such a duct. The wall materials and the air stream are closely coupled together. Heat transfer would lead to mass transfer, and mass transport would lead to heat generation or dissipation. The basic data of heat and mass transfer in such a duct are the key issue influencing system performance. The knowledge of this is the basis for the design, quantification and optimization of the adsorbent beds or desiccant wheels. For such ducts, solid materials constitute the main resistance in mass transfer, while heat transfer resistance is small.

Membrane contactors are another category of heat and mass transfer equipment. They have been extensively used in air purification, water treatment, gas separation, air dehumidification, and chemical production. They have the benefits of selective permeation of some desired substances. They are compact, and easy to maintain. The packing density can be as high as $1000\,m^2/m^3$, so the treatment capability is rather high. The membranes used for substance separation can be made into plate membranes or hollow fibers. Figure 1.8 shows a hollow fiber membrane module for air humidification [4]. It is well known that air dehumidification is a critical task, but air humidification is equally important in dry seasons. In winter when the rooms are heated, the indoor air is quite dry. Occupants would feel very uncomfortable, therefore it is necessary to have air humidification. In fuel cells, H_2 should be humidified before it is used in electrochemical reactions to generate electricity. Humidification of H_2 is a prerequisite for fuel cell applications. The module in Figure 1.8 can be used for air humidification. According to this concept, hollow fibers are packed together and assembled in a plastic shell to form a shell-and-tube structure. Water flows inside the fibers, while dry air flows across the fiber bank. Moisture permeates from the water flow to the air stream. The membranes, which are in hollow fibers, can prevent the leaking of liquid water to the air stream. However, they permit the permeation of moisture from water to air. In this way, the air is humidified. There are numerous fibers in a membrane module. Membrane-formed ducts are also the basic elements for heat and mass transfer. They provide the area for moisture permeation. Figure 1.9 shows a representative repetitive cell in a staggered

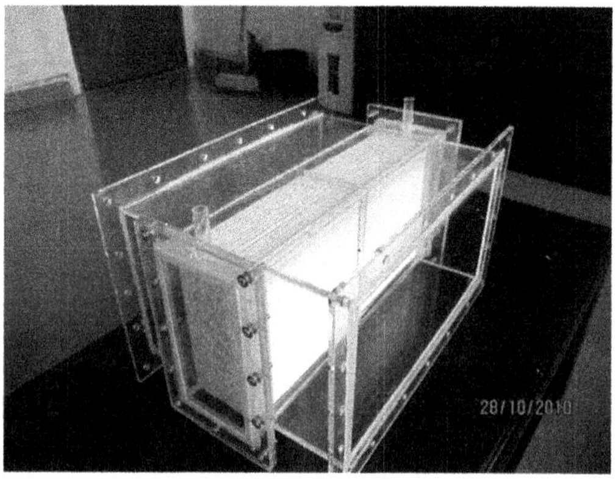

FIGURE 1.8

A hollow fiber membrane module for air humidification.

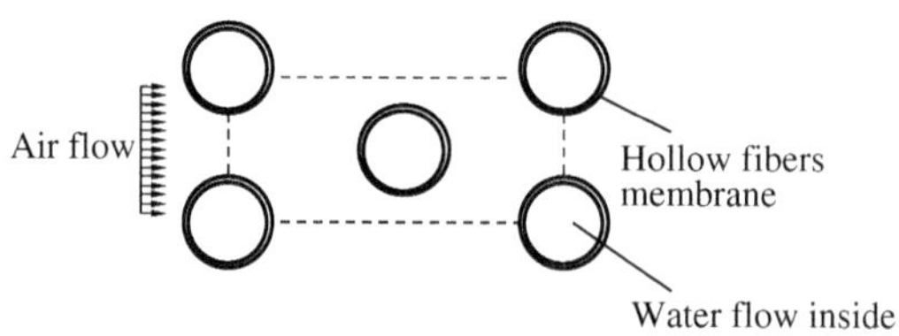

FIGURE 1.9

A representative cell in the hollow fiber membrane module. Water flows inside the fibers, and the air stream flows in the duct surrounded by the fibers. The equivalent air duct is surrounded by the dashed line.

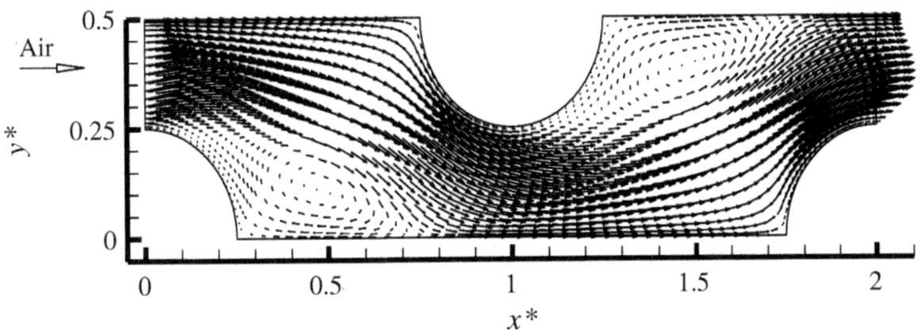

FIGURE 1.10

The velocity vectors in the hypothetical duct. Flow fields in a duct surrounded by staggered arrayed fibers. x^* and y^* are dimensionless transverse and longitudinal coordinates.

aligned hollow fiber membrane bundle [5]. As seen, it is composed of two ducts: the inside duct and the outside duct. The water flows in the inside duct. It is a common flow in a round tube. Air flows in the external hypothetical duct formed by the fibers, surrounded by the dashed lines. In this respect, it is like a representative duct in a heat transfer tube. However, the heat and mass transfer in such a duct formed by membranes is more complex. Besides the complex flow pattern, heat and mass transfer in the air duct is closely related to the membrane materials, which in turn are closely related to the water properties in the fibers. So it is a three-phase conjugate problem: water, membrane, and air. The transport phenomenon in such a cell plays a key role in humidification performance. Due to the complexity, an analytical solution of such a three-phase (or three-element) system is always impossible. Therefore numerical methodology is highly relied on to find the solution. Figure 1.10 shows the calculated velocity fields in the air duct [5].

1.2 Special features of heat and mass transfer in ducts

As mentioned, the conjugate problems in these typical heat and mass exchangers are rather complex. It is usually difficult to solve these problems with analytical methods. Numerical methodology provides a useful tool for performance predictions. The requirements in numerical methods also lie in the fact that there are special features in heat and mass transfer in the ducts of heat and mass exchangers.

1.2.1 Irregular geometry

Simple and regular cross sections such as circular and rectangular ducts are not popular in novel heat and mass exchangers. Though simple, their air-side convective heat and mass transfer coefficients are not high. To augment convective heat and mass transfer coefficients, ducts of irregular cross section are generally used. The most commonly used duct cross sections include sinusoidal, triangular, and hexagonal. Figure 1.11 shows the most commonly used duct cross sections for heat and mass exchangers. Some other novel structures are being developed. For instance, Figure 1.12 shows a novel structure for a total heat exchanger, the so-called cross corrugated triangular duct [6]. In this structure, parallel plain plates are first corrugated to form wave-like plates. The corrugated plates are then stacked together with

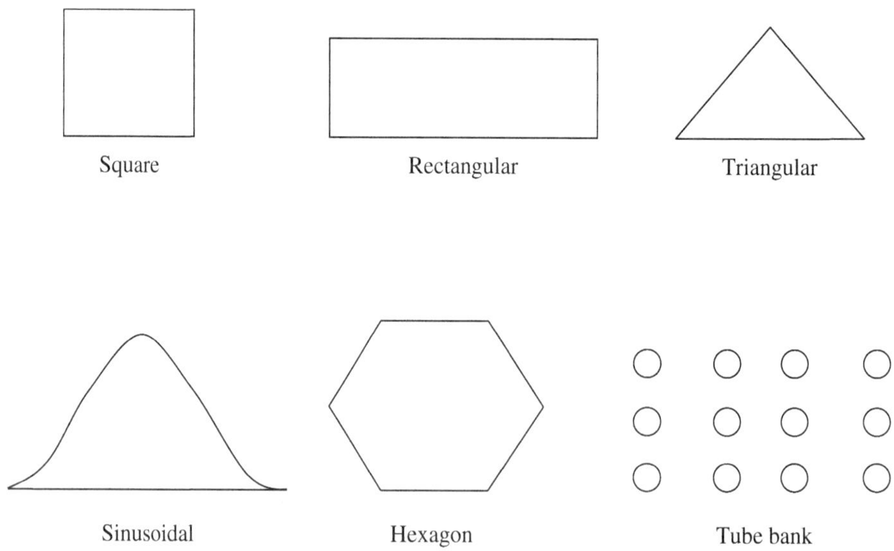

FIGURE 1.11

The most commonly used cross sections for ducts.

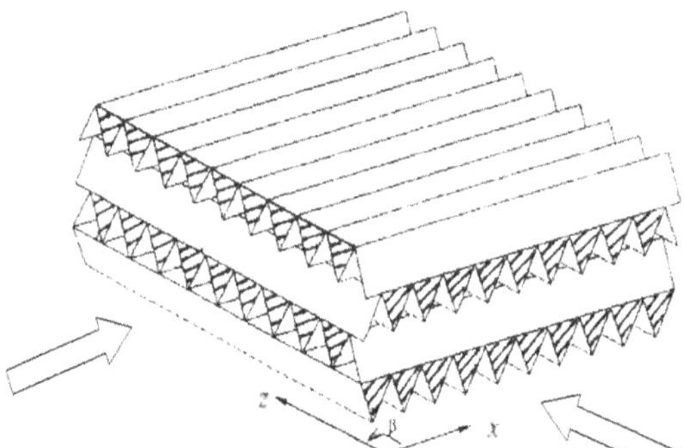

FIGURE 1.12

A cross-corrugated triangular duct for a total heat exchanger.

neighboring plates oriented 90° to each other. In this way, a contraction-expansion structured duct is formed. Air flow and heat and mass transfer in the duct are intensified. Further, the structure is compact. Its mechanical strength is rather high.

Sinusoidal ducts are very common. The sine fins are naturally formed by the mechanical pressing of plain plates into corrugations. When the plates and the corrugated fins are stacked together, sinusoidal plate-fin ducts are formed. They have high packing density. The contact area between air and solid is very high. Further, the mechanical strength of this honeycomb structure is very great, even with very thin plates and fins. They are easy to produce on a large scale by machine, and the cost is low. The reliability is very high. For these reasons, the sinusoidal plate-fin duct structure has become the most popular structure in heat and mass exchangers, including total heat exchangers, desiccant wheels, adsorbent beds, etc.

The cross sections of the ducts are irregular. It's very difficult to obtain the velocity fields, temperature fields, and concentration fields in the ducts, even with ideal walls of uniform temperature boundary conditions. Therefore, in the past, the effects of the irregular cross sections were usually ignored. Heat and mass transfer were assumed to take place between a stream and a rectangular plate. The details in the duct could not be included. Numerical methods are always the only way to solve fluid flow and heat and mass transfer problems in these ducts of irregular cross section.

1.2.2 Multi-element systems

As mentioned for Figure 1.10, a duct in a hollow fiber membrane contactor is a three-element system. Similarly, most of the ducts in heat and mass exchangers are three-element systems as well. For instance, a duct in a total heat exchanger is an air-membrane-air system. A duct in an adsorbent bed is an air-solid adsorbent two-element system. The elements in these systems are closely related. The temperature and concentration values on solid surfaces are co-determined by the fluids and the materials. Solution of these elements simultaneously is not an easy task and relies heavily on numerical techniques.

1.2.3 Coupled heat and mass transfer

Heat and mass transfer in these ducts are closely related to each other. Adsorption is an exothermic process while desorption is an endothermic process. Mass transfer would lead to heat transport, and vice versa. Equations for fluid flow, heat transfer, and mass transfer should be combined together and solved simultaneously. The set of partial differential equations is rather complex without simplifications. The problem can only be solved by numerical methods. Further, the mass fluxes on material surfaces are rather inhomogeneous, which leads to inhomogeneous distributions of both temperature boundary conditions and concentration boundary conditions. Predictions of the heat and mass transfer data in these ducts under such complex boundary conditions are quite difficult. Numerical calculations can provide qualitative and quantitative results. Figures 1.13 and 1.14 show the boundary conditions for temperature and concentration on a membrane surface in a total heat exchanger [7], respectively. As can be seen, they are quite non-uniform.

1.2.4 Closely conjugated with materials

The solid walls in heat and mass exchangers are not common metals. They are functional materials to perform certain functions, such as adsorption or permeation. A typical duct for heat and mass transfer is shown in Figure 1.15. Similar to a common heat transfer tube, velocity, temperature, and humidity fields develop as the fluid enters the tube. The difference is that the phenomenon is more complex. The fluid has close relations with the solid materials. The thermophysical and interfacial properties of solid materials, such as adsorption properties, porosities, heat conductivities, contact angles, etc., all have tremendous impacts on the convective heat and mass transfer characteristics in the duct.

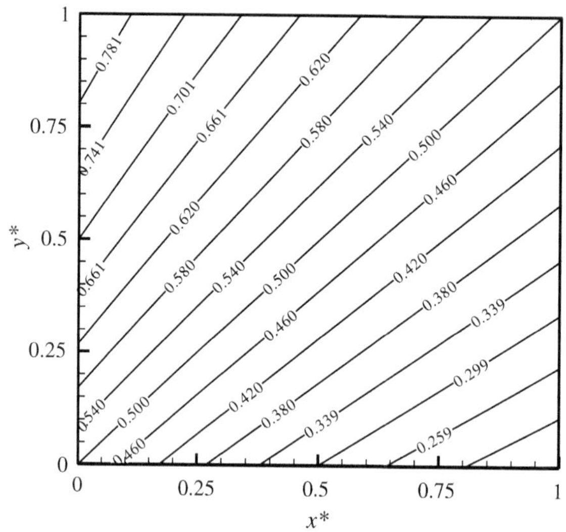

FIGURE 1.13

Temperature distributions on the membrane surface in a total heat exchanger. The values are dimensionless temperatures; x^* and y^* are dimensionless coordinates.

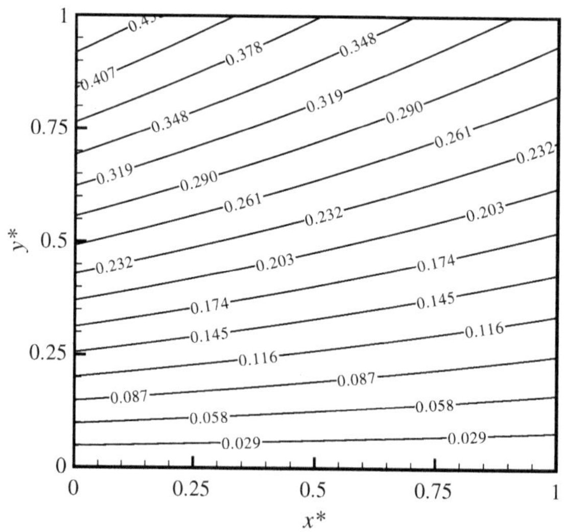

FIGURE 1.14

Humidity distributions on the membrane surface in a total heat exchanger. The values are dimensionless humidity on the exhaust air side; x^* and y^* are dimensionless coordinates.

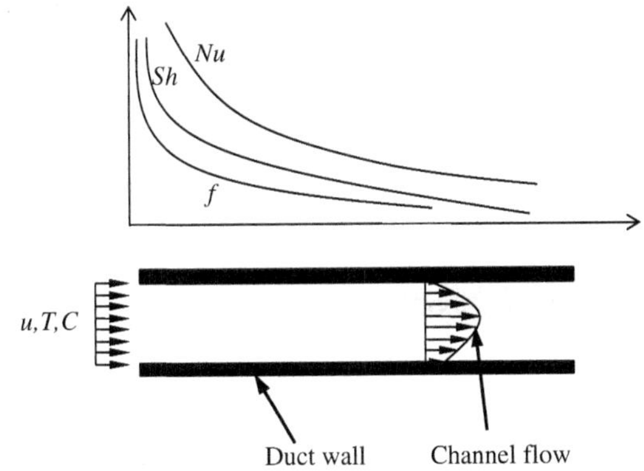

FIGURE 1.15

Development of fluid flow, heat and mass boundaries in a duct for heat and mass transfer.

The difficulties also lie in the fact that the materials in novel heat and mass exchangers are new. They have been newly fabricated or recently developed. In most cases, their thermophysical and interfacial properties are unknown to heat transfer scientists. So before the work on numerical heat transfer, many pre-investigations of the materials are necessary to discover the thermophysical and interfacial properties of the duct wall materials. This is a cross-disciplinary work which requires not only know-how and expertise, but also hard labor. Only after much time, effort and many observations using special equipment can the properties be revealed. Figures 1.16 to 1.18 show the thermophysical and interfacial properties of some duct wall materials. Figure 1.16 shows the adsorption and desorption properties of an adsorbent materials. As seen, a hysteresis phenomenon is clearly observed. The

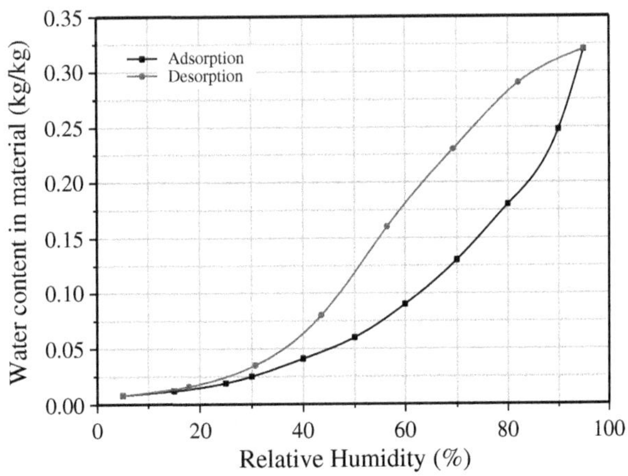

FIGURE 1.16

Adsorption and desorption isotherms of a wall material measured by an adsorption meter.

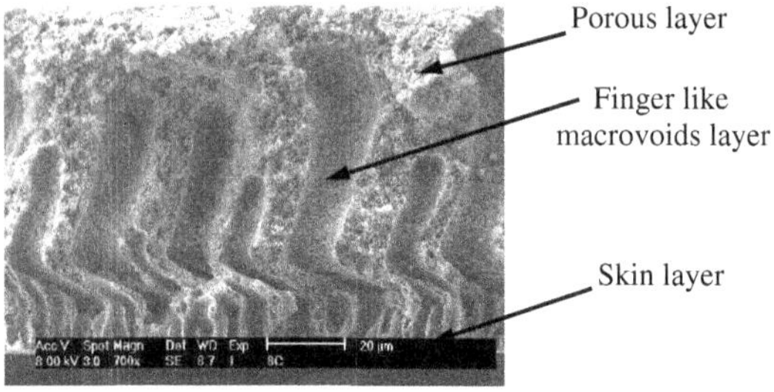

FIGURE 1.17

SEM (scanning electron micrograph) graph of the cross section of a membrane.

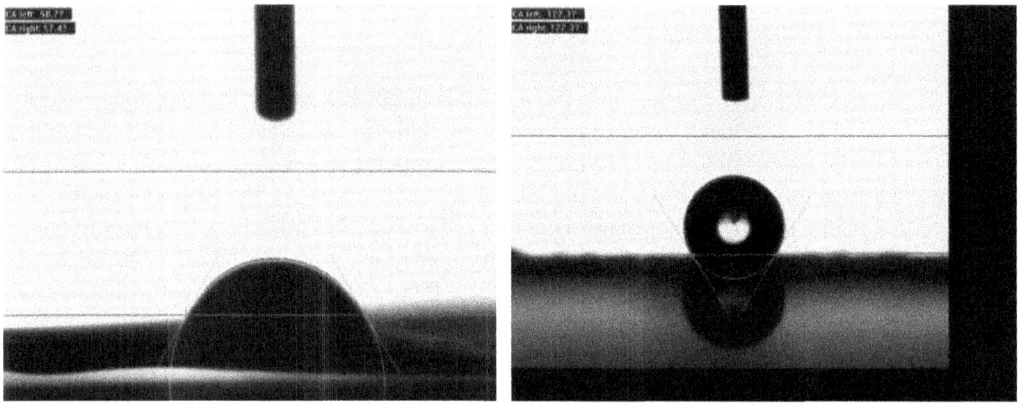

FIGURE 1.18

Contact angles measured by a contact angle meter. One is hydrophilic and the other is hydrophobic.

thermophysical properties are usually highly nonlinear. They change with temperature and operating conditions. Figure 1.17 is an SEM (scanning electron micrograph) picture of a composite membrane. The complex structures in the membranes are revealed by fractal theories. Figure 1.18 shows two pictures of contact angle measurements. One is hydrophilic (contact angle $<90°$), the other is hydrophobic (contact angle $>90°$). All these properties will influence the conjugate heat and mass transfer in the ducts.

1.2.5 Large biot number ducts

One important parameter in conjugate heat and mass transfer in ducts is the Biot number. It is defined by

$$Bi_{\mathrm{h}} = \frac{\delta/\lambda}{1/h} \tag{1.1}$$

$$Bi_{\mathrm{m}} = \frac{\delta/D_{\mathrm{vs}}}{1/k} \tag{1.2}$$

$$Nu = \frac{hD_{\mathrm{h}}}{\lambda} \tag{1.3}$$

$$Sh = \frac{kD_{\mathrm{h}}}{D_{\mathrm{va}}} \tag{1.4}$$

where δ is solid wall thickness (m), λ is thermal conductivity ($\mathrm{kW\,m^{-1}\,K^{-1}}$), D_{vs} is equivalent vapor diffusivity in solid wall ($\mathrm{m^2/s}$) calculated on the bases of air humidity differences, h ($\mathrm{kW\,m^{-2}\,K^{-1}}$) and k (m/s) are convective heat and mass transfer coefficients, respectively. Nu and Sh are Nusselt and Sherwood numbers in the duct, respectively. D_{h} is the hydrodynamic diameter of duct (m), D_{va} is mass diffusivity in air ($\mathrm{m^2/s}$). Subscripts "h" and "m" refer to heat and mass respectively.

According to the above definitions, for a membrane or adsorbent formed duct, the Bi numbers for heat and mass can be calculated. Assuming the duct is a plain plate-fin duct with a 1 m/s air flow, the properties of typical walls are [8]: $D_{\mathrm{h}}=1.66\,\mathrm{mm}$, $\delta=0.2\,\mathrm{mm}$; $\lambda_{\mathrm{s}}=0.38\ \mathrm{W\,m^{-1}\,K^{-1}}$; $\lambda_{\mathrm{a}}=0.0263\ \mathrm{W\,m^{-1}\,K^{-1}}$; $D_{\mathrm{vs}}=2.97\mathrm{e}{-7}\,\mathrm{m^2/s}$; $D_{\mathrm{va}}=2.82\mathrm{e}{-5}\,\mathrm{m^2/s}$; $Nu=0.95$, $Sh=0.71$; The calculated heat and mass transfer coefficients are: $h=15.05\ \mathrm{W\,m^{-2}\,K^{-1}}$; $k=0.012\,\mathrm{m/s}$; The calculated Biot numbers are: $Bi_{\mathrm{h}}=7.92\times10^{-3}$; $Bi_{\mathrm{m}}=8.08$.

This example shows that the Biot numbers for heat transfer are very small; however, the Biot numbers for mass transfer are much larger. The Biot numbers for mass transfer are three magnitudes higher than heat transfer. The differences are the results of the relatively lower mass diffusivities in solid materials, even with the novel "highly diffusive" materials [8]. Comparison of the Biot numbers for heat transfer and for mass transfer is shown schematically in Figure 1.19.

When $Bi<0.1$, the resistance in wall thickness can be ignored, and a lumped parameter model can be adopted for wall thickness [9]. This example then consequently implies that for heat transfer analysis, lumped parameters can be assumed for solid walls; however, for mass transfer, distributed models

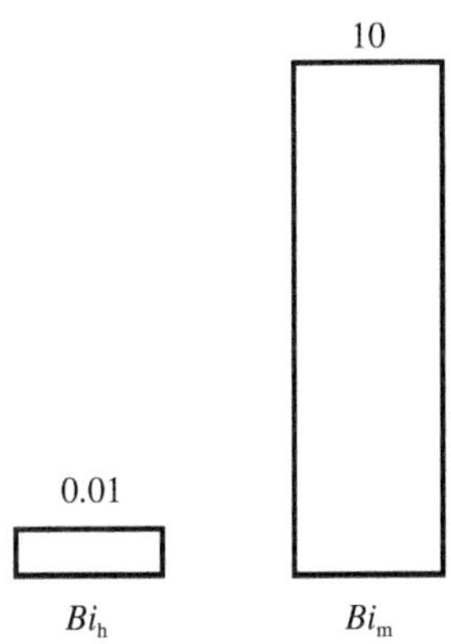

FIGURE 1.19

Comparison of the magnitudes of the Biot numbers for heat transfer and mass transfer.

of wall thickness must be considered. The heat and mass transfer ducts are called "**large Biot number ducts**", because the Biot numbers for mass transfer are much larger. For these ducts, the wall materials have a great effect on mass transfer, which in turn will have an effect on heat transfer. So a distributed model considering the parameter differences in wall thickness should be adopted. This will make the model complicated and a numerical solution is required.

Figure 1.20 shows the dimensionless humidity profiles in two plate-fin triangular ducts [10]. One duct is with low diffusivity materials ($Bi_m = 10$) and the other one is with high diffusivity materials ($Bi_m = 0.1$). As can be seen, the fin materials have changed the humidity profiles in the duct substantially.

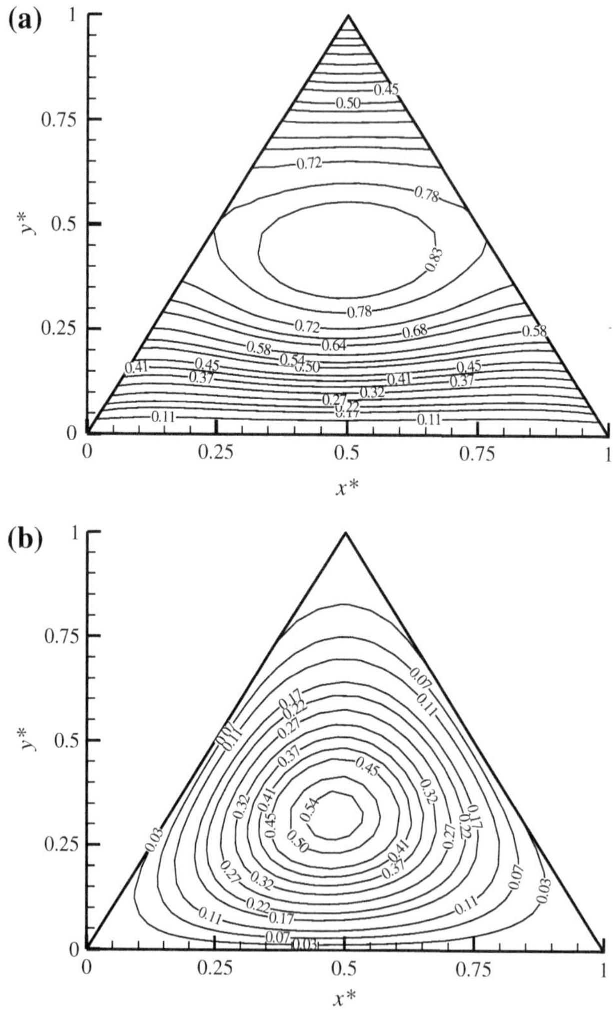

FIGURE 1.20

Dimensionless humidity profiles in plate-fin triangular ducts with different fin materials. (a) Low diffusivity fins, $Bi_m = 10$; (b) high diffusivity fins, $Bi_m = 0.1$.

1.2.6 Delayed development of boundary layers

The influence of solid wall materials is great. Because of the interactions with wall materials, compared to common heat transfer tubes, the thermal and mass boundary layers develop more slowly. The thermal and mass entry lengths account for major fractions of the duct lengths. Therefore, in most cases, the entry lengths should be considered in heat and mass transfer in the ducts of heat and mass exchangers. Similar to metal heat transfer tubes, hydrodynamic boundary layers develop very fast. So the fluid can be assumed as hydrodynamically fully developed, but developing in both thermal and mass boundaries. The Nusselt and Sherwood numbers in the entry regions are higher than the fully developed values. The effects of entry regions are great and they should be considered in performance analysis.

The situation is more complex if the system has liquid flow in the ducts. Liquid usually has much larger Schmidt numbers. So the mass entry lengths for fluids are much longer than for air streams. Figure 1.21 shows the development of mass profiles for a liquid solution in a duct with membrane walls [11]. As seen, due to the large Schmidt numbers, the mass boundary layers are still developing even at the end of the duct. However, the thermal boundary layers are fully developed shortly after the inlet [11]. These phenomena cannot be revealed without a detailed modeling of conjugate heat and mass transfer in the entry regions. To realize that goal, a numerical scheme is required because analytical methods are difficult to apply to this task.

All these special features indicate that conjugate heat and mass transfer in the ducts of heat and mass exchangers constitute a multi-element coupled heat and mass transfer system. The fluids interact closely with the wall materials. The boundary conditions for fluids are neither uniform values nor uniform fluxes. Rather, they are co-determined by the solids and the fluids. The irregularity of the duct

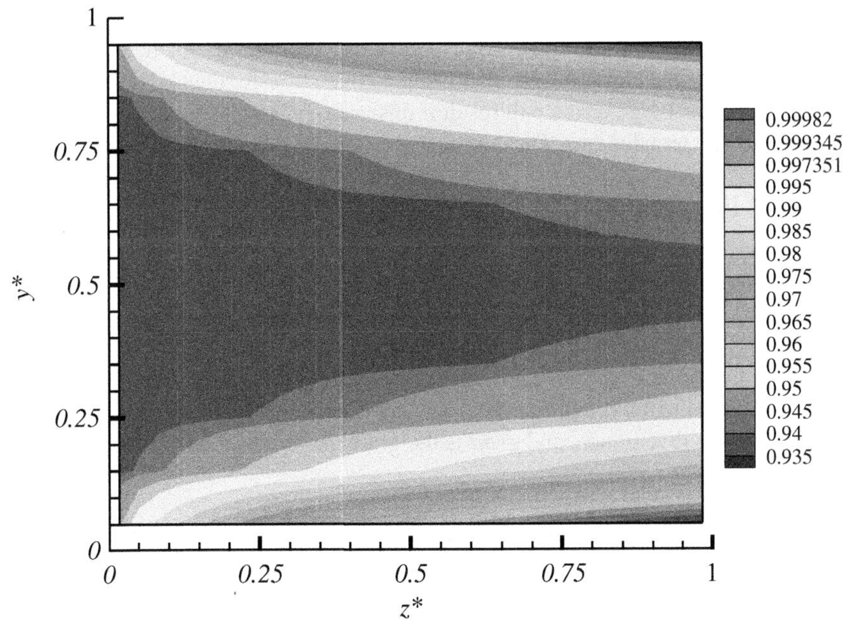

FIGURE 1.21

Dimensionless mass fraction profiles along the duct for a solution stream. The shading represents dimensionless mass friction of the solution. y^* is dimensionless longitudinal coordinates, z^* is dimensionless axial coordinates, $z^* = z/(D_h ReSc)$.

geometries, the nonlinearity of wall material properties, and the delayed entry lengths all contribute to the fact that there are tremendous differences between a heat and mass transfer duct and a common metal heat transfer tube. The transport data in modern textbooks cannot be simply adopted in the analysis of novel heat and mass exchangers. New knowledge should be built up. It is certainly a tough task considering the coupling of partial differential equations governing the transport phenomena in the duct. Fortunately, numerical techniques can help to accomplish this job.

1.2.7 Failure of the Chilton–Colburn *j* factor analogy

Traditionally, mass transfer coefficients are calculated from heat transfer coefficients by the Chilton–Colburn *j* factor analogy [8]. According to this concept, the *j* factor for mass transfer equals that for heat transfer, or

$$j_{\mathrm{m}} = j_{\mathrm{h}} \tag{1.5}$$

$$j_{\mathrm{m}} = St_{\mathrm{m}} Sc^{2/3} \tag{1.6}$$

$$j_{\mathrm{h}} = St_{\mathrm{h}} Pr^{2/3} \tag{1.7}$$

where the Stanton numbers for heat transfer and mass transfer are $St_{\mathrm{h}} = \frac{h}{\rho c_{\mathrm{p}} u}$ and $St_{\mathrm{m}} = \frac{k}{u}$ respectively.

The analogy can be used because generally the dimensionless heat transfer equations have the same forms as the dimensionless mass transfer equations. The solutions to the two set of equations are the same if the boundary conditions are also the same. Therefore, the above analogy is valid only when the dimensionless boundary conditions for heat transfer and mass transfer are the same. Regrettably, as observed from Figures 1.13, 1.14, and 1.20, the dimensionless boundary conditions for heat transfer and mass transfer are quite different for the large Biot number heat and mass transfer ducts. Therefore, the Chilton–Colburn *j* factor analogy fails. Large errors will result if this analogy is still adopted to obtain mass transfer coefficients from the heat transfer data. The conjugate problem should be addressed in a different way.

1.2.8 Summary of the problem

The general form for fluid flow, heat, and mass transport in the fluid (air or liquid) duct can be expressed by

$$\frac{\partial(\rho\phi)}{\partial t} + div(\rho \boldsymbol{U}\phi) = div(\Gamma_\phi \boldsymbol{grad}\phi) + S_\phi \tag{1.8}$$

where t is time (s), ρ is fluid density (kg/m^3), ϕ is a general variable which can be velocity, temperature, or concentration, $\boldsymbol{U}$ is the velocity vector matrix, Γ_ϕ is generalized diffusivity, and S_ϕ is a generalized source term. Function *div* is divergence, and ***grad*** is the gradient vector. It can also be expressed as

$$\frac{\partial(\rho\phi)}{\partial t} + \nabla(\rho \boldsymbol{U}\phi) = \nabla(\Gamma_\phi \nabla\phi) + S_\phi \tag{1.9}$$

If Γ_ϕ is a constant, then

$$\frac{\partial(\rho\phi)}{\partial t} + \nabla(\rho \boldsymbol{U}\phi) = \Gamma_\phi \nabla^2\phi + S_\phi \tag{1.10}$$

where ∇^2 is the Laplace operator.

For ducts in heat and mass exchangers, the solid walls are usually too dense to form a macroscopic fluid flow inside. So the macroscopic fluid flow in the solid is ignored. Heat and mass diffusion in the solid can be expressed by a general form as

$$\frac{\partial(\rho\phi)}{\partial t} = \nabla\left(\Gamma_\phi \nabla\phi\right) + S_\phi \tag{1.11}$$

where ϕ is temperature or concentration in the solid. This equation is usually conveniently expressed by the Fourier equation for heat conduction [9] and/or Fick's equation for mass diffusion [8].

Besides, the properties in the solid are governed by non-linear equations

$$\phi_i = f_i\left(\phi_1, \phi_2, \cdots, \phi_N\right) \tag{1.12}$$

The conjugate problem becomes: (1) to solve Eqs. (1.9), (1.11), and (1.12) simultaneously; (2) to solve the problem on the real duct geometries; (3) to solve the equations with variations in materials properties.

The problem is summarized in Figure 1.22.

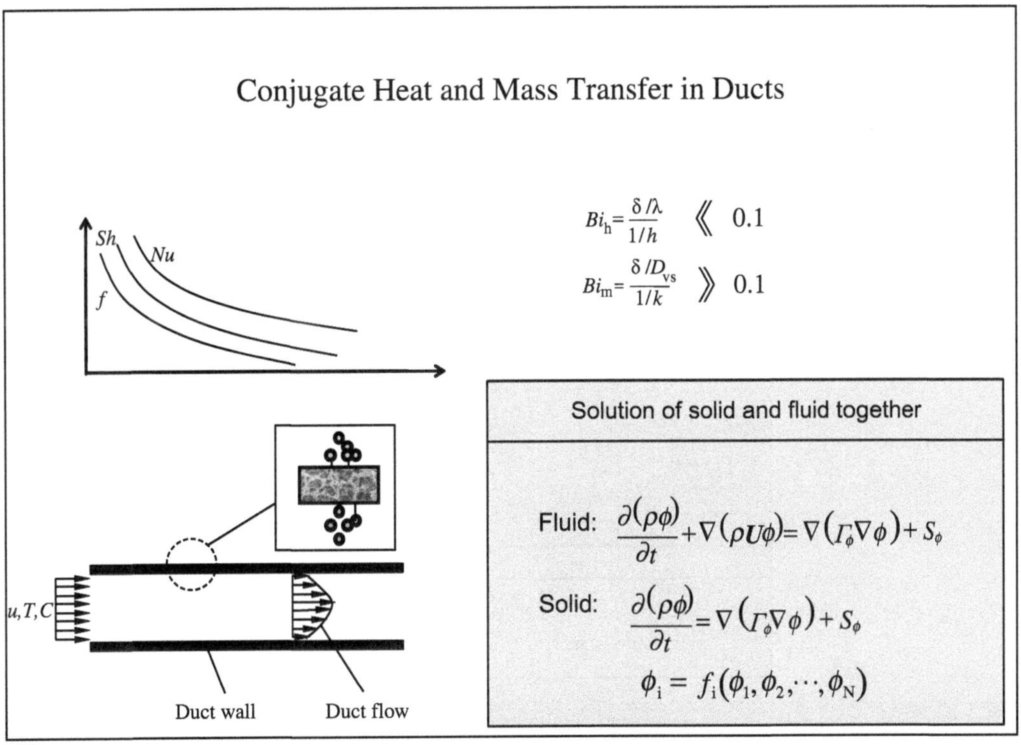

FIGURE 1.22

Schematic of the conjugate heat and mass transfer in ducts.

1.3 Current research trends

The concept of conjugate heat transfer was proposed many years ago in the 1960s by Perelman [12]. A very important piece of research was published by Luikov [13], where fluid exchanged heat with a plain plate. A typical conjugate problem is shown in Figure 1.23. A lot of theoretical work was carried out on this problem. Simultaneous convection and conduction in a flat plate were investigated by many authors with a variety of methods [14], partial difference simulations, or analytical solutions after simplification. The main investigated phenomena include: (1) convection from plain plate surfaces with known temperatures; (2) heat transfer in fins; (3) internal flows in a round tube. In case (1), with some assumptions, an analytical solution can be obtained for the temperature distribution in the solids. In case (2), fin temperatures and fin efficiencies are obtained by finite difference methods. In case (3), a quasi-steady method is used to give analytical solutions of wall and fluid temperatures [15].

Besides the above classic conjugate heat transfer problems, there were other studies of conjugate heat transfer in engineering applications; for instance, heat transfer in ribbed passages [16], heat transfer in micro-channels [17], heat transfer in fin arrays [18], heat transfer in mobile phones [14], drying of a porous flat plate [19], etc. Generally speaking, all previous investigations of conjugate heat transfer problems were confined by the following factors: (1) the geometries are rather simple; (2) multi-element systems are not considered; (3) no mass transfer is involved; (4) considerable simplifications are made concerning boundary conditions; (5) ducts are seldom studied and only plates are focused on. During these investigations, though many authors tried to find analytical solutions for the conjugate problems [14,20], numerical modeling was adopted by most of the authors [19,21,22]. The reason is that analytical solutions can only be found for simple problems. For the more engineering-oriented studies, numerical schemes are the only method. Even considering this trend, research on conjugate heat and mass transfer in ducts is very scarce. A systematic study of the conjugate heat and mass transfer in ducts, especially in heat and mass exchangers, is highly desirable, both for academics and for industry.

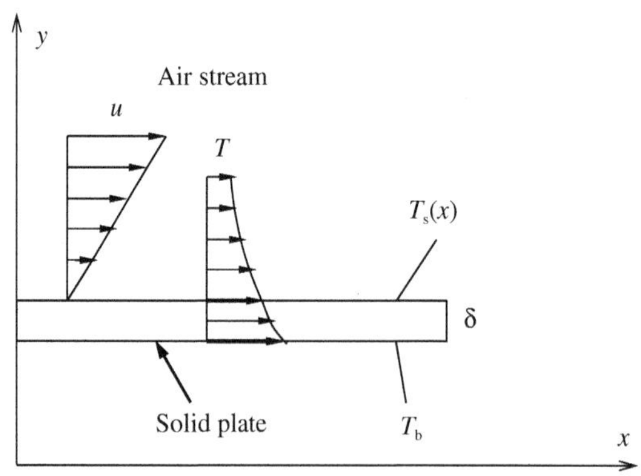

FIGURE 1.23

A typical conjugate heat transfer problem on a flat plate.

1.4 Summary

The special features encountered in heat and mass exchangers require a thorough investigation of the conjugate heat and mass transfer in ducts. However, due to the irregularities in geometries, complexities in governing equations, shortage of knowledge of materials, etc., knowledge on conjugate heat and mass transfer in ducts is rather scarce. This kind of problem can be called **the heat and mass transfer problem in large Biot number ducts**. For such ducts, the classic Chilton–Colburn j factor analogy fails to estimate mass transfer coefficients from the heat transfer data.

Over the past 20 years, the current author and his research group have conducted much research on conjugate heat and mass transfer in large Biot number ducts. The highlights are the combinations of heat and mass transfer with new functional materials. Much knowledge has been accumulated on the basic transport data in heat and mass transfer ducts. Appendix 1 at the end of this book gives a list of the publications of our group on this topic. However, knowledge of this area is still quite sporadic. It is not systematic enough. There is a need to collect it together and give a systematic introduction.

Against this background, the author began to write this book. In next chapters, the transport phenomena in conjugate heat and mass transfer in the ducts of heat and mass exchangers will be introduced in detail. They will provide the fundamental data for system design and performance optimization.

References

[1] U.S. Energy Information Administration (EIA). The Annual Energy Outlook Report 2012, June 2012, Washington DC. http://www.eia.gov/forecasts/aeo.

[2] Zhang LZ. Heat and mass transfer in a cross flow membrane-based enthalpy exchanger under naturally formed boundary conditions. Int J Heat Mass Transfer 2007;50(1–2):151–62.

[3] Zhang LZ, Niu JL. Performance comparisons of desiccant wheels for air dehumidification and enthalpy recovery. Appl Therm Eng 2002;22(12):1347–67.

[4] Zhang LZ. Coupled heat and mass transfer in an application-scale cross-flow hollow fiber membrane module for air humidification. Int J Heat Mass Transfer 2012;55(21–22):5861–9.

[5] Zhang LZ, Huang SM, Pei LX. Conjugate heat and mass transfer in a cross-flow hollow fiber membrane contactor for liquid desiccant air dehumidification. Int J Heat Mass Transfer 2012;55(25–26):8065–76.

[6] Zhang LZ, Chen ZY. Convective heat transfer in cross-corrugated triangular ducts under uniform heat flux boundary conditions. Int J Heat Mass Transfer 2011;54(1–3):597–605.

[7] Zhang LZ, Xiao F. Simultaneous heat and moisture transfer through a composite supported liquid membrane. Int J Heat Mass Transfer 2008;51(9–10):2179–89.

[8] Zhang LZ. Total heat recovery: heat and moisture recovery from ventilation air. New York: Nova Science; 2009. p. 151.

[9] Incropera FP, Dewitt DP. Introduction to heat transfer. 3rd ed. New York: John Wiley & Sons; 1996 [Chapter 5 p. 216].

[10] Zhang LZ. Heat and mass transfer in plate-fin enthalpy exchangers with different plate and fin materials. Int J Heat Mass Tranfer 2009;52(11–12):2704–13.

[11] Huang SM, Zhang LZ, Tang K, Pei LX. Fluid flow and heat mass transfer in membrane parallel-plates channels used for liquid desiccant air dehumidification. Int J Heat Mass Tranfer 2012;55(3):2571–80.

[12] Perelman TL. On conjugated problems of heat transfer. Int J Heat Mass Tranfer 1961;3:293–303.

[13] Luikov AV. Conjugate convective heat transfer problems. Int J Heat Mass Tranfer 1974;17:257–65.

[14] Karvinen R. Use of analytical expressions of convection in conjugated heat transfer problems. ASME J Heat Transfer 2012;134:031007-1-9.
[15] Jakob M. Heat transfer. New York: John Wiley & Sons; 1959.
[16] Iaccarino G, Ooi A, Durbin PA, Behnia M. Conjugate heat transfer predictions in two-dimensional ribbed passages. Int J Heat Fluid Flow 2002;23:340–5.
[17] Nonino C, Savino S, Giudice SD, Mansutti L. Conjugate forced convection and heat conduction in circular micro-channels. Int J Heat Fluid Flow 2009;30:823–30.
[18] Mereu S, Sciubba E, Bejan A. The optimal cooling of a stack of heat generating boards with fixed pressure drop, flow rate or pumping power. Int J Heat Mass Tranfer 1993;36:3677–86.
[19] Defraeye T, Blocken B, Carmeliet J. Analysis of convective heat and mass transfer coefficients for convective drying of a porous flat plate by conjugate modeling. Int J Heat Mass Tranfer 2012;55:112–24.
[20] Stein CF, Johansson P, Bergh J, Lofdahl L, Sen M, Gad-el-Hak M. An analytical asymptotic solution to a conjugate heat transfer problem. Int J Heat Mass Transfer 2002;45:2485–500.
[21] Kadem S, Lachemet A, Younsi R, Kocaefe D. 3d-transient modeling of heat and mass transfer during heat treatment of wood. Int Commun Heat Mass 2011;38:717–22.
[22] Juncu G. Unsteady conjugate forced convection heat/mass transfer from a finite flat plate. Int J Therm Sci 2008;47:972–84.

CHAPTER

2 Conjugate Heat and Mass Transfer in Adsorbent Ducts

2.1 The honeycomb type adsorbent beds

Heat and mass transfer processes are in many cases performed by fixed adsorbent beds. Mass transfer in the adsorbent beds is the necessary process for processing a stream. This mass transfer process of adsorption usually accompanies heat transfer. Further, deliberate heating and cooling of the beds is practiced to produce bed regeneration and adsorption because adsorption needs to operate at a low temperature and regeneration needs to operate at a high temperature. Temperature swings between low and high temperatures are the driving force for adsorption beds.

Adsorption processes have been widely used in many applications, such as adsorption cooling, air cleaning, organic vapor separation, etc. They have been used since the 1950s for the high-efficiency removal of a wide variety of moisture and organic vapors and several types of inorganic gases. The use of adsorption processes has been expanding recently due to innovations in the designs of the systems and to the development of new adsorbents. Adsorption systems designed for odor control and other low contaminant concentration applications (<10 ppm) are relatively simple. In these cases, the adsorbent bed is discarded as it approaches saturation with the contaminant. These systems are termed nonregenerative because the absorbent material is not reused. This type of adsorption process is also used extensively in large-scale applications having solvent vapor concentrations in the range 10 to 10,000 ppm. However, they are costly, because large quantities of adsorbent are discarded after use. To decrease the cost and re-use the adsorbent bed, regenerative beds systems are used. Prior to becoming saturated with the adsorbed adsorbates, the adsorbent is isolated from the process stream and treated with a hot regenerative stream to drive the adsorbed compounds out of the solid adsorbent and exhaust them. After regeneration, the regenerated bed is then switched to the process again to begin another cycle of use. Figure 2.1 shows a regenerative adsorption system with two beds. As seen, when valves V1 and V2 are open and valves V3 and V4 are closed; bed 1 works in adsorption phase, while bed 2 works in regeneration phase. Similarly, when valves V3 and V4 are open and valves V1 and V2 are closed, bed 1 works in regeneration phase, while bed 2 works in adsorption phase. The process stream is cool and high in adsorbate concentration, while the regeneration stream is hot and low in adsorbate concentration. The beds swing between hot and low temperatures. The exhaust stream can be further cooled and condensed to recover the adsorbate, such as organic solvents.

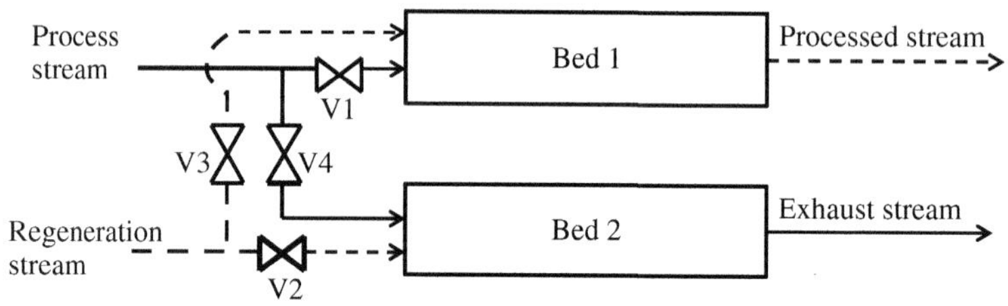

FIGURE 2.1

Schematic of an adsorbent-bed system.

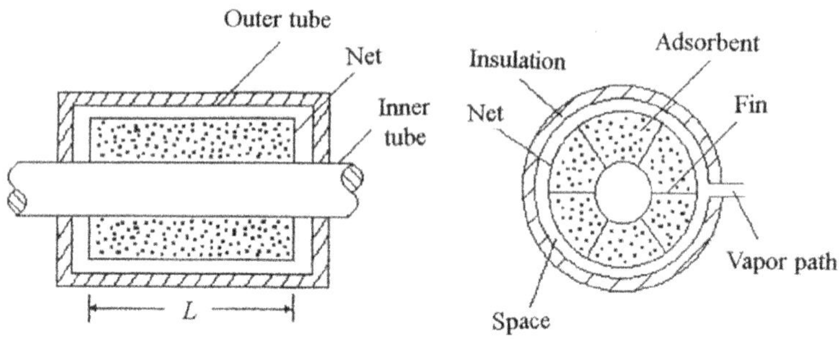

FIGURE 2.2

Schematic of a packed adsorbent bed.

There are two types of adsorbent bed. One is that packed with small adsorbent pallets. Figure 2.2 shows an adsorbent bed used for adsorption cooling [1]. As can be seen, it is composed of two heat transfer pipes. Adsorbent particles of Zeolite 13X with 3mm in diameter are packed between the outer pipe and the inner pipe. Fins are inserted inside the adsorbent bed to enhance heat transfer.

This type of adsorbent bed is easy to construct and can be used to process a wide variety of gases. Fluid flows in the cavities among the pellets and exchanges heat and mass with the pellets. Heat and mass transfer in packed adsorbent beds can be found in many text books on chemical engineering [2,3]. They are well studied and the correlations for heat and mass transfer predictions are well established. A more detailed modeling of fluid flow and heat and mass transfer in packed adsorbent beds can also be found in recent references [4,5]. In such cases, N-S equations inside the adsorbents should be solved. They are more accurate yet more complicated. Applications of packed beds are widespread. However, packed beds are bulky and low in heat and mass transfer. Because the adsorbent beds are very thick (>3 mm), the overall heat and mass transfer coefficients are very low. Large quantities of adsorbents must be packed in a bed to have a certain treatment capability during a given period. The heat conductivity and mass diffusivity of gases in adsorbent materials are very low, so adsorbents should be thin enough to have acceptable treatment power. To solve this problem, honeycomb-type adsorbent beds are constructed.

In a honeycomb-type adsorbent bed, adsorbent is fabricated on support material that is as thin as a paper and corrugated into a plate-fin honeycomb structure. The support material is usually fiber glass

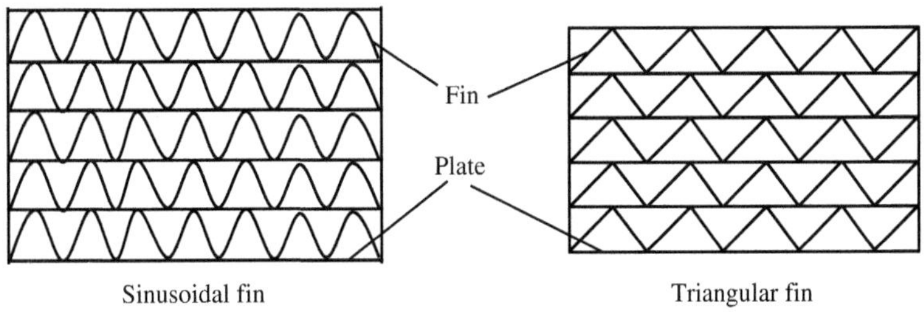

FIGURE 2.3

Two honeycomb-type adsorbent beds, one with sinusoidal fins and the other with triangular fins.

and it is very porous. It can absorb and station large quantities of adsorbent solution. After the fiber glass is corrugated to form a honeycomb structure, it is soaked with raw adsorbent solution. Then it is heat treated to form solid adsorbent walls on the fiber glass matrix. In-situ fabrication is the common method of material making. The formed solid is dense. The finally formed adsorbent bed is like a honeycomb type plate-fin heat exchanger, as depicted in Figure 2.3. However, the wall (including the plates and fins) materials are not common metals, but fiber glass with adsorbents stationed inside. The wall thickness is usually 0.1–0.6 mm. The specific contact areas of such beds are very large, of the order of 500–700 m^2/m^3, so heat and mass transfer capabilities are rather high. The problem of low heat and mass transfer coefficients in a packed bed is overcome by such thin wall thickness and large contact area. As can also be seen, the beds are formed by numerous fine ducts. The cross sections of these ducts are usually either in sinusoidal shapes or in triangular shapes. The cool process stream flows along these ducts, exchanging heat and mass with duct walls. When they are close to saturation, they are switched to regeneration mode, and the regeneration process starts. They work in a cyclic manner of cooling-adsorption-heating-regenerating.

Adsorbent ducts are the basic elements in these beds, which are composed of fluid duct and adsorbent walls. Compared to packed beds, heat and mass transfer in honeycomb-type beds is different: (1) there are no macroscopic fluid flows inside the adsorbent walls, which are usually dense compared to packed pallets. Heat conduction and mass diffusion inside the adsorbent walls are the transport mechanisms in the solids. (2) The materials are quick in response to outside conditions. (3) Fluids flow in straight ducts while exchanging heat and mass with adsorbent walls. This is similar to common plate-fin heat exchangers. However, the wall temperatures and concentrations are temporal. The boundary conditions on wall surfaces change with time. This is quite different from common plate-fin heat exchangers, which always work in steady-state conditions. For adsorbent beds, the transient behaviors should be investigated.

2.2 Current researches on honeycomb type adsorbent beds

Honeycomb-type adsorbents beds have attracted much attention in air dehumidification and enthalpy recovery. Desiccant wheels and energy wheels can be regarded as revolving honeycomb adsorbents beds. In these applications, hygroscopic materials such as zeolite, silica gel, and LiCl are fabricated on the paper or fiber glass, corrugated into honeycomb plate-fin channels. A desiccant wheel is shown in Figure 2.4.

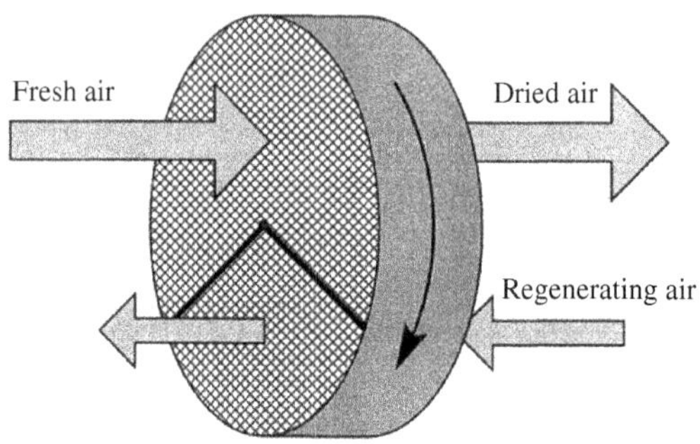

FIGURE 2.4

A schematic of a desiccant wheel.

As can be seen, it rotates between the process air (the fresh air) and the regenerating air. Heat and moisture in the fresh air are absorbed by the ducts walls. It becomes dry air after flowing out of the ducts. The adsorbents walls adsorb moisture and become saturated. Their capability to dry air is refreshed after they rotate through the regenerating zone, where hot air flows in the ducts and drives the moisture out. The wheel rotates continuously and the fresh air is dried continuously. The wheel is like many beds cycling continuously.

There are many studies of heat and moisture transfer in desiccant wheels. Zhang and Niu [6,7] investigated the heat and mass transfer in a desiccant wheel. Both the axial and the transverse heat conduction and mass diffusion in the solids were considered. Zhang et al. [8] simulated the one-dimensional heat and mass transfer in a desiccant wheel. A lumped parameter model was used to simplify heat and mass transfer in duct thickness. The different behaviors of air dehumidification and enthalpy recovery were revealed. Ruivo et al. [9–11] investigated heat and moisture transfer in a single channel in the desiccant wheel. De Antonellis et al. [12] investigated the performance of heat and mass transfer in a desiccant wheel with rotational speed. In all these models, the honeycomb ducts were assumed as parallel-plate channels. The duct geometry was not considered. The convective heat and mass transfer properties between the solid and air were directly borrowed from the data obtained for common plate-fin heat exchangers. The temporal behaviors of the honeycomb ducts were not considered. The effects of the cross-sectional shapes of the duct were not considered either. Recently several studies considered these effects. Niu and Zhang [13] studied the effects of deformed sinusoidal duct cross sections on the fully developed Nusselt and Sherwood numbers in the ducts; however, the properties of the adsorbent wall were not coupled to the air flow. Al-Sharqawi and Lior [14,15] investigated the effects of duct geometry on solid desiccant dehumidification. The wall properties were coupled to the air flow; however, only parallel plates and triangular ducts were modeled and no Nusselt or Sherwood data were obtained for future component design. Further, the model developed was only appropriate to packed beds since the macroscopic fluid flows inside the solids dominated heat and mass transfer. In contrast, mass diffusion and heat conduction are the transport mechanism in the solids, rather than macroscopic convection. Solid walls in honeycomb-type beds are much denser and no macro-scale fluid can penetrate through the walls.

In summary, there has not yet been a study that fully considers the conjugate heat and mass transfer in a duct of a honeycomb-type adsorbent wall. Conjugate heat and mass transfer in adsorbent ducts is the research objective. The following factors should be taken into account: (1) the duct geometry; (2) mass diffusion and heat conduction inside the solids; (3) conjugate heat and mass transfer between fluids and solids; (4) simultaneous heat and mass transfer; (5) the temporal behaviors of boundary conditions on wall surfaces; (6) the transient characters of Nusselt and Sherwood numbers in the ducts. In the next sections, these characteristics will be studied in detail.

2.3 The physical model for adsorbent ducts

A fixed honeycomb-type adsorbent bed is considered. The work mode is similar to Figure 2.1. The bed is switched between adsorption and regeneration modes by valves. It is made of hygroscopic material in-situ fabricated silica gel to dehumidify air. The bed structure is the same as in Figure 2.3. One section of the bed is independently shown in Figure 2.5 to plot the geometries of the duct. As can be seen, the bed is a plate-fin structure. The net air duct is surrounded by a plate and a sinusoidal fin. The duct walls (the plate and the fin) are made with silica gel with a thickness of δ. There are numerous parallel ducts in the bed. Only a representative duct is modeled here. The duct geometry is of sinusoidal shape, the most common one. A single duct in the bed is plotted in Figure 2.6. The geometries of the duct are shown in Figures 2.5 and 2.6. The mid plane of the sinusoidal fin satisfies a sin function,

$$y = a\left[1 - \cos\left(\frac{\pi}{b}x\right)\right] \tag{2.1}$$

Aspect ratio of duct

$$\alpha_{\mathrm{f}} = \frac{2a}{2b} \tag{2.2}$$

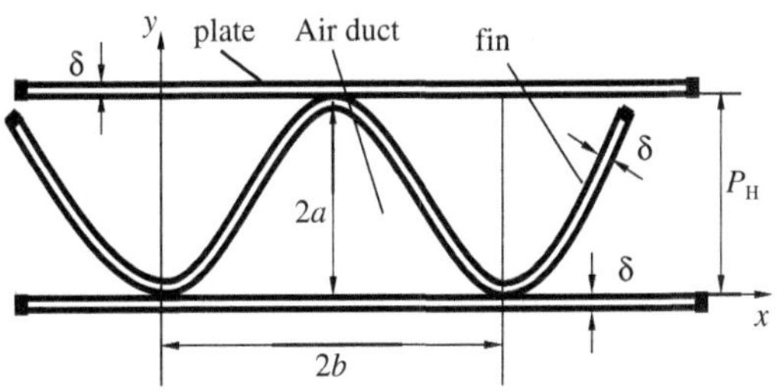

FIGURE 2.5

A section of the honeycomb adsorbent bed showing the geometry of adsorbent ducts, consisting of the net air duct and plate-fin solid adsorbent walls.

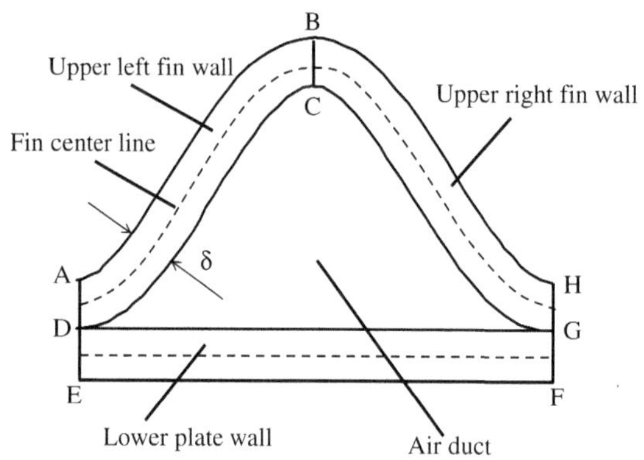

FIGURE 2.6

A representative sinusoidal adsorbent duct in the bed, showing the net air duct, the upper left fin wall, the upper right fin wall, and the lower plate wall. The dashed lines are the mid planes in the walls.

Fin thickness ratio

$$\beta_{\mathrm{f}} = \frac{\delta}{2b} \tag{2.3}$$

Width of a duct

$$L_{\mathrm{W}} = 2b \tag{2.4}$$

Channel pitch

$$P_{\mathrm{H}} = 2a + 0.5\delta \tag{2.5}$$

Overall height of the duct in Figure 2.6

$$L_{\mathrm{H}} = P_{\mathrm{H}} + \delta \tag{2.6}$$

Hydraulic diameter of the duct

$$D_{\mathrm{h}} = \frac{4A_{\mathrm{c}}}{P_{\mathrm{f}}} \tag{2.7}$$

where A_{c} is the cross-sectional area of the net air duct (m^2), and P_{f} is the perimeter of the net air duct surrounded by adsorbents (m).

Cool and moist air flows through the ducts. The duct wall is cooled from the previous regeneration stage and begins to adsorb moisture from the process fresh air. The air is heated and dried. This is the adsorption mode. After the solids adsorb a certain quantities of moisture, it is switched to regeneration mode. Hot air flows through the duct. The adsorbent wall is heated and the water content in the solid is driven off. The regeneration air becomes cool and humid. The solids become dry. This is the regeneration mode.

2.4 The mathematical model

Fluid flow and conjugate heat and mass transfer in the duct are considered. The flow in the duct is considered to be laminar and hydrodynamically fully developed, but developing in temperature and concentration in the entrance region of the duct. The fluid is Newtonian with constant thermal properties.

2.4.1 Air stream

The governing equations are summarized as follows [16,17].
Momentum

$$\mu\left(\frac{\partial^2 u}{\partial x^2}+\frac{\partial^2 u}{\partial y^2}\right)=\frac{dP}{dz} \tag{2.8}$$

where μ is dynamic viscosity of air stream (Pa s), u is fluid velocity (m/s), P is the pressure (Pa), z is the axial coordinate (m). As can be seen, this is a fully developed steady flow.

However, heat and mass conservations are transitional.

Transient energy conservation

$$\rho_a c_{Pa}\frac{\partial T}{\partial t}+\rho_a c_{Pa} u\frac{\partial T}{\partial z}=\lambda_a\left(\frac{\partial^2 T}{\partial x^2}+\frac{\partial^2 T}{\partial y^2}\right) \tag{2.9}$$

where t is time (s), T is fluid temperature (K), λ_a is thermal conductivity of air ($\mathrm{kW\,m^{-1}K^{-1}}$), ρ_a is density of air ($\mathrm{kg/m^3}$), c_{pa} is specific heat of air ($\mathrm{kJ\,kg^{-1}K^{-1}}$).

Transient mass conservation

$$\frac{\partial \omega}{\partial t}+u\frac{\partial \omega}{\partial z}=D_{va}\left(\frac{\partial^2 \omega}{\partial x^2}+\frac{\partial^2 \omega}{\partial y^2}\right) \tag{2.10}$$

where ω is the humidity ratio (kg moisture/kg dry air), D_{va} is the diffusivity of vapor in dry air ($\mathrm{m^2 s^{-1}}$).

The above three governing equations (2.8)–(2.10) can be normalized to

$$\frac{\partial^2 u^*}{\partial x^{*2}}+\frac{\partial^2 u^*}{\partial y^{*2}}+\frac{L_W^2}{D_h^2}=0 \tag{2.11}$$

$$c_1\frac{\partial \theta}{\partial \tau}+U\frac{\partial \theta}{\partial z^*}=\frac{\partial^2 \theta}{\partial x^{*2}}+\frac{\partial^2 \theta}{\partial y^{*2}} \tag{2.12}$$

$$c_2\frac{\partial \omega^*}{\partial \tau}+\frac{U}{Le}\frac{\partial \omega^*}{\partial z^*}=\frac{\partial^2 \omega^*}{\partial x^*2}+\frac{\partial^2 \omega^*}{\partial y^*2} \tag{2.13}$$

with dimensionless velocity

$$u^*=-\frac{\mu u}{(dP/dz)D_h^2} \tag{2.14}$$

and dimensionless temperature

$$\theta = \frac{T - T_{ci}}{T_{hi} - T_{ci}} \tag{2.15}$$

where in the equations, T_{hi} is the inlet temperature of the heating fluid, and T_{ci} is the inlet temperature of the cooling fluid.

The dimensionless humidity is defined by

$$\omega^* = \frac{\omega - \omega_{min}}{\omega_{max} - \omega_{min}} \tag{2.16}$$

where ω_{max} is the maximum air humidity, which is in equilibrium with the wall at temperature T_{hi} and water content w_{max}; ω_{min} is the minimum air humidity, which is in equilibrium with the wall at temperature T_{ci} and water content w_{min}; w_{max} and w_{min} are the maximum and minimum water uptake in solid, respectively.

Dimensionless time

$$\tau = \frac{t}{t_{cyc}} \tag{2.17}$$

$$t_{cyc} = \frac{z_L}{u_m} = \frac{D_h^2 z_L^*}{a_a} \tag{2.18}$$

where t_{cyc} is a hypothetic cycling time of the fluid (s), z_L is the length of the duct (m), a_a is the thermal diffusivity of air (m^2/s).

Constants in the equations

$$c_1 = \frac{\text{Re}\,\text{Pr}\,L_W^2}{D_h^2} \tag{2.19}$$

$$c_2 = \frac{\text{Re}\,\text{Pr}\,L_W^2}{LeD_h^2} \tag{2.20}$$

Dimensionless coordinates are defined by

$$x^* = \frac{x}{L_W} \tag{2.21}$$

$$y^* = \frac{y}{L_W} \tag{2.22}$$

$$z^* = \frac{z}{D_h RePr} \tag{2.23}$$

where Re is the Reynolds number and Pr is the Prandtl number, and Le is the Lewis number.

$$Le = \frac{\mathrm{Pr}}{Sc} = \frac{a_{\mathrm{a}}}{D_{\mathrm{va}}} \tag{2.24}$$

For moisture air, $Le = 0.82$–0.84.

In Eqs. (2.12), (2.13), the velocity coefficient U is defined by

$$U = \frac{u^*}{u_{\mathrm{m}}^*}\frac{L_{\mathrm{W}}^2}{D_{\mathrm{h}}^2} \tag{2.25}$$

where u_{m}^* is the average dimensionless velocity on the duct cross section, and it is calculated by

$$u_{\mathrm{m}}^* = \frac{\int\int u^* dA}{A_{\mathrm{c}}} \tag{2.26}$$

For hydrodynamically developed fluid flow in ducts, the characteristics of fluid flow in the duct can be represented by the product of the friction factor and the Reynolds number as

$$(fRe) = \left(-\frac{D_{\mathrm{h}}\frac{dP}{dz}}{\frac{1}{2}\rho_{\mathrm{a}}u_{\mathrm{m}}^2}\right)\left(\frac{\rho_{\mathrm{a}}u_{\mathrm{m}}D_{\mathrm{h}}}{\mu}\right) = \frac{2}{u_m^*} \tag{2.27}$$

Dimensionless bulk temperature

$$\theta_{\mathrm{b}} = \frac{\int\int u^*\theta dA}{\int\int u^* dA} \tag{2.28}$$

Nusselt number

$$Nu = \frac{hD_{\mathrm{h}}}{\lambda_{\mathrm{a}}} \tag{2.29}$$

Sherwood number

$$Sh = \frac{kD_{\mathrm{h}}}{D_{\mathrm{va}}} \tag{2.30}$$

Other characteristic parameters are:

Reynolds number

$$Re = \frac{u_{\mathrm{m}}D_{\mathrm{h}}}{\nu} \tag{2.31}$$

Prandtl number

$$Pr = \frac{\nu}{a_{\mathrm{a}}} \tag{2.32}$$

Schmidt number

$$Sc = \frac{\nu}{D_{\mathrm{va}}} \tag{2.33}$$

Peclet number

$$Pe = RePr \tag{2.34}$$

Kinematic viscosity of air

$$\nu = \frac{\mu}{\rho_a} \tag{2.35}$$

Thermal diffusivity of air

$$a_a = \frac{\lambda_a}{\rho_a c_{pa}} \tag{2.36}$$

where h and k are the convective heat transfer coefficient ($\text{kW m}^{-2}\text{K}^{-1}$) and the convective mass transfer coefficient (m/s) is between fluid and wall, respectively. The local Nusselt and Sherwood numbers may change from point to point on a duct surface. It is more practical to evaluate the peripherally mean local Nusselt and Sherwood numbers along the duct.

Considering the energy balance in a control volume of length Δz, heat transferred through convection

$$Q_1 = h_L P_f \Delta z (T_W - T_b) \tag{2.37}$$

where h_L is the peripherally mean and axially local heat transfer coefficient, and subscripts "w" and "b" refer to "wall" and "bulk" respectively. The enthalpy change of the fluid in the control volume

$$Q_2 = \rho_a u_m A_c c_p \Delta T_b \tag{2.38}$$

Since

$$Q_1 = Q_2 \tag{2.39}$$

and

$$\Delta z = \Delta z^* D_h RePr = \Delta z^* \frac{\rho c_p u_m D_h^2}{\lambda_a} \tag{2.40}$$

the peripherally mean local heat transfer coefficient becomes

$$h_L = \frac{1}{4(\theta_w - \theta_b)} \frac{\lambda_a}{D_h} \frac{\Delta\theta_b}{\Delta z^*} \tag{2.41}$$

Substituting the above equation to Eq. (2.29), and considering the control volume to be infinitely small, then we obtain the peripherally local Nusselt number as

$$Nu_L = \frac{1}{4(\theta_w - \theta_b)} \frac{d\theta_b}{dz^*} \tag{2.42}$$

The average Nusselt number from 0 to z^*

$$Nu_m = \frac{1}{z^*} \int_0^{z^*} Nu_L dz^* \tag{2.43}$$

Correspondingly, a mass balance on the control volume will have the following equations

$$k_{\mathrm{L}} = -\frac{1}{4(\omega_{\mathrm{w}}^{*} - \omega_{\mathrm{b}}^{*})} \frac{D_{\mathrm{va}}}{LeD_{\mathrm{h}}} \frac{\Delta\omega_{\mathrm{b}}^{*}}{\Delta z^{*}} \tag{2.44}$$

$$Sh_{\mathrm{L}} = -\frac{1}{4Le(\omega_{\mathrm{w}}^{*} - \omega_{\mathrm{b}}^{*})} \frac{d\omega_{\mathrm{b}}^{*}}{dz^{*}} \tag{2.45}$$

$$Sh_{\mathrm{m}} = \frac{1}{z^{*}} \int_{0}^{z^{*}} Sh_{\mathrm{L}} dz^{*} \tag{2.46}$$

As will be discussed later, the local Nusselt number and the local Sherwood number decrease asymptotically from very high values near the entrance of a duct to certain fully developed values at the end of thermal or mass entry length. Under a uniform temperature (uniform mass concentration) boundary condition, the fully developed value is denoted as Nu_{T} (Sh_{T} for mass). Under a uniform heat flux (mass flux) boundary condition, it is denoted as Nu_{H} (Sh_{H} for mass). In this study, under the real convective flow boundary conditions, the fully developed value is denoted as Nu_{C} (Sh_{C} for mass transfer).

2.4.2 Solid walls

Energy conservation in the solid

$$(\rho c_{\mathrm{p}})_{\mathrm{tot}} \frac{\partial T}{\partial t} - \rho_{\mathrm{s}} q_{\mathrm{st}} \frac{\partial w}{\partial t} = \frac{\partial}{\partial x}\left(\lambda_{\mathrm{s}} \frac{\partial T}{\partial x}\right) + \frac{\partial}{\partial y}\left(\lambda_{\mathrm{s}} \frac{\partial T}{\partial y}\right) + \frac{\partial}{\partial z}\left(\lambda_{\mathrm{s}} \frac{\partial T}{\partial z}\right) \tag{2.47}$$

$$(\rho c_{\mathrm{p}})_{\mathrm{tot}} = \varepsilon_{\mathrm{t}} \rho_{\mathrm{a}} c_{\mathrm{pa}} + \rho_{\mathrm{s}}(c_{\mathrm{ps}} + w c_{\mathrm{pw}}) \tag{2.48}$$

Mass conservation

$$\varepsilon_{\mathrm{t}} \rho_{\mathrm{a}} \frac{\partial \omega}{\partial t} + \rho_{\mathrm{s}} \frac{\partial w}{\partial t} = \frac{\partial}{\partial x}\left(\rho_{\mathrm{a}} D_{\mathrm{vs}} \frac{\partial \omega}{\partial x}\right) + \frac{\partial}{\partial y}\left(\rho_{\mathrm{a}} D_{\mathrm{vs}} \frac{\partial \omega}{\partial y}\right) + \frac{\partial}{\partial z}\left(\rho_{\mathrm{a}} D_{\mathrm{vs}} \frac{\partial \omega}{\partial z}\right) \tag{2.49}$$

where ε_{t} is the porosity of adsorbent, subscript "s" refers to solid, q_{st} is the adsorption heat (kJ/kg); the variable w is water uptake in the membrane (kg moisture/kg dry membrane), D_{vs} is equivalent vapor diffusivity in solid walls, and ω is the equilibrium humidity ratio of air with adsorbents.

For dense materials, the relationship between adsorbed water diffusivity and equivalent vapor diffusivity is

$$\rho_{\mathrm{s}} D_{\mathrm{ws}} \frac{\partial w}{\partial z} = \rho_{\mathrm{a}} D_{\mathrm{vs}} \frac{\partial \omega}{\partial z} \tag{2.50}$$

If a simplified equation is used to reflect the relation between adsorbed water uptake and equivalent equilibrium humidity, then

$$w = K_p \omega \tag{2.51}$$

where K_p is the partition coefficient [(kg water/kg material)/(kg vapor/kg air)]. This is the slope of adsorption curve to humidity. Substituting Eq. (2.51) into Eq. (2.50), then the relation between water diffusivity and equivalent moisture diffusivity in the solid is expressed by

$$D_{vs} = \frac{\rho_s D_{ws} K_p}{\rho_a} \tag{2.52}$$

Considering non-equilibrium adsorption, the adsorption rate is

$$\frac{\partial w}{\partial t} = k_m \left(w_{eq} - w \right) \tag{2.53}$$

where k_m is the internal mass transfer coefficient (1/s) of adsorbents, and w_{eq} is the equilibrium water content in solid at temperature T and humidity ratio ω. This can be estimated by the isothermal adsorption curve as

$$\frac{w}{w_{eq}} = \left(1 - e^{-k_m t} \right) \tag{2.54}$$

According to the above equation, when the time is $t_{0.5}$ for adsorption to reach $w = 0.5 w_{eq}$, $k_m = -\ln 0.5 / t_{0.5}$.

The equilibrium water uptake in the material can be expressed by a general sorption curve as

$$w_{eq} = \frac{W_{max}}{1 - C_s + C_s/RH} \tag{2.55}$$

where W_{max} is the maximum water uptake of the solid material (kg/kg), C_s is a constant called the shape factor for the material, and RH is air relative humidity. This sorption curve directly links water content to RH, a variable that can be directly measured. The curve can be used to calculate K_p. The basic thermal physical parameters are obtained in the laboratory.

Substituting Eq. (2.51) into (2.53), Eq. (2.53) can be written to

$$\frac{\partial w}{\partial t} = k_m' \left(\omega_{eq} - \omega \right) \tag{2.56}$$

$$k_m' = K_p k_m \tag{2.57}$$

The relative humidity is calculated by humidity ratio and temperature as [18]

$$\frac{RH}{\omega} = \frac{e^{5294/T}}{10^6} - 1.61 RH \tag{2.58}$$

where T is in K. The second term on the right side of the equation will generally have less than a 5% effect, and it can be usually ignored.

As can be seen, this model considers both the internal resistance of the adsorbents and the external resistance in the pores of the adsorbents. If the internal resistance is ignored, then equilibrium adsorption is assumed and

$$\frac{\partial w}{\partial t} = \sigma_{\mathrm{T}} \frac{\partial T}{\partial t} + \sigma_{\omega} \frac{\partial \omega}{\partial t} \tag{2.59}$$

$$\sigma_{\mathrm{T}} = \left. \frac{\partial w}{\partial T} \right|_{\omega} \tag{2.60}$$

$$\sigma_{\omega} = \left. \frac{\partial w}{\partial \omega} \right|_{T} \tag{2.61}$$

Still considering both the internal and the external resistance, substituting Eq. (2.56) into (2.47) and (2.49), the above Eqs. (2.47) and (2.49) can be normalized to

$$c_3 \frac{\partial \theta}{\partial \tau} - c_4 \frac{\partial w^*}{\partial \tau} = c_{\mathrm{x}} \frac{\partial}{\partial x^*}\left(\lambda_{\mathrm{s}} \frac{\partial \theta}{\partial x^*}\right) + c_y \frac{\partial}{\partial y^*}\left(\lambda_{\mathrm{s}} \frac{\partial \theta}{\partial y^*}\right) + c_z \frac{\partial}{\partial z^*}\left(\lambda_{\mathrm{s}} \frac{\partial \theta}{\partial z^*}\right) \tag{2.62}$$

$$c_5 \frac{\partial \omega^*}{\partial \tau} + c_6 w_{\mathrm{eq}}^* = c_{\mathrm{x}} \frac{\partial}{\partial x^*}\left(D_{\mathrm{vs}} \frac{\partial \omega^*}{\partial x^*}\right) + c_y \frac{\partial}{\partial y^*}\left(D_{\mathrm{vs}} \frac{\partial \omega^*}{\partial y^*}\right) + c_z \frac{\partial}{\partial z^*}\left(D_{\mathrm{vs}} \frac{\partial \omega^*}{\partial z^*}\right) \tag{2.63}$$

$$c_3 = \frac{(\rho c_{\mathrm{p}})_{\mathrm{tot}}}{t_{\mathrm{cyc}}} \tag{2.64}$$

$$c_4 = \frac{\rho_{\mathrm{s}} q_{\mathrm{st}} (w_{\max} - w_{\min})}{t_{\mathrm{cyc}} (T_{\mathrm{hi}} - T_{\mathrm{ci}})} \tag{2.65}$$

$$c_{\mathrm{x}} = \frac{1}{L_{\mathrm{W}}^2} \tag{2.66}$$

$$c_y = \frac{1}{L_{\mathrm{W}}^2} \tag{2.67}$$

$$c_{\mathrm{z}} = \frac{1}{(D_{\mathrm{h}} \mathrm{RePr})^2} \tag{2.68}$$

$$c_5 = \frac{\varepsilon_{\mathrm{t}} \rho_{\mathrm{a}}}{t_{\mathrm{cyc}}} - \rho_{\mathrm{s}} K'_{\mathrm{m}} \tag{2.69}$$

$$c_6 = \rho_s k_m' \tag{2.70}$$

Dimensionless water uptake

$$w^* = \frac{w - w_{min}}{w_{max} - w_{min}} \tag{2.71}$$

where w_{max} and w_{min} are the maximum and minimum water uptake in the solid. Usually, in the solid, heat conduction and mass diffusion in fin thickness are much larger than in other two directions, and the heat conduction and mass diffusion in fin height and in duct length can be ignored. However, for general purposes, this study considers the heat conduction and mass diffusion in all three dimensions.

2.4.3 Heat conductivity and moisture diffusivity

Heat conductivity

There are many equations for the estimation of thermal conductivity in porous media [19]. In this case, the porous medium has three phases: air, adsorbed water, and solid matrix. Therefore heat conductivity in the solid wall can be analyzed by

$$\lambda_s = \lambda_a^{\varepsilon_t - v_w} \cdot \lambda_w^{v_w} \cdot \lambda_{gel}^{1-\varepsilon_t} \tag{2.72}$$

where subscript "gel" denotes solid part in the porous material. Volumetric content of water in the porous media is calculated by

$$v_w = \frac{w\rho_s}{\rho_w} \tag{2.73}$$

Vapor diffusivity in porous media

The established theory considers three mechanisms for mass transfer in a pore as depicted in Figure 2.7: Poiseuille flow, ordinary molecular diffusion and Knudsen diffusion, or a combination of these.

The governing quantity that provides a guideline in determining which mechanism is operative in a given pore under given operating conditions is the ratio of the pore size to the mean free path λ_i, which is calculated using the following expression [17],

$$\lambda_i = \frac{k_B T}{\sqrt{2}\pi \sigma_i^2 P_m} \tag{2.74}$$

where σ_i is the molecular collision diameter (m), 2.641 Å and 3.711 Å for water vapor and air, respectively; k_B is the Boltzmann constant, 1.38×10^{-23} J/K, P_m is the mean total pressure within the pores in adsorbent (Pa), and T is the absolute temperature (K).

Under room temperature and atmospheric pressure, the calculated λ_i for moisture is 0.07 μm, while under vacuum conditions, the mean free path for air may be several microns to several meters. Obviously, operating conditions have a great influence on diffusion mechanisms.

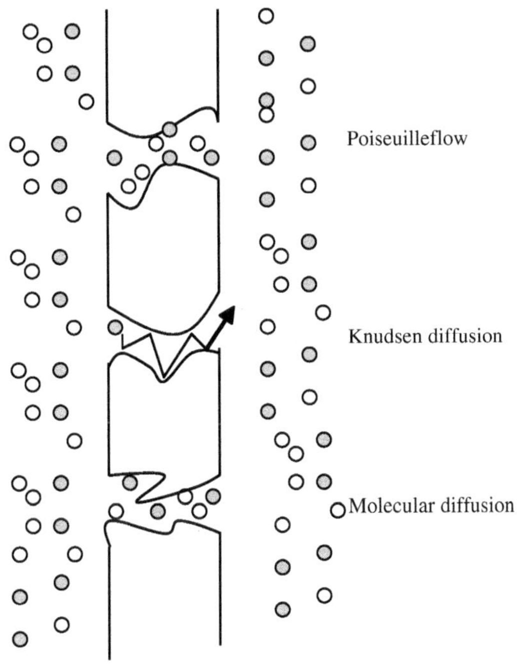

FIGURE 2.7

Diffusion mechanisms for gases in pores.

Knudsen number

$$Kn = \frac{\lambda_i}{d_p} \tag{2.75}$$

where d_p is pore diameter (m).

Poisseuille flow

When $Kn < 0.01$, i.e., the pore size is large in relation to the mean free path of gas molecules, the molecule-molecule collisions between gas molecules themselves will dominate and viscous Poisseuille flow will occur.

Knudsen diffusion

If the mean free path of the gas molecules is large in relation to the pore size, the molecule-pore wall collisions are dominant over the molecule-molecule collisions and the gas transport takes place via Knudsen flow.

Knudsen diffusion coefficient

$$D_K = \frac{d_P}{3}\sqrt{\frac{8RT}{\pi M_v}} \tag{2.76}$$

where R is the gas constant, 8.314 J/(mol K).

Molecule diffusion

If the dominant resistance is the molecular diffusion resistance caused by the virtually stagnant air trapped within the material pores, one obtains the following relationship for the gas flow in a single pore, where the ordinary diffusivity of water vapor molecule in air is expressed by [18]

$$D_O = \frac{1.86 \times 10^{-7} T^{3/2}}{P_m \sigma_{va}^2 \Omega} \sqrt{\frac{1}{M_v} + \frac{1}{M_a}} \tag{2.77}$$

where D_O is in m^2/s; T is in K; M_v and M_a are molecule weight in g/mol for vapor and air respectively; P_m is pressure of the system in atm; σ_{va} is collision diameter in angstrom (10^{-10} m), which is the mean molecular diameter of vapor and air, $\sigma_{va}=0.5(\sigma_v+\sigma_a)$; for air $\sigma_a=3.711$ Å; for vapor $\sigma_v=2.641$ Å; Ω is a function of the interactions of molecules of vapor and air, which is around 0.45 for a vapor-air mixture.

As discussed previously, when Kn (ratio of the pore size to the mean free path) ≥ 10, the Knudsen flow is dominant, and the Poisseuille mechanism may be ignored. Actually, in most cases in the air conditioning industry with microporous adsorbents, the Knudsen number is larger than 10 and Poisseuille flow can be ignored, and the flow is considered to be combined Knudsen and ordinary diffusion.

The effective diffusivity of combined Knudsen and ordinary flow is

$$D_{KO}^{-1} = \left(D_K^{-1} + D_O^{-1}\right) \tag{2.78}$$

Final effective diffusivity in the solid wall

$$D_{vs} = \frac{\varepsilon_t}{\tau_s} D_{KO} \tag{2.79}$$

where τ_s is tortuosity of pores, which is usually selected as 2.8.

2.4.4 Initial and boundary conditions

Velocity boundary conditions

For fluid, $u^*=0$, on the three walls of the duct.

Initial conditions

At $\tau = 0$, $w = w_0$; $\omega = \omega_0$; $\theta = \theta_0$; for both the air and the solid.

Inlet conditions for air stream

For adsorption mode, when $0 < \tau < \tau_{ad}$, $\omega = \omega_{ci}$; $T_a = T_{ci}$.
For desorption mode, when $0 < \tau < \tau_{de}$, $\omega_a = \omega_{hi}$; $T_a = T_{hi}$.

Conjugate boundary conditions

At the interfaces between the air stream and adsorbent walls

$$-\lambda_a \frac{\partial T}{\partial n}\bigg|_{\text{airside}} = -\lambda_s \frac{\partial T}{\partial n}\bigg|_{\text{solid side}} \tag{2.80}$$

$$-\rho_a D_{va} \frac{\partial \omega}{\partial n}\bigg|_{\text{airside}} = -\rho_a D_{vs} \frac{\partial \omega}{\partial n}\bigg|_{\text{solid side}} \tag{2.81}$$

Skew-symmetric boundary conditions

Due to the skew-symmetric placement of ducts in the honeycomb bed, as seen in Figure 2.6, there are skew-symmetric boundary conditions on the fin surfaces. As depicted schematically in Figure 2.8, the heat and mass fluxes on fin surface AB are skew-symmetric to fin surface DC. In Figure 2.8, the fin is straightened to cleanly show the symmetry of the fin length.

$$q_{AB} = -\lambda_s \left(\frac{\partial T}{\partial n}\right)_{AB} \tag{2.82}$$

$$q_{DC} = -\lambda_s \left(\frac{\partial T}{\partial n}\right)_{DC} \tag{2.83}$$

$$q_{AB}\big|_s = q_{DC}\big|_{L_f - s} \tag{2.84}$$

$$\dot{m}_{AB} = -\rho_a D_{vs} \left(\frac{\partial \omega}{\partial n}\right)_{AB} \tag{2.85}$$

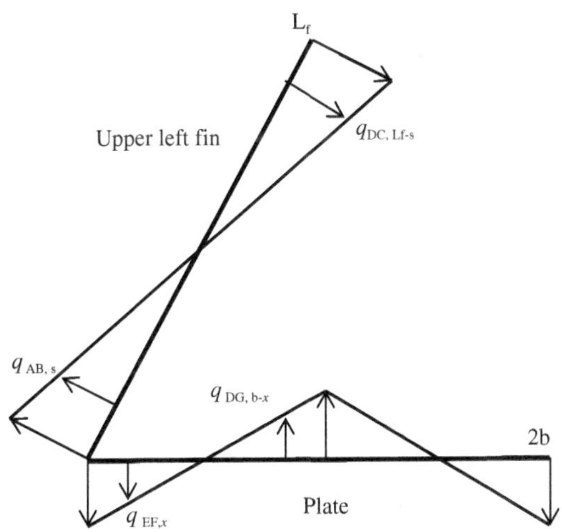

FIGURE 2.8

The skew-symmetric boundary conditions on the fins and plate. The graph shows the upper left fin and the plate, and the skew-symmetric behaviors of the heat flux on the boundaries.

$$\dot{m}_{\mathrm{DC}} = -\rho_{\mathrm{a}} D_{\mathrm{vs}} \left(\frac{\partial \omega}{\partial n} \right)_{\mathrm{DC}} \tag{2.86}$$

$$\dot{m}_{\mathrm{AB}}\big|_s = \dot{m}_{\mathrm{DC}}\big|_{L_{\mathrm{f}}-s} \tag{2.87}$$

where s is fin length from the fin root, L_{f} is the total fin length of AB or DC. Boundary conditions on fin surface BH are skew-symmetric to fin surface CG. Similarly, boundary conditions on plate surface EF are also skew-symmetric to DG, however the symmetric point is the 1/4 lower plate length. In this way, the boundary conditions on fin and plate surfaces are defined.

$$q_{\mathrm{EF}}\big|_x = q_{\mathrm{DG}}\big|_{b-x} \text{ for } x < \mathrm{b} \tag{2.88}$$

$$q_{\mathrm{EF}}\big|_x = q_{\mathrm{DG}}\big|_{3b-x} \text{ for } x > \mathrm{b} \tag{2.89}$$

$$q_{\mathrm{EF}}\big|_{x=b} = 0 \text{ for } x = \mathrm{b} \tag{2.90}$$

Similar relations exist for mass flux on the EF boundary.

Other symmetric boundary conditions

On other boundaries, such as on ADE and HGF, in Figure 2.6, symmetric boundaries are set.

$$\left(\frac{\partial T}{\partial n} \right) = \left(\frac{\partial \omega}{\partial n} \right) = 0 \tag{2.91}$$

The axial gradients of temperature and humidity in the solid at the inlet and outlet surfaces are also set to zero.

2.5 Boundary-fitted coordinates

Commonly, a duct of rectangular cross section is easy to solve under the x–y coordinate system. However, the cross section in this case is a sinusoidal one. To facilitate the solution, a boundary-fitted coordinate transformation technique is used to transfer the sinusoidal domain to a square domain. Another benefit with this methodology is that the program can be easily modified to calculate other ducts of arbitrary cross-sectional shapes, as long as the grid points on the boundaries are specified. This will provide a broad basis for program validation.

The basic idea of the boundary-fitted coordinate system is to have a coordinate system such that the body contour coincides with the coordinate lines. The transformation between the physical coordinates (x, y) and the boundary-fitted coordinates (ξ, η), which is usually a square domain, is achieved by solving two Poisson equations on the (x, y) domain [18],

$$\frac{\partial^2 \xi}{\partial x^2} + \frac{\partial^2 \xi}{\partial y^2} = P(\xi, \eta) \tag{2.92}$$

$$\frac{\partial^2 \eta}{\partial x^2} + \frac{\partial^2 \eta}{\partial y^2} = Q(\xi, \eta) \tag{2.93}$$

where $P(\xi, \eta)$ and $Q(\xi, \eta)$ are grid distribution inhomogeneous functions in the computational domain. These two equations may be more easily solved on the computational plane. Therefore, Thompson et al. [18] inverted Eqs. (2.92) and (2.93) into the transformed domain (ξ, η), where the boundary is easy to specify. At the same time, using the method proposed by Thomas and Middlecoff [18] for selecting P, Q, Eqs. (2.92) and (2.93) are inverted into

$$\alpha\left[\frac{\partial^2 x}{\partial \xi^2} + \phi\frac{\partial x}{\partial \xi}\right] - 2\beta\frac{\partial^2 x}{\partial \xi \partial \eta} + \gamma\left[\frac{\partial^2 x}{\partial \eta^2} + \psi\frac{\partial x}{\partial \eta}\right] = 0 \tag{2.94}$$

$$\alpha\left[\frac{\partial^2 y}{\partial \xi^2} + \phi\frac{\partial y}{\partial \xi}\right] - 2\beta\frac{\partial^2 y}{\partial \xi \partial \eta} + \gamma\left[\frac{\partial^2 y}{\partial \eta^2} + \psi\frac{\partial y}{\partial \eta}\right] = 0 \tag{2.95}$$

where

$$\alpha = \left[\frac{\partial x}{\partial \eta}\right]^2 + \left[\frac{\partial y}{\partial \eta}\right]^2 \tag{2.96}$$

$$\beta = \frac{\partial x}{\partial \xi}\frac{\partial x}{\partial \eta} + \frac{\partial y}{\partial \xi}\frac{\partial y}{\partial \eta} \tag{2.97}$$

$$\gamma = \left[\frac{\partial x}{\partial \xi}\right]^2 + \left[\frac{\partial y}{\partial \xi}\right]^2 \tag{2.98}$$

$$\phi = -\frac{\left[\frac{\partial x}{\partial \xi}\frac{\partial^2 x}{\partial \xi^2} + \frac{\partial y}{\partial \xi}\frac{\partial^2 y}{\partial \xi^2}\right]}{\left[\frac{\partial x}{\partial \xi}\right]^2 + \left[\frac{\partial y}{\partial \xi}\right]^2} \tag{2.99}$$

$$\psi = -\frac{\left[\frac{\partial y}{\partial \eta}\frac{\partial^2 y}{\partial \eta^2} + \frac{\partial x}{\partial \eta}\frac{\partial^2 x}{\partial \eta^2}\right]}{\left[\frac{\partial x}{\partial \eta}\right]^2 + \left[\frac{\partial y}{\partial \eta}\right]^2} \tag{2.100}$$

In the above expressions, x and y refer to the dimensionless coordinates, if normalized coordinates are used on the physical plane. The numerical value of ϕ at each grid point along the horizontal boundary $\eta = \eta_b$ in terms of boundary values x, y is computed once the differential operators are replaced by central-difference operators in Eq. (2.99). The values of ϕ at internal mesh points are computed by linear interpolation along the vertical mesh lines $\xi =$ const. Similarly, the numerical values of ψ at mesh points along vertical boundary $\xi = \xi_b$ are computed through the central difference of Eq. (2.100).

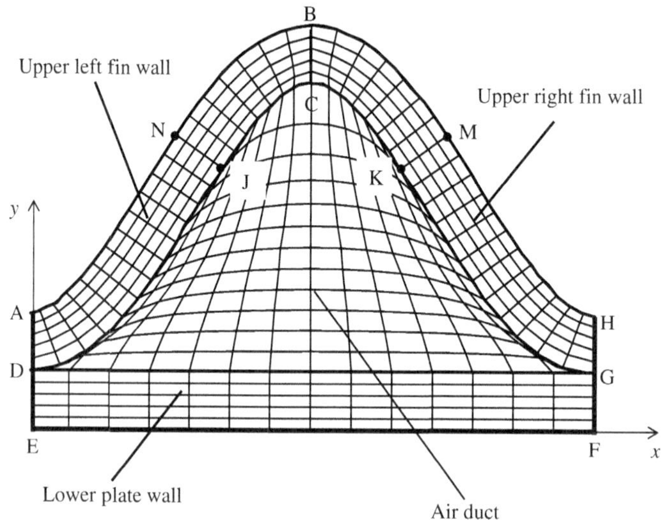

FIGURE 2.9

Grid structure of a sine duct including the net air duct and the surrounding walls, on the duct cross section, *x*–*y* plane.

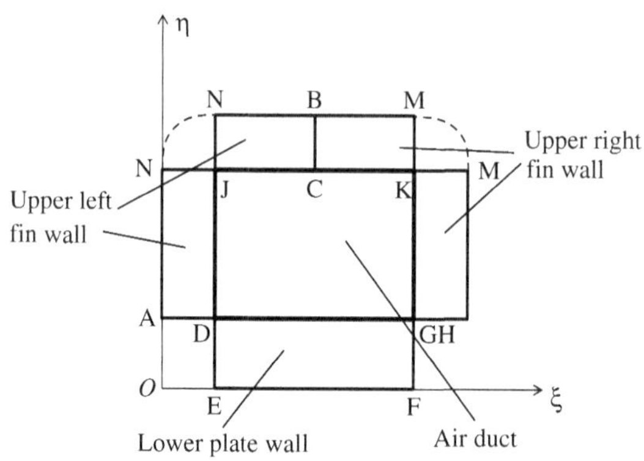

FIGURE 2.10

The computational plane corresponding to the physical plane shown in Figure 2.9.

Linear interpolations along the horizontal lines $\eta = \text{const}$ are performed to obtain the values of ψ at the internal grids. The procedure for evaluating the parameters ϕ and ψ ensures that the grid throughout the interior of the computational domain will be governed by the grid distribution that is assigned on the boundaries, and that the transverse grid lines will be locally orthogonal to the boundaries.

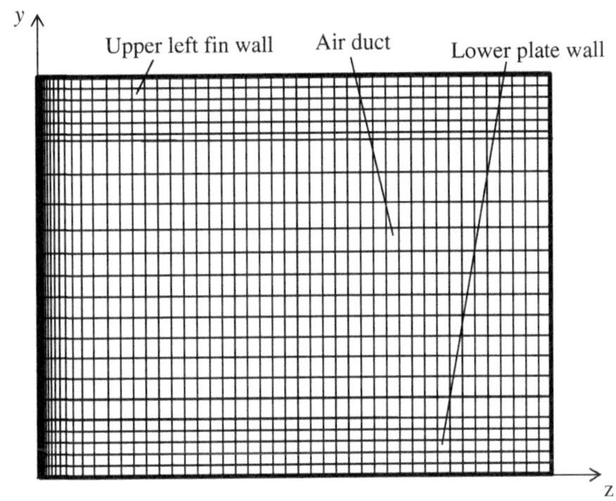

FIGURE 2.11

Grid structure on the *z–y* plane.

Once the values of ϕ and ψ are calculated, the numerical solution of Eqs. (2.94) and (2.95) by standard successive line less-relaxation on a uniform, rectangular grid $\Delta\xi$, $\Delta\eta$ results in a grid point distribution throughout the physical domain that is controlled entirely by the distribution of grid points on the boundaries.

The resulting grid constructions for the sinusoidal duct and the corresponding computational domain are shown in Figures 2.9 and 2.10, respectively.

After the setting up of boundary-fitted coordinates, the dimensionless governing equations for heat and mass transfer are transformed to the following equations on the computational plane. The grid structure on the *z–y* plane is plotted in Figure 2.11, showing the axial direction. On the computational plane, the grids in ξ (from 0 to 1) and η (from 0 to 1) are uniformly placed, but the grids are non-uniformly placed axially, to give a denser observation near the fluid inlet.

Air stream

$$\frac{\partial}{\partial \xi}\left[\frac{1}{J}\left(\alpha\frac{\partial u^*}{\partial \xi}-\beta\frac{\partial u^*}{\partial \eta}\right)\right]+\frac{\partial}{\partial \eta}\left[\frac{1}{J}\left(\gamma\frac{\partial u^*}{\partial \eta}-\beta\frac{\partial u^*}{\partial \xi}\right)\right]+J\frac{L_{\mathrm{w}}^2}{D_{\mathrm{h}}^2}=0 \tag{2.101}$$

$$Jc_1\frac{\partial \theta}{\partial \tau}+JU\frac{\partial \theta}{\partial z^*}=\frac{\partial}{\partial \xi}\left[\frac{1}{J}\left(\alpha\frac{\partial \theta}{\partial \xi}-\beta\frac{\partial \theta}{\partial \eta}\right)\right]+\frac{\partial}{\partial \eta}\left[\frac{1}{J}\left(\gamma\frac{\partial \theta}{\partial \eta}-\beta\frac{\partial \theta}{\partial \xi}\right)\right] \tag{2.102}$$

$$Jc_2\frac{\partial \omega^*}{\partial \tau}+\frac{JU}{Le}\frac{\partial \omega^*}{\partial z^*}=\frac{\partial}{\partial \xi}\left[\frac{1}{J}\left(\alpha\frac{\partial \omega^*}{\partial \xi}-\beta\frac{\partial \omega^*}{\partial \eta}\right)\right]+\frac{\partial}{\partial \eta}\left[\frac{1}{J}\left(\gamma\frac{\partial \omega^*}{\partial \eta}-\beta\frac{\partial \omega^*}{\partial \xi}\right)\right] \tag{2.103}$$

Solid

$$
\begin{aligned}
Jc_3\frac{\partial\theta}{\partial\tau}-Jc_4\frac{\partial w^*}{\partial\tau} &= c_x\frac{\partial}{\partial\xi}\left[\frac{\lambda_s}{J}\left(\alpha\frac{\partial\theta}{\partial\xi}-\beta\frac{\partial\theta}{\partial\eta}\right)\right]\\
&+c_y\frac{\partial}{\partial\eta}\left[\frac{\lambda_s}{J}\left(\gamma\frac{\partial\theta}{\partial\eta}-\beta\frac{\partial\theta}{\partial\xi}\right)\right]+Jc_z\frac{\partial}{\partial z^*}\left(\lambda_s\frac{\partial\theta}{\partial z^*}\right)
\end{aligned}
\tag{2.104}
$$

$$
\begin{aligned}
Jc_5\frac{\partial w^*}{\partial\tau}+Jc_6w^*_{eq} &= c_x\frac{\partial}{\partial\xi}\left[\frac{D_{vs}}{J}\left(\alpha\frac{\partial\omega^*}{\partial\xi}-\beta\frac{\partial\omega^*}{\partial\eta}\right)\right]\\
&+c_y\frac{\partial}{\partial\eta}\left[\frac{D_{vs}}{J}\left(\gamma\frac{\partial\omega^*}{\partial\eta}-\beta\frac{\partial\omega^*}{\partial\xi}\right)\right]+Jc_z\frac{\partial}{\partial z^*}\left(D_{vs}\frac{\partial\omega^*}{\partial z^*}\right)
\end{aligned}
\tag{2.105}
$$

where α, β, γ have the same definition as in Eqs. (2.96)–(2.98), J is the Jacobian transformation operator, which is defined as

$$J=\frac{\partial(x,y)}{\partial(\xi,\eta)}=\frac{\partial x}{\partial\xi}\frac{\partial y}{\partial\eta}-\frac{\partial x}{\partial\eta}\frac{\partial y}{\partial\xi} \tag{2.106}$$

Differential area on the physical cross section is

$$dA=Jd\xi d\eta \tag{2.107}$$

In the expressions of J, x and y here represent the dimensionless coordinates if normalized coordinates are used on the physical plane.

2.6 The finite difference forms of governing equations

Above governing differential equations on the computational plane can be transformed to finite difference equations. The grid numbering is shown in Figure 2.12 for the i–j directions and Figure 2.13 for the j–k directions. It is a three-dimensional node structure. P represents node (i, j, k). W and E represent the nodes $i-1$, and $i+1$, respectively. S and N represent the nodes $j-1$, and $j+1$, respectively. B and T represent the nodes $k-1$, and $k+1$, respectively. Position "e" refers to the east boundary of the control volume P. Similar explanations for "w," "n," "s," "b," "t." Point NE/2 refers to the center of meshes P-N-NE-E, and point SE/2 refers to the center of meshes P-S-SE-E.

Air momentum

$$
\begin{aligned}
&\frac{\partial}{\partial\xi}\left[\frac{1}{J}\left(\alpha\frac{\partial u^*}{\partial\xi}-\beta\frac{\partial u^*}{\partial\eta}\right)\right]+\frac{\partial}{\partial\eta}\left[\frac{1}{J}\left(\gamma\frac{\partial u^*}{\partial\eta}-\beta\frac{\partial u^*}{\partial\xi}\right)\right]+J\frac{L_w^2}{D_h^2}\\
&=\frac{1}{\Delta\xi}\left[\frac{1}{J}\left(\alpha\frac{\Delta u^*}{\Delta\xi}-\beta\frac{\Delta u^*}{\Delta\eta}\right)\right]_e-\frac{1}{\Delta\xi}\left[\frac{1}{J}\left(\alpha\frac{\Delta u^*}{\Delta\xi}-\beta\frac{\Delta u^*}{\Delta\eta}\right)\right]_w\\
&+\frac{1}{\Delta\eta}\left[\frac{1}{J}\left(\gamma\frac{\Delta u^*}{\Delta\eta}-\beta\frac{\Delta u^*}{\Delta\xi}\right)\right]_n-\frac{1}{\Delta\eta}\left[\frac{1}{J}\left(\gamma\frac{\Delta u^*}{\Delta\eta}-\beta\frac{\Delta u^*}{\Delta\xi}\right)\right]_s+J\frac{L_w^2}{D_h^2}
\end{aligned}
\tag{2.108}
$$

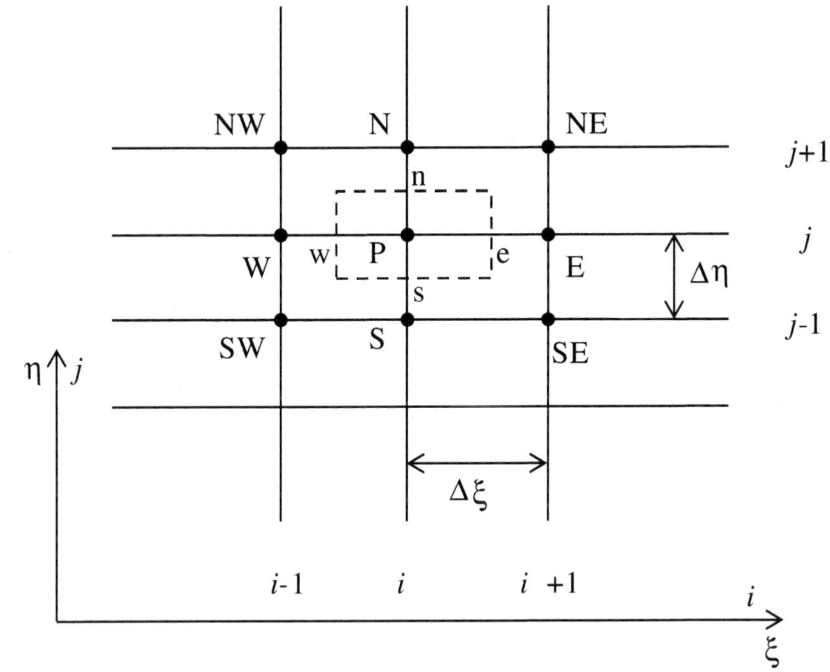

FIGURE 2.12

Grid numbering on the ξ–η plane (position index, *i*, *j*).

$$
\begin{aligned}
&= \left\{ \begin{array}{l} \frac{1}{\Delta\xi}\left(\frac{\alpha}{J}\right)_{e}\left(\frac{u_{E}^{*}-u_{P}^{*}}{\Delta\xi_{e}}\right) - \frac{1}{\Delta\xi}\left(\frac{\alpha}{J}\right)_{w}\left(\frac{u_{P}^{*}-u_{W}^{*}}{\Delta\xi_{w}}\right) + \frac{1}{\Delta\eta}\left(\frac{\gamma}{J}\right)_{n}\left(\frac{u_{N}^{*}-u_{P}^{*}}{\Delta\eta_{n}}\right) \\ -\frac{1}{\Delta\eta}\left(\frac{\gamma}{J}\right)_{s}\left(\frac{u_{P}^{*}-u_{S}^{*}}{\Delta\eta_{s}}\right) \end{array} \right\} \\
&+ \left\{ \begin{array}{l} -\frac{1}{\Delta\xi}\left(\frac{\beta}{J}\right)_{e}\left(\frac{u_{NE/2}^{*}-u_{SE/2}^{*}}{\Delta\eta_{e}}\right) + \frac{1}{\Delta\xi}\left(\frac{\beta}{J}\right)_{w}\left(\frac{u_{NW/2}^{*}-u_{SW/2}^{*}}{\Delta\eta_{w}}\right) \\ -\frac{1}{\Delta\eta}\left(\frac{\beta}{J}\right)_{n}\left(\frac{u_{NE/2}^{*}-u_{NW/2}^{*}}{\Delta\xi_{n}}\right) + \frac{1}{\Delta\eta}\left(\frac{\beta}{J}\right)_{s}\left(\frac{u_{SE/2}^{*}-u_{SW/2}^{*}}{\Delta\xi_{s}}\right) \end{array} \right\} \\
&+J\frac{L_{w}^{2}}{D_{h}^{2}} = 0
\end{aligned}
$$

The finite difference can be summarized by

$$a_{P}u_{P}^{*} = a_{E}u_{E}^{*} + a_{W}u_{W}^{*} + a_{N}u_{N}^{*} + a_{S}u_{S}^{*} + b_{1} + b_{2} \tag{2.109}$$

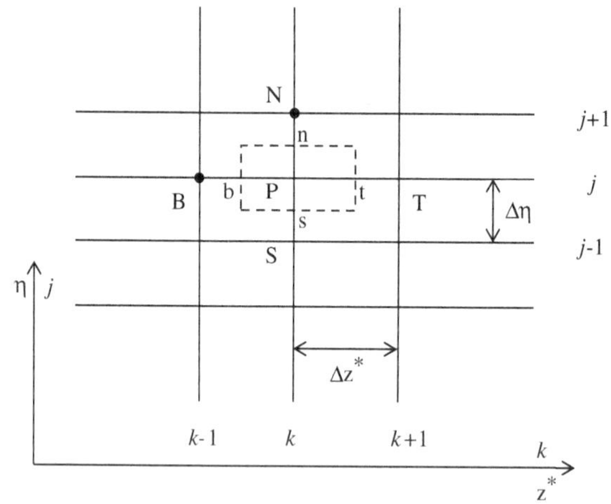

FIGURE 2.13

Grid numbering on the z*–η plane (position index, *k*, *j*).

Where

$$a_P = a_E + a_W + a_N + a_S$$

$$b_1 = -\frac{1}{(\Delta\xi)_P}\left(\frac{\beta}{J}\right)_e\left(\frac{u^*_{NE/2} - u^*_{SE/2}}{\Delta\eta_e}\right) + \frac{1}{(\Delta\xi)_P}\left(\frac{\beta}{J}\right)_w\left(\frac{u^*_{NW/2} - u^*_{SW/2}}{\Delta\eta_w}\right) - \frac{1}{(\Delta\eta)_P}\left(\frac{\beta}{J}\right)_n\left(\frac{u^*_{NE/2} - u^*_{NW/2}}{\Delta\xi_n}\right) + \frac{1}{(\Delta\eta)_P}\left(\frac{\beta}{J}\right)_s\left(\frac{u^*_{SE/2} - u^*_{SW/2}}{\Delta\xi_s}\right)$$

$$b_2 = J\frac{L_W^2}{D_h^2};\quad a_E = \frac{1}{(\Delta\xi)_P}\left(\frac{\alpha}{J}\right)_e\frac{1}{(\Delta\xi)_e};\quad a_W = \frac{1}{(\Delta\xi)_P}\left(\frac{\alpha}{J}\right)_w\frac{1}{(\Delta\xi)_w}$$

$$a_N = \frac{1}{(\Delta\eta)_P}\left(\frac{\gamma}{J}\right)_n\frac{1}{(\Delta\eta)_n};\quad a_S = \frac{1}{(\Delta\eta)_P}\left(\frac{\gamma}{J}\right)_s\frac{1}{(\Delta\eta)_s}$$

$$\frac{u^*_{NE/2} - u^*_{SE/2}}{\Delta\eta_e} = \frac{u^*_N + u^*_{NE} - u^*_S - u^*_{SE}}{4\Delta\eta_e} = \frac{u^*_{i,j+1} + u^*_{i+1,j+1} - u^*_{i,j-1} - u^*_{i+1,j-1}}{4\Delta\eta_e}$$

Air temperature

$$a_P\theta_P = a_E\theta_E + a_W\theta_W + a_N\theta_N + a_S\theta_S + a_B\theta_B + a^0_P\theta^0_P + b_1 \tag{2.110}$$

where

$$a_P = a_E + a_W + a_N + a_S + a_B + a_P^0$$
$$a_P^0 = \frac{Jc_1}{\Delta\tau}$$
$$b_1 = -\frac{1}{(\Delta\xi)_P}\left(\frac{\beta}{J}\right)_e \frac{\theta_{NE/2} - \theta_{SE/2}}{(\Delta\eta)_e} + \frac{1}{(\Delta\xi)_P}\left(\frac{\beta}{J}\right)_w \frac{\theta_{NW/2} - \theta_{SW/2}}{(\Delta\eta)_w} - \frac{1}{(\Delta\eta)_P}\left(\frac{\beta}{J}\right)_n \frac{\theta_{NE/2} - \theta_{NW/2}}{(\Delta\xi)_n} + \frac{1}{(\Delta\eta)_P}\left(\frac{\beta}{J}\right)_s \frac{\theta_{SE/2} - \theta_{SW/2}}{(\Delta\xi)_s}$$
$$a_E = \frac{1}{(\Delta\xi)_P}\left(\frac{\alpha}{J}\right)_e \frac{1}{(\Delta\xi)_e}; \quad a_W = \frac{1}{(\Delta\xi)_P}\left(\frac{\alpha}{J}\right)_w \frac{1}{(\Delta\xi)_w}; \quad a_N = \frac{1}{(\Delta\eta)_P}\left(\frac{\gamma}{J}\right)_n \frac{1}{(\Delta\eta)_n}$$
$$a_S = \frac{1}{(\Delta\eta)_P}\left(\frac{\gamma}{J}\right)_s \frac{1}{(\Delta\eta)_s}; \; a_B = \frac{JU}{(\Delta z^*)_P}$$

Air humidity

$$a_P\omega_P^* = a_E\omega_E^* + a_W\omega_W^* + a_N\omega_N^* + a_S\omega_S^* + a_B\omega_B^* + a_P^0\omega_{P0}^* + b_1 \tag{2.111}$$

where

$$a_P = a_E + a_W + a_N + a_S + a_B + a_P^0$$
$$a_P^0 = \frac{Jc_2}{\Delta\tau}$$
$$b_1 = -\frac{1}{(\Delta\xi)_P}\left(\frac{\beta}{J}\right)_e \frac{\omega_{NE/2}^* - \omega_{SE/2}^*}{(\Delta\eta)_e} + \frac{1}{(\Delta\xi)_P}\left(\frac{\beta}{J}\right)_w \frac{\omega_{NW/2}^* - \omega_{SW/2}^*}{(\Delta\eta)_w} - \frac{1}{(\Delta\eta)_P}\left(\frac{\beta}{J}\right)_n \frac{\omega_{NE/2}^* - \omega_{NW/2}^*}{(\Delta\xi)_n} + \frac{1}{(\Delta\eta)_P}\left(\frac{\beta}{J}\right)_s \frac{\omega_{SE/2}^* - \omega_{SW/2}^*}{(\Delta\xi)_s}$$
$$a_E = \frac{1}{(\Delta\xi)_P}\left(\frac{\alpha}{J}\right)_e \frac{1}{(\Delta\xi)_e}; \quad a_W = \frac{1}{(\Delta\xi)_P}\left(\frac{\alpha}{J}\right)_w \frac{1}{(\Delta\xi)_w}; \quad a_N = \frac{1}{(\Delta\eta)_P}\left(\frac{\gamma}{J}\right)_n \frac{1}{(\Delta\eta)_n}$$
$$a_S = \frac{1}{(\Delta\eta)_P}\left(\frac{\gamma}{J}\right)_s \frac{1}{(\Delta\eta)_s}; \quad a_B = \frac{JU}{Le(\Delta z^*)_P}$$

Solid wall temperature

$$a_P\theta_P = a_E\theta_E + a_W\theta_W + a_N\theta_N + a_S\theta_S + a_B\theta_B + a_T\theta_T + a_P^0\theta_P^0 + b_1 + b_2 \tag{2.112}$$

where

$$
\begin{aligned}
a_P &= a_E + a_W + a_N + a_S + a_B + a_T + a_P^0 \\
a_P^0 &= \frac{Jc_3}{\Delta\tau} \\
b_1 &= -\frac{c_x}{(\Delta\xi)_P}\left(\frac{\lambda_s\beta}{J}\right)_e \frac{\theta_{NE/2}-\theta_{SE/2}}{(\Delta\eta)_e} + \frac{c_x}{(\Delta\xi)_P}\left(\frac{\lambda_s\beta}{J}\right)_w \frac{\theta_{NW/2}-\theta_{SW/2}}{(\Delta\eta)_w} \\
&\quad -\frac{c_y}{(\Delta\eta)_P}\left(\frac{\lambda_s\beta}{J}\right)_n \frac{\theta_{NE/2}-\theta_{NW/2}}{(\Delta\xi)_n} + \frac{c_y}{(\Delta\eta)_P}\left(\frac{\lambda_s\beta}{J}\right)_s \frac{\theta_{SE/2}-\theta_{SW/2}}{(\Delta\xi)_s} \\
b_2 &= Jc_4\frac{\Delta w_P^*}{\Delta\tau}; \quad a_E = \frac{c_x}{(\Delta\xi)_P}\left(\frac{\lambda_s\alpha}{J}\right)_e \frac{1}{(\Delta\xi)_e}; \quad a_W = \frac{c_x}{(\Delta\xi)_P}\left(\frac{\lambda_s\alpha}{J}\right)_W \frac{1}{(\Delta\xi)_w} \\
a_N &= \frac{c_y}{(\Delta\eta)_P}\left(\frac{\lambda_s\gamma}{J}\right)_n \frac{1}{(\Delta\eta)_n}; \quad a_S = \frac{c_y}{(\Delta\eta)_P}\left(\frac{\lambda_s\gamma}{J}\right)_s \frac{1}{(\Delta\eta)_s}; a_B = \frac{c_zJ}{(\Delta z^*)_P}\left(\frac{\lambda_s}{\Delta z^*}\right)_b \\
a_T &= \frac{c_zJ}{(\Delta z^*)_P}\left(\frac{\lambda_s}{\Delta z^*}\right)_t; \\
\frac{\theta_{NE/2}-\theta_{SE/2}}{(\Delta\eta)_e} &= \frac{\theta_N+\theta_{NE}-\theta_S-\theta_{SE}}{4(\Delta\eta)_e}
\end{aligned}
$$

Control volume on physical plane: $\Delta V = J\Delta\xi\,\Delta\eta\,\Delta z^*$.

Solid wall humidity

$$
a_P\omega_P^* = a_E\omega_E^* + a_W\omega_W^* + a_N\omega_N^* + a_S\omega_S^* + a_B\omega_B^* + a_T\omega_T^* + a_P^0\omega_{P0}^* + b_1 + b_2 \tag{2.113}
$$

where

$$
\begin{aligned}
a_P &= a_E + a_W + a_N + a_S + a_B + a_T + a_P^0 \\
a_P^0 &= \frac{Jc_5}{\Delta\tau} \\
b_1 &= -\frac{c_x}{(\Delta\xi)_P}\left(\frac{D_{vs}\beta}{J}\right)_e \frac{\omega_{NE/2}^*-\omega_{SE/2}^*}{(\Delta\eta)_e} + \frac{c_x}{(\Delta\xi)_P}\left(\frac{D_{vs}\beta}{J}\right)_w \frac{\omega_{NW/2}^*-\omega_{SW/2}^*}{(\Delta\eta)_w} \\
&\quad -\frac{c_y}{(\Delta\eta)_P}\left(\frac{D_{vs}\beta}{J}\right)_n \frac{\omega_{NE/2}^*-\omega_{NW/2}^*}{(\Delta\xi)_n} + \frac{c_y}{(\Delta\eta)_P}\left(\frac{D_{vs}\beta}{J}\right)_s \frac{\omega_{SE/2}^*-\omega_{SW/2}^*}{(\Delta\xi)_s} \\
b_2 &= -Jc_6 w_{eq,p}^*; \quad a_E = \frac{c_x}{(\Delta\xi)_P}\left(\frac{D_{vs}\alpha}{J}\right)_e \frac{1}{(\Delta\xi)_e}; \quad a_W = \frac{c_x}{(\Delta\xi)_P}\left(\frac{D_{vs}\alpha}{J}\right)_w \frac{1}{(\Delta\xi)_w} \\
a_N &= \frac{c_y}{(\Delta\eta)_P}\left(\frac{D_{vs}\gamma}{J}\right)_n \frac{1}{(\Delta\eta)_n}; \quad a_S = \frac{c_y}{(\Delta\eta)_P}\left(\frac{D_{vs}\gamma}{J}\right)_s \frac{1}{(\Delta\eta)_s}; \quad a_B = \frac{Jc_z}{(\Delta z^*)_P}\left(\frac{D_{vs}}{\Delta z^*}\right)_b; \\
a_T &= \frac{Jc_z}{(\Delta z^*)_P}\left(\frac{D_{vs}}{\Delta z^*}\right)_t
\end{aligned}
$$

In the above equations, the finite difference forms in term b_1 are the result of transitions from the physical plane to the computational plane. They have the forms

$$\frac{\phi_{NE/2}-\phi_{SE/2}}{\Delta\eta_e}=\frac{\phi_N+\phi_{NE}-\phi_S-\phi_{SE}}{4\Delta\eta_e}=\frac{\phi_{i,j+1,k}+\phi_{i+1,j+1,k}-\phi_{i,j-1,k}-\phi_{i+1,j-1,k}}{4\Delta\eta_e} \tag{2.114}$$

$$\frac{\phi_{NW/2}-\phi_{SW/2}}{\Delta\eta_w}=\frac{\phi_N+\phi_{NW}-\phi_S-\phi_{SW}}{4\Delta\eta_w}=\frac{\phi_{i,j+1,k}+\phi_{i-1,j+1,k}-\phi_{i,j-1,k}-\phi_{i-1,j-1,k}}{4\Delta\eta_w} \tag{2.115}$$

$$\frac{\phi_{NE/2}-\phi_{NW/2}}{\Delta\xi_n}=\frac{\phi_E+\phi_{NE}-\phi_W-\phi_{NW}}{4\Delta\xi_n}=\frac{\phi_{i+1,j,k}+\phi_{i+1,j+1,k}-\phi_{i-1,j,k}-\phi_{i-1,j+1,k}}{4\Delta\xi_n} \tag{2.116}$$

$$\frac{\phi_{SE/2}-\phi_{SW/2}}{\Delta\xi_s}=\frac{\phi_E+\phi_{SE}-\phi_W-\phi_{SW}}{4\Delta\xi_s}=\frac{\phi_{i+1,j,k}+\phi_{i+1,j-1,k}-\phi_{i-1,j,k}-\phi_{i-1,j-1,k}}{4\Delta\xi_s} \tag{2.117}$$

where ϕ is a general variable, representing u^*, θ, or ω^*. For velocity, it is only two-dimensional (i, j).

Conjugate boundary conditions

The normal gradient on the physical plane boundary is transformed to the gradient on the computation plane at the boundary $\xi=\text{const}$,

$$\frac{\partial\phi}{\partial n^{(\xi)}}=\frac{\alpha\phi_\xi-\beta\phi_\eta}{J\sqrt{\alpha}}=\frac{\partial\phi}{\partial n^{(x,y)}} \tag{2.118}$$

The gradient can be written in finite difference form at point (i, j):

$$\frac{\alpha\phi_\xi-\beta\phi_\eta}{J\sqrt{\alpha}}=\frac{\alpha}{J\sqrt{\alpha}}\frac{\phi_{i,j,k}-\phi_{i-1,j,k}}{\Delta\xi}-\frac{\beta}{J\sqrt{\alpha}}\frac{\phi_{i,j,k}-\phi_{i,j-1,k}}{\Delta\eta} \tag{2.119}$$

On the computation plane at the boundary $\eta=\text{const}$,

$$\frac{\partial\phi}{\partial n^{(\eta)}}=\frac{-\beta\phi_\xi+\gamma\phi_\eta}{J\sqrt{\gamma}}=\frac{\partial\phi}{\partial n^{(x,y)}} \tag{2.120}$$

The gradient can be written in finite difference form at point (i, j):

$$\frac{-\beta\phi_\xi+\gamma\phi_\eta}{J\sqrt{\gamma}}=\frac{-\beta}{J\sqrt{\gamma}}\frac{\phi_{i,j,k}-\phi_{i-1,j,k}}{\Delta\xi}+\frac{\gamma}{J\sqrt{\gamma}}\frac{\phi_{i,j,k}-\phi_{i,j-1,k}}{\Delta\eta} \tag{2.121}$$

where ϕ is also a general variable, representing θ, or ω^*. For velocity, it is only in two dimensions (i, j). The symmetric boundary conditions at ADE and HGF in Figure 2.9 can be calculated with the above equations.

The above equations can also be used to calculate the values at conjugate boundaries if the heat and mass capacities at the nodes on the boundaries are considered infinitesimal or they are at steady state. However, in this study, the heat and mass transfer are transitional, so are the node values at boundaries. Therefore the conjugate boundary conditions are also transitional. Transitional energy and mass balance equations should be set up for the nodes at boundaries to calculate the properties of these nodes. Consequently, algebraic equations (2.112) and (2.113) should be used to calculate boundary values, for a more accurate solution.

Transient boundaries

Figure 2.14 shows a boundary node P in the solid. The neighboring node N is on the air side. The air-side properties should be used to calculate the flux from the air side. The control volume at the boundary is only half a normal internal control volume on the solid side, as shown by the node P, which is surrounded by the dashed line in Figure 2.13. The volume is $0.5J\Delta\xi\Delta\eta\Delta z^*$. Taking temperature as an example, the final heat balance equation for the boundary node P can be revised from Eq. (2.112). However, air-side properties, including thermophysical properties and grid sizes, should be used for calculating the constant a_N and the third term in b_1, which govern the fluxes from the north. They are as follows.

Solid wall temperature for the boundary node P,

$$a_P\theta_P = a_E\theta_E + a_W\theta_W + a_N\theta_N + a_S\theta_S + a_B\theta_B + a_T\theta_T + a_P^0\theta_P^0 + b_1 + b_2 \tag{2.122}$$

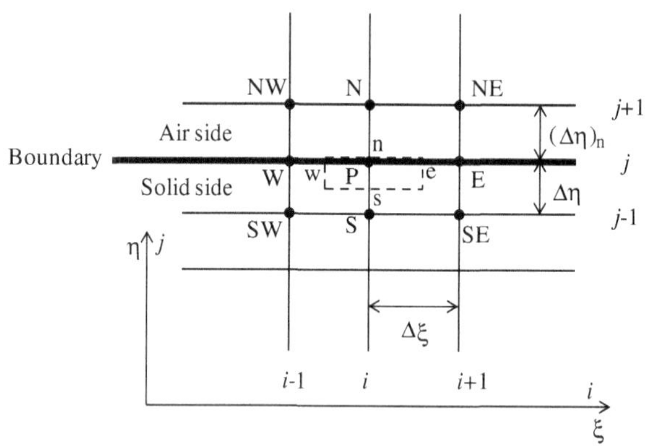

FIGURE 2.14

Grid structure at the boundary between the lower plate and air duct.

where

$$
\begin{aligned}
a_{\mathrm{P}} &= a_{\mathrm{E}} + a_{\mathrm{W}} + a_{\mathrm{N}} + a_{\mathrm{S}} + a_{\mathrm{B}} + a_{\mathrm{T}} + a_{\mathrm{P}}^{0} \\
a_{\mathrm{P}}^{0} &= \frac{Jc_3}{\Delta\tau} \\
b_1 &= -\frac{c_{\mathrm{x}}}{\Delta\xi}\left(\frac{\lambda_{\mathrm{s}}\beta}{J}\right)_{\mathrm{e}}\frac{\theta_{\mathrm{PE/2}}-\theta_{\mathrm{SE/2}}}{\Delta\eta_{\mathrm{e}}} + \frac{c_{\mathrm{x}}}{\Delta\xi}\left(\frac{\lambda_{\mathrm{s}}\beta}{J}\right)_{\mathrm{w}}\frac{\theta_{\mathrm{PW/2}}-\theta_{\mathrm{SW/2}}}{\Delta\eta_{\mathrm{w}}} \\
&\quad -\frac{c_{\mathrm{y}}}{0.5\Delta\eta}\left(\frac{\lambda_{\mathrm{a}}\beta}{J}\right)_{\mathrm{n}}\frac{\theta_{\mathrm{PE/2}}-\theta_{\mathrm{PW/2}}}{\Delta\xi_{\mathrm{n}}} + \frac{c_{\mathrm{y}}}{0.5\Delta\eta}\left(\frac{\lambda_{\mathrm{s}}\beta}{J}\right)_{\mathrm{s}}\frac{\theta_{\mathrm{SE/2}}-\theta_{\mathrm{SW/2}}}{(\Delta\xi)_{\mathrm{s}}} \\
b_2 &= Jc_4\frac{\Delta w_{\mathrm{P}}^{*}}{\Delta\tau};\quad a_{\mathrm{E}} = \frac{c_{\mathrm{x}}}{\Delta\xi}\left(\frac{\lambda_{\mathrm{s}}\alpha}{J}\right)_{\mathrm{e}}\frac{1}{\Delta\xi};\quad a_{\mathrm{w}} = \frac{c_{\mathrm{x}}}{\Delta\xi}\left(\frac{\lambda_{\mathrm{s}}\alpha}{J}\right)_{\mathrm{W}}\frac{1}{\Delta\xi} \\
a_{\mathrm{N}} &= \frac{c_{\mathrm{y}}}{0.5\Delta\eta}\left(\frac{\lambda_{\mathrm{a}}\gamma}{J}\right)_{\mathrm{n}}\frac{1}{(\Delta\eta)_{\mathrm{n}}};\quad a_{\mathrm{S}} = \frac{c_{\mathrm{y}}}{0.5\Delta\eta}\left(\frac{\lambda_{\mathrm{s}}\gamma}{J}\right)_{\mathrm{s}}\frac{1}{\Delta\eta};\quad a_{\mathrm{B}} = \frac{c_z J}{(\Delta z^{*})_{\mathrm{P}}}\left(\frac{\lambda_{\mathrm{s}}}{\Delta z^{*}}\right)_{\mathrm{b}} \\
a_{\mathrm{T}} &= \frac{c_z J}{(\Delta z^{*})_{\mathrm{P}}}\left(\frac{\lambda_{\mathrm{s}}}{\Delta z^{*}}\right)_{\mathrm{t}}; \\
\frac{\theta_{\mathrm{PE/2}}-\theta_{\mathrm{SE/2}}}{\Delta\eta_{\mathrm{e}}} &= \frac{\theta_{\mathrm{E}}+\theta_{\mathrm{P}}-\theta_{\mathrm{SE}}-\theta_{\mathrm{S}}}{2\Delta\eta};\quad \frac{\theta_{\mathrm{PW/2}}-\theta_{\mathrm{SW/2}}}{\Delta\eta_{\mathrm{w}}} = \frac{\theta_{\mathrm{W}}+\theta_{\mathrm{P}}-\theta_{\mathrm{SW}}-\theta_{\mathrm{S}}}{2\Delta\eta} \\
\frac{\theta_{\mathrm{PE/2}}-\theta_{\mathrm{PW/2}}}{\Delta\xi_{\mathrm{n}}} &= \frac{\theta_{\mathrm{E}}-\theta_{\mathrm{W}}}{2\Delta\xi};\quad \frac{\theta_{\mathrm{SE/2}}-\theta_{\mathrm{SW/2}}}{\Delta\xi_{\mathrm{s}}} = \frac{\theta_{\mathrm{E}}+\theta_{\mathrm{SE}}-\theta_{\mathrm{W}}-\theta_{\mathrm{SW}}}{4\Delta\xi}
\end{aligned}
$$

Similarly, humidity can be calculated by a revision of Eq. (2.113). It should be noted that uniform grid sizes are constructed along ξ and/or η directions. However, the grid sizes are different along the z direction. Also, the grid sizes on the air side may be different from those on the solid side, or $(\Delta\eta)_{\mathrm{n}}$ may be different from $(\Delta\eta)$.

Steady-state boundaries

If the temporal behaviors of the boundary nodes are not considered, then steady-state balances of heat fluxes can be used to calculate boundary values.

Boundary η = const

If the grids on the air side are numbered independently of the solid side, as shown in Figure 2.15 for a boundary η = const, the node numbers on each side should be clear. Note that the boundary node P is on line $j = m + 1$ for the solid side, but on line $j = 1$ for the air side, as shown in Figure 2.15.

From Eq. (2.121),

$$
q_{\mathrm{a}} = \lambda_{\mathrm{a}}\frac{-\beta}{J\sqrt{\gamma}}\frac{\theta_{i+1,1,k}^{\mathrm{a}}-\theta_{i-1,1,k}^{\mathrm{a}}}{2\Delta\xi} + \lambda_{\mathrm{a}}\frac{\gamma}{J\sqrt{\gamma}}\frac{\theta_{i,2,k}^{\mathrm{a}}-\theta_{i,1,k}^{\mathrm{a}}}{\Delta\eta_{\mathrm{a}}}
$$

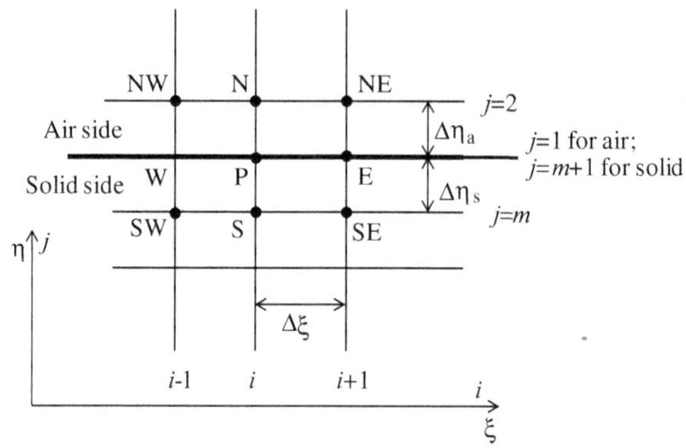

FIGURE 2.15

The nodes at the boundary η = const for heat and mass flux balance.

It can be further written to

$$q_{\mathrm{a}} = a_1\left(\theta^{\mathrm{a}}_{i+1,1,k} - \theta^{\mathrm{a}}_{i-1,1,k}\right) + a_2\left(\theta^{\mathrm{a}}_{i,2,k} - \theta^{\mathrm{a}}_{i,1,k}\right)$$
$$a_1 = \frac{-\beta}{J\sqrt{\gamma}}\frac{\lambda_{\mathrm{a}}}{2\Delta\xi}, \quad a_2 = \frac{\gamma}{J\sqrt{\gamma}}\frac{\lambda_{\mathrm{a}}}{\Delta\eta_{\mathrm{a}}}$$

where γ, β and J are calculated by air-side grids around P.

At the same time,

$$q_{\mathrm{s}} = \lambda_{\mathrm{s}}\frac{\beta}{J\sqrt{\gamma}}\frac{\theta^{\mathrm{s}}_{i+1,m+1,k} - \theta^{\mathrm{s}}_{i-1,m+1,k}}{2\Delta\xi} + \lambda_{\mathrm{s}}\frac{\gamma}{J\sqrt{\gamma}}\frac{\theta^{\mathrm{s}}_{i,m+1,k} - \theta^{\mathrm{s}}_{i,m,k}}{\Delta\eta_{\mathrm{s}}}$$
$$q_{\mathrm{s}} = b_1\left(\theta^{\mathrm{s}}_{i+1,m+1,k} - \theta^{\mathrm{s}}_{i-1,m+1,k}\right) + b_2\left(\theta^{\mathrm{s}}_{i,m+1,k} - \theta^{\mathrm{s}}_{i,m,k}\right)$$
$$b_1 = \frac{-\beta}{J\sqrt{\gamma}}\frac{\lambda_{\mathrm{s}}}{2\Delta\xi}, \quad b_2 = \frac{\gamma}{J\sqrt{\gamma}}\frac{\lambda_{\mathrm{s}}}{\Delta\eta_{\mathrm{s}}}$$

where γ, β and J are calculated by solid-side grids around P.

Since

$$q_{\mathrm{a}} = q_{\mathrm{s}}, \quad \text{and} \quad \theta^{\mathrm{a}}_{i,1,k} = \theta^{\mathrm{s}}_{i,m+1,k}$$

then the boundary temperature is calculated by combining the above equations

$$\theta^{\mathrm{s}}_{i,m+1,k} = \frac{(-a_1 + b_1)(\theta^{\mathrm{s}}_{i+1,m+1,k} - \theta^{\mathrm{s}}_{i-1,m+1,k}) - a_2\theta^{\mathrm{a}}_{i,2,k} - b_2\theta^{\mathrm{s}}_{i,m,k}}{(-a_2 - b_2)} \tag{2.123}$$

At boundary node P, the derivatives in Eqs. (2.96) to (2.98) and (2.106), are expressed specifically by:

Air side:

$$\frac{\partial x}{\partial \xi}=\frac{x_{i+1,1}-x_{i-1,1}}{2\Delta\xi},\qquad \frac{\partial x}{\partial \eta}=\frac{x_{i,2}-x_{i,1}}{\Delta\eta_{\mathrm{a}}}$$

$$\frac{\partial y}{\partial \xi}=\frac{y_{i+1,1}-y_{i-1,1}}{2\Delta\xi},\qquad \frac{\partial x}{\partial \eta}=\frac{y_{i,2}-y_{i,1}}{\Delta\eta_{\mathrm{a}}}$$

Solid side:

$$\frac{\partial x}{\partial \xi}=\frac{x_{i+1,m+1}-x_{i-1,m+1}}{2\Delta\xi},\qquad \frac{\partial x}{\partial \eta}=\frac{x_{i,m+1}-x_{i,m}}{\Delta\eta_{\mathrm{s}}}$$

$$\frac{\partial y}{\partial \xi}=\frac{y_{i+1,m+1}-y_{i-1,m+1}}{2\Delta\xi},\qquad \frac{\partial y}{\partial \eta}=\frac{y_{i,m+1}-y_{i,m}}{\Delta\eta_{\mathrm{s}}}$$

where x and y represent the dimensionless coordinates if normalized coordinates are used on the physical plane.

As can be seen, the steady-state method is simpler than the temporal method of Eq. (2.122).

Boundary $\xi=$ const

For boundary $\xi=$const, with similar techniques, Eq. (2.119) can be transformed to calculate the boundary values. Figure 2.16 shows a node P at the boundary $\xi=$const. Note that boundary node P is on line $i=n+1$ for the solid side, but on line $i=1$ for the air side, as shown in Figure 2.16.

From Eq. (2.119),

$$q_{\mathrm{a}}=\lambda_{\mathrm{a}}\frac{\alpha}{J\sqrt{\alpha}}\frac{\theta^{\mathrm{a}}_{2,j,k}-\theta^{\mathrm{a}}_{1,j,k}}{\Delta\xi_{\mathrm{a}}}-\lambda_{\mathrm{a}}\frac{\beta}{J\sqrt{\alpha}}\frac{\theta^{\mathrm{a}}_{1,j+1,k}-\theta^{\mathrm{a}}_{1,j-1,k}}{2\Delta\eta}$$

It can be further written to

$$q_{\mathrm{a}}=a_1\left(\theta^{\mathrm{a}}_{2,j,k}-\theta^{\mathrm{a}}_{1,j,k}\right)+a_2\left(\theta^{\mathrm{a}}_{1,j+1,k}-\theta^{\mathrm{a}}_{1,j-1,k}\right)$$

$$a_1=\frac{\alpha}{J\sqrt{\alpha}}\frac{\lambda_{\mathrm{a}}}{\Delta\xi_{\mathrm{a}}},\qquad a_2=\frac{-\beta}{J\sqrt{\alpha}}\frac{\lambda_{\mathrm{a}}}{2\Delta\eta}$$

where γ, β and J are calculated by air-side grids around P. It can be further written to

$$q_{\mathrm{s}}=\lambda_{\mathrm{s}}\frac{\alpha}{J\sqrt{\alpha}}\frac{\theta^{\mathrm{s}}_{n+1,j,k}-\theta^{\mathrm{s}}_{n,j,k}}{\Delta\xi_{\mathrm{s}}}-\lambda_{\mathrm{s}}\frac{\beta}{J\sqrt{\alpha}}\frac{\theta^{\mathrm{s}}_{n+1,j+1,k}-\theta^{\mathrm{s}}_{n+1,j-1,k}}{2\Delta\eta}$$

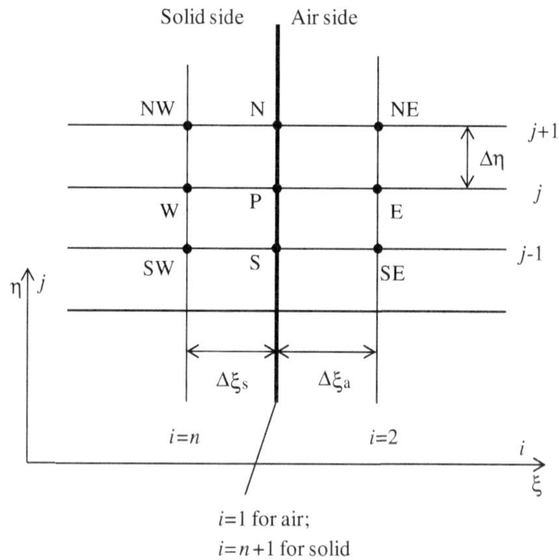

FIGURE 2.16

The nodes at the boundary ξ = const for heat and mass flux balance.

It can be further written to

$$q_s = b_1\left(\theta^s_{n+1,j,k} - \theta^s_{n,j,k}\right) + b_2\left(\theta^s_{n+1,j+1,k} - \theta^s_{n+1,j-1,k}\right)$$

$$b_1 = \frac{\alpha}{J\sqrt{\alpha}}\frac{\lambda_s}{\Delta\xi_s}; \qquad b_2 = \frac{-\beta}{J\sqrt{\alpha}}\frac{\lambda_s}{2\Delta\eta}$$

In the above equations, the parameters α, β and J are calculated by solid-side grids around P.

Since $q_a = q_s$, and $\theta^a_{1,j,k} = \theta^s_{n+1,j,k}$ then the boundary temperature is calculated by combining the above equations

$$\theta^s_{n+1,j,k} = \frac{(a_2 - b_2)\left(\theta^s_{n+1,j+1,k} - \theta^s_{n+1,j-1,k}\right) + a_1\theta^a_{2,j,k} + b_1\theta^s_{n,j,k}}{(a_1 + b_1)} \tag{2.124}$$

Superscripts or subscripts "s" mean "solid side." At boundary node P, the derivatives in Eqs. (2.96) to (2.98) and (2.106) are expressed specifically by:

Air side:

$$\frac{\partial x}{\partial \xi} = \frac{x_{2,j} - x_{1,j}}{\Delta\xi_a}, \qquad \frac{\partial x}{\partial \eta} = \frac{x_{1,j+1} - x_{1,j-1}}{2\Delta\eta}$$

$$\frac{\partial y}{\partial \xi} = \frac{y_{2,j} - y_{1,j}}{\Delta\xi_a}, \qquad \frac{\partial y}{\partial \eta} = \frac{y_{1,j+1} - y_{1,j-1}}{2\Delta\eta}$$

Solid side:

$$\frac{\partial x}{\partial \xi} = \frac{x_{n+1,j} - x_{n,j}}{\Delta \xi_s}, \quad \frac{\partial x}{\partial \eta} = \frac{x_{n+1,j+1} - x_{n+1,j-1}}{2\Delta \eta}$$
$$\frac{\partial y}{\partial \xi} = \frac{y_{n+1,j} - y_{n,j}}{\Delta \xi_s}, \quad \frac{\partial y}{\partial \eta} = \frac{y_{n+1,j+1} - y_{n+1,j-1}}{2\Delta \eta}$$

where x and y represent the dimensionless coordinates if normalized coordinates are used on the physical plane.

The skew-symmetric boundaries on the computational plane

On boundary AB, the heat or mass fluxes are skew-symmetric to boundary DC by the fin center

$$\left.\frac{\partial \phi}{\partial n}\right|_{\mathrm{AB},s^*} = \left.\frac{\partial \phi}{\partial n}\right|_{\mathrm{DC},L_f^*-s^*} \tag{2.125}$$

where s^* is normalized fin length

$$s^* = \frac{s}{L_W} \tag{2.126}$$

$$L_f^* = \frac{L_f}{L_W} \tag{2.127}$$

The boundary values on BH are symmetric to those on BA. On boundary EF, the heat or mass fluxes are skew-symmetric to those on DG by 1/4 plate length, so

$$\left.\frac{\partial \phi}{\partial n}\right|_{EF,x^*} = \left.\frac{\partial \phi}{\partial n}\right|_{DG,0.5-x^*} \quad \text{for } x^* < 0.5 \tag{2.128}$$

$$\left.\frac{\partial \phi}{\partial n}\right|_{EF,x^*} = \left.\frac{\partial \phi}{\partial n}\right|_{DG,1.5-x^*} \quad \text{for } x^* > 0.5 \tag{2.129}$$

$$\left.\frac{\partial \phi}{\partial n}\right|_{EF,x^*} = 0 \quad \text{for } x^* = 0.5 \tag{2.130}$$

The gradients on the boundaries of the computational plane can be calculated from Eqs. (2.119) and (2.121).

2.7 Numerical scheme and validation

The meshes both in the walls and in the fluid are three-dimensional. To account for the inlet influences, grids are denser near the inlet, while being relatively sparser and evenly distributed after the inlet. The grid structures are shown in Figure 2.9 and Figure 2.11.

The discretized momentum, energy and mass conservation equations for the air stream and the solids on the computational plane, Eqs. (2.109)–(2.113), are solved by the ADI (alternate direction implicit) iteration technique. They are algebraic equations. The problem is a typical conjugate one. First the fluid velocity distribution is calculated. Temperature and humidity values on the solid side are assumed. After the solution of temperature and humidity fields in the fluid, temperature and humidity values at the conjugate boundaries (between the walls and fluid) are calculated, with Eq. (2.122) for temperature. Then the temperature and humidity fields in the solid walls are solved on the generated grids, together with the skew-symmetric boundary conditions for the solid walls. Finally, other symmetric boundaries are calculated. The whole process needs iterations. The iteration loop is expressed by the following:

a. Grid generation for both the fluid and the solid walls.
b. Solve momentum equation Eq. (2.109). Get the velocity fields and resistance data for the duct.
c. Assume initial temperature and humidity fields in the solids.
d. Taking current temperature and humidity values of the solids as the boundary conditions for the fluid, get the temperature and humidity fields in the fluid, by the solution of Eqs. (2.110) and (2.111).
e. Calculate the temperature and humidity values at the conjugate boundaries, DJCKG and DG, by the solution of equations of nodes at these boundaries.
f. Taking the above solved boundary values as the default boundary values for the solids, calculate the temperature and humidity profiles in the solids, the upper sinusoidal fin walls and the lower

Table 2.1 Fully developed (fRe) and Nusselt numbers for ducts of various cross sections.

Cross sections			(fRe)			Nu_T		
			Refs [20–22]	This study	Error (%)	Refs [20–22]	This study	Error (%)
Circular			64.0	63.48	0.8	3.657	3.66	0.6
Rectan-gular	$2a/2b$	0.125	82.0	81.32	0.9	5.60	5.73	2.3
		0.25	73.0	72.64	0.6	4.44	4.55	2.5
		0.5	62.0	61.28	1.3	3.39	3.43	1.8
		1.0	56.91	56.44	1.5	2.976	3.06	2.8
Isosceles triangular	$2a/2b$	0.289	52.97	51.48	2.8	2.301	2.262	1.7
		0.5	53.20	51.76	2.9	2.359	2.451	3.9
		0.866	53.28	53.64	0.5	2.500	2.594	3.7
		1.866	52.36	51.84	1.0	2.284	2.391	4.7
Sine	$2a/2b$	0.5	44.83	44.68	0.4	2.12	2.077	2.0
		0.75	48.94	48.84	0.2	2.33	2.374	1.7
		1.0	52.01	51.81	0.3	2.45	2.377	2.9
		1.5	56.01	56.45	0.8	2.6	2.521	3.0
		2.0	58.21	58.59	0.7	—	2.718	—

plate wall. Use appropriate equations. For instance, solve Eq. (2.122) for temperatures. During this process, the skew-symmetric boundaries of ANBMH and EF should be solved simultaneously with the solution of internal grids. They are floating boundaries whose values change with the distribution of internal flows. The fluxes at these boundaries should be input to the internal grids. Otherwise, the propagation of temperature and humidity values with time cannot be generated.

g. Calculate other symmetric boundaries, such as ADE and HGF. Use Eqs. (2.123) and (2.124).

h. Go to (d), until the old values and the newly calculated values of temperature and humidity at all calculating nodes are converged.

i. Output temperature and humidity fields, and calculate Nusselt and Sherwoood numbers.

To ensure the accuracy of the results presented, a grid independence test was performed for the duct to determine the effects of the grid size. It indicated that 21×21 grids on duct cross section and $\Delta z^* = 0.001$ axially are adequate (less than 0.1% difference compared with 31×31 grids and $\Delta z^* = 0.0005$).

To further validate the numerical program, ordinary ducts of various cross sections were calculated under uniform wall temperature conditions. In this case, only the fluid flow needs to be solved. For hydrodynamically fully developed laminar flow in ducts, (fRe) is a constant. The local Nusselt numbers in the duct will decrease along the flow and reach stable values when the flow is thermally fully developed. The fully developed Nu values under uniform temperature conditions are denoted as Nu_T. The calculated values of (fRe) and Nu_T for various cross sections and aspect ratios are listed in Table 2.1. Comparisons are made with the values from well-known references [20–22]. As can be seen, the current study predicts the flow well. The results are in accordance with the published data. Maximum uncertainty is less than 5%. The code is acceptable.

2.8 Experimental work

An experimental test rig is built to study the heat and moisture transfer properties of the honeycomb desiccant ducts. The test rig is shown in Figure 2.17. The schematic is shown in Figure 2.18. The whole test rig is composed of two blowers, ductwork, the adsorbent bed, heaters, and instrumentation. Air velocities, temperatures and humidity to and from the adsorbent bed are monitored and measured by sensors. Their variations with time are recorded by a data log meter. The temperatures of the regeneration air are controlled by the heater. The temperatures and humidity of the process air are adjusted by the room air, where the whole test rig is installed. The regeneration air and the process air flow through the adsorbent bed alternately. As can be seen, when valves V2 and V3 are open, and V1 and V4 are closed, the hot regeneration air flows through the adsorbent bed. It is regenerated. After regeneration, valves V1 and V4 are opened, and V2 and V3 are closed. Cool and moist process air flows through the adsorbent bed. This is the adsorption stage.

A honeycomb-type adsorbent bed has been built and tested. The properties of the adsorbent bed are listed in Table 2.2. It is composed of 51 layers of plates separated by 50 corrugated fins, which form 50 parallel channels. Each channel is divided into 30 sinusoidal ducts by sinudoidal fins. The matrixes of plates and fins are fiberglass. When they are stacked together and form the structure, they are soaked with water glass (Na_2O·and SiO_2). Then they are dipped into HCl solution to react. After that they are

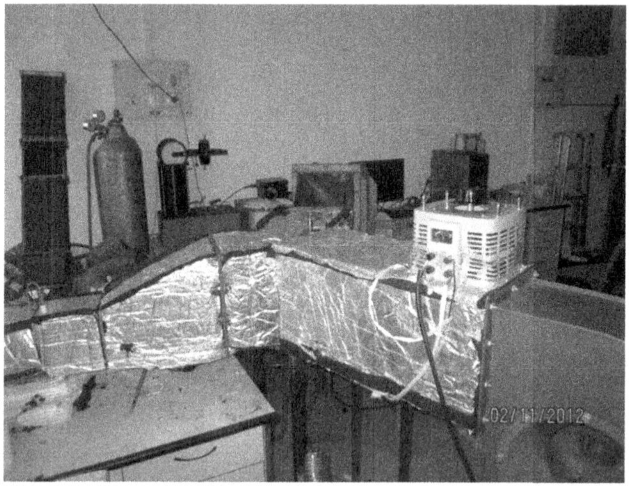

FIGURE 2.17

The test rig for the adsorbent bed.

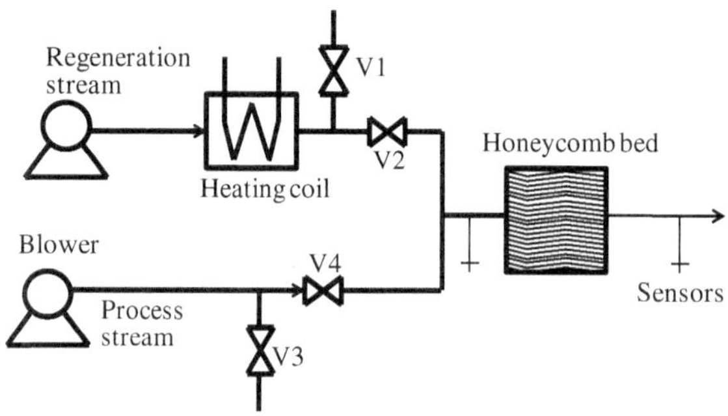

FIGURE 2.18

Schematic of the test rig.

heat treated to form composite materials on the walls. Finally, a honeycomb-type adsorbent bed is achieved. A piece of the fabricated adsorbent bed is shown in Figure 2.19. It is composed of ducts of silica gel walls.

The adsorbent bed is installed in the test rig for adsorption-regeneration cycling. The adsorbent bed is pretreated in a 60 °C hot air stream for regeneration. Before the test, the velocities, temperature and humidity of the inlet air streams are adjusted to the designed points. The adsorbent bed is by-passed during this pre-conditioning process. The cycling time is set to 6 minutes, with 3 minutes for adsorption and 3 minutes for desorption. The outlet temperatures and humidity are recorded and plotted with time in Figures 2.20 and 2.21 for temperature and humidity respectively. The inlet humidity is

Table 2.2 Physical and thermodynamic properties of the honeycomb adsorbent bed.

Item	Symbol	Unit	Value
Dimensions of bed, width × height × depth		mm × mm × mm	90 × 90.3 × 100
Duct width	$2b$	mm	3.0
Duct height (mid plane)	$2a$	mm	1.5
Wall thickness	δ	mm	0.3
Duct length	z_L	mm	100
Air velocity	u_i	m/s	1.5
Number of ducts in the adsorbent bed			1500
Mean pore diameter in the material	d_p	nm	5.2
Maximum adsorption ratio	W_{max}	kg/kg	0.35
Thermal conductivity of material	λ_s	$Wm^{-1}K^{-1}$	0.218
Density of material	ρ_s	kg/m^3	514
Specific heat of material	c_{ps}	$kJkg^{-1}K^{-1}$	0.876
Adsorption heat	q_{st}	kJ/kg	2438
Constant in Sorption curve	C_s		1.0
Porosity of material	ε_t		0.31
Tortuosity	τ_s		2.8
Moisture diffusivity in air	D_{va}	m^2/s	2.64e-5
Internal mass transfer coefficient of material	k'_m	s^{-1}	0.32

FIGURE 2.19

A segment of the tested bed.

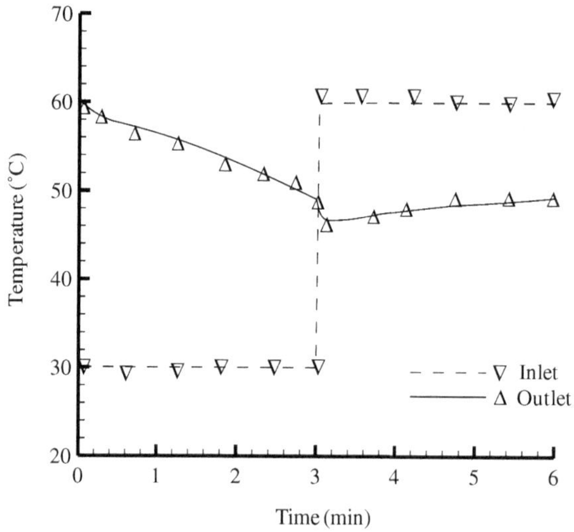

FIGURE 2.20

Variations of inlet and outlet air temperatures with time during an adsorption-regeneration cycle. The discrete dots are measured data. Initial conditions: $T_0=60\,°C$, $\omega_0=0.019$ kg/kg, $w_0=0.052$ kg/kg. Inlet conditions for adsorption: Re = 100, $T_{ai}=30\,°C$, $\omega_{ai}=0.019$ kg/kg; Inlet conditions for desorption: Re = 100, $T_{ai}=60\,°C$, $\omega_{ai}=0.019$ kg/kg.

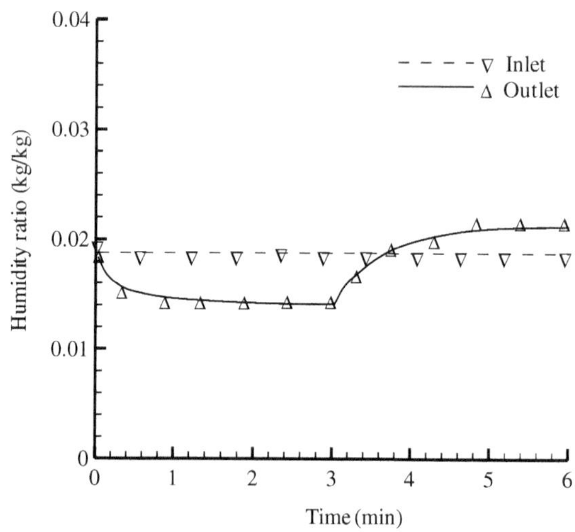

FIGURE 2.21

Variations of inlet and outlet air humidity with time. The discrete dots are measured data. Inlet conditions for adsorption: Re = 100, $T_{ai}=30\,°C$, $\omega_{ai}=0.019$ kg/kg; Inlet conditions for desorption: Re = 100, $T_{ai}=60\,°C$, $\omega_{ai}=0.019$ kg/kg.

kept unchanged while the inlet temperature is step-changed from 30 to 60 °C to realize the cooling-adsorption and heating-desorption processes. From this figure, it can be seen that the model predicts performance well. In this respect, the model is validated again. After the model validation, the numerical work will be employed to discuss the detailed transport phenomena inside the ducts.

2.9 Temperature and humidity profiles in the duct

The heat and mass transport in a single duct is modeled with various operating and duct parameters. The cooling and adsorption process is modeled for 4 minutes. Figure 2.22 shows the velocity profiles in a sine duct with aspect ratio of 0.5. The flow is hydrodynamically fully developed. The velocity profiles don't change with duct length. The shapes of the velocity contours are the same as those in common heat transfer tubes of sinusoidal cross sections. They are circular in the center but triangular near the walls. There are dead zones at the duct corners which are harmful for heat and mass transfer.

Figure 2.23 shows the temperature profiles in the duct at the cross section at $z^* = 0.4$. As can be seen, the temperature gradients in the air are higher, but lower in the solid. This is because the thermal conductivity in the solids is 10-times higher than in air. The small differences in wall temperature are also due to the simultaneous cooling of both sides by the fluids. In this respect, the wall thickness is halved. The high thermal conductivity and the small heat conduction length contribute to the fact that the temperatures in the wall are quite homogeneous. Therefore a lumped parameter assumption of wall temperature is considered reasonable in previous studies by other authors. The skew-symmetric nature of temperature distributions at wall boundaries can be observed from this figure. This is quite

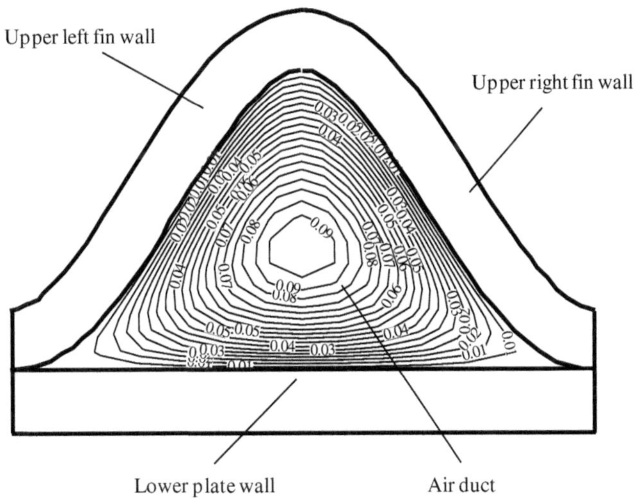

FIGURE 2.22

Profiles of dimensionless velocity (u^*) at duct cross section. Channel aspect ratio $\alpha_f = 0.5$, wall thickness ratio $\beta_f = 0.2$, Channel width $L_w = 3.0$ mm. Inlet conditions: Re = 100.

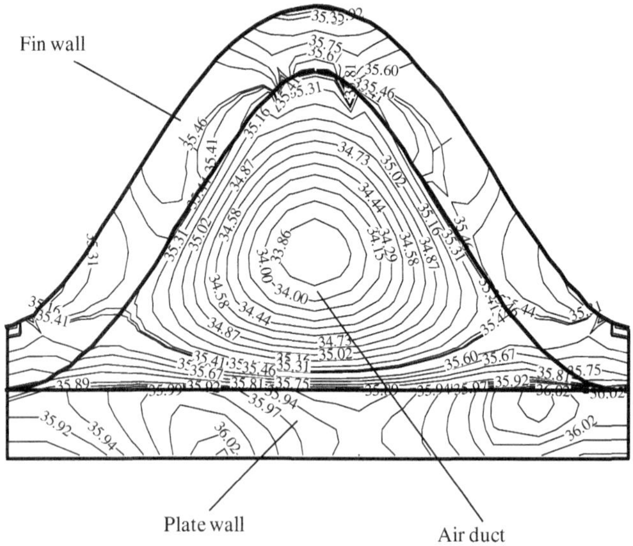

FIGURE 2.23

Profiles of temperatures (°C) at cross section $z^*=0.4$ and $t=4$ mins. Cooling and adsorption mode. Channel aspect ratio $\alpha_f=0.5$, wall thickness ratio $\beta_f=0.2$, Channel width $L_w=3.0$ mm. Inlet conditions: Re $=100$, $T_{ai}=30\,°C$, $\omega_{ai}=0.019$ kg/kg; Initial conditions: $T_0=60\,°C$, $\omega_0=0.019$ kg/kg, $w_0=0.052$ kg/kg.

characteristic for the numerous ducts in a honeycomb desiccant wall. This phenomenon could not be revealed by other previous simple models which did not reflect the effects of duct geometries. The temperature distribution in the duct is in the thermal entry length region. So temperature profiles vary with axial positions. Another point that should be noted is that due to the skew-symmetric cooling, the heat and mass transfer in the dead spaces in duct corners, which is very important when only the duct internal surfaces are cooled, is enhanced. Therefore performance is improved. The internal surfaces in the dead corners are not well cooled; however, the external surfaces are well cooled due to the boundaries' skew symmetry. Thus, the numerous honeycomb-type ducts are well placed for heat mass transfer.

The temperatures in the lower plate are somewhat higher than the upper sine fins. This is because the upper fins are cooled more quickly by the fluid. The disparities in the two walls are the result of the aspect ratios. This model shows the different heat transfer properties of the duct walls, even in the same duct and at the same duct cross section.

Figure 2.24 shows the humidity profiles at the cross section $z^*=0.4$ and $t=4$ min. As can be seen, the humidity profiles also exhibit a skew-symmetric nature. This is the same for temperature profiles. However, large gradients in wall thickness exist, which indicate high mass transfer resistance. There are large differences in humidity in wall thickness, even if the wall thickness is very small. Therefore a lumped parameter model for mass transfer in walls is no longer valid. Correspondingly, the contours of the water content in the solids are shown in Figure 2.25. The water content isolines are also mostly parallel to the solid walls, meaning high mass transfer resistance inside the walls.

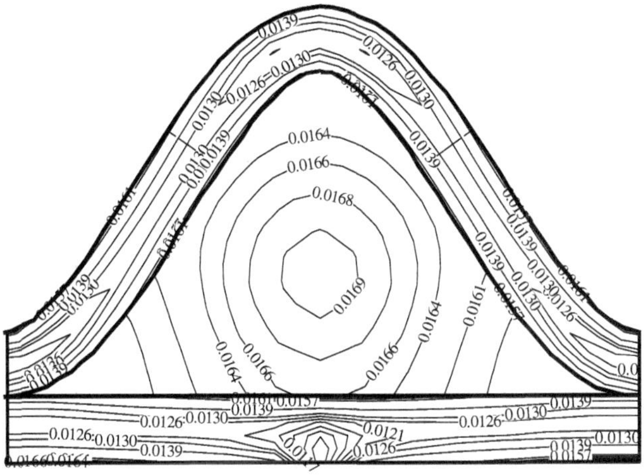

FIGURE 2.24

Profiles of humidity (kg/kg) at cross section $z^*=0.4$ and $t=4$ mins. Cooling and adsorption mode. Channel aspect ratio $\alpha_f=0.5$, wall thickness ratio $\beta_f=0.2$, Channel width $L_w=3.0$ mm. Inlet conditions: Re = 100, $T_{ai}=30\,°C$, $\omega_{ai}=0.019$ kg/kg; Initial conditions: $T_0=60\,°C$, $\omega_0=0.019$ kg/kg, $w_0=0.052$ kg/kg.

The axial distributions of temperature, humidity and water content in the upper left fin are shown in Figures 2.26 to 2.28 respectively, for the cross section at the fin center (at half fin thickness, and parallel to the upper left sin wall). The curved fins are straightened to plot these contours.

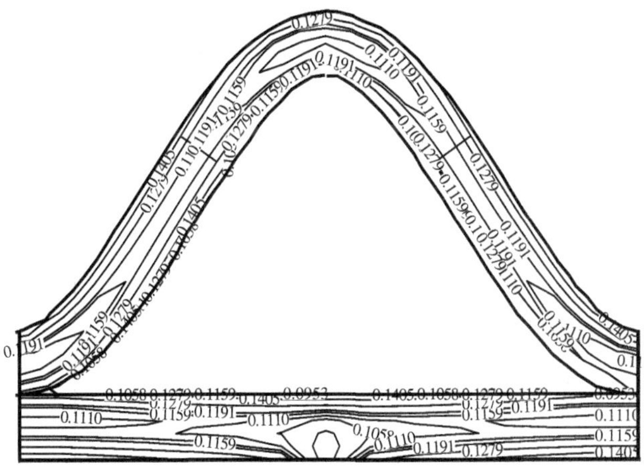

FIGURE 2.25

Profiles of water content in solid wall (kg/kg) at cross section $z^*=0.4$ and $t=4$ mins. Cooling and adsorption mode. Channel aspect ratio $\alpha_f=0.5$, wall thickness ratio $\beta_f=0.2$, Channel width $L_w=3.0$ mm. Inlet conditions: Re = 100, $T_{ai}=30\,°C$, $\omega_{ai}=0.019$ kg/kg; Initial conditions: $T_0=60\,°C$, $\omega_0=0.019$ kg/kg, $w_0=0.052$ kg/kg.

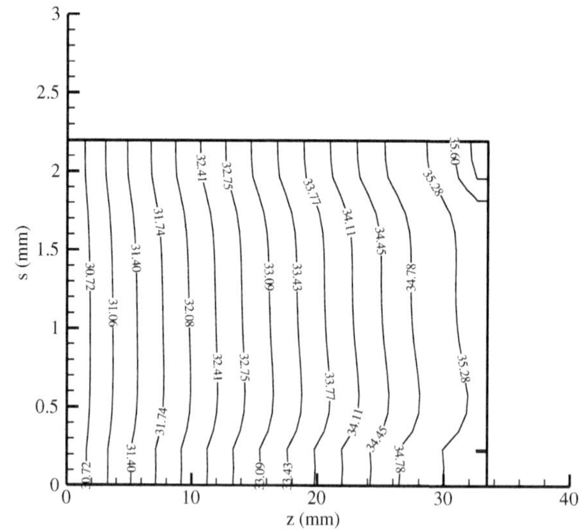

FIGURE 2.26

Profiles of temperature (°C) on the middle plane in the upper left sinusoidal fin (0.5δ, center of fin thickness) at $t = 4$ mins. Cooling and adsorption mode. Channel aspect ratio $\alpha_f = 0.5$, wall thickness ratio $\beta_f = 0.2$, Channel width $L_w = 3.0$ mm. Inlet conditions: Re = 100, $T_{ai} = 30$ °C, $\omega_{ai} = 0.019$ kg/kg; Initial conditions: $T_0 = 60$ °C, $\omega_0 = 0.019$ kg/kg, $w_0 = 0.052$ kg/kg.

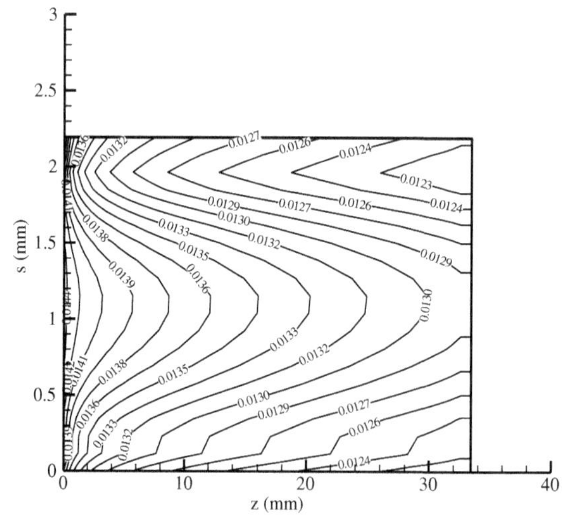

FIGURE 2.27

Profiles of humidity (kg/kg) on the middle plane in the upper left sinusoidal fin (center of fin thickness) at $t = 4$ mins and adsorption mode. Channel aspect ratio $\alpha_f = 0.5$, wall thickness ratio $\beta_f = 0.2$, Channel width $L_w = 3.0$ mm. Inlet conditions: Re = 100, $T_{ai} = 30$ °C, $\omega_{ai} = 0.019$ kg/kg; Initial conditions: $T_0 = 60$ °C, $\omega_0 = 0.019$ kg/kg, $w_0 = 0.052$ kg/kg.

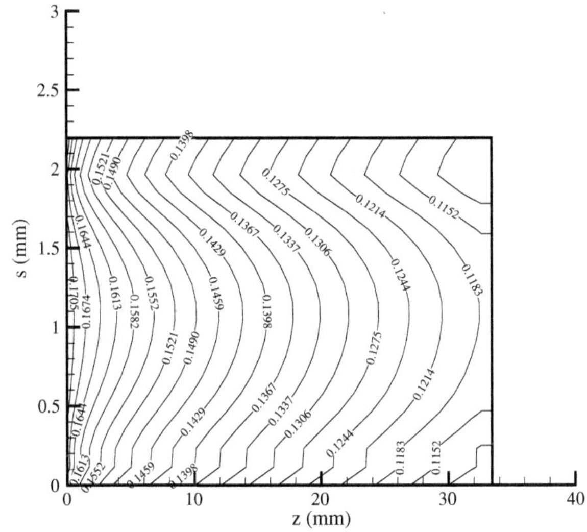

FIGURE 2.28

Profiles of water content (kg/kg) on the middle plane of the solid sinusoidal fin (center of fin thickness) at $t=4$ mins. Cooling and adsorption mode. Channel aspect ratio $\alpha_f=0.5$, wall thickness ratio $\beta_f=0.2$, Channel width $L_w=3.0$ mm. Inlet conditions: Re = 100, $T_{ai}=30\,°C$, $\omega_{ai}=0.019$ kg/kg; Initial conditions: $T_0=60\,°C$, $\omega_0=0.019$ kg/kg, $w_0=0.052$ kg/kg.

As can be seen from these figures, the heat and mass transfers in the duct are quite three-dimensional. The shapes of the contours are mostly influenced by the axial distributions of temperature, and humidity in the fluid. The wall temperatures are quite homogeneous peripherally, but inhomogeneous axially. In contrast, the humidity and water content values in the solids are quite inhomogeneous both peripherally and axially. The differences are caused by different resistances in the solids.

The axial distributions of temperature and humidity in the duct, both in the air stream and in the solids, are plotted in Figures 2.29 and 2.30 on a z–y cross section at the center of the duct. Similarly, the observed temperature differences with wall thickness are quite small; however, humidity differences with wall thickness are quite large. There are large temperature differences, but quite small humidity differences axially in the solids. There are both large temperature and humidity gradients along the axial direction in the fluid.

The temperature and humidity profiles in the duct for the heating and desorption phase are similar. Figures 2.31 and 2.32 show the temperature and humidity profiles respectively on a cross section during the heating and desorption process. As can be seen, the shapes of the contours are similar to those in the cooling process; however, the directions for heat and mass transfer are opposite. In heating and regeneration mode, moisture transfers from the solid walls to the air stream, while heat transfers from the air stream to the solid walls. In cooling and adsorption mode, heat transfers from the solid walls to the air stream, while moisture transfers from the air stream to the solid walls.

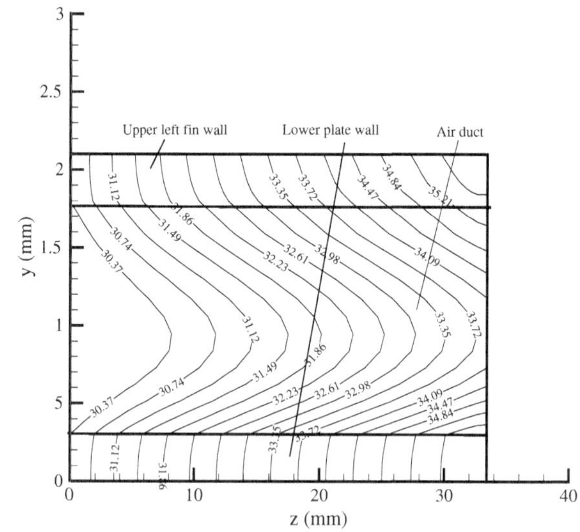

FIGURE 2.29

Profiles of temperature (°C) on the middle z-y plane in channel ($x^* = 0.5$) at $t = 4$ mins. Cooling and adsorption mode. Channel aspect ratio $\alpha_f = 0.5$, wall thickness ratio $\beta_f = 0.2$, Channel width $L_w = 3.0$ mm. Inlet conditions: $Re = 100$, $T_{ai} = 30$ °C, $\omega_{ai} = 0.019$ kg/kg; Initial conditions: $T_0 = 60$ °C, $\omega_0 = 0.019$ kg/kg, $w_0 = 0.052$ kg/kg.

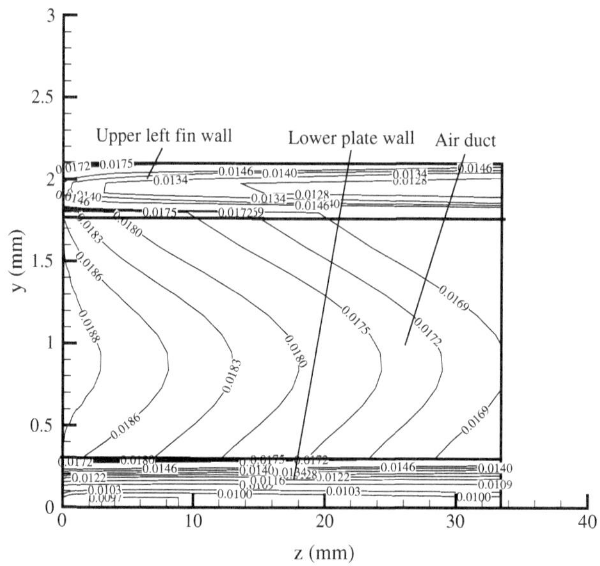

FIGURE 2.30

Profiles of humidity (kg/kg) on the middle z-y plane in channel ($x^* = 0.5$) at $t = 4$ mins. Cooling and adsorption mode. Channel aspect ratio $\alpha_f = 0.5$, wall thickness ratio $\beta_f = 0.2$, channel width $L_w = 3.0$ mm. Inlet conditions: $Re = 100$, $T_{ai} = 30$ °C, $\omega_{ai} = 0.019$ kg/kg; Initial conditions: $T_0 = 60$ °C, $\omega_0 = 0.019$ kg/kg, $w_0 = 0.052$ kg/kg.

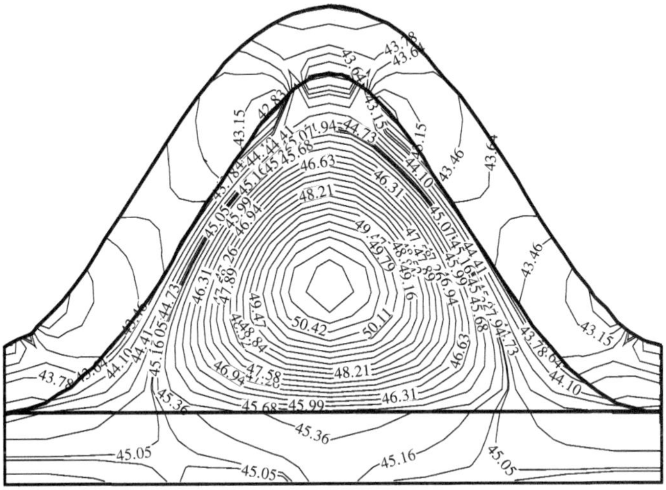

FIGURE 2.31

Profiles of temperatures (°c)at cross section $z^* = 0.4$ and $t = 4$ mins. Heating and desorption mode. Channel aspect ratio $\alpha_f = 0.5$, wall thickness ratio $\beta_f = 0.2$, Channel width $L_w = 3.0$ mm. Inlet conditions: Re = 100, $T_{ai} = 60$ °C, $\omega_{ai} = 0.019$ kg/kg; Initial conditions: $T_0 = 30$ °C, $\omega_0 = 0.019$ kg/kg, $w_0 = 0.25$ kg/kg.

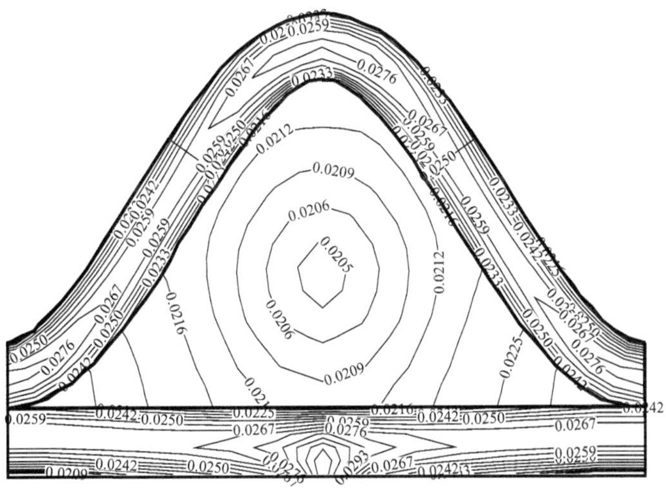

FIGURE 2.32

Profiles of humidity (kg/kg) at cross section $z^* = 0.4$ and $t = 4$ mins. Heating and desorption mode. Channel aspect ratio $\alpha_f = 0.5$, wall thickness ratio $\beta_f = 0.2$, Channel width $L_w = 3.0$ mm. Inlet conditions: Re = 100, $T_{ai} = 60$ °C, $\omega_{ai} = 0.019$ kg/kg; Initial conditions: $T_0 = 30$ °C, $\omega_0 = 0.019$ kg/kg, $w_0 = 0.25$ kg/kg.

2.10 Nusselt and Sherwood numbers in sinusoidal ducts

In ducts of common heat exchangers, temperature and humidity profiles develop along the flow. The Nusselt and Sherwood numbers develop along the flow also. They are very large at the beginning due to the very thin boundary layers. They decrease asymptotically as the boundary layers develop. They will finally reach their stable values at the end of the thermal or mass entry length. The constant values at these stable points are called fully developed values. The fully developed Nusselt numbers for the most commonly encountered ducts of various cross sections under uniform temperature conditions (denoted as Nu_T) and under uniform heat flux boundary conditions (denoted as Nu_H) can be found in many textbooks [20–22]. This book will not repeat them here. The Nusselt and Sherwood numbers in a conjugate duct surrounded by sinusoidal desiccant walls are given in this study. The heat and mass transfer in the solids would have great impacts on the Nusselt and Sherwood numbers in a real duct of conjugate heat and mass transfer.

Figure 2.33 shows the bulk temperature of the air stream, and the mean temperature of the internal walls in the duct along the flow at time $t = 2$ min. It also plots the development of Nusselt numbers, both local and mean, along the flow. As can be seen, the behaviors are very similar to common heat transfer tubes. However, much longer thermal entry lengths are observed. The thermal entry length is $z^* = 0.3$

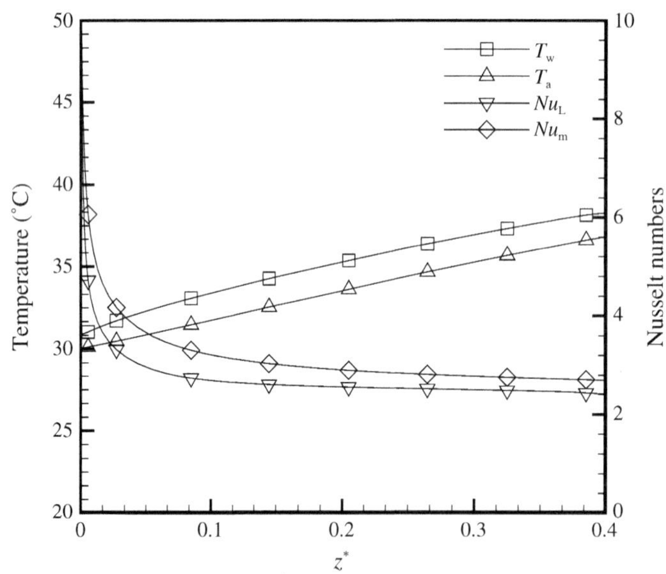

FIGURE 2.33

Variations of temperature and Nusselt numbers along duct length at $t = 2$ mins. Cooling and adsorption mode. Channel aspect ratio $\alpha_f = 0.5$, wall thickness ratio $\beta_f = 0.2$, Channel width $L_w = 3.0$ mm. Inlet conditions: Re $= 100$, $T_{ai} = 30$ °C, $\omega_{ai} = 0.019$ kg/kg; Initial conditions: $T_0 = 60$ °C, $\omega_0 = 0.019$ kg/kg, $w_0 = 0.052$ kg/kg.

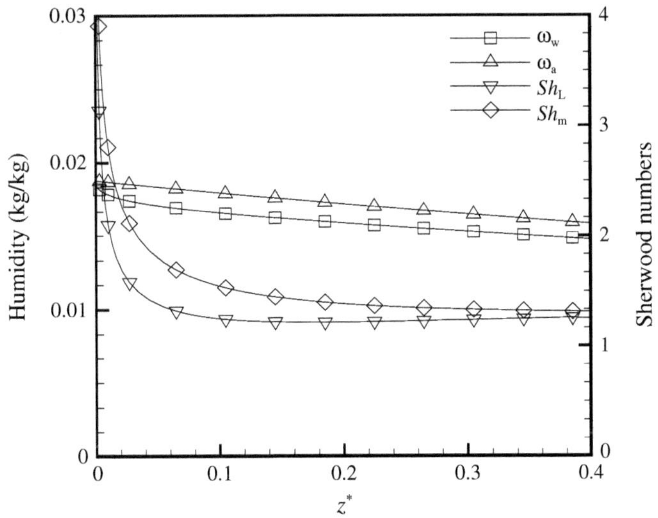

FIGURE 2.34

Variations of humidity and Sherwood numbers along duct length at $t=2$ mins. Cooling and adsorption mode. Channel aspect ratio $\alpha_f=0.5$, wall thickness ratio $\beta_f=0.2$, Channel width $L_w=3.0$ mm. Inlet conditions: Re$=100$, $T_{ai}=30\,°C$, $\omega_{ai}=0.019$ kg/kg; Initial conditions: $T_0=60\,°C$, $\omega_0=0.019$ kg/kg, $w_0=0.052$ kg/kg.

in this case. In contrast, the thermal entry length for a common duct under uniform wall temperature conditions is 0.05 [21]. The delayed entry length is due to the effects of temperature distributions on the duct walls. Figure 2.34 shows the development of the bulk humidity of the air flow and the mean humidity of the internal surfaces of the walls, along the flow at $t=2$ min. Also plotted are the variations of local and mean Sherwood numbers along the flow. As can be seen, the development of Sherwood numbers is similar to that of Nusselt numbers. After the thermal or mass entry lengths, the Nusselt and Sherwood numbers can be called "developed."

Influences of time

The operation of the adsorbent bed is a transient one. The values on the duct surfaces change with time during the operation. So the boundary conditions change with time. The Nusselt and Sherwood numbers in the duct change with time as well. Figures 2.35 and 2.36 show the variations of air temperature, humidity at the outlet, and the local and mean Nusselt and Sherwood numbers at the duct outlet. They change with time. Not only do the outlet temperature and humidity change with time, but also the Nusselt and Sherwood numbers change with time. However, after about 1–2 minutes, the outlet Nusselt and Sherwood numbers gradually reach stable values. The Nusselt and Sherwood numbers at this stage are called "fully developed" values. The finally fully developed Nusselt and Sherwood numbers under the conjugate boundary conditions are denoted as Nu_C and Sh_C respectively. If they are only fully developed axially, but not temporally, they are called "half developed."

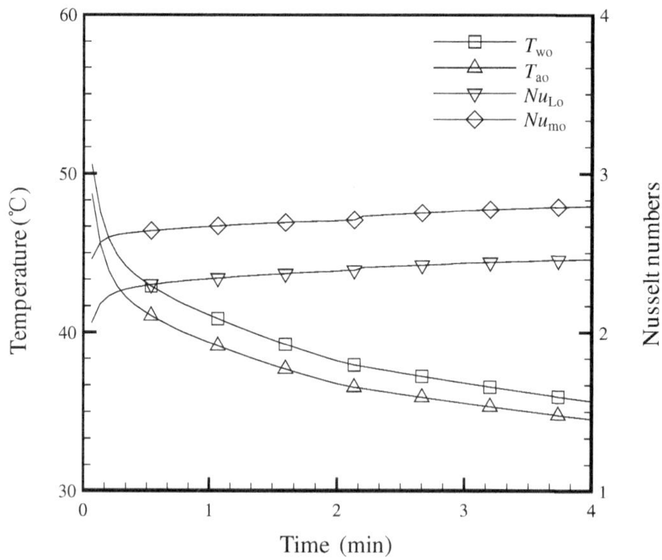

FIGURE 2.35

Variations of temperature and Nusselt numbers at duct outlet ($z^* = 0.4$) with time. Cooling and adsorption mode. Channel aspect ratio $\alpha_f = 0.5$, wall thickness ratio $\beta_f = 0.2$, Channel width $L_w = 3.0$ mm. Inlet conditions: $Re = 100$, $T_{ai} = 30$ °C, $\omega_{ai} = 0.019$ kg/kg; Initial conditions: $T_0 = 60$ °C, $\omega_0 = 0.019$ kg/kg, $w_0 = 0.052$ kg/kg.

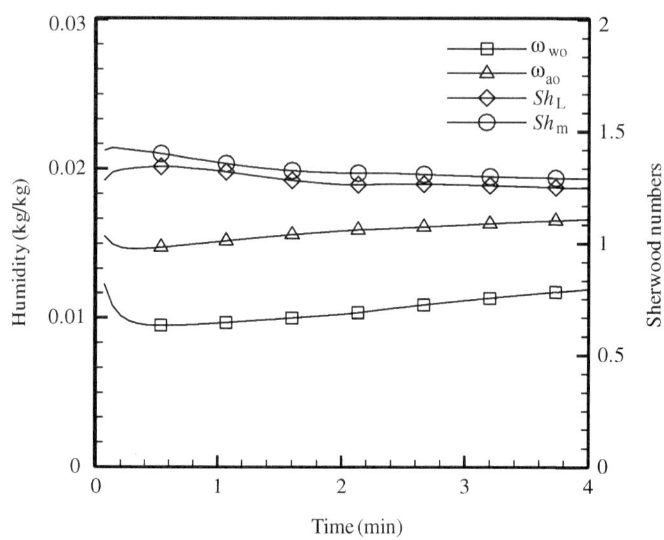

FIGURE 2.36

Variations of Humidity and Sherwood numbers at duct outlet ($z^* = 0.4$) with time. Cooling and adsorption mode. Channel aspect ratio $\alpha_f = 0.5$, wall thickness ratio $\beta_f = 0.2$, Channel width $L_w = 3.0$ mm.Inlet conditions: $Re = 100$, $T_{ai} = 30$ °C, $\omega_{ai} = 0.019$ kg/kg; Initial conditions: $T_0 = 60$ °C, $\omega_0 = 0.019$ kg/kg, $w_0 = 0.052$ kg/kg.

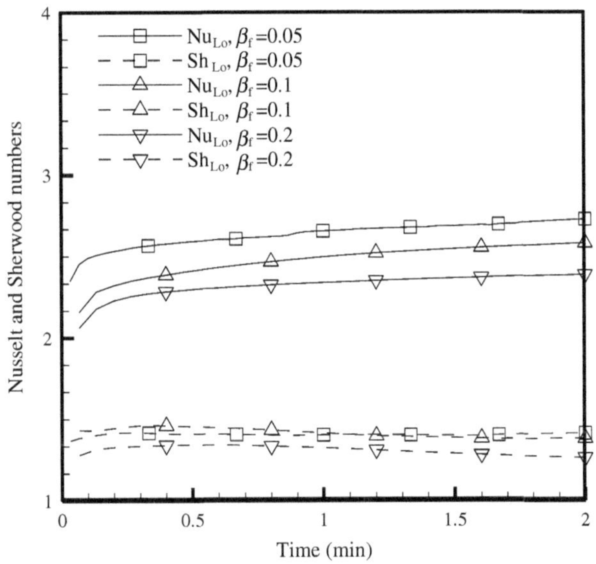

FIGURE 2.37

Variations of local Nusselt and Sherwood numbers at duct outlet with time for different wall thickness ratios. Cooling and adsorption mode. Channel aspect ratio $\alpha_f=0.5$, wall thickness ratio $\beta_f=0.05$, 0.1, 0.2; Channel width $L_w=3.0$ mm. Inlet conditions: Re = 100, $T_{ai}=30\,°C$, $\omega_{ai}=0.019$ kg/kg; Initial conditions: $T_O=60\,°C$, $\omega_O=0.019$ kg/kg, $w_O=0.052$ kg/kg.

Influence of wall thickness

Besides operation time, the thickness of the solid walls also has a great impact on heat and mass transfer in the ducts, and consequently on Nusselt and Sherwood numbers. Figure 2.37 shows the influences of wall thickness on the Nusselt and Sherwood numbers at the duct outlet. As can be seen, generally, the greater the duct wall thickness is, the lower the Nusselt and Sherwood numbers are. However, the effects of fin thickness are more complex. The fins would lead to the redistribution of temperature and humidity fields on the walls. So in some cases, the fully developed Nusselt and Sherwood numbers with wall thickness may be higher than those without wall thickness and under uniform temperature boundary conditions. The Nusselt and Sherwood numbers in the ducts of desiccant walls are quite different from those in simple traditional heat transfer tubes without considering the effects of walls.

2.11 Fully developed Nusselt and Sherwood numbers

When the Nusselt and Sherwood numbers are fully developed, both axially and temporally, they are called the fully developed values. The fully developed Nu_C and Sh_C numbers have been calculated and are listed in Table 2.3. When the fin thickness is 0, the ducts are common heat or mass transfer ducts without consideration of wall thickness. Wall effects are not considered. The values are calculated under uniform wall temperature or under wall concentration boundary conditions (T). When the fin

Table 2.3 Fully developed Nusselt and Sherwood numbers for sinusoidal adsorbent ducts (ducts with adsorbent walls).

	Fin thickness ratio, β_f			
Aspect ratio, α_f ↓	**0**		**0.05**	
	Nu_T	Sh_T	Nu_C	Sh_C
0.25	1.875	1.869	2.730	0.580
0.5	2.077	2.077	3.340	1.460
1	2.377	2.376	3.130	2.130
1.5	2.520	2.519	3.040	2.210
2	2.718	2.717	2.910	2.410
	β_f			
α_f ↓	**0.1**		**0.2**	
	Nu_C	Sh_C	Nu_C	Sh_C
0.25	2.460	0.580	2.080	0.540
0.5	2.980	1.390	3.010	1.490
1	3.010	1.910	3.110	1.340
1.5	3.390	1.520	3.630	1.120
2	3.240	1.210	3.590	1.030
	β_f			
α_f ↓	**0.3**			
	Nu_C	Sh_C		
0.25	1.830	0.540		
0.5	3.130	1.560		
1				
1.5				
2				

Notes: Nu_T, under uniform temperature wall conditions; Sh_T, under uniform concentration wall boundary conditions; Nu_C and Sh_C, under conjugate boundary conditions; β_f =0, 0.05, 0.1, 0.2, 0.3; α_f=0.25, 0.5, 1, 1.5, 2.

thickness is greater than 0, the values are calculated under conjugate boundary conditions. As can be seen from this table, the aspect ratios and duct wall thickness all have a tremendous influence on the Nusselt and Sherwood numbers. They provide the guidelines for future modeling of heat and mass transfer in adsorbent beds, as well as for desiccant wheels.

2.12 Triangular ducts

Triangular ducts are one category of ducts that are commonly used in honeycomb-type adsorbent beds. The cross sections are in isosceles triangles, as shown in Figure 2.3. The temperature and humidity profiles in a triangular duct with desiccant walls are shown in Figures 2.38 and 2.39,

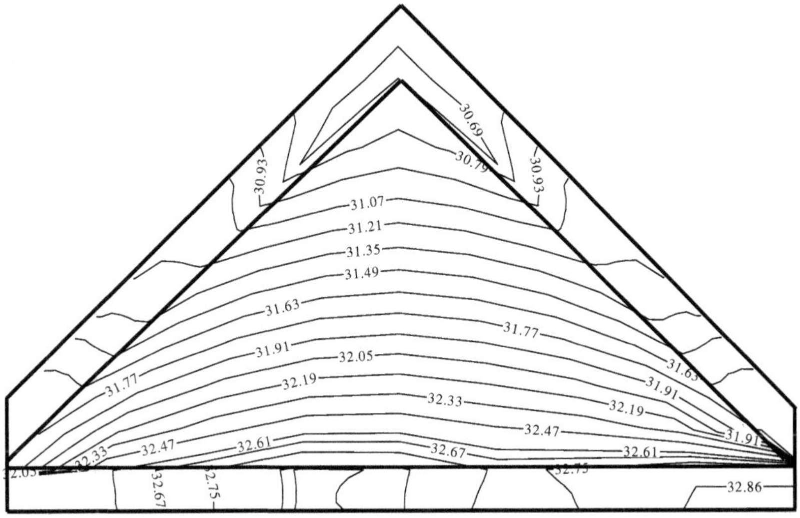

FIGURE 2.38

Profiles of temperatures (°C) at cross section $z^* = 0.9$ and $t = 2$ mins for an isosceles triangular duct. Cooling and adsorption mode. Channel aspect ratio $\alpha_f = 0.5$, wall thickness ratio $\beta_f = 0.05$, Channel width $L_w = 3.0$ mm. Inlet conditions: Re = 100, $T_{ai} = 30$ °C, $\omega_{ai} = 0.019$ kg/kg; Initial conditions: $T_0 = 60$ °C, $\omega_0 = 0.019$ kg/kg, $w_0 = 0.052$ kg/kg.

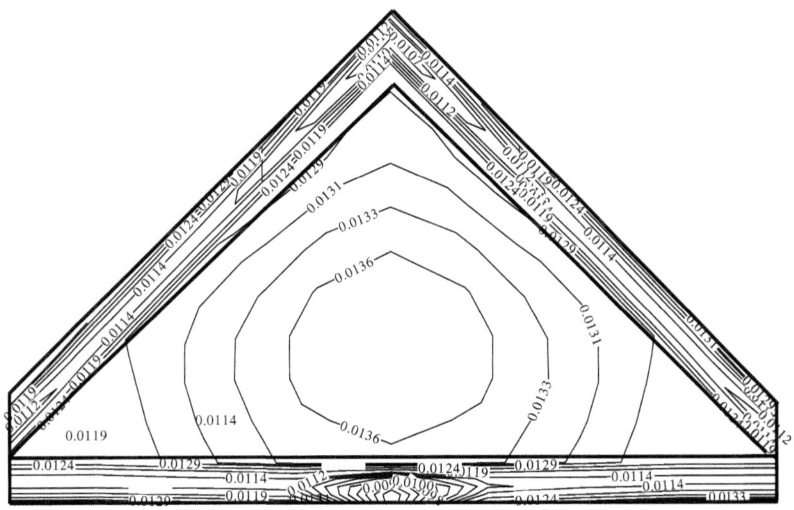

FIGURE 2.39

Profiles of humidity (kg/kg) at cross section $z^* = 0.9$ and $t = 2$ mins for an isosceles triangular duct. Cooling and adsorption mode. Channel aspect ratio $\alpha_f = 0.5$, wall thickness ratio $\beta_f = 0.05$, Channel width $L_w = 3.0$ mm. Inlet conditions: Re = 100, $T_{ai} = 30$ °C, $\omega_{ai} = 0.019$ kg/kg; Initial conditions: $T_0 = 60$ °C, $\omega_0 = 0.019$ kg/kg, $w_0 = 0.052$ kg/kg.

Table 2.4 Fully developed Nusselt and Sherwood numbers for isosceles triangular adsorbent ducts (ducts with desiccant walls).

Aspect ratio, α_f ↓	Apex angle	Fin thickness ratio, β_f			
		0		0.05	
		Nu_T	Sh_T	Nu_C	Sh_C
0.25	126.8	1.952	1.943	2.930	0.660
0.5	90	2.371	2.370	3.020	1.610
0.866	60	2.532	2.530	3.350	2.110
1	53.2	2.530	2.529	3.630	2.250
1.5	36.8	2.434	2.431	4.080	1.170
2	28	2.308	2.305	3.880	1.030
		β_f			
α_f ↓	Apex angle	0.1		0.2	
		Nu_C	Sh_C	Nu_C	Sh_C
0.25	126.8	2.530	0.630	2.470	0.610
0.5	90	3.040	1.510	3.010	1.510
0.866	60	3.440	2.160	3.460	2.160
1	53.2	3.730	2.230	3.610	2.280
1.5	36.8	4.040	1.270	4.270	1.090
2	28	3.980	1.050	4.090	1.020
		β_f			
α_f ↓	Apex angle	0.3			
		Nu_C	Sh_C		
0.25	126.8	2.360	0.580		
0.5	90	2.980	1.490		
0.866	60	3.410	2.150		
1	53.2	3.610	2.220		
1.5	36.8	3.770	1.020		
2	28	4.190	0.930		

Notes: Nu_T, under uniform temperature wall conditions; Sh_T, under uniform concentration wall boundary conditions; Nu_C and Sh_C, under conjugate boundary conditions; β_f =0, 0.05, 0.1, 0.2, 0.3; α_f=0.25, 0.5, 0.866, 1, 1.5, 2.

respectively. As can be seen, the shapes and characteristics are quite similar to those of sinusoidal ducts. The fully developed Nusselt and Sherwood numbers for a triangular duct with desiccant walls under conjugate boundary conditions are listed in Table 2.4. They can provide some reference for future system design and performance optimizations. The data can be used for the modeling of desiccant wheels.

Heat and mass transfer analogy

As can be seen from Tables 2.3 and 2.4, the ratio of Sh_C/Nu_C varies from 0.2 to 0.82. A simple analogy between heat and mass transfer no longer holds. The distributions of temperature, humidity, heat flux, and mass flux are very complicated, which leads to the difficulties in building a simple analogy between *Nu* and *Sh* values.

2.13 Conclusion

Conjugate heat and mass transfer in adsorbent ducts is a complicated phenomenon. The difficulties lie in three aspects: (1) The walls are closely conjugated with air flow; (2) heat transfer is closely coupled with mass transfer; (3) the cross sections are complex. Besides, many factors influence heat and mass transfer performance; for instance: the aspect ratios, the plate and fin thickness, the duct length, the adsorption and/or desorption processes, and even the operation time. These influences cannot be determined without a complete conjugate transient model as proposed in this chapter. Previous simple models are unable to shed light on the conjugate heat and moisture transfer problems in these ducts. More detailed temperature and humidity profiles in the ducts are disclosed here. More accurate Nusselt and Sherwood numbers specifically obtained for these ducts have been calculated. The fully developed Nusselt and Sherwood numbers obtained in this chapter will provide the basic data and guidelines for system design and performance optimization of honeycomb-type adsorbent beds, as well as for desiccant wheels. The data can be used to replace the convective Nusselt and Sherwood numbers used in the modeling of adsorbent beds and desiccant wheels, which were always borrowed from classic heat transfer tubes.

References

[1] Zhang LZ. Design and testing of an automobile waste heat adsorption cooling system. Applied Thermal Engineering 1999;20(1):103–14.

[2] Vafai K. Handbook of Porous Media. New York: Marcel Dekker; 2000.

[3] Kaviany M. Principles of Heat Transfer in Porous Media. London: Springer; 2011.

[4] Solmus I, Rees DA, Yamali C, Baker D. A two-energy equation model for dynamic heat and mass transfer in an adsorbent bed using silica gel/water pair. International Journal of Heat and Mass Transfer 2012;55:5275–88.

[5] Zhang LZ, Wang L. Momentum and heat transfer in the adsorbent of a waste heat adsorption cooling system. Energy 1999;24(7):605–24.

[6] Zhang LZ, Niu JL. Performance comparisons of desiccant wheels for air dehumidification and enthalpy recovery. Applied Thermal Engineering 2002;22(12):1347–67.

[7] Niu JL, Zhang LZ. Effects of wall thickness on the heat and moisture transfers in desiccant wheels for air dehumidification and enthalpy recovery. International Communications in Heat and Mass Transfer 2002;29(2):255–68.

[8] Zhang XJ, Dai YJ, Wang RZ. A simulation study of heat and mass transfer in a honeycombed rotary desiccant dehumidifier. Applied Thermal Engineering 2003;23:989–1003.

[9] Ruivo CR, Costa JJ, Figueiredo AR. Parametric study of the cyclic behavior of a hygroscopic matrix in a desiccant air flow system. Heat Mass Transfer 2011;47:1101–12.

[10] Ruivo CR, Costa JJ, Figueiredo AR. Validity of pseudo-gas-side-controlled models to predict the behavior of desiccant matrices. International Journal of Thermal Sciences 2009;48:2171–8.
[11] Ruivo CR, Costa JJ, Figueiredo AR. On the behavior of hygroscopic wheels: part I - channel modeling. International Journal of Heat and Mass Transfer 2007;50:4812–22.
[12] De Antonellis S, Joppolo CM, Molinaroli L. Simulation, performance analysis and optimization of desiccant wheels. Energy and Buildings 2010;42:1386–93.
[13] Niu JL, Zhang LZ. Heat transfer and friction coefficients in corrugated ducts confined by sinusoidal and arc curves. International Journal of Heat and Mass Transfer 2002;45(3):571–8.
[14] Al-Sharqawi HS, Lior N. Effect of flow-duct geometry on solid desiccant dehumidification. Industrial & Engineering Chemistry Research 2008;47:1569–85.
[15] Al-Sharqawi HS, Lior N. Transient two-dimensional dehumidification and desorption behavior of plate desiccants under humid air flow: a conjugate model that includes diffusion and adsorption/desorption within the desiccant. Industrial & Engineering Chemistry Research 2011;50:8859–80.
[16] Zhang LZ, Niu JL. A numerical study of laminar forced convection in sinusoidal ducts with arc lower boundaries under uniform wall temperature. Numerical Heat Transfer, Part A: Applications 2001;40(1):55–72.
[17] Zhang LZ. Laminar flow and heat transfer in plate-fin triangular ducts in thermally developing entry region. International Journal of Heat and Mass Transfer 2007;50(7–8):1637–40.
[18] Cussler EL. Diffusion Mass Transfer in Fluid Systems. 2nd ed.. Cambridge: Cambridge University Press; 2000.
[19] Zamel N, Li XG, Shen J, Becker J, Wiegmann A. Estimating effective thermal conductivity in carbon paper diffusion media. Chemical Engineering Science 2010;65:3994–4006.
[20] Shah RK, London AL. Laminar Flow Forced Convection in Ducts. New York: Academic Press; 1978.. pp. 253–260
[21] Incropera FP, Dewitt DP. Introduction to Heat Transfer. 3rd ed.. New York: John Wiley & Sons; 1996.. Chapter 8, p. 416
[22] Shah RK, Bhatti MS. Laminar convection heat transfer in ducts. In: Kakac S, , Shah RK, Aung W, editors. Handbook of Single-Phase Convective Heat Transfer. New York: Wiley; 1987.

CHAPTER

Heat and Mass Transfer in Revolving Adsorbent Wheels

3

3.1 Introduction

Adsorbent beds have been extensively used in various applications, such as energy recovery, air dehumidification, gas separations, air purification, etc. They work in transient mode. When the adsorbents become saturated, they need to be regenerated. To achieve continuous operation, adsorbents beds are revolved continuously between a process air and a regeneration air. For instance, Figure 3.1 shows schematically an air dehumidification and cooling system, or a desiccant cooling system. As can be seen, it is composed of a desiccant wheel, an evaporative cooler, a heater and the ducting work. The process air flows through the honeycomb ducts in the wheel and is dehumidified. The hot and dehumidified air is then cooled by an evaporative cooler and is supplied to indoors. At the same time, a regeneration air stream is heated by a heater and is then flowed through the ducts in the wheel. The saturated walls are regenerated and the capability to dehumidify air is recovered. The adsorbent wheel (desiccant wheel) is shown schematically in Figure 3.2. The wheel is composed of parallel narrow ducts of hygroscopic materials such as silica gel or LiCl-impregnated paper. When wet air passes through the channels, moisture is adsorbed to or desorbed from the solid channel walls. The surface of the wheel is also shown in Figure 3.3. The whole wheel is divided into an adsorption section and a desorption section. They are separated by plates. Obviously, sealing of the ducts is important. The cross-over of air between the two ducts is inevitably a problem, which is usually ignored in theoretical studies. The adsorption section is usually larger than desorption section.

This chapter will concentrate on the modeling of desiccant wheels. Desiccant wheels can be used in two applications: air dehumidification [1–3] and enthalpy recovery [4–6]. In the first application, process air is dried after it flows through the wheel, which rotates continuously between the process air and a hot regenerative air stream, as described in the above figure. The dried air can either be used directly or be employed to provide cooling following further psychrometric processes, i.e. so-called desiccant cooling. For this purpose the wheels are made of materials of high adsorption capability and low specific heat. The loading ratio of active materials—the adsorbents—should also be high. In the latter application, the desiccant wheel rotates between the fresh outside air and the exhaust air from room. Here the fresh air is hot and humid while the exhaust air is cool and dry. Heat and humidity would be recovered from the exhaust in winter and excess heat and moisture would be transferred to the exhaust

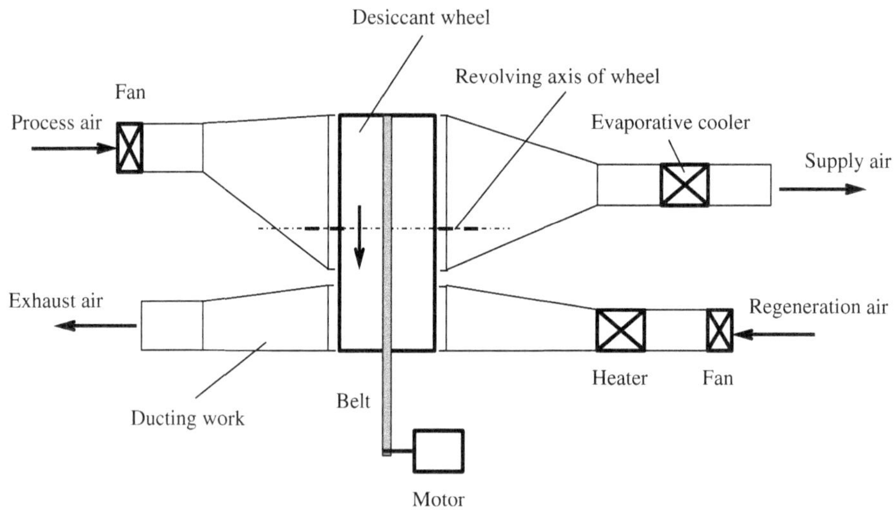

FIGURE 3.1

Schematic of a desiccant cooling system with a revolving adsorbent wheel (desiccant wheel).

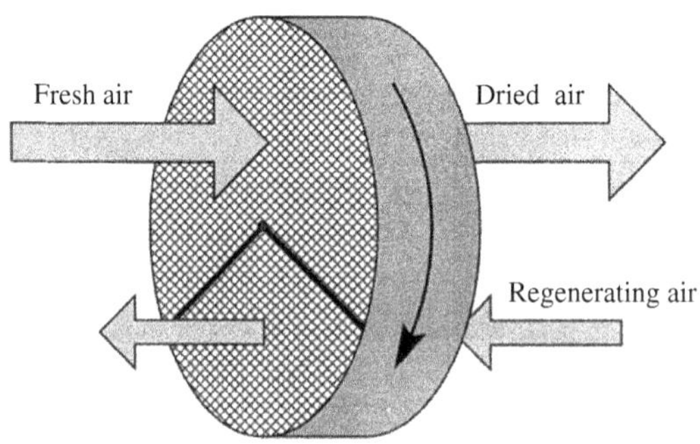

FIGURE 3.2

A schematic of an adsorbent wheel (desiccant wheel).

to cool and dehumidify the process air in summer. The latter application is also called passive dehumidification. Such wheels are termed energy wheels, and wheels of this type are made of materials of high specific heat and density. High thermal conductivity is the common requirement for both applications.

The air flow and the materials are strongly coupled in heat and mass transfer in the wheels. As described in the previous chapter, detailed conjugate heat and mass transfer in a single duct with appropriate duct geometries and boundary conditions should be studied. However, it is quite time-consuming to track the revolving wheel. The quick revolving of the wheels requires that the model be solved quickly in response to varying outside operating conditions. The solutions should also be quick for performance optimization. A good model should be simple but should consider these influencing factors.

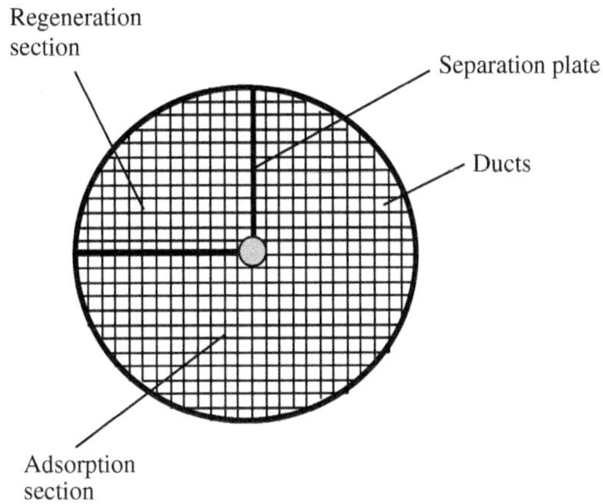

FIGURE 3.3

Surface of a desiccant wheel.

Much work has been done on the simplified modeling of heat and mass transfer in desiccant wheels. Various models have been proposed [5–11]. The common assumption is that the heat and mass transfer occurs between an air stream and a solid flat plate. This can really simplify numerical calculations. However, nearly all these models use transport data for a traditional heat transfer metal pipe as the air-side convective heat and mass transfer coefficients. In previous chapters it has been shown that this is in fact questionable. In this book, to account for the effects of duct geometries, and other operating conditions, the Nusselt and Sherwood numbers obtained in the previous chapter are adopted to calculate the air-side heat and mass transfer coefficients. In this way, the solution is fast and more accurate.

3.2 A simplified model

The adsorbent wheel is a rotating cylindrical wheel of length L and with numerous mini-channels. It is divided into two sections: Fresh air (process air) side cooling and adsorption section (angle fraction α_0) and exhaust air-side heating and desorption section (fraction $1 - \alpha_0$), where the wet fresh air and dry exhaust air streams are in a counter-flow arrangement. Duct geometries can be rectangular, hexagonal, triangular or sinusoidal shapes. The most commonly used is a sinusoidal shape due to its ease of manufacture. The flow channels that are parallel to the axis of the wheel usually have a base supporting material with an active desiccant material impregnated on their surfaces. Modern manufacturing technology (in-situ fabrication) has made it possible for the desiccant material to be closely cross-linked to the base material and distributed evenly into the macro-voids in the base material. The mixed materials can be regarded as one homogeneous material. In this model, a representative duct in the wheel is selected, as shown in Figure 3.4. The wheel duct is approximated by an air stream and two parallel adsorbent plates. Heat and mass transfer in the wheel takes place between the air stream and the two solid plates. It is a parallel-plate duct. A wheel has numerous such ducts. Because the cycles the wheel ducts undergo in rotating are identical, wheel performance can be represented by a single duct.

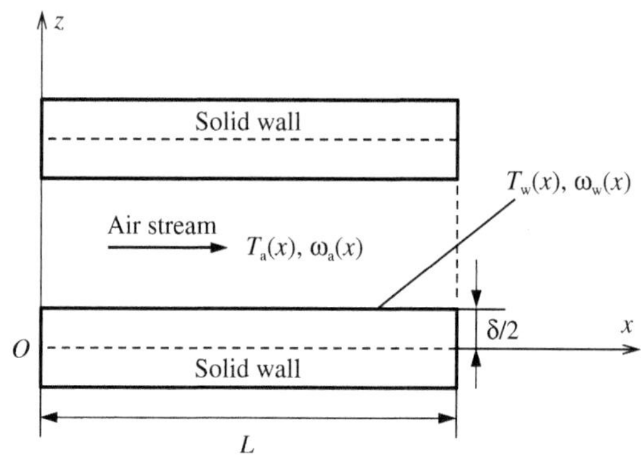

FIGURE 3.4

A representative parallel plate duct in the desiccant wheel.

The model used is transient and one-dimensional. Because of symmetry, the transverse heat conduction and mass diffusion are not considered. The previous chapter has also shown that heat and mass resistance in wall thickness are the dominant factors due to small wall thickness, so only heat conduction and mass diffusion in wall thickness are considered. The axial conduction and diffusion are ignored. Because of symmetry, the mid-plane of a solid wall can be considered to be adiabatic, and a half-size duct surrounded by dashed lines is used as the physical model, as shown in Figure 3.3. For air flow, heat and mass transfers are one-dimensional and in the axial direction.

Governing equations

The air stream in the assumed parallel-plates duct is assumed to be a slug flow. It is fully developed and the axial diffusion terms are ignored. Then the heat and mass conservation for the air stream

$$\frac{1}{u_a}\frac{\partial T_a}{\partial t} + \frac{\partial T_a}{\partial x} = \frac{4h}{D_h u_a \rho_a c_{pa}}\left(T_s - T_a\right) \tag{3.1}$$

$$\frac{1}{u_a}\frac{\partial \omega_a}{\partial t} + \frac{\partial \omega_a}{\partial x} = \frac{4k}{D_h u_a}\left(\omega_s - \omega_a\right) \tag{3.2}$$

where u_a is the air bulk velocity (m/s), T_a and ω_a are temperature (°C) and humidity ratio (kg/kg) of air respectively, t is time (s), x is axial coordinate (m), D_h is the hydrodynamic diameter of the duct (m), ρ is density (kg/m^3) and c_p is specific heat (kJ kg^{-1} K^{-1}), h and k are the convective heat transfer (kW m^{-2} K^{-1}) and mass transfer (m/s) coefficients between the air stream and the solid surface, respectively. In the equations, subscripts "s" and "a" refer to "surface" and "air" respectively. Heat and mass transfer coefficients are calculated by Nusselt number

$$Nu = \frac{hD_h}{\lambda_a} \tag{3.3}$$

Sherwood number

$$Sh = \frac{kD_h}{D_{va}} \tag{3.4}$$

where λ and D_{va} are heat conductivity (kW m^{-2} K^{-1}) and moisture diffusivity in air (m^2/s), respectively. The Nusselt and Sherwood numbers are obtained in Chapter 2 for various duct cross sections and fin thickness. The friction factors for pressure drop calculations can also be found in Chapter 2.

Hydraulic diameter of the duct

$$D_h = \frac{4A_c}{P_f} \tag{3.5}$$

where A_c is the cross-sectional area of a real air duct in the wheel (m^2), and P_f is the perimeter of a real air duct in the wheel (m).

For cooling and adsorption section, i.e., $0 \le \alpha^* < \alpha_0$, the inlet conditions,

$$\begin{cases} T_a\big|_{x=0} = T_{ci} \\ \omega_a\big|_{x=0} = \omega_{ci} \end{cases} \tag{3.6}$$

For heating and desorption section, i.e., $\alpha_0 \le \alpha^* < 1$:

$$\begin{cases} T_a\big|_{x=L} = T_{hi} \\ \omega_a\big|_{x=L} = \omega_{hi} \end{cases} \tag{3.7}$$

where subscripts "c," "h," "i" and "o" refer to cooling air, heating air, inlet and outlet, respectively. The dimensionless angle $\alpha^* = \alpha/2\pi$, and α is angle in the wheel (rad). The adsorption section in the wheel accounts for ($\alpha_0^* 2\pi$) angles of the whole wheel (2π).

Enthalpy conservation in the desiccant

$$\rho_d c_{tot} \frac{\partial T}{\partial t} = \frac{\partial}{\partial z}\left(\lambda_d \frac{\partial T}{\partial z}\right) + q_{st}\rho_d \frac{\partial w}{\partial t} \tag{3.8}$$

where ρ_d is the density of the desiccant wall (kg m^{-3}), λ_d is the thermal conductivity of the wall (kW m^{-1} K^{-1}), w is the water content in the wall (kg/kg), z is the coordinate in the thickness (m), q_{st} is the adsorption heat (kJ/kg), c_{tot} is the total heat capacity of moist desiccant, which includes two parts (dry desiccant and adsorbed water) and is calculated by

$$c_{tot} = c_{pd} + wc_{pw} \tag{3.9}$$

where c_{pd} and c_{pw} are the specific heats (kJ kg^{-1} K^{-1}) of dry desiccant and liquid water respectively. The heat capacity of moist air in the solids is ignored.

The mechanisms of moisture diffusion in desiccant are very complicated. Two phases of water, namely, gas and adsorbed state, co-exist and diffuse in the pores of the solid. There are three dominant diffusion mechanisms [5]: surface diffusion, ordinary diffusion, and Knudsen diffusion. In this model, all these mechanisms are considered and are summarized together. The equivalent moisture diffusivity in solids, based on air humidity ratio differences, is denoted by $D_{vs}(m^2/s)$. The moisture conservation in the solid can be expressed as

$$\varepsilon_t \rho_a \frac{\partial \omega}{\partial t} + \rho_d \frac{\partial w}{\partial t} = \rho_a \frac{\partial}{\partial z}\left(D_{vs}\frac{\partial \omega}{\partial z}\right) \tag{3.10}$$

where ε_t is the total porosity of the desiccant. The right-hand side of Eq. (3.10) is the moisture transfer by diffusion. The diffusivity can be calculated using the equations in Chapter 2. For engineering applications, the equivalent diffusivity is evaluated based on the equilibrium humidity ratio differences of air in the solids, rather than the concentration differences of water in the solids.

From the previous chapter, the adsorption rate is

$$\frac{\partial w}{\partial t} = k_m'\left(\omega_{eq} - \omega\right) \tag{3.11}$$

where k_m' is termed as the equivalent internal mass transfer coefficient (1/s) of adsorbents, ω_{eq} is the equilibrium humidity of solid at temperature T and uptake w.

Equilibrium water content in the desiccant is governed by a general sorption isotherm as

$$w_{eq} = \frac{W_{max}}{1 - C_s + C_s/\phi} \tag{3.12}$$

where W_{max} is the maximum equilibrium water content (kg/kg), C_s is a constant that determines the isotherm shape, ϕ is the relative humidity. Selecting T and ω as two independent variables, a differential form of adsorption content can be written in terms of humidity and temperature as

$$dw_{eq} = \sigma_\omega d\omega + \sigma_T dT \tag{3.13}$$

where

$$\sigma_\omega = \left(\frac{\partial w_{eq}}{\partial \omega}\right)_T \tag{3.14}$$

$$\sigma_T = \left(\frac{\partial w_{eq}}{\partial T}\right)_\omega \tag{3.15}$$

The above equation is for the equilibrium adsorption state. Using the Clapeyron equation to represent the saturation vapor pressure and assuming a standard atmospheric pressure of 101,325 Pa gives the relation between humidity and relative humidity as

$$\frac{\phi}{\omega} = 10^{-6}e^{5294/T} - 1.61\phi \tag{3.16}$$

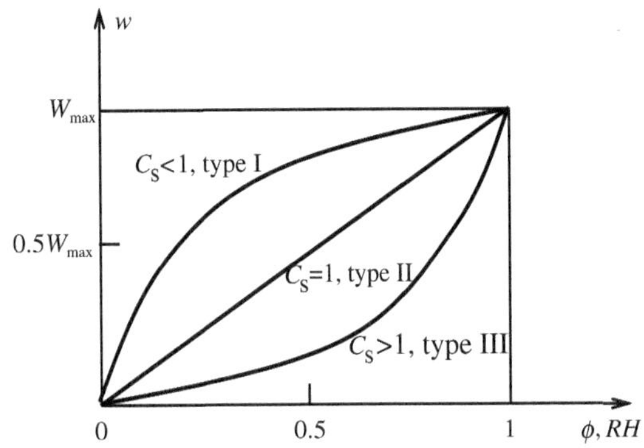

FIGURE 3.5

Sorption curves of desiccant wheel materials.

where T is in K. The second term on the right side of the equation will generally have less than a 5% effect, thus it can be ignored. Therefore

$$\sigma_\omega = 10^{-6} e^{5294/T} \frac{W_{max} C_s}{(1 - C_s + C_s/\phi)^2 \phi^2} \tag{3.17}$$

$$\sigma_T = -\frac{5294\phi}{T^2} \frac{W_{max} C_s}{(1 - C_s + C_s/\phi)^2 \phi^2} \tag{3.18}$$

where σ_ω reflects the slope of w to ω, and it is a dimensionless variable. The partial differential σ_T reflects the slope of w to T, and its unit is K^{-1}. For an enthalpy recovery wheel, the greater the σ_ω, the better the performance is; for an air dehumidification wheel, the greater the σ_T, the better the performance is. The characteristics can be found in the sorption curves of the desiccant materials. As plotted in Figure 3.5, for desiccant materials, there are three typical sorption curves. For some materials such as silica gel, $C_s = 1$ (type II); for most polymer materials, $C_s > 1$ (type III); for most inorganic materials such as zeolite molecule sieves, $C_s < 0.1$ (type I). The curve shape parameter C_s has a clear impact on the performance of materials. In most cases, non-equilibrium adsorption is applied, so Eq. (3.11) is still used.

3.2.1 Normalization

Introducing the dimensionless temperature, humidity, and water uptake

$$\theta = \frac{T - T_{ci}}{T_{hi} - T_{ci}} \tag{3.19}$$

$$\omega^* = \frac{\omega - \omega_{min}}{\omega_{max} - \omega_{min}} \tag{3.20}$$

$$w^* = \frac{w}{W_{max}} \tag{3.21}$$

and the dimensionless time

$$\tau = \frac{tN}{60} \tag{3.22}$$

where N is the rotary speed in rpm. The dimensionless coordinates

$$x^* = \frac{x}{L} \tag{3.23}$$

$$z^* = \frac{2z}{\delta} \tag{3.24}$$

Adsorption rate can be normalized to

$$\frac{\partial w^*}{\partial \tau} = k_\mathrm{m}^* \left(\omega_\mathrm{eq}^* - \omega^*\right) \tag{3.25}$$

where

$$k_\mathrm{m}^* = \frac{60k_\mathrm{m}' \left(\omega_\mathrm{max} - \omega_\mathrm{min}\right)}{NW_\mathrm{max}} \tag{3.26}$$

Energy equation (3.8) can be normalized to

$$\frac{\partial \theta}{\partial \tau} = \frac{\partial}{\partial z^*}\left(\lambda_\mathrm{z}^* \frac{\partial \theta}{\partial z^*}\right) + c_1 \left(\omega_\mathrm{eq}^* - \omega^*\right) \tag{3.27}$$

where

$$\lambda_\mathrm{z}^* = \frac{240\lambda_\mathrm{d}}{N\delta^2 \rho_\mathrm{d} c_\mathrm{tot}} \tag{3.28}$$

$$c_1 = \frac{60k_\mathrm{m}' q_\mathrm{st}}{c_\mathrm{tot} N} \frac{\left(\omega_\mathrm{max} - \omega_\mathrm{min}\right)}{\left(T_\mathrm{hi} - T_\mathrm{ci}\right)} \tag{3.29}$$

The mass equation (3.10) can be normalized to

$$\frac{\partial \omega^*}{\partial \tau} + c_2 \left(\omega_\mathrm{eq}^* - \omega^*\right) = \frac{\partial}{\partial z^*}\left(D_\mathrm{vz}^* \frac{\partial \omega^*}{\partial z^*}\right) \tag{3.30}$$

where

$$c_2 = \frac{60\rho_\mathrm{d} k_\mathrm{m}'}{\varepsilon_\mathrm{t} \rho_\mathrm{a} N} \tag{3.31}$$

$$D_\mathrm{vz}^* = \frac{240 D_\mathrm{vs}}{N\delta^2 \varepsilon_\mathrm{t}} \tag{3.32}$$

Boundary conditions for the solid phase become:

$$\left.\frac{\partial \theta}{\partial z^*}\right|_{z^*=0} = \left.\frac{\partial \theta}{\partial x^*}\right|_{x^*=0} = \left.\frac{\partial \theta}{\partial x^*}\right|_{x^*=1} = 0 \tag{3.33}$$

$$\left.\frac{\partial \omega}{\partial z^*}\right|_{z^*=0} = \left.\frac{\partial \omega}{\partial x^*}\right|_{x^*=0} = \left.\frac{\partial \omega}{\partial x^*}\right|_{x^*=1} = 0 \tag{3.34}$$

$$-\left.\frac{\partial \theta}{\partial z^*}\right|_{z^*=1} = Bi\left(\theta - \theta_{\mathrm{a}}\right) \tag{3.35}$$

$$-\left.\frac{\partial \omega}{\partial z^*}\right|_{z^*=1} = Bi_{\mathrm{m}}\left(\omega - \omega_{\mathrm{a}}\right) \tag{3.36}$$

where Bi is the Biot number for heat transfer

$$Bi = \frac{h\delta}{2\lambda_{\mathrm{d}}} \tag{3.37}$$

and Bi_{m} is the Biot number for mass transfer

$$Bi_{\mathrm{m}} = \frac{k\delta}{2D_{\mathrm{vs}}} \tag{3.38}$$

The heat and mass transfer equations for the air streams can be normalized as

$$c_3\frac{\partial \theta_{\mathrm{a}}}{\partial \tau} + \frac{\partial \theta_{\mathrm{a}}}{\partial x^*} = NTU\left(\theta_{\mathrm{s}} - \theta_{\mathrm{a}}\right) \tag{3.39}$$

$$c_3\frac{\partial \omega_{\mathrm{a}}^*}{\partial \tau} + \frac{\partial \omega_{\mathrm{a}}^*}{\partial x^*} = NTU_{\mathrm{m}}\left(\omega_{\mathrm{s}}^* - \omega_{\mathrm{a}}^*\right) \tag{3.40}$$

where

$$c_3 = \frac{NL}{60u_{\mathrm{a}}}; \qquad NTU = \frac{hA_{\mathrm{t}}}{\dot{m}_{\mathrm{a}}c_{\mathrm{pa}}}$$

$$NTU_{\mathrm{m}} = \frac{\rho_{\mathrm{a}}kA_{\mathrm{t}}}{\dot{m}_{\mathrm{a}}}$$

$$A_{\mathrm{t}} = P_{\mathrm{f}}Ln_{\mathrm{se}}; \quad \dot{m}_{\mathrm{a}} = \rho_{\mathrm{a}}u_{\mathrm{a}}A_{\mathrm{c}}n_{\mathrm{se}}$$

where A_{t} is the contact area between the solid walls and air (m^2), which is equal to the surface area of a single duct multiplied by the total number of channels in the adsorption or desorption sections of the wheel; $\dot{m}_{\mathrm{a}}$ is mass flow rate of process air or exhaust air (kg/s); n_{se} is the number of channels in the adsorption or desorption sections. The *NTU* of a single duct is the same as the *NTU* for the whole section in the wheel.

The wheel is revolving, so the initial conditions for the adsorption section are the end conditions of the regeneration section, and vice versa. The above Eqs. (3.39), (3.40), (3.27) and (3.30) are the final governing equations for the air stream and the solid.

Performance indices

The above equations are for a single duct in the wheel. The adsorption time (s) for each single duct is $t = 60\alpha^*/N$, and the desorption time is $t = 60(1-\alpha^*)/N$. The mean outlet temperature and humidity for the process air (from angle $0 \le \alpha^* < \alpha_0$, or $0 \le \tau < \alpha_0$) can be calculated by the integration of the outlet values over time τ,

$$\overline{T}_{\text{ao}} = \frac{\int_0^{\alpha_0} T_{\text{ao}} d\tau}{\alpha_0} \tag{3.41}$$

$$\overline{\omega}_{\text{ao}} = \frac{\int_0^{\alpha_0} \omega_{\text{ao}} d\tau}{\alpha_0} \tag{3.42}$$

The mean outlet temperature and humidity for the exhaust air (from angle $\alpha_0 \le \alpha^* < 1$, or $\alpha_0 \le \tau < 1$) are calculated by

$$\overline{T}_{\text{ao}} = \frac{\int_{\alpha_0}^{1} T_{\text{ao}} d\tau}{1-\alpha_0} \tag{3.43}$$

$$\overline{\omega}_{\text{ao}} = \frac{\int_{\alpha_0}^{1} \omega_{\text{ao}} d\tau}{1-\alpha_0} \tag{3.44}$$

If the wheel is used for enthalpy recovery, two measures of effectiveness are defined:

Sensible effectiveness, ε_s

$$\varepsilon_{\text{s}} = \frac{\dot{m}_{\text{c}}(T_{\text{ci}} - \overline{T}_{\text{co}})}{\dot{m}_{\text{min}}(T_{\text{ci}} - \overline{T}_{\text{hi}})} \tag{3.45}$$

Latent effectiveness, ε_L

$$\varepsilon_{\text{L}} = \frac{\dot{m}_{\text{c}}(\omega_{\text{ci}} - \overline{\omega}_{\text{co}})}{\dot{m}_{\text{min}}\omega_{\text{ci}} - \overline{\omega}_{\text{hi}})} \tag{3.46}$$

where $\dot{m}_{\text{min}}$ is the least value of process and exhaust mass flows.

Dehumidification efficiency, ε_d

$$\varepsilon_{\text{d}} = \frac{\omega_{\text{ci}} - \overline{\omega}_{\text{co}}}{\omega_{\text{ci}}} \tag{3.47}$$

Specific dehumidification power (*SDP*) on the basis of unit mass of desiccant (kg kg^{-1} s^{-1})

$$SDP = \frac{\dot{m}_c \left(\omega_{ci} - \overline{\omega}_{co}\right)}{m_d} \tag{3.48}$$

Energy coefficient of performance (*COP*)

$$COP = \frac{\dot{m}_c \left(\omega_{ci} - \overline{\omega}_{co}\right) H_v}{\dot{m}_h c_{pa} \left(T_{hi} - T_0\right)} \tag{3.49}$$

where the denominator is the energy used to heat the regeneration air from the ambient state to the inlet state, the numerator is the latent energy removed from the dehumidified air; H_v is latent heat of water vapor (kJ/kg). For the alternate cycling adsorbent beds, they are calculated for every complete cycle, including adsorption and regeneration.

3.3 Performance analysis

The one-dimensional heat and mass transfer equations of the desiccant are numerically solved by means of the ADI (alternating direction implicit) method [5,6]. Because the equations are strongly coupled and nonlinear, iterations are necessary to get converged values for each time step. This may be the only method to solve these problems. Before numerical analysis can be performed, the physical domain of the problem as well as the equations must be discretized. The whole calculating domain is divided into a number of equal-step discrete elements. Each element is identified as a control volume by a nodal point. The numbers of nodes are: 50 in axial, 10 in thickness, and 120 in time (angle). The temperature and humidity fields in the solid and in the air stream are calculated at each time interval. Then the mean outlet values are obtained by the integration of outlet values over each adsorption or desorption period. Finally, the effectiveness is calculated.

To validate the model, a test rig has been built. The schematic is shown in Figure 3.6. As seen, it is composed of: two blowers, two heaters, one humidifier, one cooler, the ducting work, and the sensors for temperature, humidity, and air velocity measurement. The desiccant wheel is used to perform heat and moisture exchange between the process air (hot and humid) and the exhaust air (cool and dry). Measurement uncertainties are: temperature 0.2 °C, humidity 4%, velocity 2%, and the uncertainties for sensible and latent effectiveness are 4.5% and 6.5% respectively. Inlet and outlet temperature, humidity and velocity, are monitored and recorded by a computer. The data sampling time is 2 s. The variations of inlet and outlet properties are recorded with time.

For enthalpy recovery, the inlet temperature and humidity are set to: fresh air, 35 °C, 0.021 kg/kg; exhaust air 24 °C, 0.012 kg/kg. The thermophysical and transport properties for the tested desiccant wheel are listed in Table 3.1. The desiccant material is a newly developed hygroscopic material similar to silica gel. It is in-situ fabricated on fiber glass, which is first corrugated to sinusoidal plate-fin honeycomb structures. The geometry of the channels in the wheel is sinusoidal with a width of 3.0 mm and a height of 1.8 mm. The effective cross-sectional area for each duct is 1.36 mm^2, and the perimeter for each duct is 3.92 mm. The wall thickness is 0.3 mm.

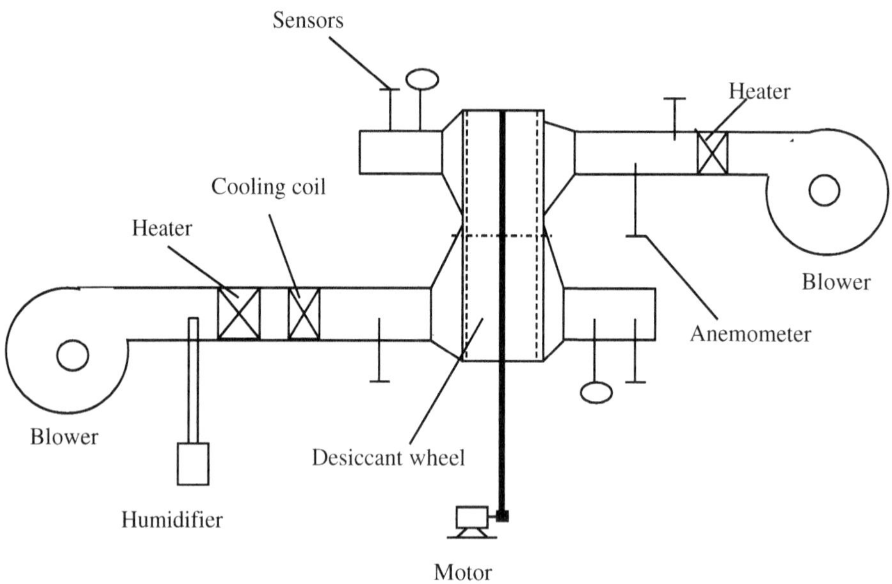

FIGURE 3.6

Test rig for the desiccant wheel.

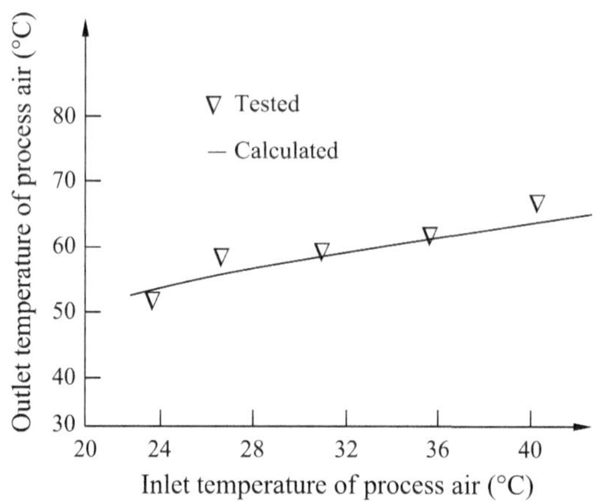

FIGURE 3.7

Comparison of the outlet temperatures of the process air. Other operating conditions are fixed.

The inlet temperature and humidity of the process air are changed, while other operating conditions are kept unchanged. The outlet temperature and humidity are measured. The measured data are compared with the data calculated with the model. The results are plotted in Figures 3.7 and 3.8. As can be seen, the agreements are generally good. The model, though only one-dimensional, can be used to estimate the

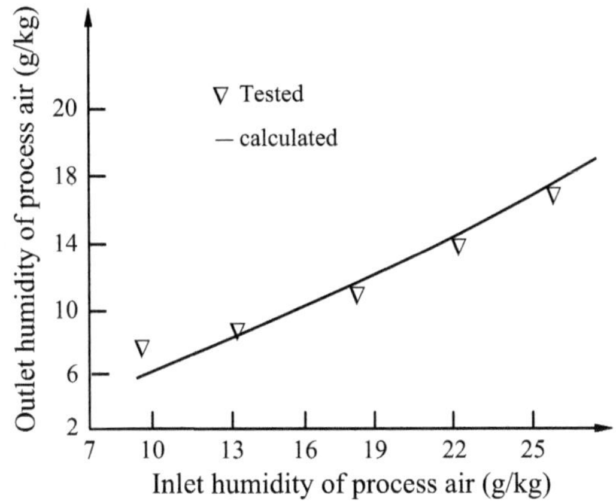

FIGURE 3.8

Comparison of the outlet humidity of the process air. Other operating conditions are fixed.

Table 3.1 Transport and thermophysical properties of the adsorbent wheel tested.

Symbol	Unit	Value
T_{ci}	°C	24.0
T_{hi}	°C	35.0
ω_{ci}	kg/kg	0.012
ω_{hi}	kg/kg	0.021
m_a	kg/h	30.6
C_s		1.1
q_{st}	kJ/kg	2650
W_{max}	kg/kg	0.35
α_0		0.50
D_{vs}	$m^2\ s^{-1}$	3.5×10^{-7}
λ_d	$W\ m^{-1}\ K^{-1}$	0.24
L	m	0.1
m_d	kg	3.85
ρ_d	$kg\ m^{-3}$	630
δ	mm	0.3
ε_t		0.46
n_{ad}		5200
n_{de}		5200
A_t (for one stream)	m^2	2.04
N	rpm	0.5

performance of desiccant wheels. In the following modeling, the operating conditions are set to the nominal operating conditions. For air dehumidification applications, the inlet conditions for the regenerating air are: inlet temperature, 90 °C, humidity 0.021 kg/kg. The process air is the same as in the Table 3.1.

Temperature and humidity profiles

If the wheel is assumed to be stationary, then both temperature and humidity vary with wheel angle. The profiles of temperature and humidity of air mean and desiccant mean with dimensionless time (or the wheel angle) during enthalpy recovery and air dehumidification are shown in Figures 3.9 and 3.10 respectively. The values at two local positions in the channel wall (one is near the cool air inlet, the other is near the warm air inlet) are also plotted in the figures. The coordinates for the two points are: Point 1, $x^* = 0.01, z^* = 0.5$; Point 2, $x^* = 0.99, z^* = 0.5$.

From Figure 3.9 for the air dehumidification application, it can be seen that in dehumidification, desiccant near the warm air inlet has a larger temperature wave than that near the cool air inlet, but the amplitudes of the humidity waves are almost the same at these two positions. The mean temperatures at positions near the warm air inlet are higher than those near the cool air inlet. However, the mean humidity near the warm air inlet is lower than that near the cool air inlet. These facts reveal that the higher the temperature is, the lower the humidity ratio for desiccant. During the heating period, the mean humidity of desiccant is higher than that of the air inlet, causing moisture to be desorbed from desiccant to air; while during the cooling period, the mean humidity of the desiccant is lower than that of air inlet, causing moisture to be adsorbed from the air to the desiccant.

From Figure 3.10 for the energy recovery application, it is seen that during the cooling period, humidity in the desiccant (both local and mean) is higher than that in the air stream and moisture desorbs from the desiccant to the cooling air, while during the heating period humidity in the desiccant is lower than that in the air stream and moisture adsorbs from the air to the desiccant. These phenomena in enthalpy recovery are just the opposite to those in air dehumidification. Furthermore, the amplitudes of the temperature and humidity waves at different locations are not in much disparity, which shows that temperature and humidity fields in the wheel are more homogeneous in enthalpy recovery than in dehumidification.

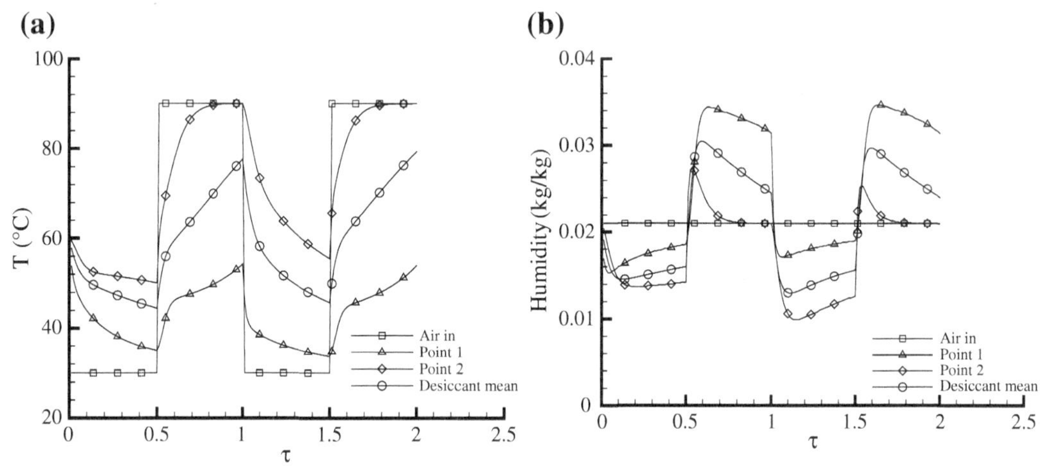

FIGURE 3.9

Variations of the temperature (a) and humidity (b) of two local points, wheel mean and air inlet, across the wheel angle for dehumidification. Point 1: near inlet; point 2: near outlet.

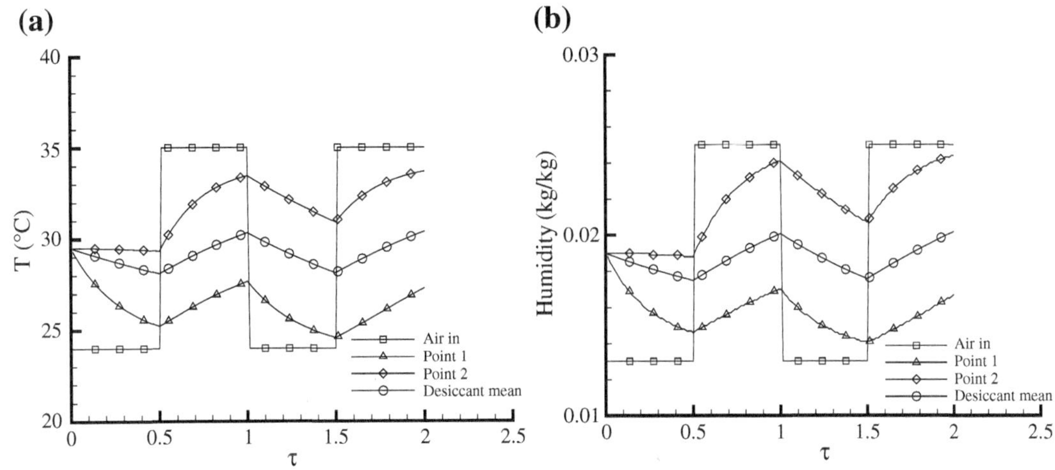

FIGURE 3.10

Variations of the temperature (a) and humidity (b) of two local points, wheel mean and air inlet, across the wheel angle for enthalpy recovery. Point 1: near inlet; point 2: near outlet.

The cyclic behaviors are observed from these temperature and humidity profiles. With the model, the properties in the solids walls can be analyzed and the performance can be investigated.

Effects of NTU

As in sensible-only heat exchangers, the number of transfer units, *NTU*, has a great influence on the system performance. There are many ways to modify *NTU*. When the wheel mass and channel size are fixed, changing the wall thickness will change the channel numbers, and consequently the contact area and *NTU*. When the wheel volume and wall thickness are fixed, changing channel sizes will change the number of channels and the packing density, and consequently the *NTU*. Figure 3.11 shows the influence of NTU_h and NTU_m on total heat recovery efficiencies. An NTU_h or NTU_m of 3.0 is needed for sensible and latent efficiencies higher than 0.60. It is noted that the latent effectiveness is usually less than the sensible effectiveness, because the moisture transfer resistance is usually larger than the heat transfer resistance. In other words, mass diffusion in the solid is far less than thermal diffusion. The Biot numbers for mass transfer are much larger than those for heat transfer.

For desiccant wheels, the effects of NTU_h and NTU_m on dehumidification efficiencies and specific dehumidification power are shown in Figure 3.12. As can be seen, they have the greatest influence on system performances. To produce acceptable performance, *NTU* of 3.0–4.0 is required.

For desiccant wheels, when the total volume of the wheel is fixed, a higher specific area can be achieved by constructing with smaller size channels and thinner channel walls. The specific contact area or the packing density is a key parameter influencing *NTU*, and consequently the performance. The honeycomb-type adsorbent wheels are beneficial because they have high packing densities.

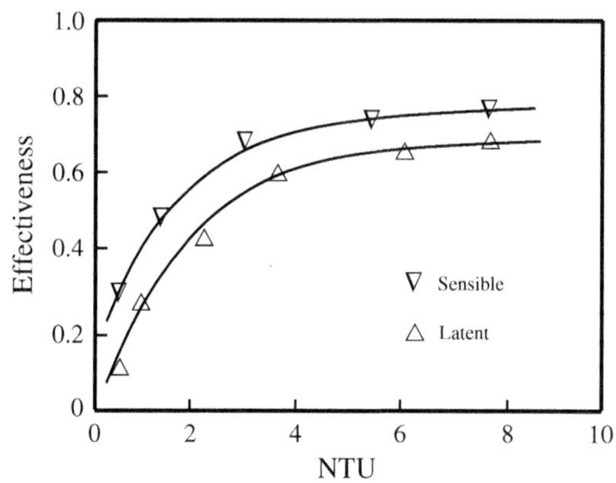

FIGURE 3.11

Effects of number of transfer units (NTU_h or NTU_m) on sensible and latent effectiveness of the wheel for energy recovery.

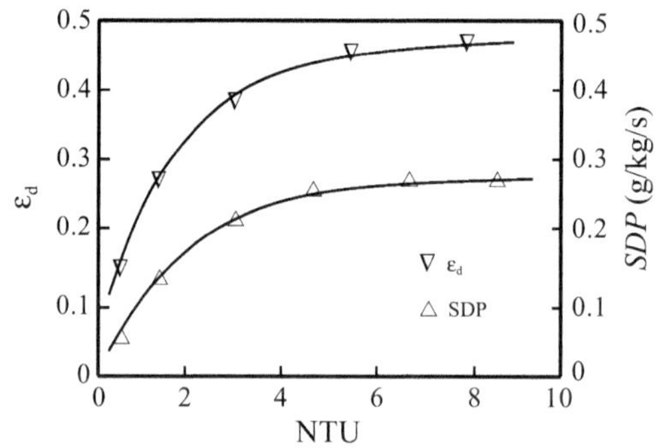

FIGURE 3.12

Effects of number of transfer units (NTU_h or NTU_m) on dehumidification efficiencies and specific dehumidification power of the wheel for dehumidification.

3.3.1 Effects of rotary speed

There exists an optimum rotary speed at which the efficiency reaches a peak. When a desiccant wheel rotates much faster than the optimum speed, the adsorption and desorption processes become too short, which results in poor performance. On the other hand, when the rotary speed is lower than the optimum, the adsorption and desorption processes become too long, which leads to more energy wasted in sensible heating/cooling rather than in sorption processes. They are less effective in either case. The variations of performance with rotary speed are plotted in Figure 3.13 for energy recovery purposes. The figure indicates that the optimum rotary speed for sensible and latent heat recovery is higher. In the following analysis, the wheels are operated at the optimum speeds.

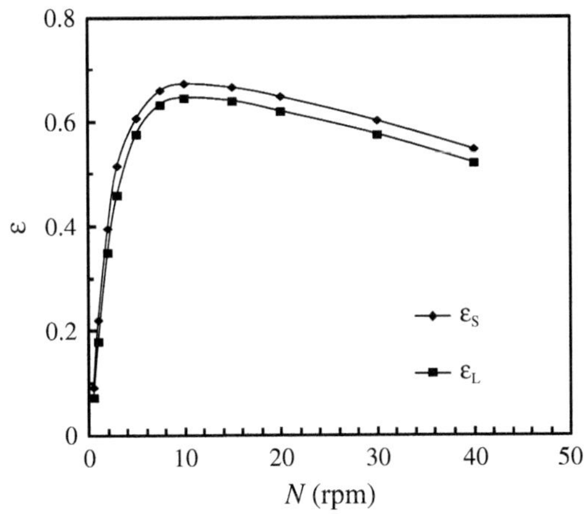

FIGURE 3.13

Effects of rotary speed on performance of the wheel for energy recovery applications.

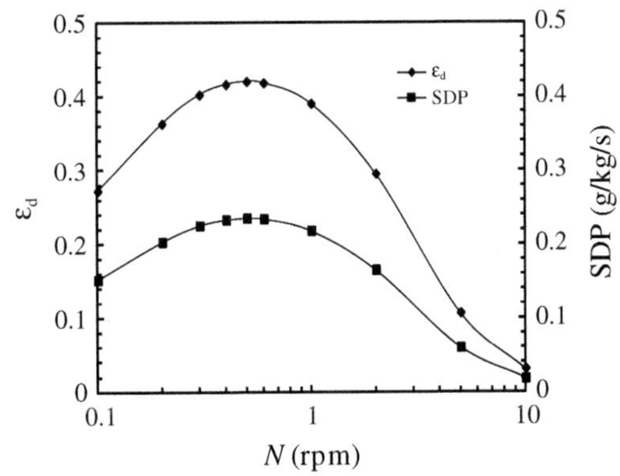

FIGURE 3.14

Effects of rotary speed on dehumidification efficiencies and specific dehumidification power for air dehumidification applications.

Similarly, for air dehumidification applications, the effects of rotary speed on dehumidification efficiencies and specific dehumidification power are plotted in Figure 3.14. There is an optimum rotary speed for the wheel and for air dehumidification it is lower than or energy recovery, with the same wheel. This is why wheels for enthalpy recovery applications should be rotated much faster than those for dehumidification to produce optimized performance.

With the model, the effects of other operating conditions and wheel properties can be investigated as well. The modeling of the conjugate heat and mass transfer in a revolving adsorbent wheel provides a tool for systems design and optimization. The thermodynamic psychrometric processes of the wheel for various applications can also be studied with the proposed model [5,6].

3.4 Conclusions

To model the conjugate heat and mass in a revolving adsorbent wheel accurately and swiftly, a one-dimensional, transient heat and mass transfer model is presented in the chapter. The advantages of such a model lie in the fact that it considers the dominant mechanisms for heat conduction and mass diffusion in the wall thickness, while using the air-side Nusselt and Sherwood numbers developed specially for adsorbent ducts. With the model, the performance of a desiccant wheel is analyzed. Many other structural and operating parameters of the wheel can also be optimized.

References

[1] Zhang LZ. Total heat recovery: Heat and moisture recovery from ventilation air. New York: Nova Science; 2009.

[2] Zheng W, Worek WM. Numerical simulation of combined heat and mass transfer processes in a rotary dehumidifier. Numer Heat Transfer A: Appl 1993;23:211–32.

[3] Tauscher R, Dinglreiter U, Durst B, Mayinger F. Transport processes in narrow channels with application to rotary exchangers. Heat and Mass Transfer 1999;35:123–31.

[4] Simonson CJ, Besant RW. Energy wheel effectiveness: Part I—development of dimensionless groups. Int J Heat Mass Transfer 1999;42:2161–70.

[5] Zhang LZ, Niu JL. Performance comparisons of desiccant wheels for air dehumidification and enthalpy recovery. Appl Therm Eng 2002;22:1347–67.

[6] Niu JL, Zhang LZ. Effects of wall thickness on the heat and moisture transfers in desiccant wheels for air dehumidification and enthalpy recovery. Int Commun Heat Mass Transfer 2002;29:255–68.

[7] Simonson CJ, Besant RW. Heat and moisture transfer in energy wheels during sorption, condensation, and frosting conditions. ASME J Heat Transfer 1998;120:699–708.

[8] Majumdar P. Heat and mass transfer in composite desiccant pore structures for dehumidification. Solar Energy 1998;62:1–10.

[9] Zhang XJ, Dai YJ, Wang RZ. A simulation study of heat and mass transfer in a honeycomb rotary desiccant dehumidifier. Appl Therm Eng 2003;23:989–1003.

[10] Antonellis SD, Joppolo CM, Molinaroli L. Simulation, performance analysis and optimization of desiccant wheels. Energy Buildings 2010;42:1386–93.

[11] Ruivo CR, Costa JJ, Figueiredo AR. Parametric study of the cyclic behavior of a hygroscopic matrix in a desiccant air flow system. Heat Mass Transfer 2011;47:1101–12.

CHAPTER

4 Heat and Mass Transfer in Air-to-Air Parallel-Plate Membrane Ducts

4.1 Introduction

Membranes that are selectively permeable to a certain substance are increasingly being used in various industries. One typical application is total heat recovery [1]. Parallel-plate membranes are the most commonly used structure due to their simplicity and ease of construction. Heat and mass transfer in parallel-plate membrane ducts are the key parameters for system design and component optimization. The geometry, similar to a rectangular duct, is simple. However, due to the complexities in the close conjugation between the membrane materials and fluid flow, a simple analytical solution is not easily available. The problem solution is highly reliant on numerical methods. Figure 4.1 shows a schematic of a parallel-plate membrane total heat exchanger. It is in a cross-flow arrangement.

In most industrialized countries, energy consumption by the HVAC sector accounts for 1/3 of the total energy consumption of the whole society. Cooling and dehumidifying fresh ventilation air constitutes 20–40% of the total energy load for HVAC in hot and humid regions. Total heat recovery-heat and moisture recovery from ventilation air has become a must-do in recent years. In many countries, building codes stipulate that total heat exchangers must be installed with air conditioning systems, for both energy conservation and indoor air quality control purposes.

Normally the water vapor content of atmospheric air is small, some tens of grams per kilogram of air. Nevertheless, due to the very high heat of vaporization, the latent heat content in air conditioning is of the same order as the sensible one. Due to the fact that the fresh air latent load accounts for 20–40% of the total load for air conditioning and air conditioning accounts for 1/3 of the total energy use in society, to reduce energy use in treating fresh air is a crucial step towards the whole of society's sustainable development. Total heat exchangers (or the so-called energy recovery ventilators, or enthalpy exchangers) could save a large fraction of energy for cooling and dehumidifying the fresh air since coolness and dryness would be recovered from the exhaust stream to the fresh air in summer. With total heat exchangers, the efficiency of the existing HVAC systems can be improved also. The reason is that normally the fresh air is dehumidified by a cooling coil through condensation followed by a re-heating process, which is very energy intensive. This part of the energy can be saved if total heat exchangers are installed to reduce the dehumidification load. Besides energy conservation, the total heat exchangers

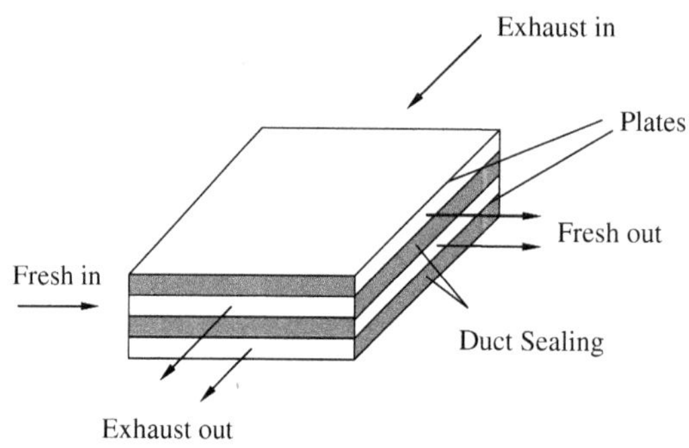

FIGURE 4.1

Schematic of a parallel-plate membrane total heat exchanger; the plates are vapor-permeable membranes.

have the additional benefits of ensuring sufficient fresh air supply, which is crucial for the prevention of epidemic respiratory diseases such as SARS and bird flu.

Energy wheels have the ability to recover both sensible heat and moisture from ventilation air [2,3]. However, membrane-based total heat exchangers have many benefits. They have no moving parts and their reliability is quite high. The duct sealing is easy and crossover problems can be prevented. The difference between a total heat exchanger and a common sensible-only heat exchanger is that water-permeable plates are used instead of metal plates. Currently there are two types of vapor-permeable materials: paper and hydrophilic polymer membranes. Paper is cheap, but its moisture efficiency is lower than 0.4. Membranes have higher latent effectiveness, above 0.65. The shortcoming is that they are more expensive than paper. As far as heat and mass transfer are concerned, both paper and the membranes can be used in heat and mass transfer studies. The transport phenomena in ducts are similar, but the moisture permeability through membranes is much higher.

Fluid flow and the Nusselt numbers for parallel-plate ducts of common metal foils under uniform temperature or uniform heat flux boundary conditions are classic heat transfer problems. The values can be found in textbooks [4–6]. However, for conjugate problems, the boundary conditions are totally different. The traditional transport data are not appropriate. On the contrary, the data should be calculated in combination with materials properties.

Parallel-plate membrane ducts can be used in air-to-air heat and mass exchangers. They can also be used in liquid-to-air heat and mass exchangers (membrane contactors). This chapter will concentrate on air-to-air ducts. The liquid-to-air ducts will be discussed in Chapter 6.

4.2 Cross-flow arrangement

4.2.1 The mathematical model

A cross-flow arrangement is the most common arrangement for air-to-air heat exchangers, mostly because of the convenience in duct sealing and the separation of the two flows. Figure 4.2 shows a practical total heat exchanger, showing the flow directions. As can be seen, the core, made of membranes, can be inserted into the exchanger shell conveniently. Replacement and the cleaning of the cores are also very convenient.

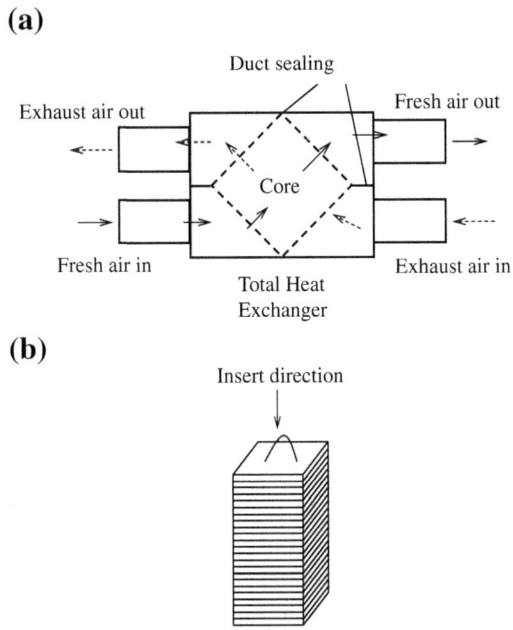

FIGURE 4.2

A schematic of a practical total heat exchanger with cross-flow arrangement. (a) The exchanger structure with flow arrangement; (b) the core that is inserted into the exchanger shell.

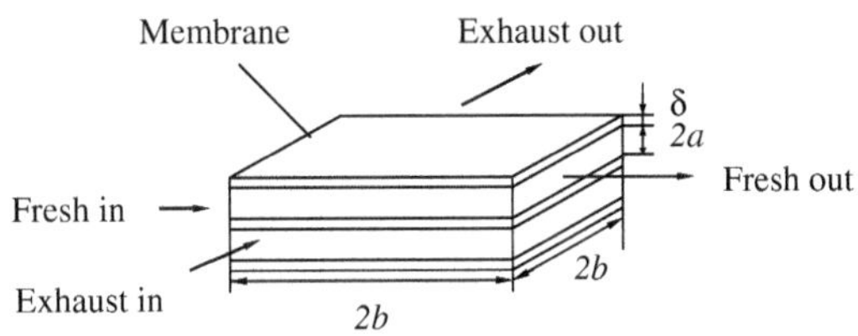

FIGURE 4.3

Schematic of a cross-flow membrane core for total heat exchangers.

The symmetric arrangement of the ducting work ensures a balanced flow between fresh air and exhaust air. In this section, heat and mass transfer in a cross flow parallel membranes duct will be studied.

Figure 4.3 shows a schematic of the membrane ducts. The fresh air and the exhaust air flow through the ducts in a cross-flow arrangement. Geometries of ducts: height $2a$, width $2b$, membrane thickness δ. The coordinate system for the model set up is shown in Figure 4.4. As shown in the figure, the fresh air, which is usually hot and humid, flows along the z axis, while the exhaust air, which is usually cool and dry, flows inversely along the x axis. Even though the coordinate system, inlet and boundary conditions are different, the equations governing the fluid flow, heat and mass transfer are in the same form for the two ducts.

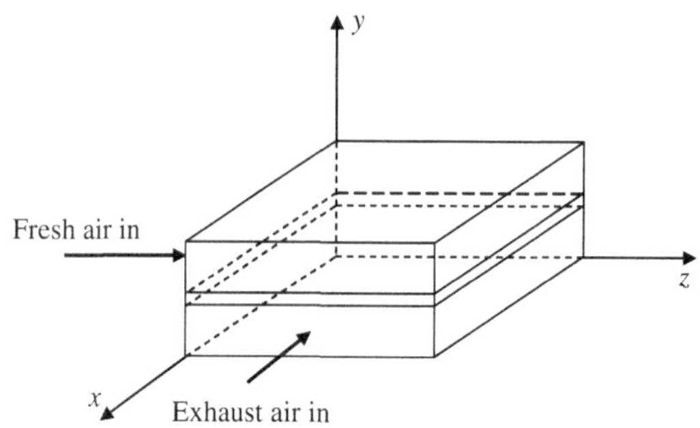

FIGURE 4.4

The coordinate system for the model set-up showing one membrane and two neighboring ducts.

In most applications, the Reynolds numbers are far below 2000, therefore laminar flow can be assumed. It is considered to be hydrodynamically fully developed, but thermally developing. The fluid is Newtonian with constant thermal properties. Additional assumptions are: (1) Adsorption of water vapor and membrane material is in an equilibrium adsorption-state; (2) the heat of sorption is assumed constant and equal to the heat of vaporization; (3) both the heat conductivity and the water diffusivity in the membrane are constants.

Momentum conservation

For fully developed laminar flow in ducts, the Navier-Stokes equations reduce to [7]

$$\mu\left(\frac{\partial^2 u}{\partial x^2}+\frac{\partial^2 u}{\partial y^2}\right)=\frac{dP}{dz} \tag{4.1}$$

where μ is dynamic viscosity (Pa.s), u is fluid velocity (m/s), P is the pressure (Pa), z is the axial coordinate (m). It is the flow direction.

Energy conservation

$$\rho c_p u\frac{\partial T}{\partial z}=\lambda\left(\frac{\partial^2 T}{\partial x^2}+\frac{\partial^2 T}{\partial y^2}\right) \tag{4.2}$$

where T is fluid temperature (K), λ is thermal conductivity ($\mathrm{kW\,m^{-1}\,K^{-1}}$), ($\rho c_p$) is heat capacity of moist air. It is calculated by the following equation as

$$\rho c_p=\rho_a\left(c_{pa}+\omega c_{pv}\right) \tag{4.3}$$

where ρ_a is density of dry air ($\mathrm{kg\,m^{-3}}$), c_{pa} and c_{pv} are specific heats of dry air and water vapor, respectively ($\mathrm{kJ\,kg^{-1}\,K^{-1}}$), ω is humidity ratio (kg vapor/kg dry air).

Mass conservation

$$u\frac{\partial \omega}{\partial z} = D_{va}\left(\frac{\partial^2 \omega}{\partial x^2} + \frac{\partial^2 \omega}{\partial y^2}\right) \tag{4.4}$$

where D_{va} is vapor diffusivity in dry air ($m^2 s^{-1}$).

Normalization of equations

The above three governing equations can be normalized to

$$\frac{\partial^2 u^*}{\partial x^{*2}} + \left(\frac{b}{a}\right)^2 \frac{\partial^2 u^*}{\partial y^{*2}} + \frac{4b^2}{D_h^2} = 0 \tag{4.5}$$

$$U\frac{\partial \theta}{\partial z_h^*} = \frac{\partial^2 \theta}{\partial x^{*2}} + \left(\frac{b}{a}\right)^2 \frac{\partial^2 \theta}{\partial y^{*2}} \tag{4.6}$$

$$U\frac{\partial \xi}{\partial z_m^*} = \frac{\partial^2 \xi}{\partial x^{*2}} + \left(\frac{b}{a}\right)^2 \frac{\partial^2 \xi}{\partial y^{*2}} \tag{4.7}$$

where the dimensionless velocity is

$$u^* = -\frac{\mu u}{(dP/dz)D_h^2} \tag{4.8}$$

and the dimensionless temperature is

$$\theta = \frac{T - T_{ci}}{T_{hi} - T_{ci}} \tag{4.9}$$

where T_{hi} is the inlet temperature of the fresh air (hot and humid), and T_{ci} is the inlet temperature of the exhaust air (cool and dry).

Dimensionless humidity is

$$\xi = \frac{\omega - \omega_{ci}}{\omega_{hi} - \omega_{ci}} \tag{4.10}$$

where ω_{hi} is the inlet humidity of the fresh air, and ω_{ci} is the inlet humidity of the exhaust air.

Dimensionless coordinates

$$x^* = \frac{x}{2b} \tag{4.11}$$

$$y^* = \frac{y}{2a} \tag{4.12}$$

$$z_h^* = \frac{z}{D_h Re Pr} \tag{4.13}$$

$$z_m^* = \frac{z}{D_h Re Sc} \tag{4.14}$$

where hydraulic diameter

$$D_h = \frac{4A_c}{P_f} \tag{4.15}$$

where A_c is the cross-sectional area of the duct (m^2), P_f is the perimeter of the duct (m). Pr is the Prandtl number, and Sc is the Schmidt number. The dimensionless axis z^* has a different definition for heat transfer and mass transfer.

In Eqs. (4.6) and (4.7), velocity coefficient U is defined by

$$U = \frac{u^*}{u_m^*} \frac{4b^2}{D_h^2} \tag{4.16}$$

where u_m^* is the average dimensionless velocity on a cross section, and it is calculated by

$$u_m^* = \frac{\iint u^* dA}{A_c} \tag{4.17}$$

The characteristics of fluid flow in the duct can be represented by the product of the friction factor and the Reynolds number as

$$(fRe) = \left(-\frac{D_h \frac{dP}{dz}}{\frac{1}{2}\rho u_m^2}\right)\left(\frac{\rho u_m D_h}{\mu}\right) = \frac{2}{u_m^*} \tag{4.18}$$

Dimensionless bulk temperature

$$\theta_b(z_h^*) = \frac{\iint u^* \theta dA}{\iint u^* dA} \tag{4.19}$$

Dimensionless bulk humidity

$$\xi_b(z_m^*) = \frac{\iint u^* \xi dA}{\iint u^* dA} \tag{4.20}$$

Nusselt number

$$Nu = \frac{hD_h}{\lambda} \tag{4.21}$$

Sherwood number

$$Sh = \frac{kD_h}{D_{va}} \tag{4.22}$$

where h and k are convective heat transfer coefficient ($\mathrm{kW\,m^{-2}\,K^{-1}}$) and convective mass transfer coefficient (m/s) between fluid and wall, respectively. As mentioned previously, the local Nusselt and Sherwood numbers may change from point to point on a duct surface. It is more practical to evaluate the peripherally mean local Nusselt and Sherwood numbers along the duct.

Through heat and mass balances on a control volume [7], the peripherally mean local heat and mass transfer coefficients can be calculated by

$$h_{\mathrm{L}} = -\frac{1}{4\left(\theta_{\mathrm{w}} - \theta_{\mathrm{b}}\right)} \frac{\lambda}{D_{\mathrm{h}}} \frac{\Delta\theta_{\mathrm{b}}}{\Delta z_{\mathrm{h}}^{*}} \tag{4.23}$$

$$k_{\mathrm{L}} = -\frac{1}{4\left(\xi_{\mathrm{w}} - \xi_{\mathrm{b}}\right)} \frac{D_{\mathrm{va}}}{D_{\mathrm{h}}} \frac{\Delta\xi_{\mathrm{b}}}{\Delta z_{\mathrm{m}}^{*}} \tag{4.24}$$

The peripherally local Nusselt and Sherwood numbers are

$$Nu_{\mathrm{L}} = -\frac{1}{4\left(\theta_{\mathrm{w}} - \theta_{\mathrm{b}}\right)} \frac{d\theta_{\mathrm{b}}}{dz_{\mathrm{h}}^{*}} \tag{4.25}$$

$$Sh_{\mathrm{L}} = -\frac{1}{4\left(\xi_{\mathrm{w}} - \xi_{\mathrm{b}}\right)} \frac{d\xi_{\mathrm{b}}}{dz_{\mathrm{m}}^{*}} \tag{4.26}$$

Average Nusselt number from 0 to z_{h}^{*}

$$Nu_{\mathrm{m}} = \frac{1}{z_h^{*}} \int_0^{z_h^{*}} Nu_{\mathrm{L}} dz_{\mathrm{h}}^{*} \tag{4.27}$$

Average Sherwood number from 0 to z_{m}^{*}

$$Sh_{\mathrm{m}} = \frac{1}{z_m^{*}} \int_0^{z_{\mathrm{m}}^{*}} Sh_{\mathrm{L}} dz_{\mathrm{m}}^{*} \tag{4.28}$$

The local Nusselt and local Sherwood numbers decrease asymptotically from very high values near the entrance of a tube to certain fully developed values at the end of thermal or mass entry lengths. Under uniform temperature (uniform mass concentration) boundary conditions, the fully developed value is denoted as Nu_{T} (Sh_{T} for mass). Under uniform heat flux (mass flux) boundary conditions, it is denoted as Nu_{H} (Sh_{H}). As a new definition, under the real conjugate boundary conditions, the fully developed value is denoted as Nu_{C} (Sh_{C} for mass transfer).

4.2.2 Boundary conditions

The governing equations for the exhaust air stream are in the same forms as for fresh air. The only differences are the coordinate systems for the model set-up. The membrane plate is a square one. If we use

x', y', z' to replace the coordinates x, y, z, in Eqs. (4.1)–(4.4) , respectively, then we get the equations for the exhaust stream. The relations of the two coordinates systems are

$$\begin{cases} x' = z \\ y' = y \\ z' = 2b - x \end{cases} \tag{4.29}$$

Inlet conditions for fresh air

$$z_h^* = 0, \quad \theta = 1 \tag{4.30}$$

$$z_m^* = 0, \quad \xi = 1 \tag{4.31}$$

Inlet conditions for exhaust air

$$z_h^* = 0, \quad \theta = 0 \tag{4.32}$$

$$z_m^* = 0, \quad \xi = 0 \tag{4.33}$$

Boundary conditions of two air streams for velocity

$$u^* = 0, \quad \text{on all walls of the duct.} \tag{4.34}$$

Adiabatic boundary conditions for fresh air

$$x^* = 0 \quad \text{or} \quad x^* = 1, \quad \frac{\partial \theta}{\partial x^*} = \frac{\partial \xi}{\partial x^*} = 0 \tag{4.35}$$

and for exhaust air

$$x'^* = 0 \quad \text{or} \quad x'^* = 1, \quad \frac{\partial \theta}{\partial x'^*} = \frac{\partial \xi}{\partial x'^*} = 0 \tag{4.36}$$

Conjugate boundary conditions on membrane surfaces, for fresh air

$$y^* = 0 \quad \text{or} \quad y^* = 1, \quad \theta\left(x^*, z_G^*\right) = \theta_{m1} \tag{4.37}$$

and for exhaust air

$$y'^* = 0 \quad \text{or} \quad y'^* = 1, \quad \theta\left(x'^*, z_G'^*\right) = \theta_{m2} \tag{4.38}$$

where term $\theta\left(x^*, z_G^*\right)$ refers to air temperature adjacent to the membrane surface at point $\left(x^*, z_G^*\right)$, subscript "m" refers to the membrane, and "1" and "2" refer to the fresh side and the exhaust side at the same point of the membrane, respectively. The variable z_G^* is the dimensionless geometric position

$$z_G^* = \frac{z}{2b} \tag{4.39}$$

Due to the small thickness in membrane (100 μm, thermal conductivity 0.127 W m^{-1}K^{-1}), temperature differences between the two sides of a membrane are rather small. Heat released on the adsorption side of the membrane (fresh-air side) could be balanced by the heat absorbed on the desorption side (exhaust side) of the membrane. Temperature differences between the two sides are of the order of 10^{-4} °C.

$$\theta_{m1} = \theta_{m2} = \theta_m \tag{4.40}$$

The value of θ_m is not a fixed value. Rather, it is a result of local couplings between the two streams on the membrane surface. In this way, the temperatures of the two fluids are conjugated. Heat flux $q_h = \lambda(T_{m1} - T_{m2})/\delta$.

Mass boundary conditions on membrane surfaces, for fresh air

$$y^* = 0 \quad \text{or} \quad y^* = 1, \quad q_m\left(x^*, z_G^*\right) = \dot{m}_v \tag{4.41}$$

For exhaust air

$$y'^* = 0 \quad \text{or} \quad y'^* = 1, \quad q_m\left(x'^*, z_G'^*\right) = \dot{m}_v \tag{4.42}$$

where $\dot{m}_v$ (kg m^{-2}s^{-1}) is the moisture emission rate through the membrane at point (x^*, z_G^*), and it is determined by a diffusion equation in the membrane as

$$\dot{m}_v = \rho_m D_{wm} \frac{w_{m1} - w_{m2}}{\delta} \tag{4.43}$$

where ρ_m and D_{wm} are density of membrane and water diffusivity in membrane, respectively. As seen, here D_{wm} is evaluated by water content difference in membranes. Variable w is water uptake in membrane (kg moisture/kg dry membrane), and it is expressed by a general sorption equation as

$$w = \frac{W_{max}}{1 - C_s + C_s/RH} \tag{4.44}$$

where W_{max} is the maximum water uptake of membrane material (kg/kg); C_s is a constant named the shape factor for the material; RH is air relative humidity. This sorption curve directly links water content to RH, a variable that can be directly measured. The basic thermal physical parameters are obtained in the laboratory. RH can be calculated from humidity ratio ω.

On the air side, moisture emission adjacent to the membrane surface on the fresh-air or exhaust-air side is calculated by

$$q_m = -\rho_a D_{va} \left.\frac{\partial \omega}{\partial y}\right|_{y=0,2a} \tag{4.45}$$

Heat and mass fluxes q_h and q_m are not constant values. They are co-determined by the membrane and the two flows. In this way, the heat and mass transfer are conjugated.

4.2.3 Numerical methods

The partial differential equations for momentum, energy, and mass transport developed together with the boundary conditions are discretized by means of a finite volume method. For the convective terms, an upwind differencing scheme is used, by replacing the boundary value by the values of the neighboring upstream nodes. For the Peclet numbers involved in the present problem, this approximation is close to reality. The diffusive fluxes at the nodes are approximated by assuming a linear variation of variables between two grid points. With this procedure, an algebraic tri-diagonal matrix will be formed to find the values at each node [7].

To ensure the accuracy of the results presented, numerical tests were performed for the duct to determine the effects of the grid size. It indicates that 42×42 grids on cross section and $\Delta z^* = 0.00025$ axially are adequate (less than 0.1% difference compared with 52×52 grids and $\Delta z^* = 0.000125$). The final numerical uncertainty is 0.1%.

There are in total three components, fresh air, exhaust air, and membrane, to be calculated. Iterations are required to find the solutions for the three components. When solving equations for one component, the values in other components are assumed to be already known values and will give assumed boundary conditions to this component. The solution of the three components is repeated until the solutions are converged, i.e., differences of old and new temperature and humidity values at each grid node are within a lower limit.

After these procedures, all the governing equations are satisfied simultaneously. The temperature and humidity fields on membrane surfaces are then obtained. Then the mean outlet values are calculated for the Nusselt and Sherwood number calculations.

4.2.4 Experimental work

A compressive test rig for heat and moisture exchange has been built in SCUT (South China University of Technology). With this test rig, various membrane total heat exchanger cores which are inserted into the shells when experimented upon, are tested. The test rig is shown in Figure 4.5. It consists of the ducting work, air controlling units, and instrumentation. Two parallel air ducts with a 210×210 mm cross section are assembled to form the test rig. Each duct is composed of a variable-speed blower, a wind tunnel, a set of nozzles, wind straighteners, electric heating coils, steam humidification tubes supplied by electric boilers, and temperature and humidity sensors. An exchanger shell is designed to hold the core. The small converging wind tunnels produce steady, homogeneous, fully developed air flow to the exchanger. The heating power and the steam generation currents can be adjusted according to temperature and humidity set points. After the air temperature and humidity are adjusted to the set points, the two ducts are connected to the two inlets of the exchanger shell respectively. The exchanger shell is designed to station the exchanger core and separate the two air streams in cross-flow. The cores can be inserted into the quadrate cavity in the center of the shell before the test. The whole test rig is built in a constant temperature and constant humidity room, so the inlet temperature and humidity can be controlled and maintained very well even under very hot and humid ambient weather conditions. A 15 mm thick plastic foam insulation layer is pasted onto the outer surfaces of the ducts and the shells to prevent heat dissipation from the system to the surroundings. Moisture dissipation from the air stream to the surroundings is negligible since the duct and shell materials are highly hydrophobic and they can adsorb little moisture. The heat loss from the system is below 0.6%, and moisture loss is less than 0.5%.

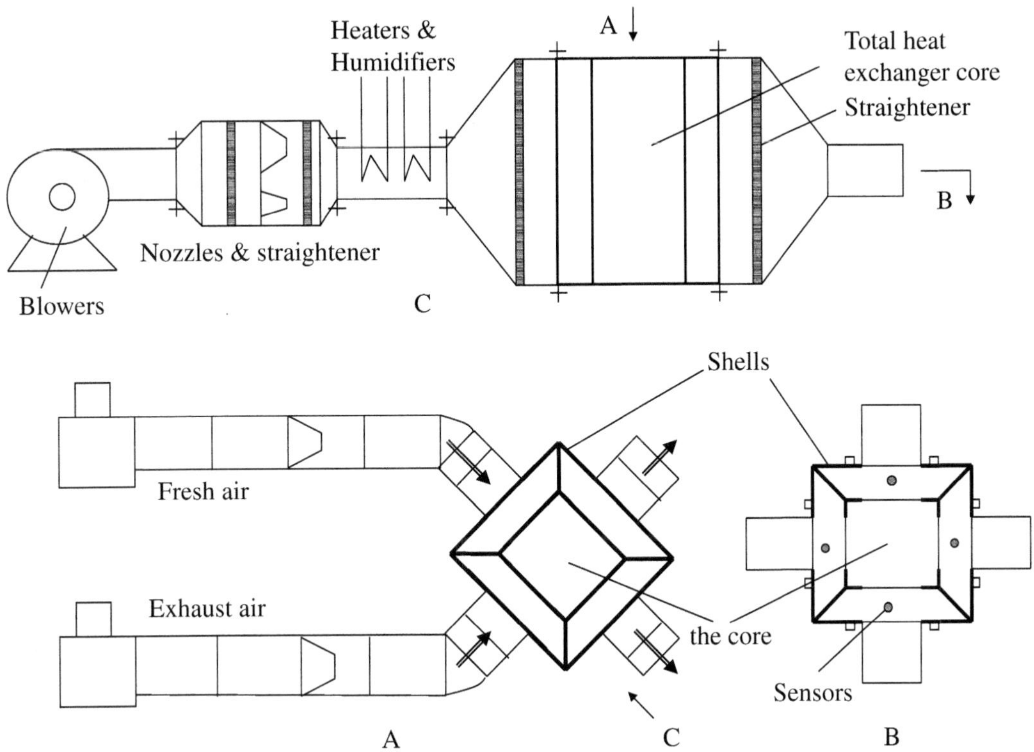

FIGURE 4.5

The comprehensive test rig for heat and moisture exchange of membrane cores.

Heat and mass balance requires a long period. The software is designed to record the data and to balance the heat and moisture transfer in the whole test rig. The nominal operating conditions are: fresh air inlet 35 °C and 0.021 kg/kg; exhaust air inlet 27 °C and 0.012 kg/kg. The corresponding inlet relative humidity (RH) is 59% and 54% for fresh air and exhaust air respectively. During the experiment, equal air flow rates are kept for the two ducts. The design air flow rates are from 50 to 300 m^3/h. In the test, they are changed by variable-speed blowers, to produce different air velocities. Humidity, temperature, and volumetric flow rates are monitored at the inlet and outlet of the exchanger. Before and after each test, temperature and humidity sensors are calibrated with a Pt-100 temperature sensor and a chilled-mirror dew-point meter. Hot-wire anemometers that are used to measure the wind speed before and after the exchanger are compared with the air flow rates measured by nozzles. The offset is controlled to within a 1% limit. Frontal air velocities vary from 0.2 m/s to 3 m/s, which are typical. Air flow under such conditions is laminar, with Reynolds numbers not exceeding 200. A digital pressure differential gauge is used to measure the pressure drop across the tested core. Six sensors are uniformly positioned on the inlet and outlet surfaces of the shell to have a mean value of measurements. The uncertainties are: temperature ±0.1 °C; humidity ±2%; volumetric flow rate ±1%, pressure drop ±1%. The final uncertainty is ±4.5% for sensible and latent effectiveness. Heat and mass balance between the fresh air and the exhaust air is checked. Their differences are controlled to be less than 0.1%.

After the measurement of mean inlet and outlet temperature and humidity, the sensible and latent effectivenesses are calculated. These are the experimentally obtained data. For balanced air flow,

$$\varepsilon_s = \frac{T_{hi} - T_{ho}}{T_{hi} - T_{ci}} \tag{4.46}$$

$$\varepsilon_L = \frac{\omega_{hi} - \omega_{ho}}{\omega_{hi} - \omega_{ci}} \tag{4.47}$$

The sensible and latent effectiveness can also be calculated with numerical outlet values. These are the numerical data. Figures 4.6 and 4.7 show the calculated and tested sensible and latent effectiveness,

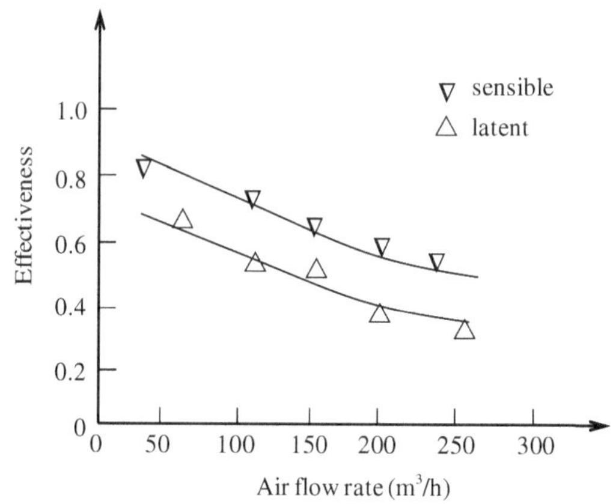

FIGURE 4.6

Sensible and latent effectiveness for Core 1. The discrete symbols are experimental data and the solid lines are numerical data.

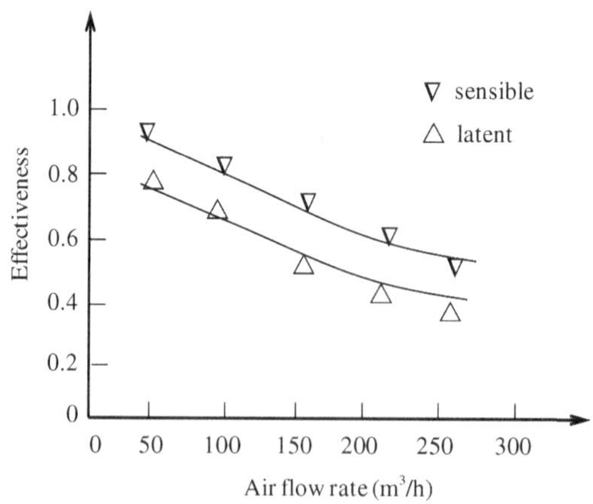

FIGURE 4.7

Sensible and latent effectiveness for Core 2. The discrete symbols are experimental data and the solid lines are numerical data.

Table 4.1 Properties of the membrane cores used.

Properties	Unit	Core 1	Core 2
ρ_m	kg/m^3	836	836
D_{wm}	m^2/s	3.2×10^{-11}	3.6×10^{-11}
λ_m	Wm^{-1}K^{-1}	0.42	0.42
δ	μm	100	80
W_{max}	kg/kg	0.27	0.32
C_s		0.81	0.94
n_{ch}		116	125
x_F, y_F, $(2b)$	mm	185	185
$2a$	mm	1.88	1.79
Core depth	mm	460	460

Note: D_{wm} here is evaluated by water content differences in membranes. n_{ch} is the number of channels for each flow.

respectively. The properties of the membrane core considered and tested are listed in Table 4.1. As observed from the figures, the numerical data predict the test satisfactorily. The maximum deviations are less than 15%. As can be seen, both the sensible and the latent effectiveness for Core 2 are higher than those for Core 1. The increase in latent effectiveness for Core 2 is much higher than that in sensible effectiveness. This is because the membranes for Core 2 are better in moisture permeability. Membranes' properties have more influence on latent effectiveness than on sensible effectiveness.

4.2.5 Temperature and humidity profiles

The temperature, humidity, heat flux, and moisture flux across the membrane are calculated and are plotted in Figures 4.8–4.12. From these figures, it can be shown that the variables on the membrane surfaces are neither uniform values nor uniform fluxes. The temperature, humidity, heat flux, and mass flux exhibit a two-dimensional distributed nature. The differences in humidity between the two surfaces of the membrane are demonstrably large, indicating high internal moisture resistance.

4.2.6 Nusselt and Sherwood numbers

Figures 4.13 and 4.14 show the local and mean Nusselt and Sherwood numbers along the duct respectively [7]. In the thermal and concentrational developing region, the local Nusselt and Sherwood numbers decrease sharply, from very high values to two lower limiting values denoted by Nu_C and Sh_C. The flow is called thermally or/and concentrationally developed after that period. The calculated Nu_C and Sh_C for various aspect ratios are listed in Table 4.2 [7]. It is found that $Sh_C \approx Nu_C$ for most aspect ratios, indicating that heat mass analogy is satisfied in this case. For comparison, the values of Nu_T and Nu_H from published references are also listed. Further, for hydrodynamically fully developed laminar flow in ducts, (fRe) is a constant. The calculated values of (fRe) for various aspect ratios are also listed in Table 4.2. These values can be used for performance analysis of total heat exchangers.

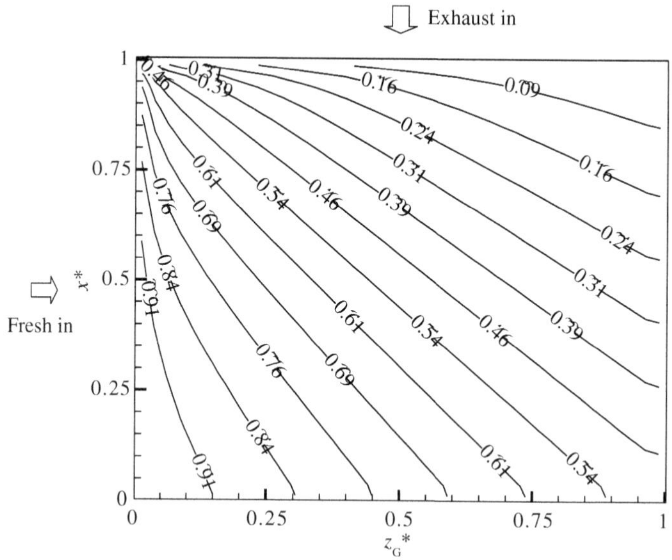

FIGURE 4.8

Dimensionless surface temperature on membrane surfaces.

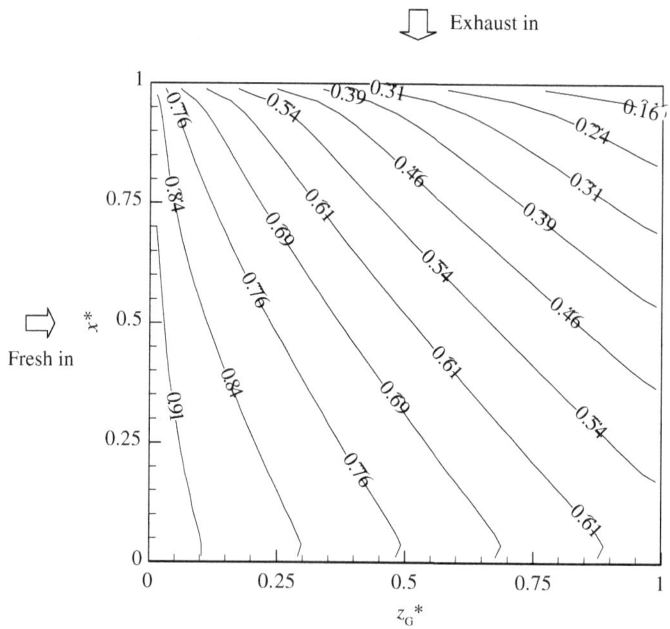

FIGURE 4.9

Dimensionless humidity on the membrane surface on the fresh-air side.

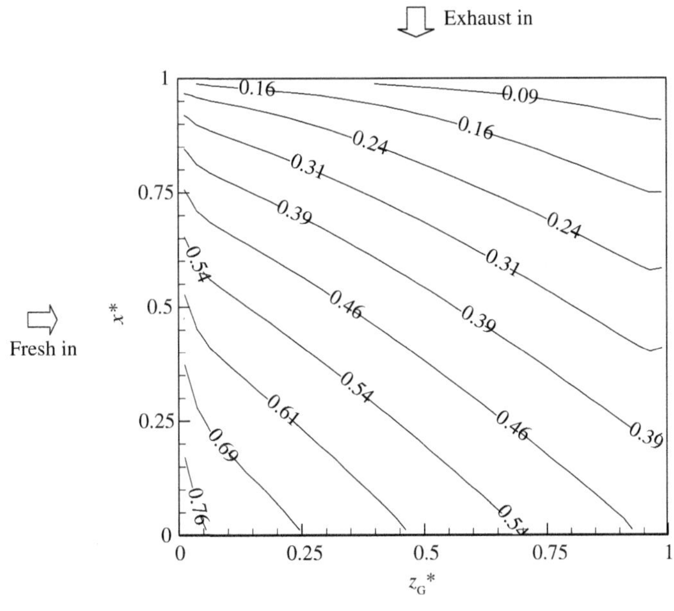

FIGURE 4.10

Dimensionless humidity on the membrane surface on the exhaust-air side.

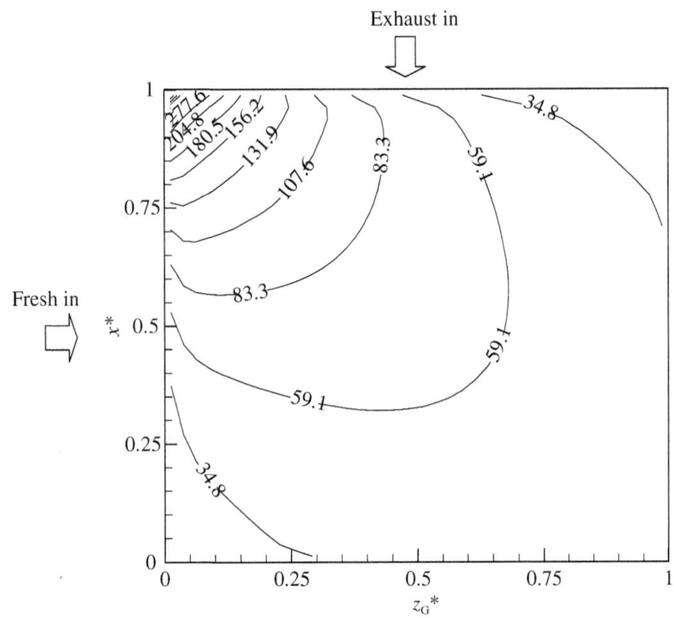

FIGURE 4.11

Profiles of heat flux q_s on the membrane surface (W/m^2).

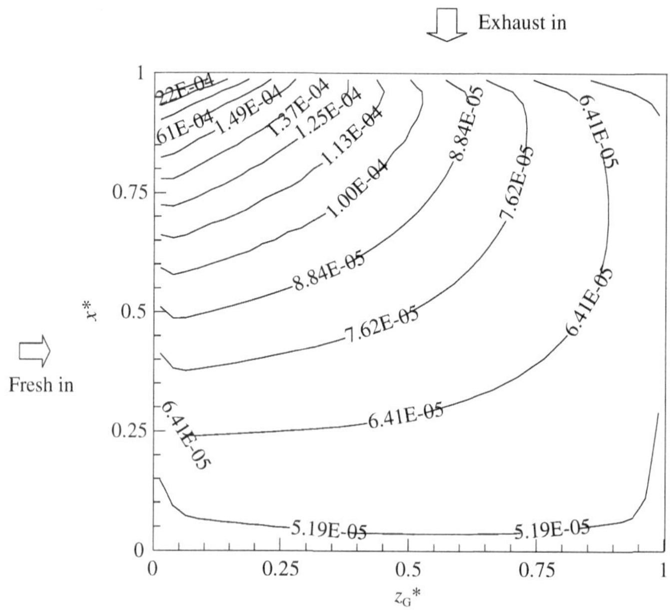

FIGURE 4.12

Profiles of mass flux q_L on the membrane surface ($kg\,m^{-2}s^{-1}$).

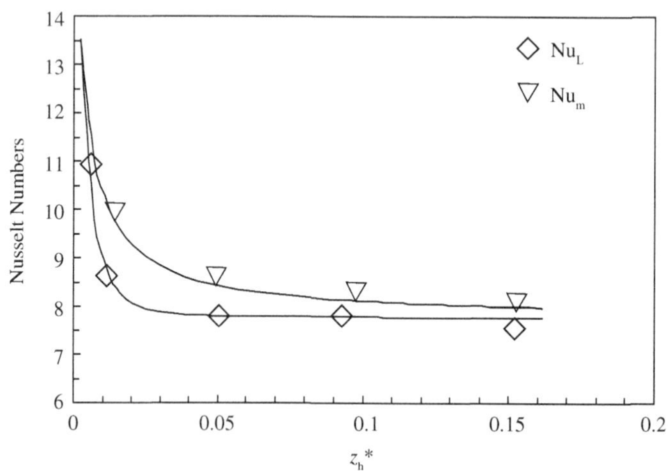

FIGURE 4.13

Local and mean Nusselt numbers along flow direction, $b/a=50$.

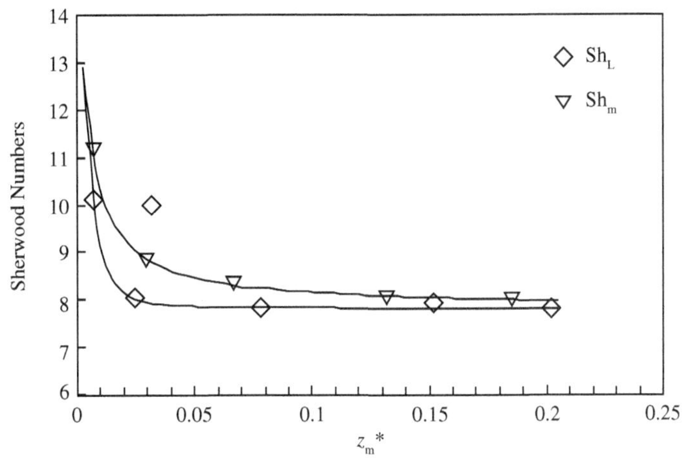

FIGURE 4.14

Local and mean Sherwood numbers along flow direction, $b/a=50$.

Table 4.2 Fully developed (fRe) and Nusselt, Sherwood numbers for various aspect ratios of parallel-plate membrane ducts.

b/a	Nu_H	Nu_T	Nu_C	Sh_C	(fRe)	(fRe)
Sources	**Ref. [4–6]**	**Ref. [4–6]**	**(conjugate)**	**(conjugate)**	**Ref. [4–6]**	**(conjugate)**
1.0	3.61	2.98	1.88	1.89	57.0	56.58
1.43	3.73	3.08	2.49	2.52	59.0	58.09
2.0	4.12	3.39	3.06	3.10	62.0	61.83
3.0	4.79	3.96	4.06	4.00	69.0	68.09
4.0	5.33	4.44	4.64	4.52	73.0	72.45
8.0	6.49	5.60	6.06	6.03	82.0	81.66
50.0			7.78	7.81		92.39
100.0			8.07	8.05		93.92
∞	8.23	7.54			96.0	

4.3 Quasi-counter flow arrangement

Cross flow has been the predominant flow arrangement for total heat exchangers due to the simplicity of duct sealing. Though they are simple total heat exchangers from an engineering point of view, their effectiveness is not as high as with counter-flow arrangements. Therefore counter flow should be considered. It should be noted that because of duct sealing, a pure counter-flow arrangement is not likely to separate the two air streams from each other. A quasi-counter-flow arrangement, as shown in Figure 4.15, is thus the best possible choice.

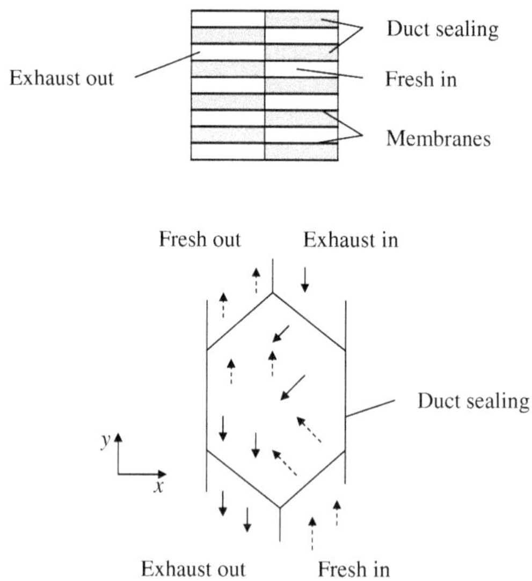

FIGURE 4.15

A quasi-counter-flow parallel-plate membrane duct structure.

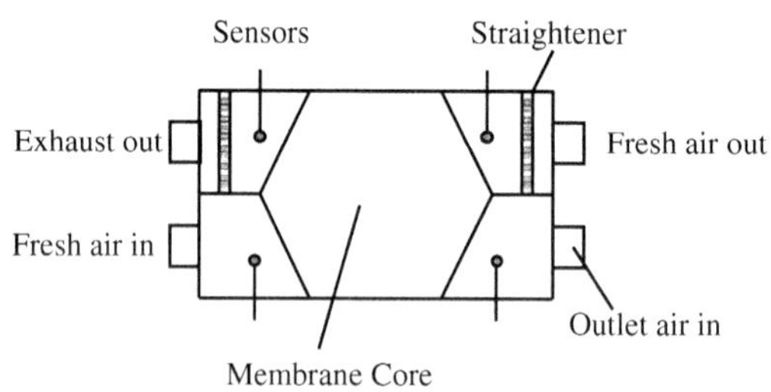

FIGURE 4.16

A total heat exchanger with a quasi-counter-flow membrane core.

In this section, the conjugate heat and mass transfer in the quasi-counter-flow parallel-plate membrane duct is modeled [8]. The Nusselt and Sherwood numbers for this structure are obtained. First, a quasi-counter-flow total heat exchanger is constructed. The structure of the whole exchanger with the core inserted is shown in Figure 4.16. The geometry of the constructed membrane is shown in Figure 4.17. The apex angle (2β) is 90°. A plastic frame is designed to station and glue the membranes together to form ducts. The duct pitch is 4 mm and in total there are 57 channels for each flow. The exchanger is designed to satisfy the ventilation needs of a 100 m^2 office, with a nominal balanced air flow rate of 150 m^3/h. The parameters of the exchanger are measured and summarized in Table 4.3. The constructed total heat exchanger core is experimented upon in the test rig in the laboratory described above. The exchanger shells are modified to hold the core. The measured outlet temperature and humidity are compared to those calculated. In this way the model and the methodology are verified.

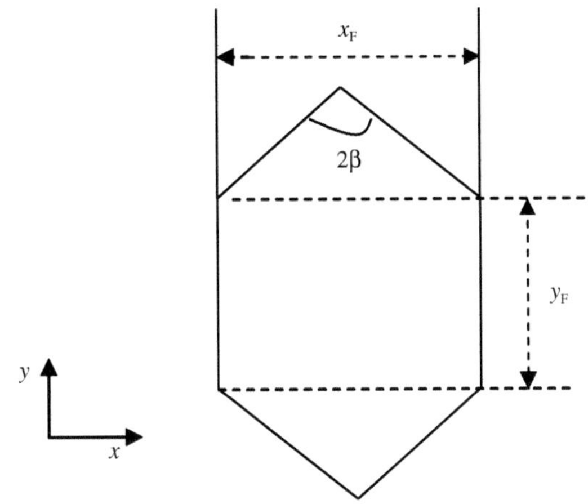

FIGURE 4.17

Geometry of a single plate cut for the exchanger.

Table 4.3 Structural and physical parameters of the quasi-counter flow membrane exchanger.

Symbol	Unit	Value	Symbol	Unit	Value
n_{ch}		57	$2a$	mm	4
x_F, y_F, $(2b)$	mm	185	β	degrees	45
δ	μm	102	ρ_m	kgm^{-3}	866
A_{tot}	m^2	6.7	D_{vm}	m^2/s	8×10^{-6}
λ_m	Wm^{-1}K^{-1}	0.43	D_{va}	m^2/s	2.82×10^{-5}
W_{max}	kg/kg	0.29	C_s		0.94

Note: D_{vm} is evaluated by equilibrium air humidity differences in membranes.

4.3.1 The mathematical model

Conservation equations in fluids

The calculating domain, surrounded by the dashed lines in Figure 4.18, consists of a membrane plate and two neighboring half air ducts. The mid-planes in the air streams are symmetric planes. Fresh air, membrane, and the exhaust air are the three components of this conjugate system. The governing equations are the same for the two flows. The coordinate for flow direction for each flow is different: y and $-y$ respectively.

The Reynolds numbers are below 2300, therefore laminar flow can be assumed. The flow structures are more complicated than pure cross flow. The duct is not straight, so the velocities are three-dimensional. The entry regions for both the fluid flow and thermal and mass boundaries are considered.

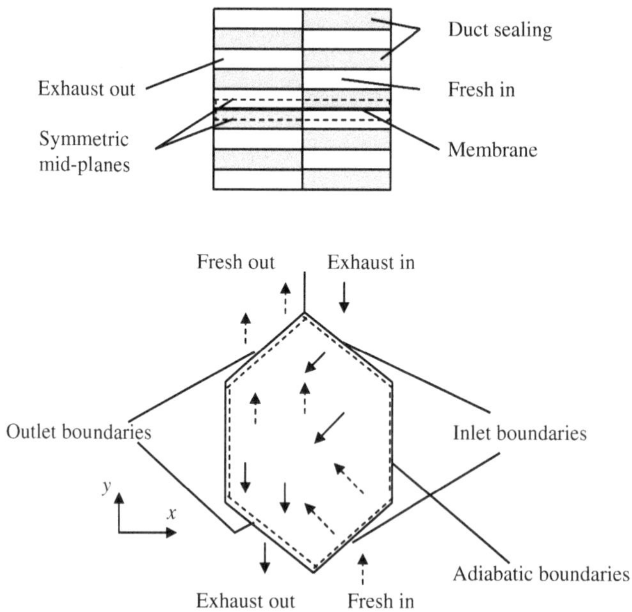

FIGURE 4.18

Calculating domain in the exchanger consists of a membrane plate and two neighboring half ducts, surrounded by dashed lines.

The hydrodynamic and heat and mass transfer problem can be described by Navier-Stokes equations in Cartesian coordinates as [8]:

Conservation of mass

$$\frac{\partial u^*}{\partial x^*} + \frac{\partial v^*}{\partial y^*} + \frac{\partial w^*}{\partial z^*} = 0 \tag{4.48}$$

where x, y and z are spanwise, streamwise, and normal coordinates for flow, respectively; u, v, and w are velocities in x, y and z directions (m/s), respectively; superscript "*" in this and the following equations represents dimensionless form. It should be noted that the coordinates system is different from the above section because velocities here are three-dimensional.

Conservation of momentum in three coordinates

$$u^* \frac{\partial u^*}{\partial x^*} + v^* \frac{\partial u^*}{\partial y^*} + w^* \frac{\partial u^*}{\partial z^*} = -\frac{\partial P^*}{\partial x^*} + \left[\frac{\partial^2 u^*}{\partial x^{*2}} + \frac{\partial^2 u^*}{\partial y^{*2}} + \frac{\partial^2 u^*}{\partial z^{*2}} \right] \tag{4.49}$$

$$u^* \frac{\partial v^*}{\partial x^*} + v^* \frac{\partial v^*}{\partial y^*} + w^* \frac{\partial v^*}{\partial z^*} = -\frac{\partial P^*}{\partial y^*} + \left[\frac{\partial^2 v^*}{\partial x^{*2}} + \frac{\partial^2 v^*}{\partial y^{*2}} + \frac{\partial^2 v^*}{\partial z^{*2}} \right] \tag{4.50}$$

$$u^*\frac{\partial w^*}{\partial x^*}+v^*\frac{\partial w^*}{\partial y^*}+w^*\frac{\partial w^*}{\partial z^*}=-\frac{\partial P^*}{\partial z^*}+\left[\frac{\partial^2 w^*}{\partial x^{*2}}+\frac{\partial^2 w^*}{\partial y^{*2}}+\frac{\partial^2 w^*}{\partial z^{*2}}\right] \tag{4.51}$$

where P represents pressure (Pa).

Conservation of heat

$$u^*\frac{\partial \theta}{\partial x^*}+v^*\frac{\partial \theta}{\partial y^*}+w^*\frac{\partial \theta}{\partial z^*}=\frac{1}{Pr}\left[\frac{\partial^2 \theta}{\partial x^{*2}}+\frac{\partial^2 \theta}{\partial y^{*2}}+\frac{\partial^2 \theta}{\partial z^{*2}}\right] \tag{4.52}$$

Conservation of water vapor

$$u^*\frac{\partial \xi}{\partial x^*}+v^*\frac{\partial \xi}{\partial y^*}+w^*\frac{\partial \xi}{\partial z^*}=\frac{1}{Sc}\left[\frac{\partial^2 \xi}{\partial x^{*2}}+\frac{\partial^2 \xi}{\partial y^{*2}}+\frac{\partial^2 \xi}{\partial z^{*2}}\right] \tag{4.53}$$

where θ and ξ are the dimensionless temperature and humidity ratio, respectively. Pr and Sc are Prandtl and Schmidt numbers for vapor-air mixture, respectively. The characteristic diameter is selected as two times the channel pitch (characteristic length, channel pitch $H_z=2a$, $D_h=4a$). The dimensionless forms for the variables are expressed as

$$x^* = \frac{x}{2H_z} \tag{4.54}$$

$$y^* = \frac{y}{2H_z} \tag{4.55}$$

$$z^* = \frac{z}{2H_z} \tag{4.56}$$

$$u^* = \frac{2uH_z}{\nu} \tag{4.57}$$

$$v^* = \frac{2vH_z}{\nu} \tag{4.58}$$

$$w^* = \frac{2wH_z}{\nu} \tag{4.59}$$

where ν is the kinematic viscosity of air (m^2/s).

The dimensionless pressure is defined as

$$P^* = \frac{4P(H_z)^2}{\rho_a \nu^2} \tag{4.60}$$

where ρ_a is air density (kg/m^3).

The Reynolds number used to characterize the airflow rate is given by

$$Re = \frac{u_{\mathrm{m}} D_{\mathrm{h}}}{\nu} \tag{4.61}$$

where u_{m} is the mean air velocity at the channel inlet. A uniform velocity distribution is assumed at the channel inlets.

Similarly, like all the ducts, the characteristics of fluid flow in the duct can be represented by the product of the friction factor and the Reynolds number as

$$(fRe) = \left(-\frac{D_{\mathrm{h}} \frac{dP}{dx}}{\frac{1}{2}\rho_{\mathrm{a}} u_{\mathrm{m}}^{2}}\right)\left(\frac{u_{\mathrm{m}} D_{\mathrm{h}}}{\nu}\right) \tag{4.62}$$

The dimensionless temperature θ and humidity ratio ξ are defined by Eqs. (4.9) and (4.10).

The Prandtl number is

$$Pr = \frac{\nu}{a_{\mathrm{a}}} \tag{4.63}$$

where a_{a} is thermal diffusivity of air (m^2/s).

The Schmidt number is

$$Sc = \frac{\nu}{D_{\mathrm{va}}} \tag{4.64}$$

Mean Nusselt number

$$Nu_{\mathrm{m}} = \frac{\bar{h} D_{\mathrm{h}}}{\lambda_{\mathrm{a}}} \tag{4.65}$$

Mean heat transfer coefficient

$$\bar{h} = \frac{\rho_{\mathrm{a}} V c_{\mathrm{pa}} \left(T_{\mathrm{hi}} - T_{\mathrm{ho}}\right)}{\Delta T_{\mathrm{lg}} A_{\mathrm{t}}} \tag{4.66}$$

Mean Sherwood number is

$$Sh_{\mathrm{m}} = \frac{\bar{k} D_{\mathrm{h}}}{D_{\mathrm{va}}} \tag{4.67}$$

Mean mass transfer coefficient

$$\bar{k} = \frac{\rho_{\mathrm{a}} V \left(\omega_{\mathrm{hi}} - \omega_{\mathrm{ho}}\right)}{\Delta \omega_{\mathrm{lg}} A_{\mathrm{t}}} \tag{4.68}$$

where ΔT_{lg} is log mean temperature difference between membrane and fluid. Membrane and fluid temperature and humidity can be calculated directly. The temperature and humidity here are mass-weighted average values. A_{t} is the heat transfer area of a single plate.

Conservation equations in membranes

Heat and mass transfer inside membranes are assumed to be isotropic. Since the thickness of membranes is rather low (of the order of 100 μm), heat and moisture transfer in the streamwise and spanwise directions are negligible. Heat and mass transfer in membranes can be simplified to one-dimensional equations (in normal direction). Steady-state heat transfer in membrane is given by

$$\frac{\partial^2 \theta_m}{\partial z^{*2}} = 0 \tag{4.69}$$

Steady-state moisture transfer in membrane:

$$\frac{\partial^2 \xi_m}{\partial z^{*2}} = 0 \tag{4.70}$$

Conjugate boundaries

At the positions where fluid contacts with membrane (on two surfaces), velocity is set to zero (no-slip boundary condition). Temperature and humidity in fluid and on the membrane surface are coupled by heat mass flux expressions as follows:

$$q_h = -\lambda_a \frac{\partial T}{\partial z}\bigg|_{\text{surface}} = -\lambda_m \frac{\partial T_m}{\partial z}\bigg|_{\text{surface}} \tag{4.71}$$

$$q_m = -\rho_a D_{va} \frac{\partial \omega}{\partial z}\bigg|_{\text{surface}} = -\rho_a D_{vm} \frac{\partial \omega_m}{\partial z}\bigg|_{\text{surface}} \tag{4.72}$$

In the above equation, D_{vm} is evaluated by air humidity differences, which are the equivalent equilibrium humidity in membranes. With these equations, the properties in the fluid are connected with those in the membranes.

At the inlets, the velocity, temperature, and humidity are set to the set point inlet values for fresh air and exhaust air, respectively. At the symmetric planes,

$$\frac{\partial \psi}{\partial z} = 0 \tag{4.73}$$

where ψ stands for variables such as velocity, temperature or humidity, and z is the normal direction to the plane. These are the boundary conditions.

4.3.2 Solution procedure

The partial differential equations for momentum, energy, and mass transport developed above are discretized by means of a finite volume method. Since three components (two air streams and the membrane) are closely interacting, iterative techniques are needed to solve these equations. The three-dimensional computational domain is created and meshed. In total 30,000 grids are generated for fresh air and exhaust air, and 18,000 grids are generated for the membrane. Then a grid independence test is conducted. It was found that a finer mesh will have little merit in further increasing the accuracy of the calculation. Therefore the current mesh number is considered sufficient.

Discretization of the equations is applied using second-order central differencing except for the convection terms, which are discretized using the bounded second-order upwind scheme (BSOU). Solution of the N-S equations is based on the SIMPLE algorithm for pressure correction.

4.3.3 Velocity, temperature and humidity profiles

The quasi-counter-flow arrangement is more complex than the counter-flow arrangement. Figures 4.19 and 4.20 show the velocity vectors at the center of the fresh air stream and the exhaust air stream, respectively. The inlet velocity is 2 m/s. As observed, the whole plate can be divided into three distinct zones: the pure counter-flow zone B, and cross-like flow zones A and C. The cross-like zones A and C are located just after the inlets and before the outlets. The incoming flow is turned around by the duct walls. It changes direction in zone A from cross flow to counter flow. Then it changes direction again in zone C from counter flow to cross flow. The velocity turnaround would lead to swirls. The smaller the areas of these cross-like zones, the more predominant the pure counter flow arrangement is, and the better the performance is. However, these cross-like zones are inescapable due to duct sealing.

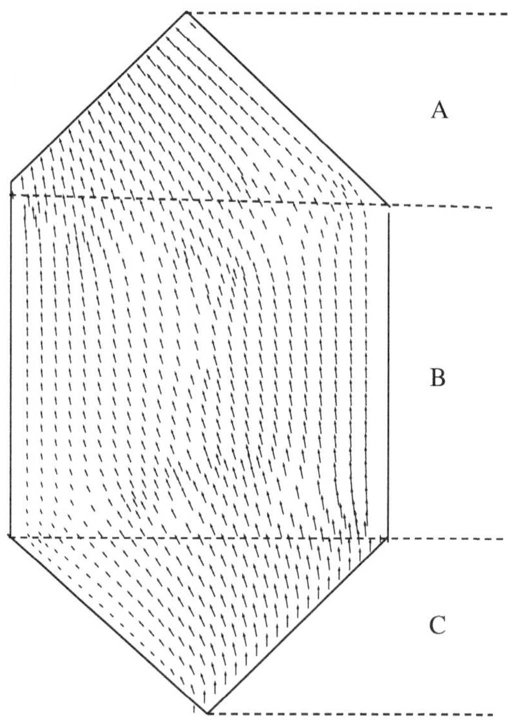

FIGURE 4.19

Velocity vectors at the center of the fresh air stream. Inlet $u_m = 2$ m/s.

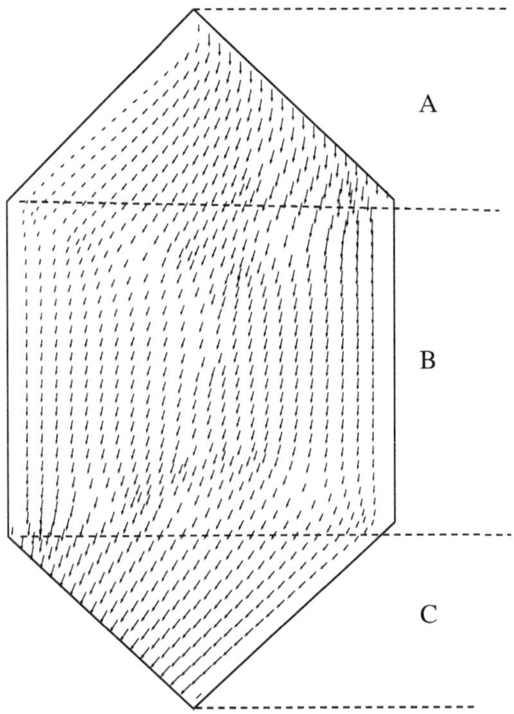

FIGURE 4.20

Velocity vectors at the center of the exhaust air. Inlet $u_m = 2\,m/s$.

In other words, a cross-flow zone is necessary to separate the two streams at duct inlets and outlets. A balance should be sought between decreasing the areas of the cross-like zones and ease of duct sealing.

The temperature and humidity profiles in the fluids and on the membrane surface are calculated. Figure 4.21 shows the temperature distributions on membrane surfaces. Due to the small membrane thickness, the temperature differences with membrane thickness are small. So the plotted values are nearly the same for membrane temperature on the exhaust-air side. The temperature distributions on membrane surfaces are very two-dimensional. They change from point to point from 0.005 to 0.889.

The calculated dimensionless humidity profiles on membrane surfaces are plotted in Figures 4.22 and 4.23 for the fresh-air side and the exhaust-air side, respectively. The humidity profiles exhibit similar trends to temperature, indicating an analogy between heat and mass transfer. However, the resistance of the membrane to mass transfer is much greater than the resistance to heat transfer. The fresh air humidity decreases from 1.0 to 0.467, while the exhaust air humidity increases from 0 to 0.633. Further, due to the higher mass transfer resistance through the membrane, there are demonstrable humidity differences between the two surfaces. For instance, humidity decreases from 0.929 to 0.331 on the surface on the fresh-air side, while it increases from 0.141 to 0.734 on the surface on the exhaust-air side. Generally, both heat transfer and moisture transfer can be divided into three distinct zones: one

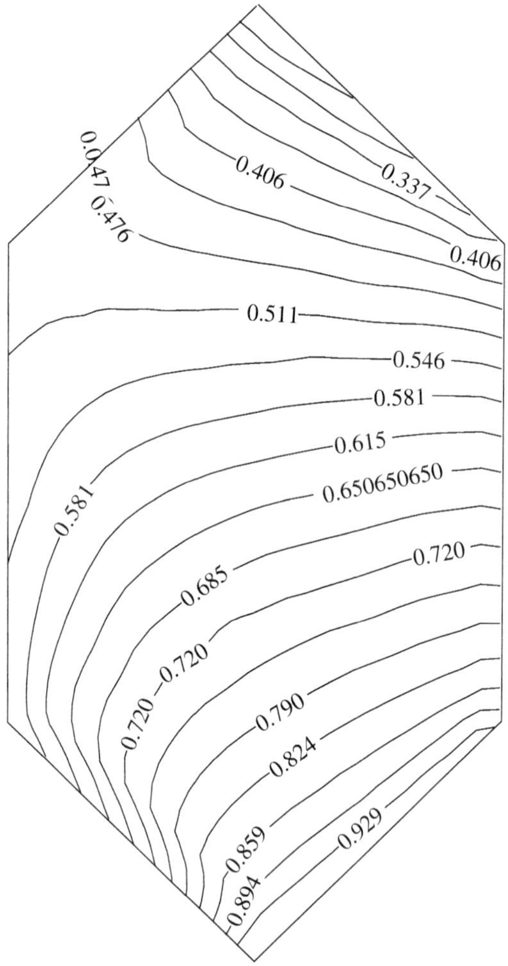

FIGURE 4.21

Dimensionless temperature profiles on the membrane surface. Inlet $u_m = 2$ m/s.

pure counter-flow regime and two cross-like regimes. This quasi-counter-flow exchanger improves the performance of the pure-counter flow regime, both for sensible heat transfer and for moisture transfer.

4.3.4 Nusselt and Sherwood numbers

Three other total heat exchangers are considered. They have the same membrane areas as the studied quasi-counter-flow exchanger. However, the flow arrangements are different: one concurrent, one cross flow and one counter flow. The calculated mean Nusselt and Sherwood numbers for the quasi-counter-flow exchanger under various air flow rates are listed in Table 4.4. The different flow arrangements have

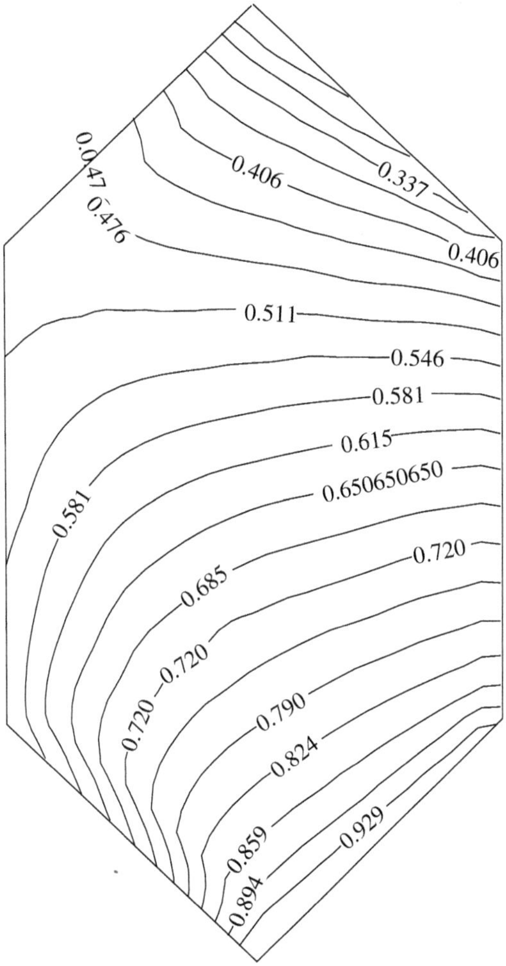

FIGURE 4.22

Dimensionless humidity profiles on the membrane surface adjacent to the fresh air stream. Inlet $u_m = 2$ m/s.

been compared. Of the flow arrangements, the cocurrent flow has the least sensible and latent effectiveness. The counter flow has the largest effectiveness. The quasi-flow has better performance than the cross flow due to improved heat mass transfer by the counter-flow zones. Generally, sensible and latent effectivenesses are improved by 5%. As can be seen, since most of the channel length is in the thermal or mass entry regions; the mean Nusselt and Sherwood numbers are larger than the fully developed values for parallel-plate ducts: $Nu = 7.54$. The higher the air velocities are, the longer the thermal and mass entry regions are. Because most of the exchanger is in the entry regions [9], a simple assumption of fully developed flow would underestimate the performance. A direct solution of the governing equations in the entry regions is a more accurate method.

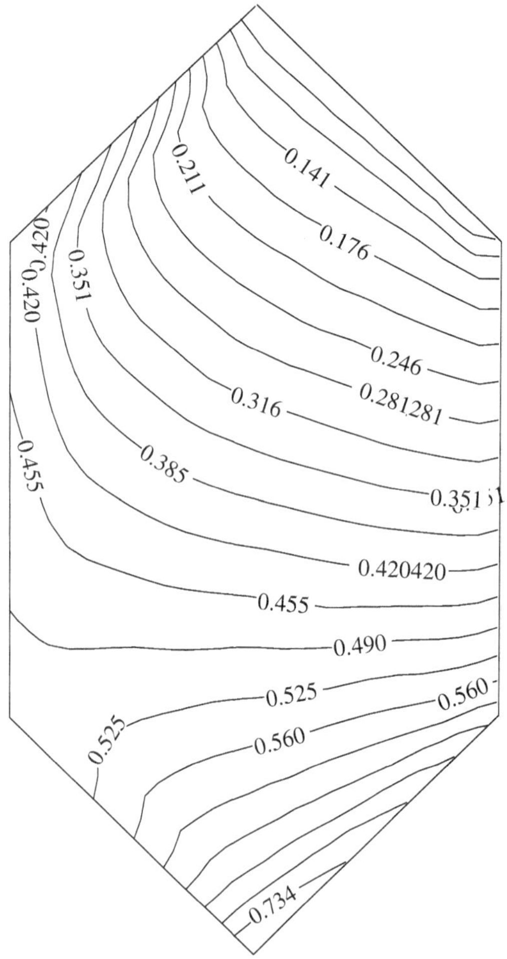

FIGURE 4.23

Dimensionless humidity profiles on the membrane surface adjacent to the exhaust air stream. Inlet $u_m = 2$ m/s.

4.4 Hydrodynamically developing regions

As mentioned, the entry regions are very important. Heat transfer in parallel-plate channels is one of the fundamental heat transfer problems in the classic textbooks. For simplicity, usually, a uniform temperature (**T** condition) or a uniform heat flux (**H** condition) boundary condition was assumed. Besides, the flow was usually assumed as hydraulically fully developed, for simplicity. With the assumed boundary values, the Nusselt numbers can be calculated.

In membrane-formed parallel-plate ducts, the phenomenon becomes more complex. Two difficulties should be overcome: (1) The boundary values are neither uniform temperature nor uniform heat

Table 4.4 Nusselt and Sherwood numbers and the effectiveness of three total heat exchangers under various flow arrangements.

V (m^3/h)	Cocurrent flow	Cross flow	Counter flow	Quasi counter flow	
		Sensible effectiveness			Nu_m
40	0.501	0.759	0.847	0.801	7.62
80	0.498	0.672	0.734	0.700	7.73
120	0.487	0.600	0.648	0.620	7.84
160	0.469	0.540	0.580	0.560	8.03
200	0.445	0.491	0.525	0.510	8.24
		Latent effectiveness			Sh_m
40	0.501	0.741	0.823	0.761	6.85
80	0.495	0.643	0.699	0.661	6.94
120	0.477	0.565	0.608	0.572	7.07
160	0.451	0.502	0.537	0.513	7.21
200	0.422	0.452	0.482	0.464	7.44

flux on the membrane surface. Rather, the boundary conditions are formed by the coupling between the two air flows: the fresh air and the exhaust air. The resistance of the membrane also has an influence on boundary conditions. It is a conjugate problem in nature. In addition, besides heat transfer, there is simultaneous mass transfer. (2) The lengths for boundary developing (or entry region) can be estimated by $\frac{x}{ReD_h} = 0.05$ for hydraulic and $\frac{x}{Re\,Pr\,D_h} = 0.05$ for thermal [5]. According to this criterion, in a 4 mm pitch channel under an inlet velocity of 1 m/s, the hydraulic entry length is $x = 0.2$ m and the thermal entry length is $x = 0.14$ m. Considering the above-mentioned typical commercial total heat exchanger of 150 m^3/h, the channel length is 0.185 m. As can be seen, a large proportion of the channel length is in the entry regions. Since the Nusselt and Sherwood numbers in the entry regions are higher than in the developed regions, a fully developed flow assumption would lead to substantial underestimations.

With the model proposed in the above section for all entry regions including flow velocity development, the Nusselt and Sherwood numbers in the entry regions can be calculated.

Figure 4.24 shows the calculated local (fRe) and Nusselt and Sherwood numbers along the dimensionless axial coordinates for fresh air or exhaust air under counter-flow arrangements. As seen, the local friction factor, the Nusselt number and the Sherwood number are very large at the air inlets. They decrease steadily and reach stable values after the boundaries are fully developed. The entry lengths are about $z^* = 0.05$, in accordance with Ref. [4]. The friction factor, Nusselt and Sherwood numbers, are rather high in the entry regions. When the heat exchanger length is short, these effects should be taken into account. As mentioned, the fully developed (fRe) is a constant: It is 96 for parallel-plate channels.

The trends shown in Figure 4.24 are similar for different flow arrangements and membrane resistance. However, the final fully developed Nusselt and Sherwood numbers are slightly different. Table 4.5 lists the calculated fully developed Nusselt and Sherwood numbers for three flow arrangements with various membrane to air conductivity (or diffusivity) ratios τ. Certainly a membrane heat

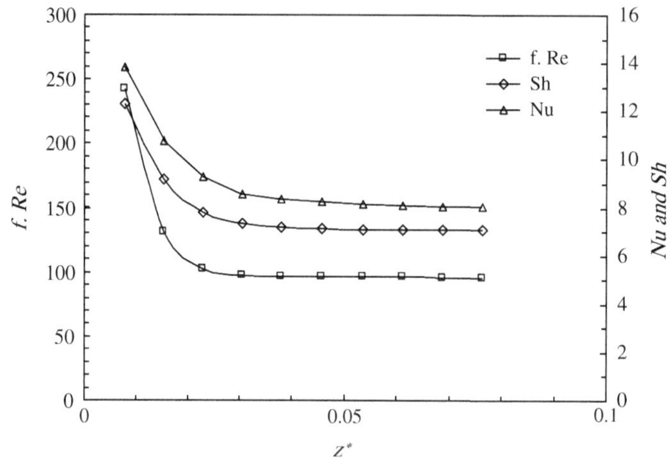

FIGURE 4.24

Variations of local (*fRe*), *Nu* and *Sh* along duct length for a counter-flow membrane duct. Parallel-plate duct.

conductivity ratio less than 1.0 or membrane diffusivity ratios greater than 0.3 only have numerical meanings. They are used just to show the effects of various membrane resistance values on fluid heat mass transfer numerically.

As can be seen in Table 4.5, the effects of membrane to air conductivity ratios on fully developed Nusselt and Sherwood numbers in fluids are negligible. The Sherwood numbers are usually less than the Nusselt numbers, due to the different values of *Pr* and *Sc*. Generally, flow arrangements and membrane resistance have some degree of influence on Nusselt and Sherwood numbers; however, this is very slight. The counter-flow arrangement has the largest Nusselt and Sherwood numbers. The cocurrent flow has the lowest values. The fully developed Nusselt number under uniform temperature conditions is $Nu_T = 7.54$, and that under uniform heat flux conditions is $Nu_H = 8.23$ [4]. As indicated, the Nusselt numbers in the counter flow are close to Nu_H, and those in the cocurrent flow are close to Nu_T. The values in the cross flow are in the middle.

Table 4.5 Fully developed Nusselt and Sherwood numbers for parallel-plate membrane ducts with different membrane to fluid conductivity ratios, under three typical arrangements.

	Cocurrent		**Counter**		**Cross**	
τ	*Nu*	*Sh*	*Nu*	*Sh*	*Nu*	*Sh*
0.1	8.01	7.03	8.17	7.22	8.12	7.13
1	8.04	7.05	8.19	7.51	8.14	7.23
10	8.06	7.06	8.21	7.33	8.16	7.26

$Nu_T = 7.54$; $Nu_H = 8.23$.

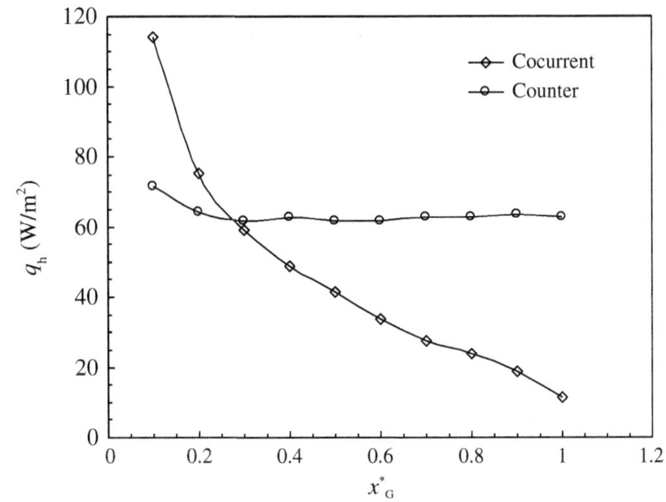

FIGURE 4.25

Heat flux distributions on the membrane surface for cocurrent flow and counter-flow total heat exchangers, with parallel-plate duct.

The distributions of heat flux on membrane surfaces are shown in Figure 4.25 for these two arrangements. As can be seen, they are inhomogeneous across the membranes, ignoring the flow arrangements. Therefore the values on the membrane surface are neither uniform temperature (humidity) nor uniform heat (mass) flux boundary conditions. The result is that the Nusselt and Sherwood numbers lie between those under uniform temperature (humidity) and uniform heat (mass) flux boundary conditions. This is shown in Table 4.5.

4.5 Summary

Parallel-plate membrane ducts, which are used extensively, are investigated. Conjugate heat and mass transfer in these ducts are considered. The flow arrangements are either cross flow or quasi-counter flow. Fluid flow and heat and mass equations governing the transport phenomena in the ducts are summarized. The entry regions for fluid flow, heat and mass boundaries, are studied. The results show that the entry regions play important roles in heat and mass transfer in the ducts. The fully developed Nusselt and Sherwood numbers under conjugate boundary conditions are obtained, which reveals heat and mass transfer behaviors different from those of traditional heat transfer tubes. The values on the membrane surface are neither uniform temperature (humidity) nor uniform heat (mass) flux boundary conditions. The fully developed Nusselt and Sherwood numbers lie between those under uniform temperature (humidity) and uniform heat (mass) flux boundary conditions. Flow arrangements and membrane properties all have some influence on Nusselt and Sherwood numbers.

References

[1] Zhang LZ. Total heat recovery: heat and moisture recovery from ventilation air. New York: Nova Science; 2009.

[2] Zhang LZ, Niu JL. Performance comparisons of desiccant wheels for air dehumidification and enthalpy recovery. Appl Thermal Eng 2002;22:1347–67.

[3] Niu JL, Zhang LZ. Effects of wall thickness on the heat and moisture transfers in desiccant wheels for air dehumidification and enthalpy recovery. Int Commun Heat Mass Transfer 2002;29:255–68.

[4] Shah RK, London AL. Laminar flow forced convection in ducts. New York: Academic Press; 1978. p. 253–60.

[5] Incropera FP, Dewitt DP. Introduction to heat transfer. 3rd ed. New York: John Wiley & Sons; 1996 (Chapter 8, p. 416).

[6] Shah RK, Bhatti MS. Laminar convection heat transfer in ducts. In: Kakac S, Shah RK, Aung W, editors. Handbook of single-phase convective heat transfer. New York: Wiley; 1987.

[7] Zhang LZ. Heat and mass transfer in a cross flow membrane-based enthalpy exchanger under naturally formed boundary conditions. Int J Heat Mass Transfer 2007;50:151–62.

[8] Zhang LZ. Heat and mass transfer in a quasi-counter flow membrane-based total heat exchanger. Int J Heat Mass Transfer 2010;53(23–24):5478–86.

[9] Zhang LZ, Liang CH, Pei LX. Conjugate heat and mass transfer in membrane-formed channels in all entry regions. Int J Heat Mass Transfer 2010;53(5–6):815–24.

CHAPTER

Heat and Mass Transfer in Plate-Fin Membrane Ducts

5

5.1 Introduction

Air-side convective heat transfer coefficients are usually low, so fins are used as extended surfaces to augment heat transfer. This leads to the invention of the plate-fin heat exchanger. The structure is shown in Figure 5.1. The cross-sections for ducts can be triangular, sinusoidal, or rectangular. Certainly other shapes are also possible, as long as they are convenient to manufacture. As for other air-to-air heat exchangers, plate-fin ducts are the most popular structure for total heat (heat and moisture) exchangers. Usually a cross flow between fresh air and exhaust air is used, due to the convenience in duct sealing.

The plate-fin structure has many benefits: (1) It is easy to construct. Especially the corrugated fins are easy to manufacture on a large scale by machine; (2) the mechanical strength is very high even with very thin plates. They are honeycomb-like, which gives high mechanical strength. (3) For air-to-air heat exchange, heat transfer intensification is needed on both sides of a plate, therefore equal structures and areas are provided on both sides. (4) Duct sealing is easy to achieve because of the multi-point contact between a fin and the neighboring plate. (5) Packing density is rather high because the channel height can be very small (1–2 mm) as a result of plates being separated by a corrugated fin. The fins are packed inside ducts. They are often called honeycomb-type heat exchangers.

Fluid flow and heat transfer in ducts are classic heat transfer problems [1–4]. They are relatively simple. Heat transfer in plate-fin ducts has been studied by several investigators [5], and can be called a conjugate heat transfer problem, which is more complicated. However, the most complicated ones are the conjugate heat and mass transfer in membrane-formed ducts, which were not addressed until 5 years ago [6–9]. These are the basic data for the design of heat and mass exchangers [10]. In this chapter a systematic discussion of the conjugate heat and mass transfer in these ducts will be presented.

5.2 Triangular ducts

Taking a total heat exchanger as an example, the conjugate heat and mass transfer in the plate-fin ducts of triangular cross sections is analyzed. The methodology is also applicable to other duct cross sections. Heat and mass transfer is accomplished by the plates and the extended surfaces—the fins.

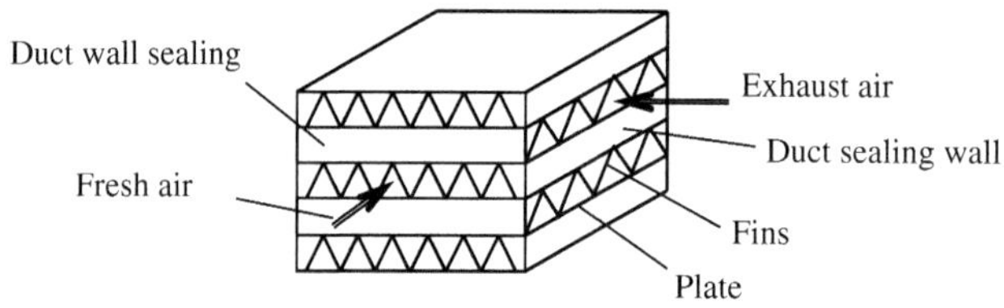

FIGURE 5.1

A plate-fin structure for air-to-air heat exchangers.

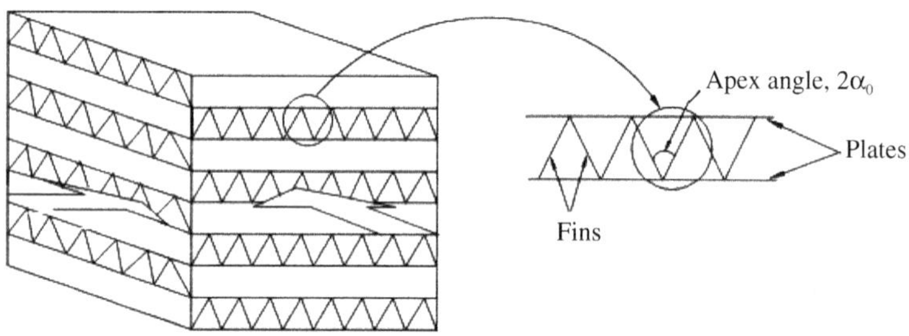

FIGURE 5.2

Schematic of a cross-flow total heat exchanger with plate-fin triangular ducts.

Heat and mass transfer in plate-fin ducts in a total heat exchanger can be separated into three steps: (1) Convective heat mass transfer from fresh air to plates and fins; (2) conductive heat and mass transfer in plates; (3) convective heat mass transfer from plates and fins to exhaust air. Of the heat mass transfer parameters, convective heat mass transfer coefficients in plate-fin ducts are the key parameter to analyze performance and product design.

In a heat exchanger, plates contact the air streams directly, but fins don't. Heat is transferred by conduction from plates to fins first, then it is transferred from fins to the air stream by convection. Fin efficiency is the key factor. It should be noted that real fins have limited conductance, which may compromise fin efficiencies. For membrane-formed total heat exchangers, as shown in Figure 5.2, when there is simultaneous heat and mass transfer, fins of finite conductance have more influence on fin efficiencies. The resulting problems of heat transfer and mass transfer are new and more challenging.

5.2.1 Governing equations

The problem considered here is that of representing duct, shown in Figure 5.3. The representative geometry for the triangular plate-fin duct is: duct height $2a$; width $2b$; duct length, L; apex angle $2\alpha_0$. The coordinate system for the two fins is x_1 and y_1, which is tangential and normal to the fins respectively.

The flow in the duct is considered to be hydrodynamically fully developed and laminar, but with both heat and mass developing in the entrance region of the duct. The fluid is Newtonian with constant

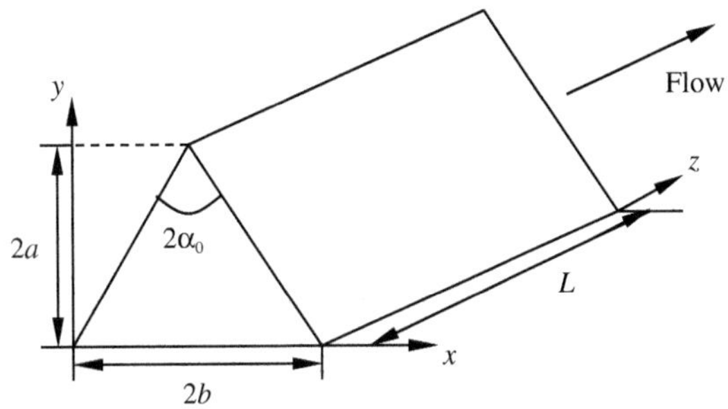

FIGURE 5.3

Geometry and coordinates of a plate-fin triangular duct.

physical properties. Additionally, uniform plate temperature and concentration boundary conditions are assumed.

For fully developed laminar flow in ducts, the Navier-Stokes equations reduce to

$$\mu\left(\frac{\partial^2 u}{\partial x^2}+\frac{\partial^2 u}{\partial y^2}\right)=\frac{dP}{dz} \tag{5.1}$$

where μ is dynamic viscosity (Pa s), z is the axial coordinate (m), P is pressure (Pa).

The above equation can be normalized to

$$\frac{\partial^2 u^*}{\partial x^{*2}}+\left(\frac{b}{a}\right)^2\frac{\partial^2 u^*}{\partial y^{*2}}+\frac{4b^2}{D_h^2}=0 \tag{5.2}$$

with a dimensionless velocity

$$u^*=-\frac{\mu u}{(dP/dz)D_h^2} \tag{5.3}$$

Dimensionless coordinates are defined by

$$x^*=\frac{x}{2b} \tag{5.4}$$

$$y^*=\frac{y}{2a} \tag{5.5}$$

where hydraulic diameter

$$D_h=\frac{4A_c}{P_f} \tag{5.6}$$

where A_c is the cross-section area of the duct (m^2), P_f is the perimeter of the duct (m).

The characteristics of fluid flow in the duct can be represented by the product of the friction factor and the Reynolds number as

$$(fRe) = \left(-\frac{D_h \frac{dP}{dz}}{\frac{1}{2}\rho_a u_m^2}\right)\left(\frac{\rho_a u_m D_h}{\mu}\right) = \frac{2}{u_m^*} \tag{5.7}$$

where u_m^* is the average dimensionless velocity on a cross section, and it is calculated by

$$u_m^* = \frac{\iint u^* dA}{A_c} \tag{5.8}$$

Heat transfer in air stream

Energy conservation in the fluid can be expressed by

$$\rho_a c_{pa} u \frac{\partial T}{\partial z} = \lambda_a \left(\frac{\partial^2 T}{\partial x^2} + \frac{\partial^2 T}{\partial y^2}\right) \tag{5.9}$$

where T is fluid temperature (K).

The above equation can be normalized to

$$U \frac{\partial \theta}{\partial z^*} = \frac{\partial^2 \theta}{\partial x^{*2}} + \left(\frac{b}{a}\right)^2 \frac{\partial^2 \theta}{\partial y^{*2}} \tag{5.10}$$

with a dimensionless temperature

$$\theta = \frac{T - T_w}{T_i - T_w} \tag{5.11}$$

where T_i is the inlet temperature of the fluid, and T_w is the wall temperature. Dimensionless axial position is defined by

$$z^* = \frac{z}{D_h Re Pr} \tag{5.12}$$

$$Le = \frac{\mathrm{Pr}}{Sc} = \frac{\lambda_a}{\rho_a c_{pa} D_{va}} \tag{5.13}$$

Le is around 0.8 at atmospheric conditions. Velocity coefficient U is defined by

$$U = \frac{u^*}{u_m^*} \frac{4b^2}{D_h^2} \tag{5.14}$$

Dimensionless bulk temperature

$$\theta_b = \frac{\iint u^* \theta dA}{\iint u^* dA} \tag{5.15}$$

Taking the plate temperature as the characteristic temperature, an energy balance in a control volume in the duct as described in previous chapters will give the equation for estimation of the local Nusselt number as

$$Nu_L = -\frac{1}{4\theta_b}\frac{d\theta_b}{dz^*} \tag{5.16}$$

and the mean Nusselt number from $z^* = 0$ to z^* by

$$Nu_m = -\frac{1}{4z^*}\ln\theta_b \tag{5.17}$$

Convective heat transfer coefficient

$$h_L = -\frac{1}{4\theta_b}\frac{\lambda_a}{D_h}\frac{d\theta_b}{dz^*} \tag{5.18}$$

Mass transfer in air stream

Air travels in the duct while exchanging moisture directly with duct walls. Mass conservation in the air stream can be expressed by

$$u\frac{\partial \omega}{\partial z} = D_{va}\left(\frac{\partial^2 \omega}{\partial x^2} + \frac{\partial^2 \omega}{\partial y^2}\right) \tag{5.19}$$

where ω is the humidity ratio of air (kg vapor/kg dry air).

The above equation can be normalized to

$$\frac{U}{Le}\frac{\partial \omega^*}{\partial z^*} = \frac{\partial^2 \omega^*}{\partial x^{*2}} + \left(\frac{b}{a}\right)^2 \frac{\partial^2 \omega^*}{\partial y^{*2}} \tag{5.20}$$

where dimensionless humidity ratio

$$\omega^* = \frac{\omega - \omega_w}{\omega_i - \omega_w} \tag{5.21}$$

where ω_i is the inlet air humidity, and ω_w is the humidity on wall surface.

Dimensionless bulk humidity

$$\omega_b^* = \frac{\iint u^* \omega^* dA}{\iint u^* dA} \tag{5.22}$$

Similarly to heat transfer, taking the plate humidity as the characteristic humidity, a mass balance in a control volume in the duct will give the equation for estimation of the local Sherwood number as

$$Sh_L = -\frac{1}{4\omega_b^* Le} \frac{d\omega_b^*}{dz^*} \tag{5.23}$$

and the mean Sherwood number from $z^* = 0$ to z^* by

$$Sh_m = -\frac{1}{4z^* Le} \ln \omega_b^* \tag{5.24}$$

Convective mass transfer coefficient

$$k_L = -\frac{1}{4\omega_b^*} \frac{D_{va}}{D_h Le} \frac{d\omega_b^*}{dz^*} \tag{5.25}$$

Heat and mass conduction in fins

At any location along the fins, there is a balance between the net heat conduction along the fins and the heat transfer from the surface of the fins to the fluid. Heat conduction in fins is governed by the following one-dimensional model

$$\lambda_f \delta \frac{d^2 T_f}{ds^2} = q_u + q_L \tag{5.26}$$

$$q_u = -\lambda_a \left(\frac{\partial T}{\partial y_1} \right)_u \tag{5.27}$$

$$q_L = \lambda_a \left(\frac{\partial T}{\partial y_1} \right)_L \tag{5.28}$$

where λ_f is fin thermal conductivity, $(\partial T/\partial y_1)$ is the normal gradient of fluid temperature on the lower or upper surface of fin. The skew-symmetric nature of boundary conditions on fins has been discussed in detail in Chapter 2. The nature is formed by the skew-symmetric fluid distributions in the structure. The heat fluxes at the lower surface and the upper surface are skew-symmetric [6], as schematically depicted in Figure 5.4. The relation is mathematically expressed by

$$q_u\big|_s = -q_L\big|_{L_f - s} \tag{5.29}$$

where L_f is the length of a single fin.

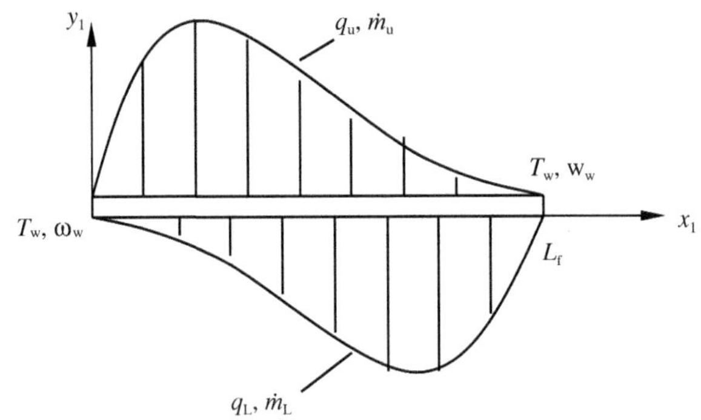

FIGURE 5.4

Skew-symmetric heat and mass fluxes distributions on the upper and lower fin-fluid interfaces.

Similarly, for mass diffusion in fins, at any location, there is a mass balance between the net water diffusion along the fin and the mass transfer from the surface of the fin to the fluid. The governing equation is expressed by

$$D_{wf}\rho_f\delta\frac{d^2w}{ds^2} = \dot{m}_u + \dot{m}_L \tag{5.30}$$

$$\dot{m}_u = -D_{va}\rho_a\left(\frac{\partial\omega}{\partial y_1}\right)_u \tag{5.31}$$

$$\dot{m}_L = D_{va}\rho_a\left(\frac{\partial\omega}{\partial y_1}\right)_L \tag{5.32}$$

where D_{wf} is diffusivity of water in fin materials (m^2/s); ρ_f is density of fin materials (kg/m^3); w is water content in fin materials (kg water/kg dry material); $\dot{m}_u$ is moisture flux from upper surface of fin, and $\dot{m}_L$ is moisture flux from lower surface. $(\partial\omega/\partial y_1)$ is the normal gradient of vapor concentration on the lower or upper surface of fin. Due to symmetry in fluid fields, the mass fluxes at the lower surface and the upper surface are also skew-symmetric. The relation is mathematically expressed by

$$\dot{m}_u\big|_s = -\,\dot{m}_L\big|_{L_f-s} \tag{5.33}$$

Equations (5.26)–(5.29) can be combined and normalized to

$$\Omega_s\frac{d^2\theta_f}{ds^{*2}} = -\left(\frac{\partial\theta}{\partial y_1^*}\right)_{s^*} - \left(\frac{\partial\theta}{\partial y_1^*}\right)_{L_f^*-s^*} \tag{5.34}$$

where Ω_s is a dimensionless parameter named the fin heat conductance parameter. It is defined by

$$\Omega_s = \frac{\lambda_f \delta}{\lambda_a (2a)} \tag{5.35}$$

where dimensionless fin temperature is also defined by Eq. (5.11)

The dimensionless coordinates

$$s^* = \frac{s}{2a} \tag{5.36}$$

$$y_1^* = \frac{y_1}{2a} \tag{5.37}$$

$$L_f^* = \frac{L_f}{2a} \tag{5.38}$$

Similar to heat conduction in the fin, equations for mass diffusion in fins, Eqs. (5.30)–(5.33), can be normalized to the following dimensionless equation:

$$\Omega_L \frac{d^2 \omega_f^*}{ds^{*2}} = -\left(\frac{\partial \omega^*}{\partial y_1^*}\right)_{s^*} - \left(\frac{\partial \omega^*}{\partial y_1^*}\right)_{L_f^* - s^*} \tag{5.39}$$

where Ω_L is defined as the dimensionless fin mass conductance parameter. It is calculated by

$$\Omega_L = \frac{\delta \rho_f D_{wf} K_p}{\rho_a D_{va} (2a)} \tag{5.40}$$

where the dimensionless humidity ratio is defined by Eq. (5.21).

Temperature differences between the two surfaces of a fin are quite small due to the small fin thickness [6]. Water content in fins is a function of relative humidity. For the air conditioning industry, heat mass exchangers always work between 40% RH and 80% RH. Considering these factors, the relative humidity can be a linear expression of the humidity ratio, and the sorption isotherm for the material can also be approximated by a linear equation as

$$w = K_p \omega \tag{5.41}$$

where K_p is defined as the partition coefficient, with the same definition as in Chapter 2. This is a simplified model for moisture adsorption.

Obviously, Eqs. (5.19) and (5.39) are in the same forms as Eqs. (5.10) and (5.34) respectively. Therefore, the solution of the dimensionless humidity ratio can be an analogy to the solution of the dimensionless temperature, if the fin mass conductance parameters are the same values as the fin heat conductance parameters. Consequently, it is only necessary to solve either the heat transfer equations or the mass transfer equations. The forms are in analogy, but the coefficients are different.

Boundary conditions

The boundary conditions for fluid

$$u^* = 0, \text{on all the walls of the duct} \tag{5.42}$$

$$\theta = 0 \text{ and } \xi = 0, \text{ at } y^* = 0 \tag{5.43}$$

Inlet condition

$$\theta = 1 \text{ and } \xi = 1, \text{ at } z^* = 0 \tag{5.44}$$

Boundary conditions for fins

$$\theta_f = 0 \text{ and } \xi_f = 0 \text{ at } s^* = 0, L_f^* \tag{5.45}$$

Conjugate conditions on fins, or fin-fluid coupling

$$\theta_f = \theta, \text{ and } \xi_f = \xi, \text{ at fin-fluid interfaces} \tag{5.46}$$

5.2.2 Numerical scheme

The duct geometry is a triangular one. For such non-rectangular shapes, a boundary-fitted coordinate transformation technique is used to transfer the irregular physical domain to a square computational domain. As introduced in previous chapters, the basic idea of the boundary-fitted coordinate system is to have a coordinate system such that the body contour coincides with the coordinate lines. The transformation between the physical coordinates and the boundary-fitted coordinates, which is usually a square domain, is achieved by solving two Poisson equations on the physical domain. For a detailed introduction of a boundary-fitted coordinate system, please see Chapter 2.

The momentum, heat, and mass Eqs. (5.2), (5.10), and (5.20) are transformed to:

$$\frac{\partial}{\partial \xi}\left[\frac{1}{J}\left(\alpha\frac{\partial u^*}{\partial \xi} - \beta\frac{\partial u^*}{\partial \eta}\right)\right] + \frac{\partial}{\partial \eta}\left[\frac{1}{J}\left(\frac{b}{a}\right)^2\left(\gamma\frac{\partial u^*}{\partial \eta} - \beta\frac{\partial u^*}{\partial \xi}\right)\right] + J\frac{4b^2}{D_h^2} = 0 \tag{5.47}$$

$$JU\frac{\partial \theta}{\partial z^*} = \frac{\partial}{\partial \xi}\left[\frac{1}{J}\left(\alpha\frac{\partial \theta}{\partial \xi} - \beta\frac{\partial \theta}{\partial \eta}\right)\right] + \frac{\partial}{\partial \eta}\left[\frac{1}{J}\left(\frac{b}{a}\right)^2\left(\gamma\frac{\partial \theta}{\partial \eta} - \beta\frac{\partial \theta}{\partial \xi}\right)\right] \tag{5.48}$$

$$\frac{JU}{Le}\frac{\partial \omega^*}{\partial z^*} = \frac{\partial}{\partial \xi}\left[\frac{1}{J}\left(\alpha\frac{\partial \omega^*}{\partial \xi} - \beta\frac{\partial \omega^*}{\partial \eta}\right)\right] + \frac{\partial}{\partial \eta}\left[\frac{1}{J}\left(\frac{b}{a}\right)^2\left(\gamma\frac{\partial \omega^*}{\partial \eta} - \beta\frac{\partial \omega^*}{\partial \xi}\right)\right] \tag{5.49}$$

where α, β, γ have the same definition as in Chapter 2, and J is the Jacobian transformation operator also defined in Chapter 2.

The discrete form for air momentum:

$$\begin{aligned}
&\frac{\partial}{\partial \xi}\left[\frac{1}{J}\left(\alpha\frac{\partial u^*}{\partial \xi}-\beta\frac{\partial u^*}{\partial \eta}\right)\right]+\frac{\partial}{\partial \eta}\left[\frac{1}{J}\left(\frac{b}{a}\right)^2\left(\gamma\frac{\partial u^*}{\partial \eta}-\beta\frac{\partial u^*}{\partial \xi}\right)\right]+J\frac{4b^2}{D_h^2}\\
&=\frac{1}{\Delta\xi}\left[\frac{1}{J}\left(\alpha\frac{\Delta u^*}{\Delta \xi}-\beta\frac{\Delta u^*}{\Delta \eta}\right)\right]_e-\frac{1}{\Delta\xi}\left[\frac{1}{J}\left(\alpha\frac{\Delta u^*}{\Delta \xi}-\beta\frac{\Delta u^*}{\Delta \eta}\right)\right]_w\\
&\quad+\frac{1}{\Delta\eta}\left[\frac{1}{J}\left(\frac{b}{a}\right)^2\left(\gamma\frac{\Delta u^*}{\Delta \eta}-\beta\frac{\Delta u^*}{\Delta \xi}\right)\right]_n-\frac{1}{\Delta\eta}\left[\frac{1}{J}\left(\frac{b}{a}\right)^2\left(\gamma\frac{\Delta u^*}{\Delta \eta}-\beta\frac{\Delta u^*}{\Delta \xi}\right)\right]_s+J\frac{4b^2}{D_h^2}\\
&=\left\{\begin{array}{l}\frac{1}{\Delta\xi}\left(\frac{\alpha}{J}\right)_e\left(\frac{u_E^*-u_P^*}{\Delta\xi_e}\right)-\frac{1}{\Delta\xi}\left(\frac{\alpha}{J}\right)_w\left(\frac{u_P^*-u_W^*}{\Delta\xi_w}\right)+\frac{1}{\Delta\eta}\left(\frac{\gamma}{J}\right)_n\left(\frac{b}{a}\right)^2\left(\frac{u_N^*-u_P^*}{\Delta\eta_n}\right)\\-\frac{1}{\Delta\eta}\left(\frac{\gamma}{J}\right)_s\left(\frac{b}{a}\right)^2\left(\frac{u_P^*-u_S^*}{\Delta\eta_s}\right)\end{array}\right\}\\
&\quad+\left\{\begin{array}{l}-\frac{1}{\Delta\xi}\left(\frac{\beta}{J}\right)_e\left(\frac{u_{NE/2}^*-u_{SE/2}^*}{\Delta\eta_e}\right)+\frac{1}{\Delta\xi}\left(\frac{\beta}{J}\right)_w\left(\frac{u_{NW/2}^*-u_{SW/2}^*}{\Delta\eta_w}\right)\\-\frac{1}{\Delta\eta}\left(\frac{\beta}{J}\right)_n\left(\frac{b}{a}\right)^2\left(\frac{u_{NE/2}^*-u_{NW/2}^*}{\Delta\xi_n}+\right)\frac{1}{\Delta\eta}\left(\frac{\beta}{J}\right)_s\left(\frac{b}{a}\right)^2\left(\frac{u_{SE/2}^*-u_{SW/2}^*}{\Delta\xi_s}\right)\end{array}\right\}\\
&\quad+J\frac{4b^2}{D_h^2}=0
\end{aligned} \tag{5.50}$$

The finite difference form can be summarized by

$$a_P u_P^* = a_E u_E^* + a_W u_W^* + a_N u_N^* + a_S u_S^* + b_1 + b_2 \tag{5.51}$$

where

$$a_P = a_E + a_W + a_N + a_S$$

$$\begin{aligned}
b_1 =& -\frac{1}{(\Delta\xi)_P}\left(\frac{\beta}{J}\right)_e\left(\frac{u_{NE/2}^*-u_{SE/2}^*}{\Delta\eta_e}\right)+\frac{1}{(\Delta\xi)_P}\left(\frac{\beta}{J}\right)_w\left(\frac{u_{NW/2}^*-u_{SW/2}^*}{\Delta\eta_w}\right)\\
&-\frac{1}{(\Delta\eta)_P}\left(\frac{\beta}{J}\right)_n\left(\frac{b}{a}\right)^2\left(\frac{u_{NE/2}^*-u_{NW/2}^*}{\Delta\xi_n}\right)+\frac{1}{(\Delta\eta)_P}\left(\frac{\beta}{J}\right)_s\left(\frac{b}{a}\right)^2\left(\frac{u_{SE/2}^*-u_{SW/2}^*}{\Delta\xi_s}\right)
\end{aligned}$$

$$b_2 = J\frac{4b^2}{{D_h}^2};\quad a_E = \frac{1}{(\Delta\xi)_P}\left(\frac{\alpha}{J}\right)_e\frac{1}{(\Delta\xi)_e};\quad a_W = \frac{1}{(\Delta\xi)_P}\left(\frac{\alpha}{J}\right)_w\frac{1}{(\Delta\xi)_w}$$

$$a_N = \frac{1}{(\Delta\eta)_P}\left(\frac{\gamma}{J}\right)_n\left(\frac{b}{a}\right)^2\frac{1}{(\Delta\eta)_n};\quad a_S = \frac{1}{(\Delta\eta)_P}\left(\frac{\gamma}{J}\right)_s\left(\frac{b}{a}\right)^2\frac{1}{(\Delta\eta)_s}$$

$$\frac{u_{NE/2}^*-u_{SE/2}^*}{\Delta\eta_e}=\frac{u_N^*+u_{NE}^*-u_S^*-u_{SE}^*}{4\Delta\eta_e}=\frac{u_{i,j+1}^*+u_{i+1,j+1}^*-u_{i,j-1}^*-u_{i+1,j-1}^*}{4\Delta\eta_e}$$

Air temperature

$$a_P\theta_P = a_E\theta_E + a_W\theta_W + a_N\theta_N + a_S\theta_S + a_B\theta_B + b_1 \tag{5.52}$$

where

$$a_P = a_E + a_W + a_N + a_S + a_B$$

$$\begin{aligned} b_1 = & -\frac{1}{(\Delta\xi)_P}\left(\frac{\beta}{J}\right)_e \frac{\theta_{NE/2} - \theta_{SE/2}}{(\Delta\eta)_e} + \frac{1}{(\Delta\xi)_P}\left(\frac{\beta}{J}\right)_w \frac{\theta_{NW/2} - \theta_{SW/2}}{(\Delta\eta)_w} \\ & -\frac{1}{(\Delta\eta)_P}\left(\frac{\beta}{J}\right)_n \left(\frac{b}{a}\right)^2 \frac{\theta_{NE/2} - \theta_{NW/2}}{(\Delta\xi)_n} + \frac{1}{(\Delta\eta)_P}\left(\frac{\beta}{J}\right)_s \left(\frac{b}{a}\right)^2 \frac{\theta_{SE/2} - \theta_{SW/2}}{(\Delta\xi)_s} \end{aligned}$$

$$a_E = \frac{1}{(\Delta\xi)_P}\left(\frac{\alpha}{J}\right)_e \frac{1}{(\Delta\xi)_e}; \quad a_W = \frac{1}{(\Delta\xi)_P}\left(\frac{\alpha}{J}\right)_w \frac{1}{(\Delta\xi)_w}; \quad a_N = \frac{1}{(\Delta\eta)_P}\left(\frac{\gamma}{J}\right)_n \left(\frac{b}{a}\right)^2 \frac{1}{(\Delta\eta)_n}$$

$$a_S = \frac{1}{(\Delta\eta)_P}\left(\frac{\gamma}{J}\right)_s \left(\frac{b}{a}\right)^2 \frac{1}{(\Delta\eta)_s}; \quad a_B = \frac{JU}{(\Delta z^*)_P}$$

Air humidity

$$a_P\overset{*}{\omega}_P = a_E\overset{*}{\omega}_E + a_W\overset{*}{\omega}_W + a_N\overset{*}{\omega}_N + a_S\overset{*}{\omega}_S + a_B\overset{*}{\omega}_B + b_1 \tag{5.53}$$

where

$$a_P = a_E + a_W + a_N + a_S + a_B$$

$$\begin{aligned} b_1 = & -\frac{1}{(\Delta\xi)_P}\left(\frac{\beta}{J}\right)_e \frac{\overset{*}{\omega}_{NE/2} - \overset{*}{\omega}_{SE/2}}{(\Delta\eta)_e} + \frac{1}{(\Delta\xi)_P}\left(\frac{\beta}{J}\right)_w \frac{\overset{*}{\omega}_{NW/2} - \overset{*}{\omega}_{SW/2}}{(\Delta\eta)_w} \\ & -\frac{1}{(\Delta\eta)_P}\left(\frac{\beta}{J}\right)_n \left(\frac{b}{a}\right)^2 \frac{\overset{*}{\omega}_{NE/2} - \overset{*}{\omega}_{NW/2}}{(\Delta\xi)_n} + \frac{1}{(\Delta\eta)_P}\left(\frac{\beta}{J}\right)_s \left(\frac{b}{a}\right)^2 \frac{\overset{*}{\omega}_{SE/2} - \overset{*}{\omega}_{SW/2}}{(\Delta\xi)_s} \end{aligned}$$

$$a_E = \frac{1}{(\Delta\xi)_P}\left(\frac{\alpha}{J}\right)_e \frac{1}{(\Delta\xi)_e}; \quad a_W = \frac{1}{(\Delta\xi)_P}\left(\frac{\alpha}{J}\right)_w \frac{1}{(\Delta\xi)_w}; \quad a_N = \frac{1}{(\Delta\eta)_P}\left(\frac{\gamma}{J}\right)_n \left(\frac{b}{a}\right)^2 \frac{1}{(\Delta\eta)_n}$$

$$a_S = \frac{1}{(\Delta\eta)_P}\left(\frac{\gamma}{J}\right)_s \left(\frac{b}{a}\right)^2 \frac{1}{(\Delta\eta)_s}; \quad a_B = \frac{JU}{(\Delta z^*)_P Le}$$

In the above equations, the nodes names and the grid numbering scheme are the same as in Chapter 2, as shown in Figure 2.12.

For fins, the left fin in Figure 5.3 is $\xi = \text{const}$ on the computational plane. So heat or mass flux is transformed to

$$\frac{\partial\phi}{\partial y_1^*} = \frac{\alpha\phi_\xi - \beta\phi_\eta}{J\sqrt{\alpha}} \tag{5.54}$$

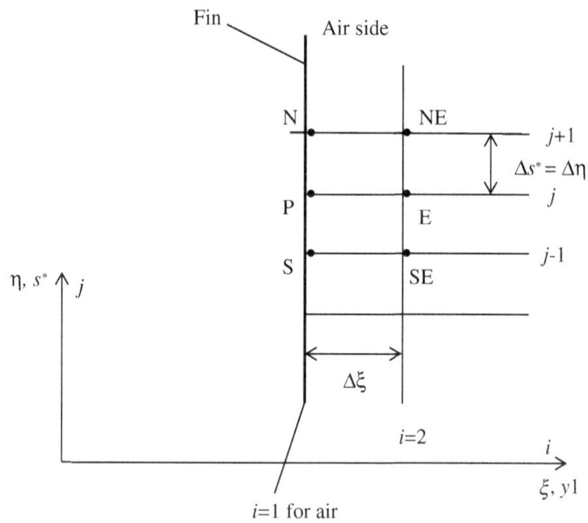

FIGURE 5.5

The grid numbering for the left fin and neighboring air duct.

The right fin in Figure 5.3 is $\eta = \text{const}$ on the computational plane. So heat or mass flux is transformed to

$$\frac{\partial \phi}{\partial y_1^*} = \frac{-\beta \phi_\xi + \gamma \phi_\eta}{J\sqrt{\gamma}} \tag{5.55}$$

where ϕ is a universal variable that can be dimensionless temperature or humidity. Both the variables and the coordinates refer to air-side values.

The discrete forms of fins in Eqs. (5.34) and (5.39) can be obtained by combining the above equations into the governing equations for fins. For instance, for the left fin, temperature or humidity is calculated by

$$a_P \phi_P = a_N \phi_N + a_S \phi_S + \left.\frac{\partial \phi}{\partial y_1^*}\right|_{s^*} + \left.\frac{\partial \phi}{\partial y_1^*}\right|_{L_f^* - s^*} \tag{5.56}$$

where

$$a_P = a_N + a_S$$

$$a_N = a_S = \frac{\Omega_\phi}{(\Delta s^*)_P^2}$$

For temperature calculations, ϕ is θ, and Ω_ϕ is Ω_S; for humidity calculations, ϕ is ω^*, and Ω_ϕ is Ω_L. The values are for fins. Nodes "N" and "S" refer to the northern and southern neighboring nodes to node "P" on the fin, as shown in Figure 5.5 for the grid structure of the fin.

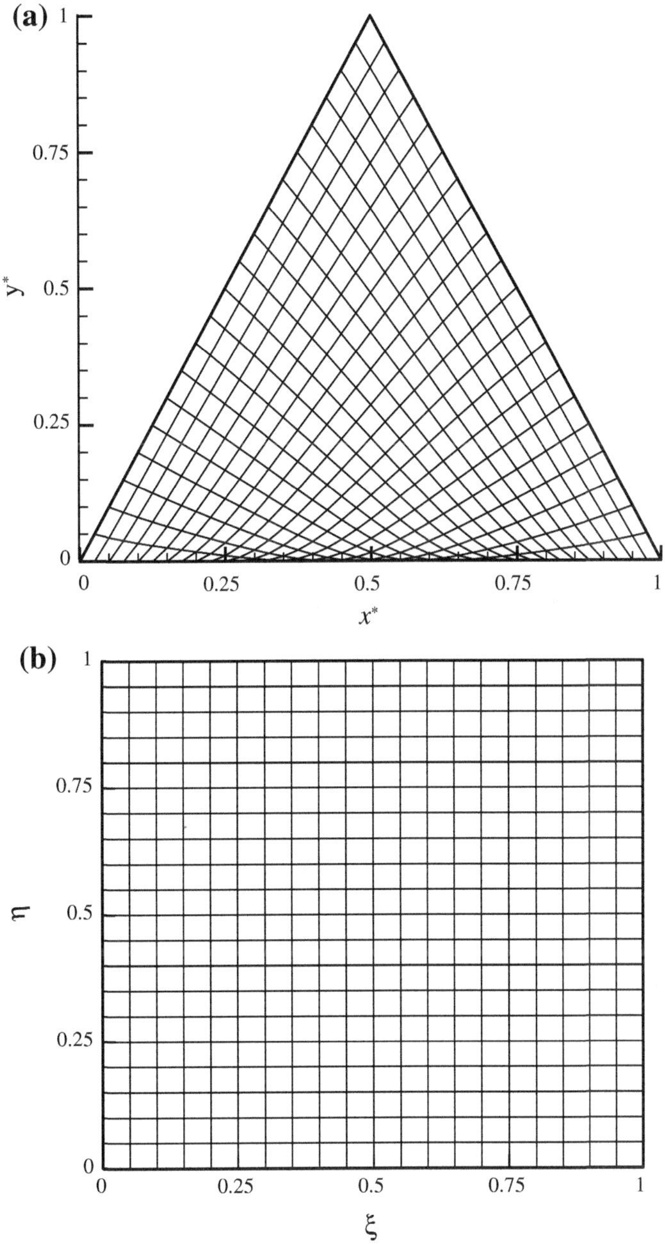

FIGURE 5.6

Grid configurations for the duct cross section. (a) Physical plane; (b) the computational domain.

The grid constructions for the duct are shown in Figure 5.6. To account for the drastic variations of temperature and mass concentration at the duct inlet, grids are dense near the inlet, while sparsely populated after the thermal entry length, as seen in Figure 5.7 for the axial grids.

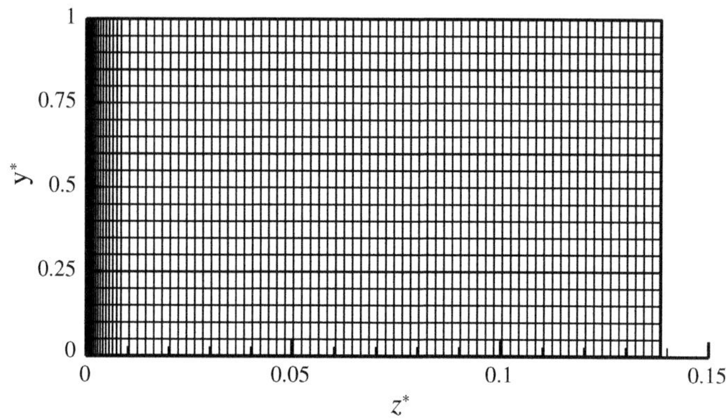

FIGURE 5.7

Mesh structure in the duct length direction.

The governing dimensionless partial differential equations are transformed from the physical domain to the boundary-fitted coordinate system. Then they are reduced to a set of algebraic equation systems as described above, which can be solved by ADI (alternate direction implicit) techniques. It is clear that iteration is needed to obtain the solution. In each direction, a tri-diagonal matrix is solved, by treating the non-linear cross-derivative as a source term and using the values of the last iteration. Although the momentum equation in the finite difference form is only solved in two directions for the determination of the velocity distribution, the energy and mass equations must be solved at every step change in the flow direction to determine the temperature and concentration distribution. The proposed finite difference schemes are implicit numerical schemes. The following convergence criterion was chosen for the study.

$$\forall i,j,k,\ \frac{\left|\phi_{i,j,k}^{m}-\phi_{i,j,k}^{m-1}\right|}{\left|\phi_{i,j,k}^{m}\right|}\leq 10^{-5} \tag{5.57}$$

where ϕ refers to the dependent variable u^*, θ or ω^*, and m stands for the mth iteration.

The problem is a typical conjugate one. After the solution of fluid velocity and humidity, humidity on two fins, is solved on the generated grids, taking the current fluid humidity as the default boundary condition. The whole calculating procedure can be summarized as the following:

a. Grid generation for both the fluid and the fins.
b. Solve the momentum equation. Get the velocity fields and resistance data for the duct.
c. Assume initial temperature and humidity fields in the fluid.
d. Taking current fluid temperature and humidity as the boundary conditions for fins, get the temperature and moisture uptake fields on two fins.
e. Taking the current values of temperature and humidity on fins as the default values, get the temperature and humidity profiles in the fluid.
f. Go to (d), until the old values and the newly calculated values of temperature and humidity at all calculating nodes are converged.

As can be seen, the solution of velocity is only executed once before the calculation of temperature and humidity. After these procedures, all the governing equations are satisfied simultaneously. Then the local and mean Nusselt and Sherwood numbers are calculated. The fully developed Nusselt and Sherwood numbers under uniform temperature or concentration boundary conditions are denoted as Nu_T and Sh_ω.

From the dimensionless governing equations, it is seen that if the value of Ω_L equals Ω_S, then the fully developed Sherwood numbers equal the fully developed Nusselt numbers. Further, the dimensionless temperature profiles at position z^* for heat transfer are the same as the humidity profiles at position $(Le\,z^*)$.

5.2.3 Validation of numerical scheme

Numerical validation

To ensure the accuracy of the results presented, a grid independence test was performed for the duct to determine the effects of the grid size. It indicated that 21 × 21 grids on the duct cross section and $\Delta z^* = 0.001$ axially are adequate (less than 0.15% difference compared with 31 × 31 grids and $\Delta z^* = 0.0005$).

To further validate the numerical program, ordinary ducts of various cross sections are calculated under uniform temperature conditions for all walls. In this case, the plate-fin duct is changed to a duct of uniform wall temperatures. For hydrodynamically fully developed laminar flow in ducts, (fRe) is a constant. The local Nusselt numbers in the duct will decrease along the flow and reach stable values when the flow is thermally fully developed. The fully developed Nu values under uniform temperature conditions are denoted as Nu_T. The calculated values of (fRe) and Nu_T for these ordinary ducts are listed in Table 5.1. Values from the literature [1–4] are shown for comparison. As can be seen, the

Table 5.1 Fully developed (*fRe*) and Nusselt numbers for ducts of various cross sections under uniform temperature conditions for all walls.

Cross Sections			(fRe)			Nu_T		
			Refs [1–4]	This Study	Error (%)	Refs [1–4]	This Study	Error (%)
Circular			64.0	63.44	0.8	3.657	3.68	0.6
Rectangular	*a/b*	0.125	82.0	81.28	0.9	5.60	5.73	2.3
		0.25	73.0	72.56	0.6	4.44	4.55	2.5
		0.5	62.0	61.20	1.3	3.39	3.45	1.8
		1.0	56.91	56.04	1.5	2.976	3.06	2.8
Isosceles triangular	*a/b*	0.289	52.97	51.48	2.8	2.301	2.262	1.7
		0.5	53.20	51.64	2.9	2.359	2.451	3.9
		0.866	53.28	53.56	0.5	2.500	2.596	3.7
		1.866	52.36	51.84	1.0	2.284	2.391	4.7
Sine	*a/b*	0.5	44.83	44.68	0.4	2.12	2.181	2.8
		0.75	48.94	48.85	0.2	2.33	2.372	1.7
		1.0	52.01	51.86	0.3	2.45	2.521	2.9
		1.5	56.01	56.46	0.8	2.6	2.573	1.1
		2.0	58.21	58.59	0.7	—	2.888	—

current study predicts the flow and heat transfer well. They are in accordance with the published data. Maximum uncertainty is less than 4.7%. This indicates that the numerical scheme is reliable.

Experimental validation

It is difficult to directly measure the local Nusselt and Sherwood numbers in small ducts. Instead, the inlet and outlet temperatures and humidities to and from the exchanger can be easily monitored. At this step, the mean Nusselt and Sherwood numbers for the whole heat mass exchanger can be estimated. A membrane total heat exchanger with plate-fin triangular ducts is constructed, as shown in Figure 5.8. The properties of the core are listed in Table 5.2. The core is then tested for heat and moisture recovery at the test rig built in SCUT, as described in the previous chapter. Two air streams, representing fresh air and exhaust air respectively, flow through the channels in the core while exchanging heat and moisture. Before they flow into the exchanger, their temperatures and humidities are adjusted to the desired set points. The air flow rates are controlled by the blowers and monitored by two flow meters. Equal air flow rates for the two streams are maintained and heat mass balances are checked.

The heat exchanged

$$Q = \rho_a V c_p (T_{fi} - T_{fo}) = \rho_a V c_p (T_{eo} - T_{ei}) \tag{5.58}$$

where V is the volumetric flow rate (m^3/s). The total heat transfer coefficient is calculated by

$$h_{tot} = \frac{Q}{A_t n_{ch} \Delta T_{lg}} \tag{5.59}$$

FIGURE 5.8

A membrane total heat exchanger core with plate-fin triangular ducts.

Table 5.2 Structural and physical parameters of the plate-fin core.

Name	Symbol	Unit	Value
Number of channels for each flow	n_{ch}		115
Half apex angle	α_0	degrees	35
Half duct width	b	mm	2.5
Hydrodynamic diameter	D_h	mm	1.66
Exchanger length	x_F, y_F	mm	185
Plate thickness	δ	μm	55
Vapor diffusivity in membrane fin	D_{wf}	m^2/s	1.27e-9
Heat conductivity	λ_f	$Wm^{-1}K^{-1}$	0.44
Heat conductivity of air	λ_a	$Wm^{-1}K^{-1}$	0.0263
Density of air	ρ_a	kg/m^3	1.18
Plate and fin density	ρ_f	kg/m^3	800
Moisture diffusivity in air	D_{va}	m^2/s	2.8210^{-5}
Partition coefficient	K_P		0.62
Kinematic viscosity	ν	m^2/s	15.89e-6

where A_t is the transfer area of a single channel (m), n_{ch} is the number of channels for each stream, ΔT_{lg} is the log mean temperature difference between two streams (K), calculated by

$$\Delta T_{lg} = \psi_s \frac{(T_{fi} - T_{eo}) - (T_{fo} - T_{ei})}{\ln \frac{(T_{fi} - T_{eo})}{(T_{fo} - T_{ei})}} \tag{5.60}$$

where Ψ_s is a correction factor for the temperature difference for the cross flow. Its specific value, which in this case is around 0.7–0.8, can be found graphically from Ref. [4].

The total heat transfer resistance includes the convective heat transfer resistance on both sides and the membrane itself, namely

$$\frac{1}{h_{tot}} = \frac{2}{h} + \frac{\delta}{\lambda_f} \tag{5.61}$$

From refabove equation, the convective heat transfer coefficient h on each side can be calculated. Then the mean Nusselt number for a single channel can be obtained from the definition of Nu. This is the experimentally obtained Nusselt number.

Similarly, moisture transferred

$$m = \rho_a V(\omega_{fi} - \omega_{fo}) = \rho_a V(\omega_{eo} - \omega_{ei}) \tag{5.62}$$

$$k_{tot} = \frac{m}{A_t n_{ch} \Delta \omega_{lg}} \tag{5.63}$$

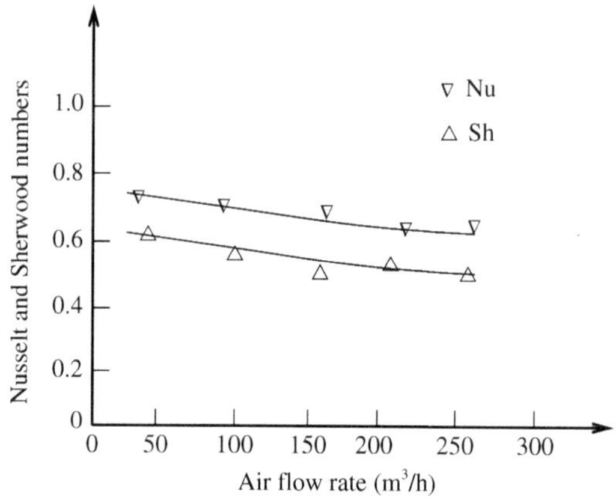

FIGURE 5.9

Comparison of tested and calculated Nusselt and Sherwood numbers for the core with plate-fin triangular ducts. The discrete symbols are the test data, and the solid lines are the numerical data.

where $\Delta\omega_{lg}$ is the log mean humidity difference between two streams (kg/kg), calculated by

$$\Delta\omega_{lg} = \psi_L \frac{(\omega_{fi} - \omega_{eo}) - (\omega_{fo} - \omega_{ei})}{\ln \frac{(\omega_{fi} - \omega_{eo})}{(\omega_{fo} - \omega_{ei})}} \tag{5.64}$$

where ψ_L is a correction factor for mass difference for the cross flow. The total moisture transfer resistance includes the convective mass transfer resistance on both sides and the membrane itself, namely

$$\frac{1}{k_{tot}} = \frac{2}{k} + \frac{\rho_a}{\rho_f} \cdot \frac{\delta}{D_{wf}} \tag{5.65}$$

From the above equation, the convective mass transfer coefficient k can be calculated. Then the mean Sherwood number can be obtained. This value is the experimentally obtained Sherwood number.

The experimentally obtained Nusselt and Sherwood numbers for different flow rates are plotted in Figure 5.9 and compared with the numerically obtained mean values. Generally, the obtained experimental values are in accordance with the numerical values. This is another validation of the program.

5.2.4 Flow, temperature and humidity fields

After the program is validated, triangular ducts with one plate of uniform temperature and two fins coupling with fluid, as shown in Figure 5.2, are modeled. Velocity, temperature and humidity fields in the fluid are obtained. Figure 5.10 shows the velocity profile distribution on the duct cross section for $\alpha_0 = 30°$. The values in the figure are dimensionless velocity u^*. As can be seen, in the center the contours are close to circular, but in places near the boundaries they are close to triangular. In the corners, there are dead spaces with little fluid flowing, which will ultimately affect heat transfer. The flow shows the same pattern as that of a common sinusoidal duct.

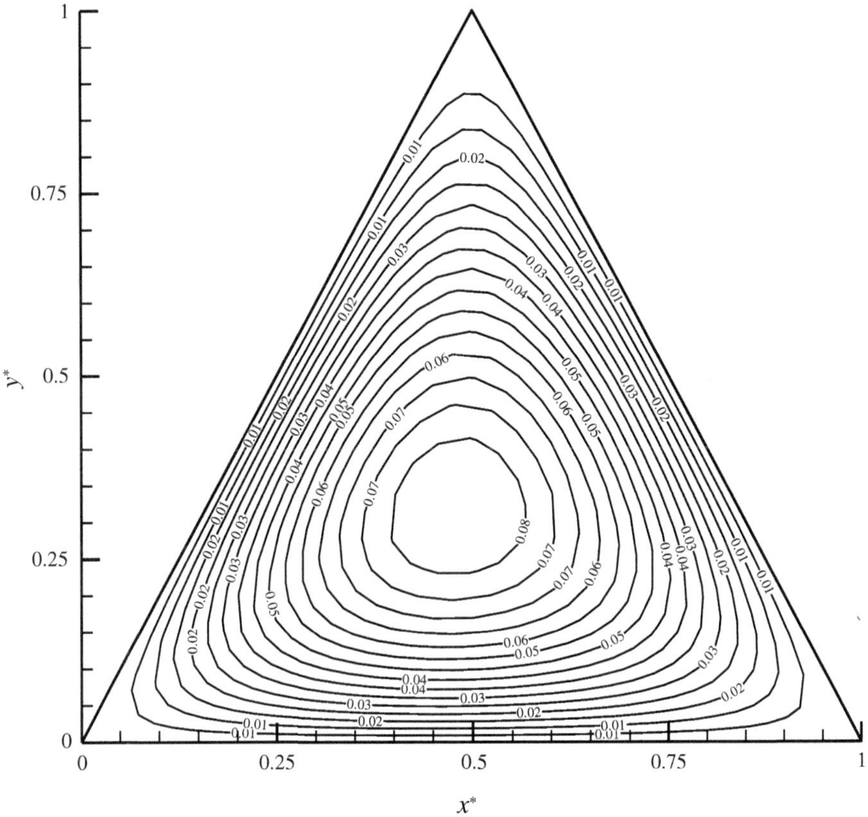

FIGURE 5.10

Dimensionless velocity profiles on duct cross section, $\alpha_0 = 30°$.

The dimensionless temperature profiles in the fluid are shown in Figures 5.11 and 5.12 for $\Omega_s = 0.1$ and 10 respectively. As can be seen, at higher fin heat conductance parameters the temperature contours are more similar to those in ducts with three walls of uniform temperature. At lower fin conductance parameters, temperature contours mostly show a pattern of straight lines parallel to the plate, which is similar to temperature profiles in parallel-plate ducts. In the center, isotherms are in elliptical shapes. When Ω_s increases, the contours transform from parallel horizontal lines to circles, implying increased heat transfer from fins to fluid.

Figure 5.13 shows the temperature profiles on the y–z plane at the center of the duct. As can be seen, heat is indeed transferred from the two ends of the plate to the fin center. Dimensionless temperature profiles on the fins are shown in Figures 5.14 and 5.15 for two fin conductance parameters $\Omega_s = 0.1$ and 10 respectively. Temperatures at locations where fins connect with the plate are equal to the plate temperature. Temperatures at other locations vary with fin heat conductance parameters. The higher the fin heat conductance parameters, the more homogeneous is the temperature distributions on the fins. The lower the fin heat conductance parameters, the less heat is transferred from fins to fluid. The isotherms on fins show a skew-symmetric behavior as expected.

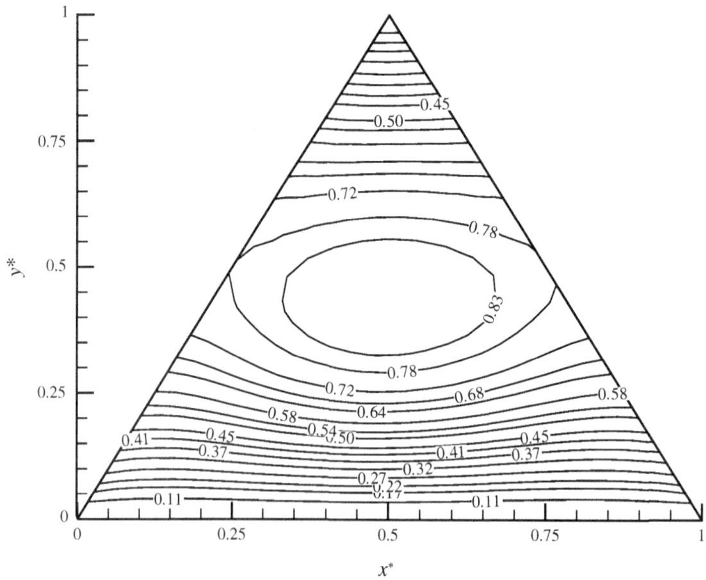

FIGURE 5.11

Dimensionless temperature contours on duct cross section at $z^* = 0.1$ for $\alpha_0 = 30°$ and $\Omega_s = 0.1$.

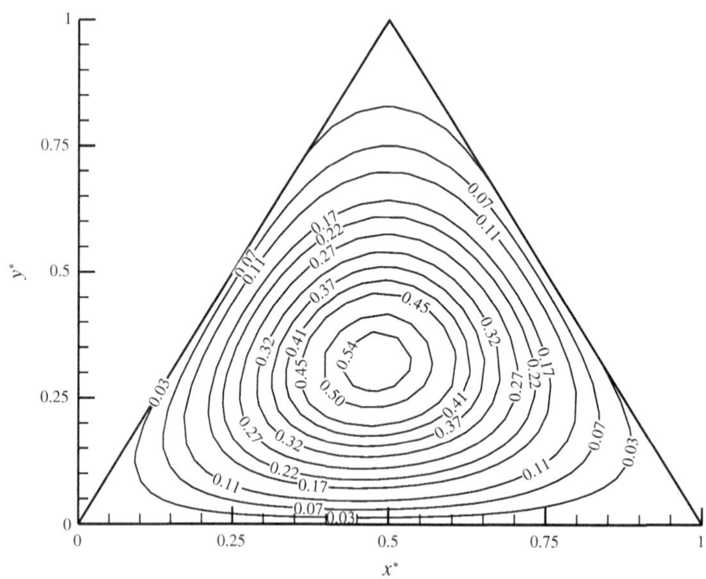

FIGURE 5.12

Dimensionless temperature contours on duct cross section at $z^* = 0.1$ for $\alpha_0 = 30°$ and $\Omega_s = 10$.

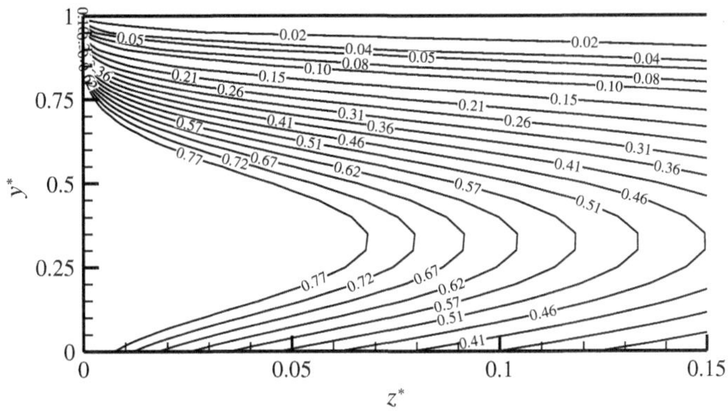

FIGURE 5.13

Dimensionless temperature contours on *y*–*z* plane at $x^* = 0.5$ for $\alpha_0 = 30°$ and $\Omega_S = 1.0$.

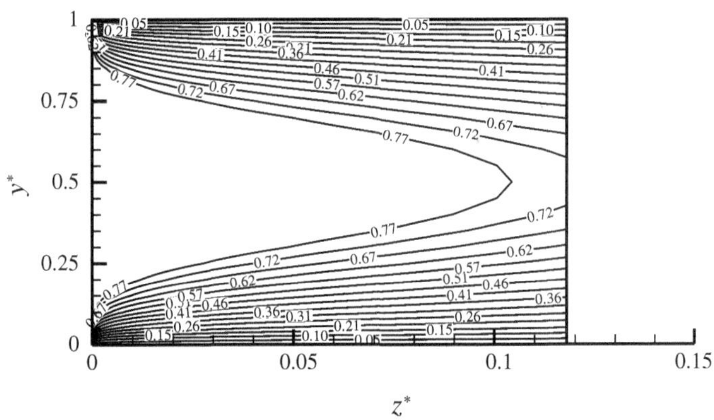

FIGURE 5.14

Dimensionless temperature contours on fins for ducts of $\alpha_0 = 30°$ and $\Omega_S = 0.1$.

The above temperature profiles are also the dimensionless humidity profiles if the fin conductance parameter for heat transfer is replaced by the fin conductance parameter for moisture transfer, and axial position z^* is replaced by $(Le\, z^*)$.

5.2.5 Nusselt and Sherwood numbers

Once the temperature and humidity fields in fluid have been determined, the dimensionless bulk temperature, bulk humidity, the local Nusselt and Sherwood numbers, and the mean Nusselt and Sherwood numbers can be calculated. Figure 5.16 shows the axial variations of local Sherwood numbers with a fin mass conductance parameter of $\Omega_L = 0.1$. The axial position z^* is a dimensionless parameter reflecting the duct length. Figure 5.17 shows the axial variations of local Nusselt numbers with a fin heat conductance parameter of $\Omega_S = 10$. The local Nusselt and Sherwood numbers decrease from high values at the

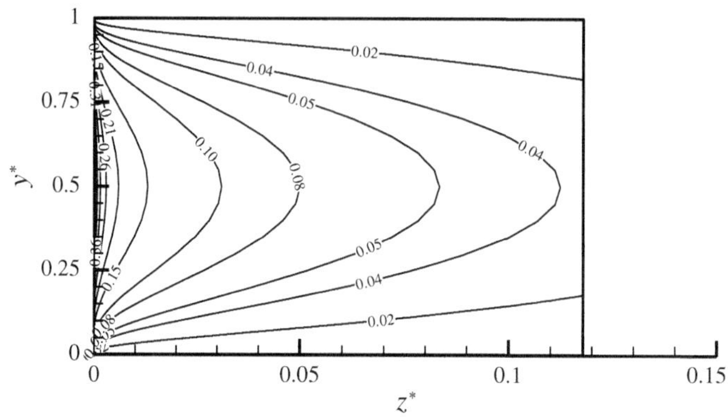

FIGURE 5.15

Dimensionless temperature contours on fins for ducts of $\alpha_0=30°$ and $\Omega_s=10.0$.

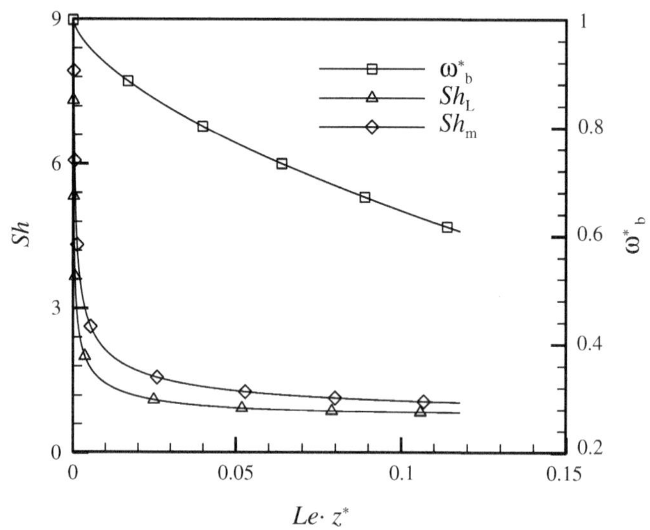

FIGURE 5.16

Variations of bulk humidity, Sh_L and Sh_m along the flow axis for duct of $\alpha_0=30°$, $\Omega_L=0.1$.

inlet, then they reach stable values Nu_T and Sh_ω. These fully developed values vary with apex angles and fin conductance parameters. The higher the fin conductance parameters, the higher are Nu_T and Sh_ω. When z^* is greater than 0.05, the flow can be considered fully developed.

Fin efficiencies for heat and mass transfer are defined as the ratios of Nu_T or Sh_ω at finite fin conductance parameters to those at Ω_s or $\Omega_L=\infty$, or

$$\eta_{fin,s} = \frac{Nu_{T,\Omega_s}}{Nu_{T,\Omega_s=\infty}} \tag{5.66}$$

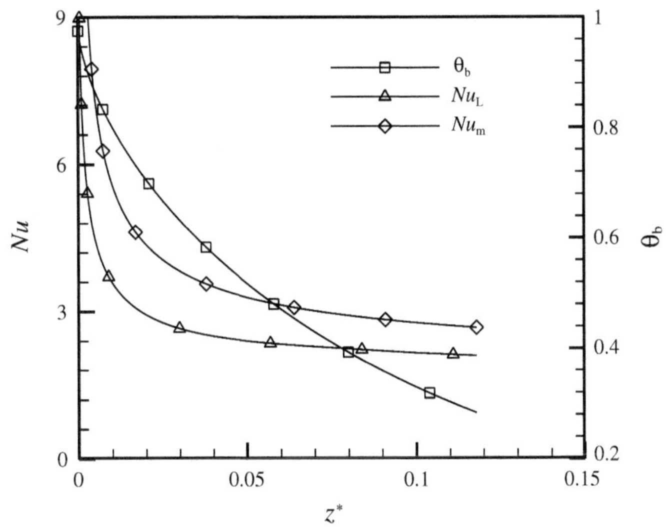

FIGURE 5.17

Variations of bulk temperature, Nu_L and Nu_m along the flow axis for duct of $\alpha_0 = 30°$, $\Omega_S = 10$.

$$\eta_{fin,L} = \frac{Sh_{\omega,\Omega_L}}{Sh_{\omega,\Omega_L=\infty}} \tag{5.67}$$

As the mass transfer equations are in the same forms as the heat transfer equations, the fin efficiency for mass transfer will equal that for heat transfer, if the fin mass conductance parameter is equal to the fin heat conductance parameter.

Table 5.3 lists the calculated Nu_T and/or Sh_ω and fin efficiencies for various apex angles and fin conductance parameters [6]. They are the basic data that can be used for compact heat exchanger or heat mass exchanger design. In the table, aspect ratio $\alpha_f = a/b$.

5.2.6 Comparisons of heat and mass transfer

As defined by Eq. (5.35), fin heat conductance parameters are determined by the duct geometry and heat conductivity of fin materials. Table 5.4 lists the values of fin heat conductance parameters for some frequently encountered materials, including metals and non-metals, for a triangular duct of height 2 mm, aspect ratio 0.5, and fin thickness 0.1 mm, as mentioned above. As can be seen, for almost all the metals the fin heat conductance parameters are larger than 100, and the resulting fin heat efficiencies can be as high as 0.90–0.98. For such traditional metal compact heat exchangers, the influence of finite fin heat conductance on heat transfer is negligible. Direct utilization of heat transfer properties for a common duct is acceptable. For the recently introduced heat mass exchangers which use non-metals to simultaneously transfer heat and moisture, the fin heat conductance parameters are usually less than 1.0, and the resulting fin efficiencies for heat transfer can be as low as 0.40. Though it's true that the

Table 5.3 Fully developed Nusselt and Sherwood numbers and fin efficiencies for plate-fin triangular ducts.

α_0	Ω_s or Ω_L	Nu_T or Sh_ω	$\eta_{fin,s}$ or $\eta_{fin,L}$	α_0	Ω_s or Ω_L	Nu_T or Sh_ω	$\eta_{fin,s}$ or $\eta_{fin,L}$
15°	∞	2.391	1.0	30°	∞	2.596	1.0
	25	2.287	0.96		25	2.516	0.97
	10	2.196	0.92		10	2.449	0.94
	5	2.062	0.86		5	2.305	0.89
	2	1.599	0.67		2	1.937	0.75
	1	1.288	0.54		1	1.604	0.62
	0.5	1.024	0.43		0.5	1.276	0.49
	0.1	0.725	0.30		0.1	0.842	0.32
	0	0.651	0.27		0	0.655	0.25
45°	∞	2.451	1.0	60°	∞	2.262	1.0
	25	2.393	0.98		25	2.105	0.93
	10	2.299	0.94		10	2.025	0.90
	5	2.182	0.89		5	1.929	0.85
	2	2.021	0.82		2	1.796	0.79
	1	1.744	0.71		1	1.618	0.72
	0.5	1.444	0.59		0.5	1.413	0.62
	0.1	0.978	0.40		0.1	1.067	0.47
	0	0.659	0.27		0	0.595	0.26

Table 5.4 Values of fin heat conductance parameter with some fin materials, duct height, 2 mm; fin thickness, 0.1 mm; fluid, air.

Fin materials	λ_f ($Wm^{-1}K^{-1}$)	Ω_s
Pure copper	401	762
Bronze	52	98.8
Iron	80.2	152.4
Steel	60.5	115
Aluminum	237	450.3
Carbon	1.6	3.04
Plywood	0.12	0.23
Wood	0.16	0.30
Clay	1.3	2.47
Glass	1.4	2.66
Paper	0.18	0.36
Teflon	0.35	0.67
Polymer membrane	0.13	0.25

fins still participate in the heat transfer enhancement, at least partially, at this stage the effects of finite fin heat conductance on heat transfer will be substantial.

Similarly, as defined by Eq. (5.40), fin mass conductance parameters are determined by duct geometry and the mass diffusivity of moisture in the fin materials. Moisture diffusivity in air is around 2.5×10^{-5} m^2/s. Moisture diffusivities in most non-metal hygroscopic materials are of the order of 10^{-10} m^2/s or less [10]. For example, water diffusivities are from 0.5×10^{-12} to 1.5×10^{-12} m^2/s in a PSS-Na/Al_2O_3 composite silicate, from 8.0×10^{-10} m^2/s to 5.0×10^{-12} m^2/s in a Nafion polymer membrane, and from 1.0×10^{-13} m^2/s to 1.0×10^{-12} m^2/s in a methylcellulose film. The densities of these materials are of the order of 1000 kg/m^3. The partition coefficients are of the order of 10% or 15–20% water uptake under 60% RH. Based on these properties, the values of fin mass conductance parameters are of the order of 10^{-3}. With such low mass conductance parameters, the fin efficiencies for mass transfer will be below 0.1 to 0.2. Under such circumstances, nearly all the mass transfer between the two air streams will be accomplished by the plate, rather than by the fins. The fins behave only like supporting materials, if excluding their role in partial participation in heat transfer. In engineering, sometimes supporting materials or spacers are necessary to separate the two streams because the plates are thin and soft.

In designing novel heat mass exchangers, it is simple to emulate a common compact heat exchanger with such a plate-fin configuration. However, the effectiveness of fins in heat transfer will be strongly compromised. Even worse, they are of little use in enhancing mass transfer. Consequently, novel configurations such as the so-called cross-corrugated triangular ducts may provide a promising choice [10], since in this structure the two fluids directly exchange heat and moisture through all the three walls, in addition to intensification by flow turbulence.

5.3 Sinusoidal ducts

When the duct cross section is changed to sinusoidal, as schematically shown in Figure 5.18, the values of the Nusselt and Sherwood numbers will change with cross sections. However, the governing equations and the solution methodology will not change. Figure 5.19 shows the coordinate system of a representative sinusoidal duct. The duct curve satisfies

$$y = a\left[1 - \cos\left(\frac{\pi}{b}x\right)\right] \tag{5.68}$$

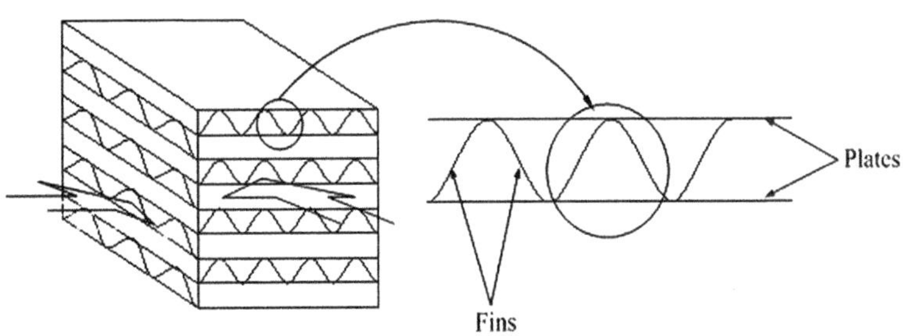

FIGURE 5.18

Schematic of a cross-flow heat exchanger with plate-fin sinusoidal ducts.

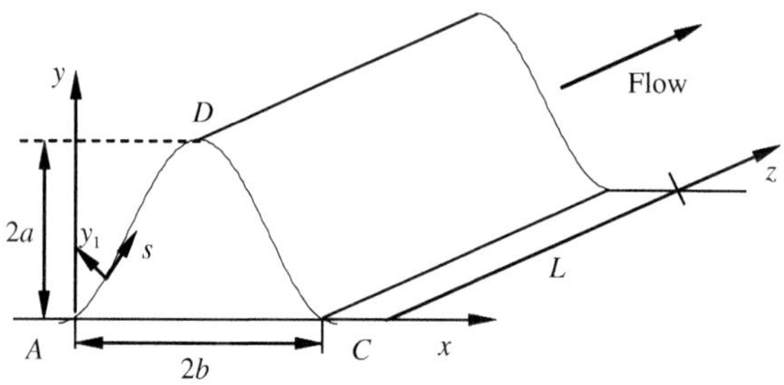

FIGURE 5.19

Geometry and coordinates of a plate-fin sinusoidal duct.

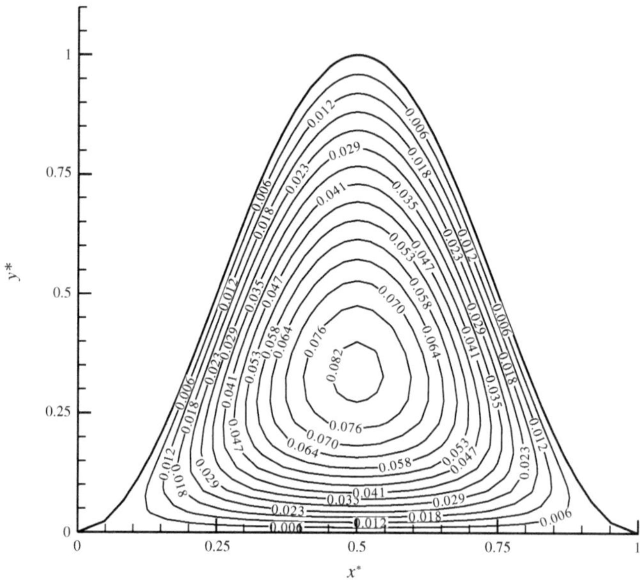

FIGURE 5.20

Dimensionless velocity profiles (u^*) on sinusoidal duct cross section, $\alpha_f = 1.0$.

For sinusoidal ducts, the duct aspect ratios are usually used to define duct geometries. The aspect ratio of the duct is defined by

$$\alpha_f = \frac{2a}{2b} \tag{5.69}$$

The solution of momentum, energy, and mass equations will provide the velocity, temperature and humidity profiles [7,8]. Figures 5.20 to 5.21 show the velocity and humidity profiles at the cross

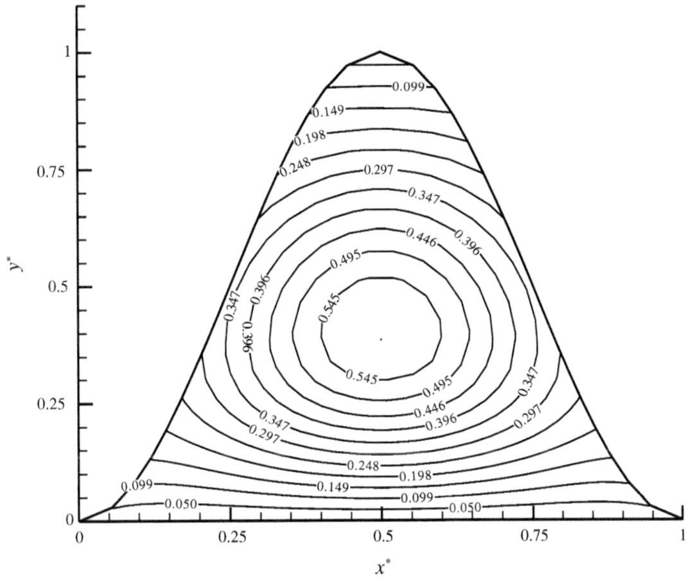

FIGURE 5.21

Dimensionless humidity contours on sinusoidal duct cross section at $z^* = 0.1$ for $\alpha_f = 1.0$ and $\Omega_L = 1.0$.

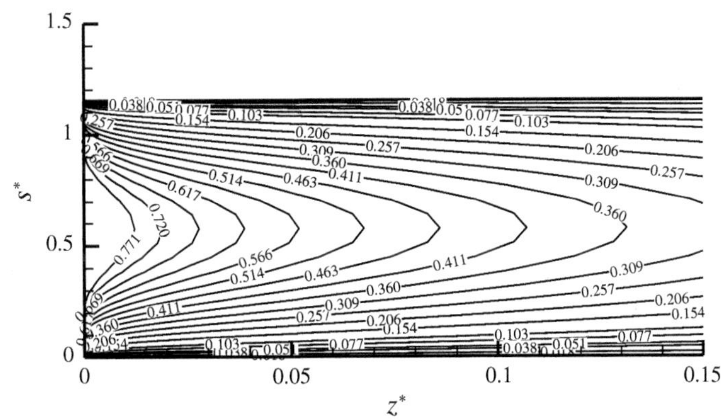

FIGURE 5.22

Dimensionless humidity contours on fins for sinusoidal ducts of $\alpha_f = 1.0$ and $\Omega_L = 1.0$.

sections of a duct, respectively. Figure 5.22 shows the humidity profiles on the fins. As can be seen, the shapes are very similar to those for triangular ducts.

The fully developed Nusselt and Sherwood numbers and fin efficiencies for plate-fin sinusoidal ducts are listed in Table 5.5. The definitions of the fin conductance parameters are the same as for triangular ducts, Eqs. (5.35) and (5.40). They provide the basic data for this kind of duct.

Table 5.5 Fully developed Nusselt and Sherwood numbers and fin efficiencies for plate-fin sinusoidal ducts.

α_f	Ω_S or Ω_L	Nu_T or Sh_ω	$\eta_{fin,s}$ or $\eta_{fin,L}$	α_f	Ω_S or Ω_L	Nu_T or Sh_ω	$\eta_{fin,s}$ or $\eta_{fin,L}$
0.2	∞	1.551	1.0	0.5	∞	2.181	1.0
	25	1.523	0.982		25	2.133	0.978
	10	1.201	0.774		10	2.107	0.966
	5	1.033	0.666		5	1.908	0.875
	2	0.863	0.556		2	1.524	0.699
	1	0.791	0.510		1	1.260	0.578
	0.5	0.739	0.476		0.5	1.055	0.484
	0.1	0.694	0.447		0.1	0.824	0.378
	0	0.641	0.413		0	0.737	0.338
1.0	∞	2.521	1.0	2.0	∞	2.888	1.0
	25	2.492	0.988		25	2.714	0.940
	10	2.369	0.940		10	2.651	0.918
	5	2.167	0.860		5	1.922	0.666
	2	1.721	0.683		2	1.511	0.523
	1	1.379	0.547		1	1.194	0.413
	0.5	1.081	0.429		0.5	0.926	0.321
	0.1	0.751	0.298		0.1	0.631	0.218
	0	0.576	0.228		0	0.358	0.124
5.0	∞	2.586	1.0	5.0	1	1.015	0.392
	25	2.390	0.924		0.5	0.833	0.322
	10	2.170	0.839		0.1	0.661	0.256
	5	1.654	0.640		0	0.173	0.067
	2	1.274	0.493				

5.4 Rectangular and parallel ducts

A schematic of a heat exchanger with plate-fin rectangular ducts is shown in Figure 5.23. This is also one of the most commonly encountered total heat exchanger structures. A representative duct is shown in Figure 5.24. The duct height is $2a$, the duct width is $2b$, and the duct aspect ratio $\alpha_f = a/b$. When the aspect ratio is infinitely small, the ducts become parallel-plate ducts. The definition of the fin heat conductance parameter is similar to that for a triangular duct.

The solution of the momentum, energy, and mass equations will provide the velocity, temperature and humidity profiles [9]. Taking the humidity profiles as examples, Figures 5.25 and 5.26 show the velocity and humidity profiles at the cross sections of a rectangular duct, respectively. Figure 5.27 shows the humidity profiles on the fins. As can be seen, the shapes are very different from those for triangular ducts.

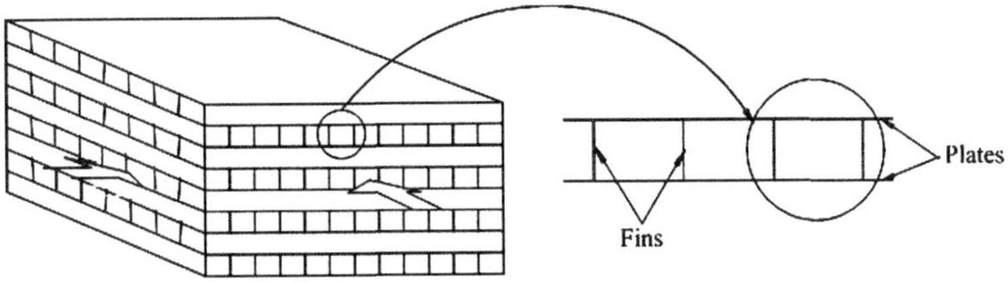

FIGURE 5.23

Schematic of a heat exchanger with plate-fin rectangular ducts.

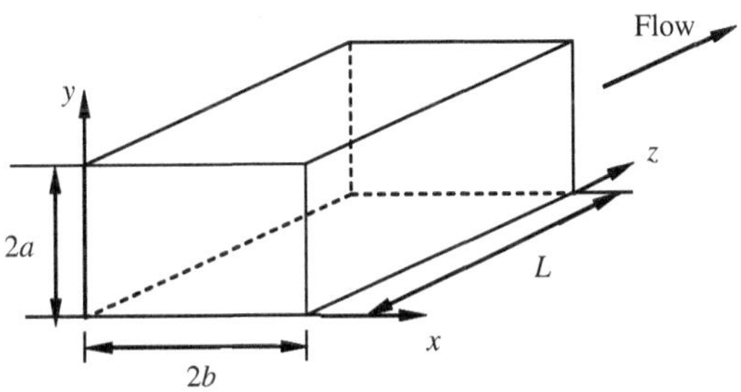

FIGURE 5.24

Geometry and coordinates of a plate-fin rectangular duct.

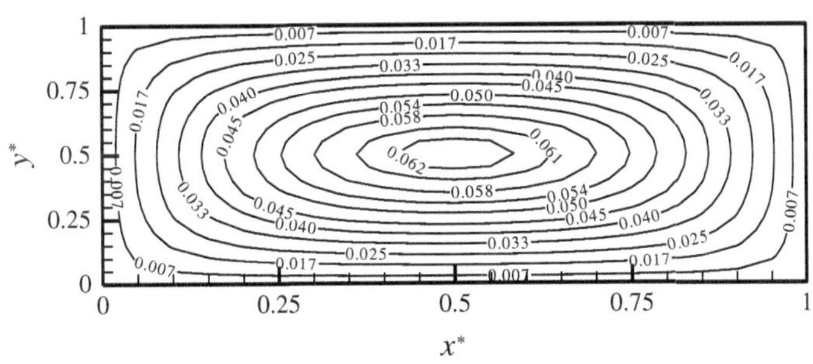

FIGURE 5.25

Dimensionless velocity profiles (u^*) on rectangular duct cross section, $\alpha_f = 0.5$.

The fully developed Nusselt and Sherwood numbers and fin efficiencies for plate-fin sinusoidal ducts are listed in Table 5.6 [9]. They provide the basic data for this kind of duct.

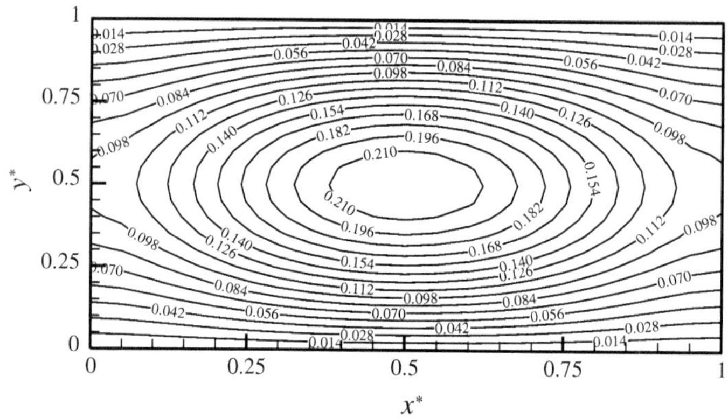

FIGURE 5.26

Dimensionless humidity contours on rectangular duct cross section at $z^* = 0.1$ for $\alpha_f = 0.5$ and $\Omega_L = 0.1$.

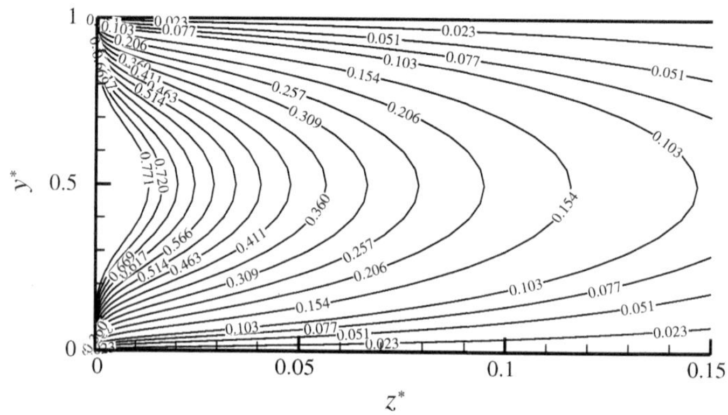

FIGURE 5.27

Dimensionless humidity contours on fins of a rectangular duct for $\alpha_f = 0.5$ and $\Omega_L = 0.1$.

Fin efficiencies deteriorate with fin conductance parameters. When fin conductance parameters increase from 0 to infinitely large, the duct's behavior changes from that of a duct with two well-conductive wall and two adiabatic walls to that of a duct with four well-conductive walls. The consequent fully developed Nusselt numbers increase accordingly. For ducts with certain limited conductance parameters, the Nusselt numbers lie between these two limiting values. For fins made with low-conductivity materials such as paper, polymer, glass, carbon, plywood, etc., the nature of the limited heat transfer from fins to fluid should be taken into account for a better estimation of the exchanger performance.

Table 5.6 Fully developed Nusselt and Sherwood numbers and fin efficiencies for plate-fin rectangular ducts.

α_f	Ω_s or Ω_L	Nu_T or Sh_ω	$\eta_{fin,s}$ or $\eta_{fin,L}$	α_f	Ω_s or Ω_L	Nu_T or Sh_ω	$\eta_{fin,s}$ or $\eta_{fin,L}$
0.125	∞	5.746	1.0	0.25	∞	4.537	1.0
	25	5.684	0.989		25	4.479	0.987
	10	5.677	0.988		10	4.464	0.984
	5	5.576	0.970		5	4.378	0.965
	2	5.560	0.968		2	4.366	0.962
	1	5.545	0.965		1	4.349	0.959
	0.5	5.539	0.964		0.5	4.341	0.957
	0.1	5.522	0.961		0.1	4.302	0.948
	0	5.502	0.958		0	4.297	0.947
0.5	∞	3.504	1.0	1.0	∞	3.021	1.0
	25	3.487	0.995		25	3.016	0.998
	10	3.481	0.993		10	3.015	0.998
	5	3.454	0.986		5	3.014	0.998
	2	3.440	0.982		2	2.686	0.889
	1	3.350	0.956		1	2.487	0.823
	0.5	3.290	0.939		0.5	2.274	0.753
	0.1	3.151	0.899		0.1	1.940	0.642
	0	3.125	0.892		0	1.815	0.601
2.0	∞	3.500	1.0	4.0	∞	4.527	1.0
	25	3.473	0.992		25	4.460	0.985
	10	3.468	0.991		10	3.394	0.750
	5	3.354	0.958		5	3.039	0.671
	2	2.254	0.644		2	2.046	0.452
	1	1.894	0.541		1	1.391	0.307
	0.5	1.493	0.427		0.5	0.938	0.207
	0.1	1.010	0.289		0.1	0.487	0.108
	0	0.872	0.249		0	0.398	0.088

5.5 Conclusion

Laminar flow and conjugate heat mass transfer in plate-fin ducts with various geometries such as sinusoidal, rectangular, triangular and parallel-plate are investigated by considering finite fin conductance in heat and mass transfer. Finite fin conductance parameters in heat and mass have determining effects on convective heat and mass transfer in the ducts.

Fins with finite conductance parameters are considered in the modeling of a conjugate plate-fin duct heat transfer problem. The results show that, besides aspect ratios, the fin conductance parameters have an equal tremendous impact on heat and mass transfer phenomena in plate-fin ducts. When fin conductance parameters increase from 0 to infinitely large, the duct's behavior changes from that of a duct with two well-conductive wall and two adiabatic walls to that of a duct with four well-conductive walls. The consequent fully developed Nusselt and Sherwood numbers increase accordingly. For ducts with certain limited conductance parameters, the Nusselt and Sherwood numbers lie between these two limiting values. For fins made with low-conductivity materials such as paper, polymer, glass, carbon, plywood, etc., the nature of the limited heat transfer from fins to fluid should be taken into account, for a better estimation of the exchanger performance. The problem is more serious for mass transfer. The fin efficiencies would be seriously diminished.

References

[1] Shah RK, London AL. Laminar flow forced convection in ducts. New York: Academic Press; 1978.

[2] Incropera FP, Dewitt DP. Introduction to heat transfer. 3rd ed. New York: John Wiley & Sons; 1996.

[3] Shah RK, Bhatti MS. Laminar convection heat transfer in ducts. In: Kakac S, Shah RK, Aung W, editors. Handbook of single-phase convective heat transfer. New York: Wiley; 1987.

[4] Kays WM, Crawford ME. Convective heat and mass transfer. 3rd edn. New York: McGraw-Hill, Inc.; 1990.

[5] Baliga BR, Azrak RR. Laminar fully developed flow and heat transfer in triangular plate-fin ducts. ASME J Heat Transfer 1986;108:24–32.

[6] Zhang LZ. Laminar flow and heat transfer in plate-fin triangular ducts in thermally developing entry region. Int J Heat Mass Transfer 2007;50:1637–40.

[7] Zhang LZ. Heat and mass transfer in plate-fin sinusoidal passages with vapor-permeable wall materials. Int J Heat Mass Transfer 2008;51:618–29.

[8] Zhang LZ. Heat and mass transfer in plate-fin enthalpy exchangers with different plate and fin materials. Int J Heat Mass Transfer 2009;52:2704–13.

[9] Zhang LZ. Thermally developing forced convection and heat transfer in rectangular plate-fin passages under uniform plate temperature. Numer. Heat Transfer Part A: Appl 2007;52(6):549–64.

[10] Zhang LZ. Total heat recovery: heat and moisture recovery from ventilation air. New York: Nova Science; 2009.

CHAPTER

6 Heat and Mass Transfer in Liquid-to-Air Parallel-Plate Membrane Ducts

6.1 Introduction

In Chapter 4, parallel-plate membrane ducts used for air-to-air heat and mass exchangers were introduced. In this chapter, parallel-plate membrane ducts used for liquid-to-air heat and mass exchangers will be introduced. Membranes are used to separate an air stream and a liquid stream. These exchangers are also called membrane contactors. The air stream and the liquid stream exchange heat and substances simultaneously. Liquid absorption is the mechanism for mass transfer. Only gases can permeate through the membrane, while the liquid solution droplets are prevented by the membranes from permeating. The membranes are highly perm-selective, therefore they are called the functional materials. Liquid gas contactors have been used in many applications: gas purification, gas absorption, CO_2 collection and storage, toxic gas disposal, gas separation, etc. Though the application fields are different, the mechanisms are the same. Transport phenomena in the ducts are also similar. This chapter will use air dehumidification as an example.

Liquid desiccant air dehumidification is one of the most widely used dehumidification technologies for air dehumidification [1–6], due to its advantages of high efficiency, absence of liquid condensate droplets and its capability for energy storage. Packed beds or packed columns are the common way of processing air, as shown in Figure 6.1. According to this method, the liquid desiccant is spread down the packings in the column. The packings are various, including structured packings, random packings, corrugated packings, Rasching rings, Pall rings, Dixon nets, etc. They provide the large contact areas between the liquid and air stream on which the liquid is spread. The air stream flows across the packing surfaces. The two flows meet on the surfaces of the packings, as schematically depicted in Figure 6.2. They exchange heat and moisture in direct contact. Since the liquid desiccant and the process air are in direct contact, small liquid desiccant droplets may be carried over by the process air to the indoor environment, which is believed to be rather harmful to occupant health, building structure and furniture. In order to overcome these drawbacks, membrane contactors are used to replace packed columns. Membranes are used to separate the liquid and the air streams, so it is also called non-contacting air dehumidification. The membranes can prevent the liquid desiccant from crossing over to the process air, while selectively permitting the transport of heat and moisture between the liquid stream and the air stream [7,8].

Figure 6.3 shows a schematic of a parallel-plate-type liquid desiccant air dehumidification module. As seen, parallel-plate ducts are formed by the membranes stacked together. Equal spacings are

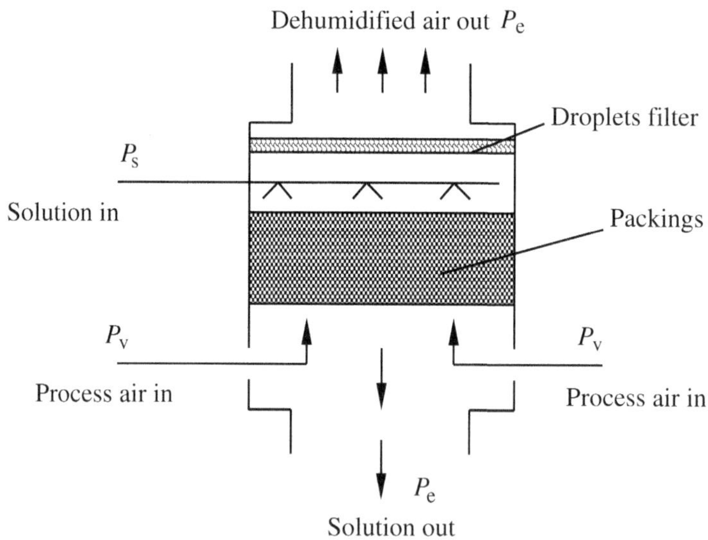

FIGURE 6.1

Schematic of a packing column for liquid air dehumidification.

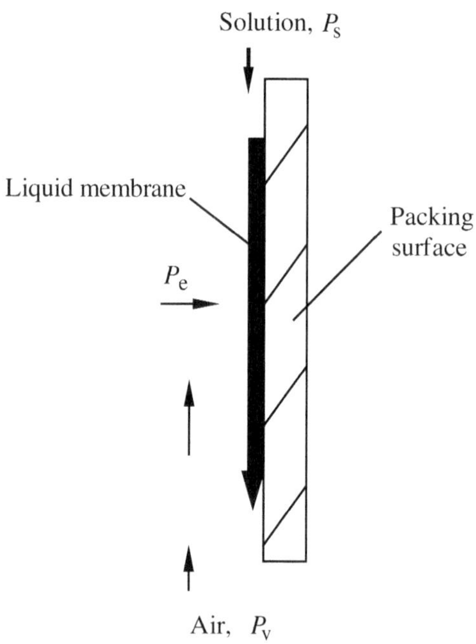

FIGURE 6.2

Mechanisms for liquid-air heat and moisture exchange.

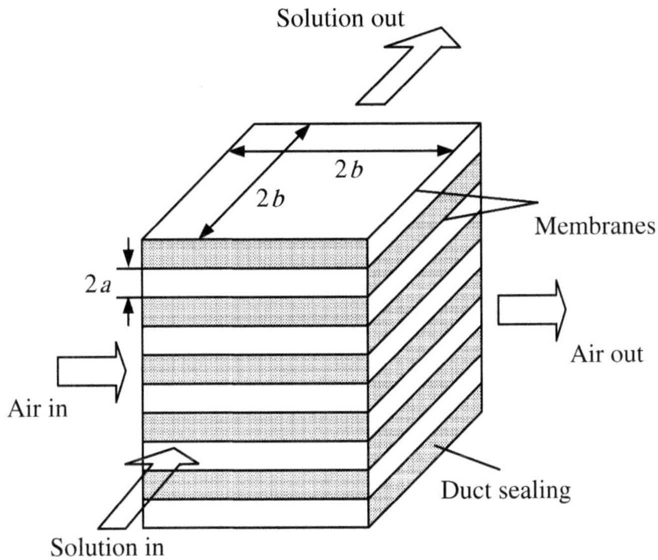

FIGURE 6.3

Structure of a parallel-plate membrane module for liquid desiccant air dehumidification.

kept between neighboring membranes to form flow channels. The solution stream and the air stream flow alternately through the parallel ducts, in a cross-flow arrangement, for ease of duct sealing. This structure is the most reliable in terms of preventing the liquid fluids from leaking. The process air is dehumidified by the liquid desiccant, while absorbing moisture from air through the membranes.

Conjugate heat and mass transfer in the above parallel-plate membrane ducts is the focus for investigation, and the basis for performance analysis and structural optimization. In the past, transport data for the common metal-formed parallel-plate heat transfer ducts were used for the analysis of membrane contactors, which leads to large discrepancies. In this chapter, the fluid flow and heat and mass transfer for these liquid-air membrane ducts will be discussed.

6.2 Mathematical model

Governing equations

Previous chapters have discussed the fluid flow and heat and mass transfer in air-to-air membrane ducts. For the fluid flow and heat and mass transfer in liquid-to-air ducts, the governing equations are similar. As depicted in Figure 6.3, the air stream and the solution stream flow through the ducts in a cross-flow arrangement. For reasons of symmetry and simplicity in calculation, one membrane and two neighboring flow channels are selected as the calculation domain. The coordinates of the unit cell are shown in Figure 6.4. As can be seen, the air stream flows along the z axis, while the solution stream flows along the $-x$ axis. Heat and moisture can be exchanged through the membrane effectively. When water vapor is absorbed by the solution, absorption heat is released on the interface between the solution and the membrane.

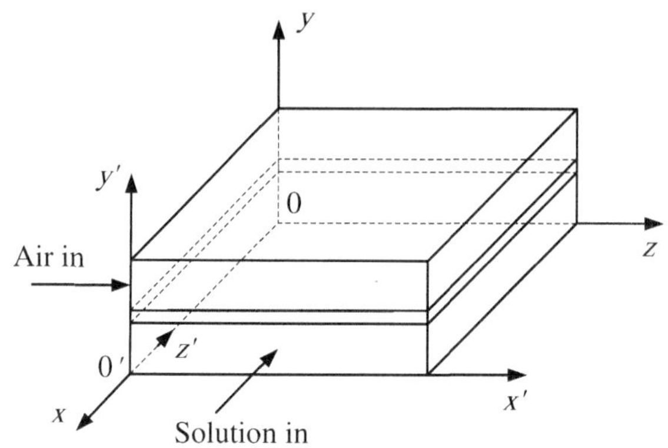

FIGURE 6.4

Coordinate system of the unit cell showing one membrane and two neighboring flow channels.

As in air-to-air ducts, Reynolds numbers are usually much less than 2300, therefore both the air and the liquid solution can be considered to be laminar. The fluids are Newtonian with constant thermo-physical properties. Other assumptions are:

(1) The flows are considered to be hydrodynamically fully developed while developing both thermally and in concentration.

(2) Axial heat and mass diffusions of the two streams are negligible compared with heat and mass convection by bulk flows. This is reasonable because Peclet numbers (*Pe*) for both the streams are much greater than 10 in this study.

For the air stream, the governing momentum, energy and mass conservation equations are [7,9,10]

$$\mu_a\left(\frac{\partial^2 u_a}{\partial x^2}+\frac{\partial^2 u_a}{\partial y^2}\right)=\frac{dP_a}{dz} \tag{6.1}$$

$$\lambda_a\left(\frac{\partial^2 T_a}{\partial x^2}+\frac{\partial^2 T_a}{\partial y^2}\right)=\rho_a c_{pa} u_a\frac{\partial T_a}{\partial z} \tag{6.2}$$

$$D_{va}\left(\frac{\partial^2 \omega_a}{\partial x^2}+\frac{\partial^2 \omega_a}{\partial y^2}\right)=u_a\frac{\partial \omega_a}{\partial z} \tag{6.3}$$

where μ_a is dynamic viscosity of air (Pa s); u_a is air velocity (m/s); P_a is pressure of the air stream (Pa); λ_a is the heat conductivity of air (W m^{-1} K^{-1}); T_a is the temperature of air (K); c_{pa} is the specific heat of air (kJ kg^{-1} K^{-1}); D_{va} is the moisture diffusivity in air (m^2/s); ω_a is the humidity ratio of air (kg moisture/kg dry air).

For the liquid, the governing momentum, energy and mass diffusion equations for the solution stream are

$$\mu_s\left(\frac{\partial^2 u_s}{\partial x'^2}+\frac{\partial^2 u_s}{\partial y'^2}\right)=\frac{dP_s}{dz'} \tag{6.4}$$

$$\lambda_s\left(\frac{\partial^2 T_s}{\partial x'^2}+\frac{\partial^2 T_s}{\partial y'^2}\right)=\rho_s c_s u_s\frac{\partial T_s}{\partial z'} \tag{6.5}$$

$$D_{ws}\left(\frac{\partial^2 X_s}{\partial x'^2}+\frac{\partial^2 X_s}{\partial y'^2}\right)=u_s\frac{\partial X_s}{\partial z'} \tag{6.6}$$

where superscript "′" means coordinates for the solution stream; μ_s is the dynamic viscosity of solution (Pa s); u_s is the solution fluid velocity (m/s); P_s is the pressure of the solution stream (Pa); λ_s is the heat conductivity of the solution (W m^{-1} K^{-1}); T_s is the temperature of the solution (K); c_s is the specific heat of the solution (kJ kg^{-1} K^{-1}); D_{ws} is the water diffusivity in solution (m^2/s); X_s is the mass fraction of the solution in the form of kg water/kg solution.

Normalization of equations

The governing equations for the air stream, Eqs. (6.1) to (6.3), can be normalized as

$$\frac{\partial^2 u_a^*}{\partial x^{*2}}+\left(\frac{b}{a}\right)^2\frac{\partial^2 u_a^*}{\partial y^{*2}}=-\frac{4b^2}{D_h^2} \tag{6.7}$$

$$\frac{\partial^2 \theta_a}{\partial x^{*2}}+\left(\frac{b}{a}\right)^2\frac{\partial^2 \theta_a}{\partial y^{*2}}=U_a\frac{\partial \theta_a}{\partial z_h^*} \tag{6.8}$$

$$\frac{\partial^2 \xi_a}{\partial x^{*2}}+\left(\frac{b}{a}\right)^2\frac{\partial^2 \xi_a}{\partial y^{*2}}=U_a\frac{\partial \xi_a}{\partial z_m^*} \tag{6.9}$$

Similarly, the governing equations for the solution stream, Eqs. (6.4) to (6.6), can be normalized as

$$\frac{\partial^2 u_s^*}{\partial x'^{*2}}+\left(\frac{b}{a}\right)^2\frac{\partial^2 u_s^*}{\partial y'^{*2}}=-\frac{4b^2}{D_h^2} \tag{6.10}$$

$$\frac{\partial^2 \theta_s}{\partial x'^{*2}}+\left(\frac{b}{a}\right)^2\frac{\partial^2 \theta_s}{\partial y'^{*2}}=U_s\frac{\partial \theta_s}{\partial z_h'^*} \tag{6.11}$$

$$\frac{\partial^2 \Theta_s}{\partial x'^{*2}}+\left(\frac{b}{a}\right)^2\frac{\partial^2 \Theta_s}{\partial y'^{*2}}=U_s\frac{\partial \Theta_s}{\partial z_m'^*} \tag{6.12}$$

where in Eqs. (6.7) to (6.12) subscripts "a" and "s" denote the air and the solution fluids, respectively. The dimensionless velocity is defined by

$$u^* = -\frac{\mu u}{D_h^2 \frac{dP}{dz}} \tag{6.13}$$

and the dimensionless temperature is defined by

$$\theta = \frac{T - T_{ai}}{T_{si} - T_{ai}} \tag{6.14}$$

where T_{ai} is the inlet temperature of air (K), and T_{si} is the inlet temperature of solution (K).

The dimensionless humidity is defined by

$$\xi = \frac{\omega - \omega_{ai}}{\omega_{si} - \omega_{ai}} \tag{6.15}$$

where ω_{ai} is the inlet humidity of air (kg/kg), and ω_{si} is the equilibrium air humidity with solution at the inlet temperature (T_{si}) and mass fraction (X_{si}).

The dimensionless mass fraction of solution is defined by

$$\Theta = \frac{X - X_{ei}}{X_{si} - X_{ei}} \tag{6.16}$$

where X_{si} is the inlet mass fraction of solution (kg water/kg solution), and X_{ei} is the equilibrium mass fraction of the solution with air stream at the inlet temperature (T_{ai}) and humidity (ω_{ai}).

Dimensionless coordinates

$$x^* = \frac{x}{2b} \tag{6.17}$$

$$y^* = \frac{y}{2a} \tag{6.18}$$

$$z_h^* = \frac{z}{RePrD_h} \tag{6.19}$$

$$z_m^* = \frac{z}{ReScD_h} \tag{6.20}$$

where hydraulic diameter

$$D_h = \frac{4A_c}{2(2a + 2b)} \tag{6.21}$$

where A_c is the cross-sectional area of the channel (m^2); Pr is the Prandtl number, and Sc is the Schmidt number. The dimensionless z^* has a different definition for heat and mass transfer. Both z_h^* and z_m^* have different values for the air and the solution streams.

In Eqs. (6.8), (6.9), (6.11) and (6.12), dimensionless velocity coefficient U is defined by

$$U = \frac{u^*}{u_m^*}\frac{4b^2}{D_h^2} \tag{6.22}$$

where u_m^* is the average dimensionless velocity on a cross-section, which is calculated by

$$u_m^* = \frac{\iint u^* \mathrm{d}A}{A_c} \tag{6.23}$$

The characteristics of the fluids flow in the channels can be represented by the product of the friction factor and the Reynolds number as

$$(fRe) = \left(\frac{-D_h \frac{\mathrm{d}P}{\mathrm{d}z}}{\rho u_m^2/2}\right)\left(\frac{\rho D_h u_m}{\mu}\right) = \frac{2}{u_m^*} \tag{6.24}$$

Dimensionless bulk temperature

$$\theta_b(z_h^*) = \frac{\iint u^* \theta \mathrm{d}A}{\iint u^* \mathrm{d}A} \tag{6.25}$$

Dimensionless bulk humidity

$$\xi_b(z_m^*) = \frac{\iint u^* \xi \mathrm{d}A}{\iint u^* \mathrm{d}A} \tag{6.26}$$

Dimensionless bulk mass fraction of solution

$$\Theta_b(z_m^*) = \frac{\iint u^* \Theta \mathrm{d}A}{\iint u^* \mathrm{d}A} \tag{6.27}$$

Nusselt number

$$Nu = \frac{hD_h}{\lambda} \tag{6.28}$$

Sherwood number

$$Sh = \frac{kD_h}{D_f} \tag{6.29}$$

where D_f is diffusivity, which represents D_{va} (moisture diffusivity in air), and D_{ws} (water diffusivity in solution); h and k are the convective heat transfer coefficient ($kW\,m^{-2}\,K^{-1}$) and convective mass transfer coefficient (m/s) between fluids and wall, respectively. As discussed previously for air-to-air ducts, the local Nusselt number and the Sherwood number are not fixed values, and will change from point to point on a channel surface. Therefore it is more reasonable to evaluate the peripherally mean local Nusselt and Sherwood numbers.

The peripherally mean local heat transfer coefficient can be calculated by [9,10]

$$h_L = -\frac{1}{4(\theta_w - \theta_b)} \frac{\lambda}{D_h} \frac{d\theta_b}{dz_h^*} \tag{6.30}$$

where subscripts "w" and "b" mean "wall" and "bulk," respectively.

Substituting Eq. (6.30) to Eq. (6.28), we can obtain the peripherally local Nusselt number as

$$Nu_L = -\frac{1}{4(\theta_w - \theta_b)} \frac{d\theta_b}{dz_h^*} \tag{6.31}$$

Averaging the Nusselt number from 0 to z_h^*

$$Nu_m = \frac{1}{z_h^*} \int_0^{z_h^*} Nu_L \, dz_h^* \tag{6.32}$$

Correspondingly, a mass balance on the control volume for the air stream will be the following equations

$$k_L = -\frac{1}{4(\xi_w - \xi_b)} \frac{D_{va}}{D_h} \frac{d\xi_b}{dz_m^*} \tag{6.33}$$

$$Sh_L = -\frac{1}{4(\xi_w - \xi_b)} \frac{d\xi_b}{dz_m^*} \tag{6.34}$$

$$Sh_m = \frac{1}{z_m^*} \int_0^{z_m^*} Sh_L \, dz_m^* \tag{6.35}$$

Similarly, a mass balance on the control volume for the solution stream will be the following equations

$$k_L = -\frac{1}{4(\Theta_w - \Theta_b)} \frac{D_{ws}}{D_h} \frac{d\Theta_b}{dz_m^*} \tag{6.36}$$

$$Sh_L = -\frac{1}{4(\Theta_w - \Theta_b)} \frac{d\Theta_b}{dz_m^*} \tag{6.37}$$

$$Sh_m = \frac{1}{z_m^*} \int_0^{z_m^*} Sh_L \, dz_m^* \tag{6.38}$$

It should be noted that both z_h^* and z_m^* in Eqs. (6.30) to (6.38) have different values for the air and the solution streams.

Boundary conditions

The coordinate systems for the two streams are different. According to Figure 6.4, the air stream and the solution stream flow in a cross-flow configuration, therefore the relationship between the two coordinates systems is

$$\begin{cases} x' = z \\ y' = y + 2a \\ z' = 2b - x \end{cases} \tag{6.39}$$

Boundary conditions of the two streams for velocity (no slip boundary condition)

$$u^* = 0, \text{ on all walls of the channels} \tag{6.40}$$

Inlet conditions for the air stream

$$z_{h,a}^* = 0, \quad \theta_a = 0 \tag{6.41}$$

$$z_{m,a}^* = 0, \quad \xi_a = 0 \tag{6.42}$$

Inlet conditions for the solution stream

$$z_{h,s}^* = 0, \quad \theta_s = 1 \tag{6.43}$$

$$z_{m,s}^* = 0, \quad \Theta_s = 1 \tag{6.44}$$

where in Eqs. (6.41) to (6.44), as mentioned, subscripts "a" and "s" stand for the air stream and the solution stream, respectively.

Adiabatic boundary conditions for the air stream

$$x^* = 0 \text{ and } x^* = 1, \quad \frac{\partial \theta_a}{\partial x^*} = \frac{\partial \xi_a}{\partial x^*} = 0 \tag{6.45}$$

and for the solution stream

$$x'^* = 0 \text{ and } x'^* = 1, \quad \frac{\partial \theta_s}{\partial x'^*} = \frac{\partial \Theta_s}{\partial x'^*} = 0 \tag{6.46}$$

Conjugate boundary conditions on membrane surfaces

Since the membrane are rather thin (of the order of 100 μm), temperature differences with membrane thickness are so small that they can be ignored. When water vapor is absorbed by the liquid desiccant, absorption heat is released on the membrane surface on the solution side. Based on these conditions, the heat balance equation on the membrane surface between the air stream and the solution stream is

$$\lambda_a \frac{\partial T_a}{\partial y}\bigg|_{y=0} + \rho_a D_{va} h_{abs} \frac{\partial \omega_a}{\partial y}\bigg|_{y=0} = \lambda_s \frac{\partial T_s}{\partial y'}\bigg|_{y'=2a} \tag{6.47}$$

where h_{abs} is absorption heat (kJ/kg).

Equation (6.47) can be normalized as

$$\lambda^* \cdot \frac{\partial \theta_a}{\partial y^*}\bigg|_{y^*=0} + h^*_{abs} \cdot \frac{\partial \xi_a}{\partial y^*}\bigg|_{y^*=0} = \frac{\partial \theta_s}{\partial y'^*}\bigg|_{y'^*=1} \tag{6.48}$$

where the dimensionless absorption heat and the dimensionless heat conductivity are defined by

$$h^*_{abs} = \frac{\rho_a D_{va} h_{abs}}{\lambda_s}\left(\frac{\omega_{si} - \omega_{ai}}{T_{si} - T_{ai}}\right) \tag{6.49}$$

and

$$\lambda^* = \frac{\lambda_a}{\lambda_s} \tag{6.50}$$

Heat flux on membrane surfaces on the air and the solution sides is equal to

$$q_h = -\lambda \frac{\partial T}{\partial y}\bigg|_{y=0,2a} \tag{6.51}$$

Since temperature and humidity are strongly coupled between the two streams and the membrane, the temperature of the membrane surfaces is not a fixed value across the surface.

Heat flux on the solution side is a sum of heat flux on the air side and latent heat flux

$$q_{h,s} = q_{h,a} + q_{Lat} \tag{6.52}$$

where q_{Lat} is latent heat flux induced by latent or absorption heat on the solution side. Equation (6.52) is the same as Eq. (6.47).

Mass boundary conditions on membrane surfaces, for the air side

$$y^* = 0 \text{ and } y^* = 1, \quad q_m(x^*, z^*_G) = \dot{m}_v \tag{6.53}$$

For the solution side

$$y'^* = 0 \text{ and } y'^* = 1, \quad q_m(x'^*, z'^*_G) = \dot{m}_v \tag{6.54}$$

where z^*_G is the dimensionless geometric position

$$z^*_G = \frac{z}{2b} \tag{6.55}$$

and $\dot{m}_v$ is the moisture emission rate through membrane at point (x^*, z_G^*), and it is determined by the diffusion equation in the membrane as

$$\dot{m}_v = \rho_a D_{vm} \frac{\omega_{m,a} - \omega_{m,s}}{\delta} \tag{6.56}$$

where D_{vm} is moisture diffusivity in the membrane (m^2/s); δ is membrane thickness (m).

On the other hand, moisture emission from the membrane surface on the air side is calculated by

$$q_m = -\rho_a D_{va} \frac{\partial \omega_a}{\partial y}\bigg|_{y=0,2a} \tag{6.57}$$

Correspondingly, moisture emission from the membrane surface on the solution side is obtained by

$$q_m = -\rho_s D_{ws} \frac{\partial X_s}{\partial y'}\bigg|_{y'=0,2a} \tag{6.58}$$

After the calculation of the outlet conditions, sensible heat effectiveness and latent effectiveness for the dehumidification process can be calculated:

$$\varepsilon_h = \frac{T_{ao} - T_{ai}}{T_{si} - T_{ai}} \tag{6.59}$$

$$\varepsilon_{Lat} = \frac{\omega_{ao} - \omega_{ai}}{\omega_{si} - \omega_{ai}} \tag{6.60}$$

The latent effectiveness is also the air dehumidification effectiveness. The sensible effectiveness reflects a temperature change in the dehumidification process. Different from sensible-only heat exchangers, absorption heat is released in dehumidification. As a result, sensible effectiveness may be negative in some cases due to the absorption heat generated. The transfer of sensible heat and moisture are in different directions now.

6.3 Solution state equations

Salt solutions can be used to absorb moisture from air. The most commonly used salt solutions are LiCl solution, LiBr solution, and $CaCl_2$ solution. The equilibrium states of the solution are determined by temperature, humidity and mass fractions of the solute. The state graphs for the three solutions are plotted in Figures 6.5 to 6.7 [11]. As can be seen, the vapor pressures rise with temperature, for a given solution concentration. This is the driving force for moisture absorption and desorption.

In calculations, it is more convenient to write the state of the solution into equations. Taking LiCl solution as an example, water vapor partial pressure, temperature, and LiCl solution concentration are governed by a thermodynamic equation [12]

$$\log p_v = A(m_s) + \frac{B(m_s)}{T} + \frac{C(m_s)}{T^2} \tag{6.61}$$

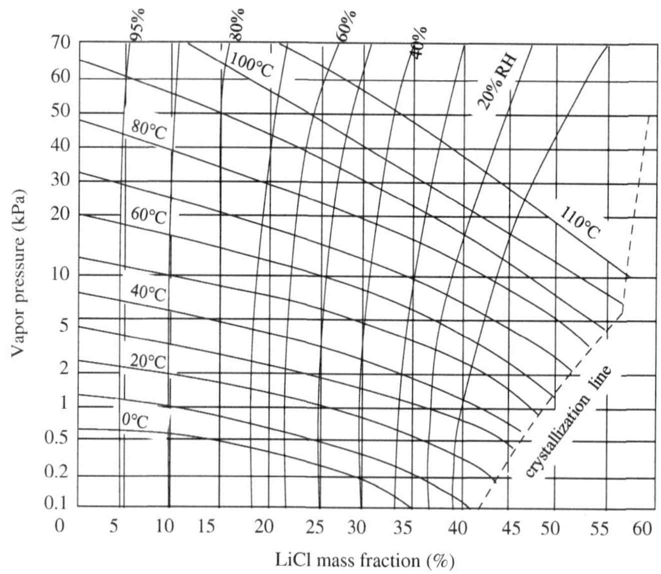

FIGURE 6.5

State graph of LiCl solution, the relations between temperature, humidity, and LiCl solute fractions.

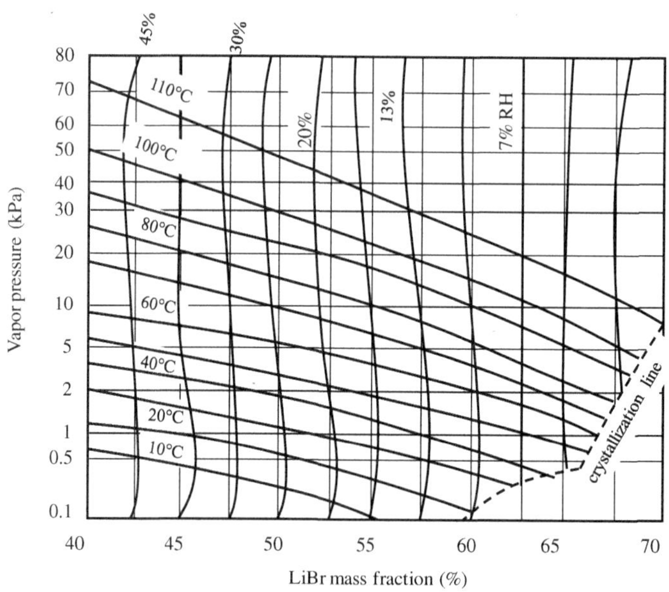

FIGURE 6.6

State graph of LiBr solution, the relations between temperature, humidity, and LiBr solute fractions.

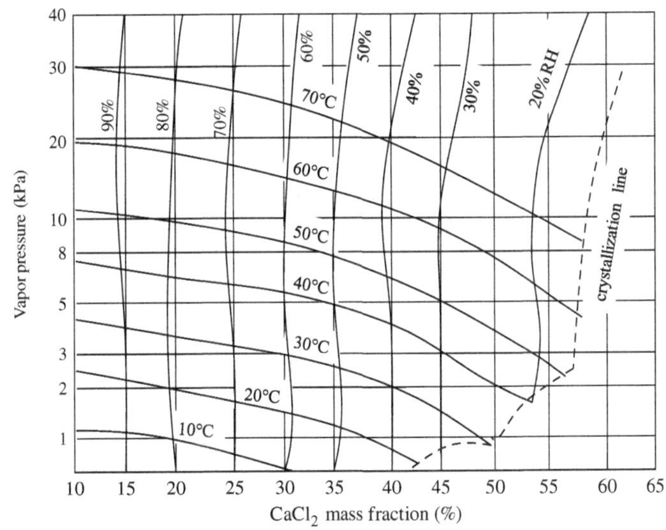

FIGURE 6.7

State graph of $CaCl_2$ solution, the relations between temperature, humidity, and $CaCl_2$ solute fractions.

$$A(m_s) = A_0 + A_1 m_s + A_2 m_s^2 + A_3 m_s^3 \quad (6.62)$$

$$B(m_s) = B_0 + B_1 m_s + B_2 m_s^2 + B_3 m_s^3 \quad (6.63)$$

$$C(m_s) = C_0 + C_1 m_s + C_2 m_s^2 + C_3 m_s^3 \quad (6.64)$$

where p_v is vapor partial pressure in kPa, T is in K, and m_s is molality of the electrolyte (mol LiCl/kg water).

$$m_s = \frac{1 - X_s}{0.0425 X_s} \quad (6.65)$$

The constants in Eqs. (6.62) to (6.64) are given by [12]

$$\begin{aligned}
A_0 &= 7.3233550 & A_1 &= -0.0623661 \\
A_2 &= 0.0061613 & A_3 &= -0.0001042 \\
B_0 &= -1718.1570 & B_1 &= 8.2255 \\
B_2 &= -2.2131 & B_3 &= 0.0246 \\
C_0 &= -97575.680 & C_1 &= 3839.979 \\
C_2 &= -421.429 & C_3 &= 16.731
\end{aligned}$$

Water concentration in solution is

$$C_w = X_s \rho_s \tag{6.66}$$

Solution density is calculated by the following Eq. [12]

$$\rho_s = \rho_w \sum_{i=0}^{3} \rho_i \left(\frac{1 - X_s}{X_s} \right)^i \tag{6.67}$$

where ρ_w is pure water density at temperature T, and ρ_i are given below [12]:

$$\rho_0 = 1.0, \quad \rho_1 = 0.540966$$
$$\rho_2 = -0.303792, \quad \rho_3 = 0.100791$$

In moist air, water vapor partial pressure is calculated by [12]

$$p_v = \frac{\omega P}{\omega + 0.622} \tag{6.68}$$

The humidity ratio in ambient air is in the range 0.005 to 0.035 kg/kg, therefore the above equation can be simplified to

$$p_v = 1.608 \omega P \tag{6.69}$$

As can be seen, the relations between the air humidity and water concentration in solution are rather complicated and need iterations to find the solution. Calculations of the thermodynamic equilibrium chart of LiCl solution with Eqs. (6.59) to (6.62) showed that under isothermal conditions a linear equation similar to Henry's sorption law can be used to express the water concentration in the solution as

$$C_w = K_p p_v + C_{w0} \tag{6.70}$$

where K_p is the Henry coefficient (kg m^{-3} Pa^{-1}), and C_{w0} is a constant (kg/m^3). K_p here has the same meaning as the partition coefficient defined in previous chapters, but with a different unit. Table 6.1 lists the curve-regressed values of K_p and C_{w0} under different temperatures.

This table gives the following correlations to estimate K_p and C_{w0} from temperature:

$$K_p = 0.2625 - 0.0093T + 9.0 \times 10^{-5} T^2 \tag{6.71}$$

$$C_{w0} = 739.9 - 1.1T \tag{6.72}$$

Table 6.1 Values of K_p and C_{w0} for LiCl solution.

T (°C)	K_p (kgm^{-3}Pa^{-1})	C_{w0} (kgm^{-3})
5	0.1434	727.8
25	0.0865	706.4
35	0.0493	700.2
45	0.0293	693.2

6.4 Numerical scheme

The governing equations deduced above with the relevant boundary conditions are solved by a finite difference numerical scheme. Since the two streams and the membrane closely interact, and temperature and humidity are also related to each other, iterative techniques are needed to solve these equations. The numerical scheme is:

a. Solve the momentum equations. Get the velocity fields and resistance data for the two neighboring ducts.
b. Assume the initial temperature and humidity in the two streams for the whole domain. Get the membrane temperature.
c. Solve the energy equations for the two streams. Get the temperature fields in the two ducts.
d. Assume an initial moisture emission rate through the membrane.
e. Based on the boundary conditions of the two streams, solve the mass equations for the two streams. Get the humidity fields in the two channels.
f. Calculate the new emission rates through the membrane, from the newly calculated humidity fields for the two streams.
g. Give the new emission rates as the default values for the emission rates. Return to step (e), until the old and the new values of emission rates are converged.
h. Based on the obtained temperature and humidity fields, calculate the new membrane temperature from the heat conduction equations for the membranes.
i. Give the new membrane temperature as the default value. Return to step (c), then re-calculate the temperature fields of the two streams, until the old values and the newly calculated values for all temperature are converged.
j. Check whether all the new temperature, humidity, mass fraction and emission rates values are converged with the old values for the whole domain. If not, return to step (c).

To ensure the accuracy of the results calculated, numerical tests were performed to determine the effects of the grid size. They indicate that 35 $\times$ 25 grids for the cross section and 32 girds for the z axis are adequate, showing less than 1.5% difference compared with 52 $\times$ 52 $\times$ 52 grids. The final numerical uncertainty is 1.0%.

After these procedures, all the governing equations are satisfied simultaneously. The temperature and humidity profiles on the membrane surfaces are the membrane-formed boundary conditions.

6.5 Experimental and numerical validation

In order to study the heat and mass transfer in the membrane-formed parallel-plate ducts, a continuous air dehumidification system has been designed. The whole test set-up is schematically depicted in Figure 6.8. As can be seen, there are two parallel-plate membrane modules, one for air dehumidification and the other for solution regeneration. The structures of the two modules are the same, as shown in Figure 6.3. The former is called a dehumidifier, while the latter is called a regenerator.

Heat and mass transfer in the dehumidifier is the focus in this study. Liquid desiccant and air flow alternately in the parallel-plate ducts. The membranes are self-made PVDF porous membranes. The porous membrane is modified by daubing a layer of silica gel on the membrane surface. The purpose

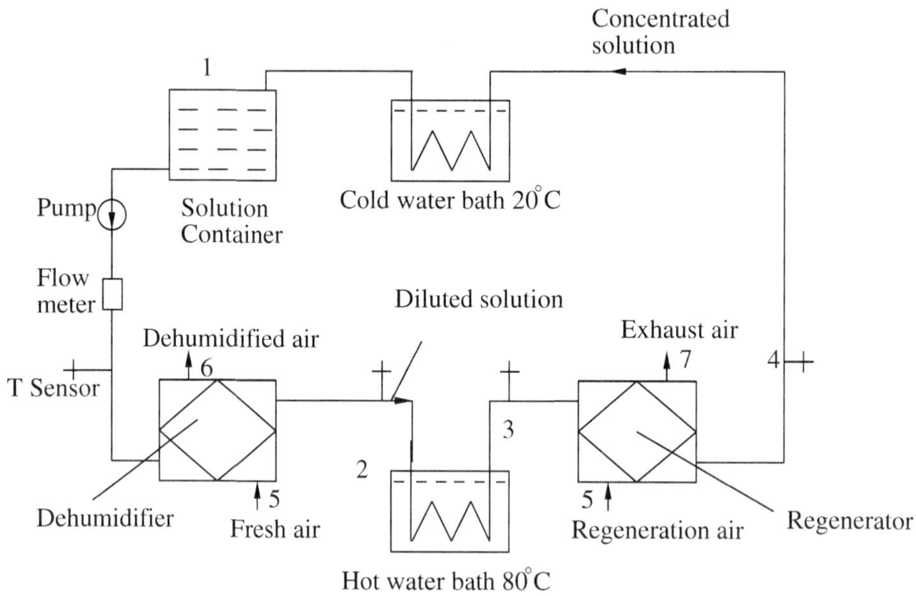

FIGURE 6.8

The schematic of the liquid air dehumidification system.

Table 6.2 Membrane properties and some transport properties.

Symbol	Unit	Value	Symbol	Unit	Value
D_{vm}	m^2/s	3.0×10^{-6}	Sc_a		0.564
D_{va}	m^2/s	2.82×10^{-5}	Sc_s		1390
D_{ws}	m^2/s	3.0×10^{-9}	T_{ai}	°C	30.0
δ	μm	120	T_{si}	°C	25.0
λ_a	$Wm^{-1}K^{-1}$	0.0263	ω_{ai}	kg/kg	0.02
λ_s	$Wm^{-1}K^{-1}$	0.46	ω_{si}	kg/kg	0.0055
c_s	$kJ\,kg^{-1}K^{-1}$	2.8	X_{si}	kg/kg	0.65
Pr_a		0.7	X_{ei}	kg/kg	0.82
Pr_s		28.36	ρ_s	kg/m^3	1215
n_{ch}		10	a	mm	1.0
ρ_m	kg/m^3	864	b	mm	100

of this modification is to prevent the liquid desiccant from leaking. The tested geometric and thermodynamic properties of the membrane are summarized in Table 6.2. In total there are 10 ducts for each flow. Some transport properties have been measured in the laboratory and are also listed in the table.

The whole set-up is placed in an air-conditioned room. The temperature and humidity of the indoor air can be adjusted, representing varying ambient air states. Both the dehumidifier and the regenerator use indoor air as working air. In this study, LiCl solution is used as the liquid desiccant. The solution flows in the sequence 1-2-3-4-1 as in Figure 6.8. Four processes are involved: dehumidification, solution heating, regeneration and solution cooling. This is a complete dehumidification and regeneration cycle.

In order to obtain the performance of the dehumidifier, air stream temperature, humidity and volumetric flow rates to and from the dehumidifier are measured. Solution stream temperature and volumetric flow rates are measured at the inlet and the outlet of the dehumidifier as well. K-type thermocouples and humidity meters (center 313, Taiwan) are used to measure the temperature and humidity, respectively. The volumetric flow rates to and out of the dehumidifier for the air and the solution streams are measured by floating-ball flow meters. A thick plastic foam layer is pasted onto the outer surface of the shells to prevent heat dissipation from the air and the solution streams to the surroundings. Moisture dissipation from the two streams to the surroundings is negligible since the shell materials are strongly hydrophobic. Heat and moisture balances are checked for the dehumidifier. The heat loss from the system is below 0.4%, and moisture loss is less than 0.2%. The uncertainties for measurements are: temperature $\pm 0.1\,°C$; humidity $\pm 2\%$; volumetric flow rate $\pm 0.5\%$.

Experimental results are used to validate the calculated values. The test set-up is operated under different operating conditions. The ranges for the operating conditions are: air velocity, 50–150 L/min; air temperature: 25–35 °C; air humidity, 40–70% *RH*; solution velocity, 1–3 L/min; temperature, 20–25 °C; mass fraction of LiCl (1-X_S), 0.3–0.33. Then the outlet values are measured. Sensible effectiveness and latent effectiveness (air dehumidification efficiencies) are calculated. The calculated results are compared to the experimentally obtained results, as shown in Figure 6.9. As can be seen, generally the model predicts the heat and mass transport processes well. Discrepancies between the numerical and

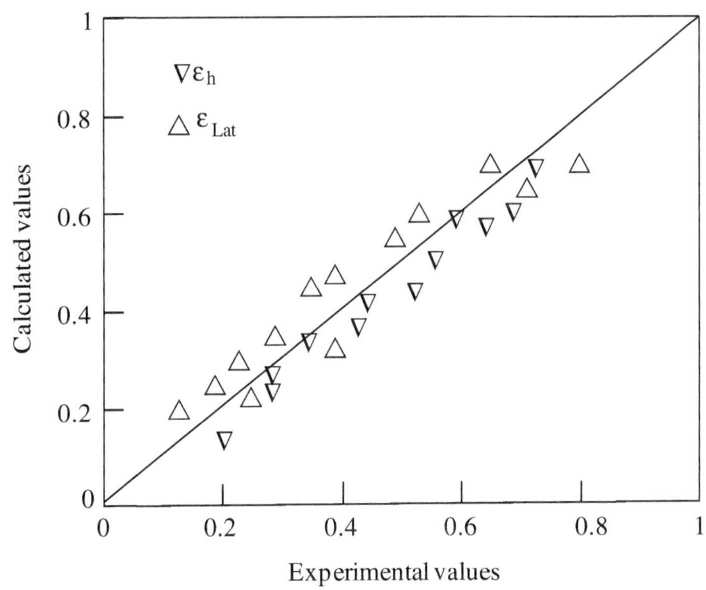

FIGURE 6.9

Comparison of numerical and experimental data for sensible and latent effectiveness.

tested data are below 9.0%. So the calculated values are in good agreement with the experimentally obtained data. Most calculated sensible effectiveness is less than the test data, while most calculated moisture effectiveness is greater than the test data. This is because heat dissipation is greater than moisture losses from the test rig.

As done previously, the code is also validated numerically. For fully developed laminar flows in the channels, (*fRe*) is a constant. The calculated fully developed (*fRe*) values for various aspect ratios are compared to those in Refs. [13,14]. Agreement is reached. Further, the fully developed Nusselt numbers under uniform temperature conditions are calculated and compared with those in the references. Agreement is also reached. Maximum differences are controlled to within 1.5%, otherwise grid structures are optimized. In this way, the code is verified. For the detailed methodology of verification, please consult the previous chapters.

6.6 Nusselt numbers

The temperature and humidity profiles under normal operating conditions are calculated. The peripherally local and mean Nusselt numbers for the two fluids along the two flow directions are shown in Figure 6.10. Since absorption heat is generated on the solution side, Nusselt numbers for the two streams are different. Though the Nusselt number values are not the same for the two streams, the trends

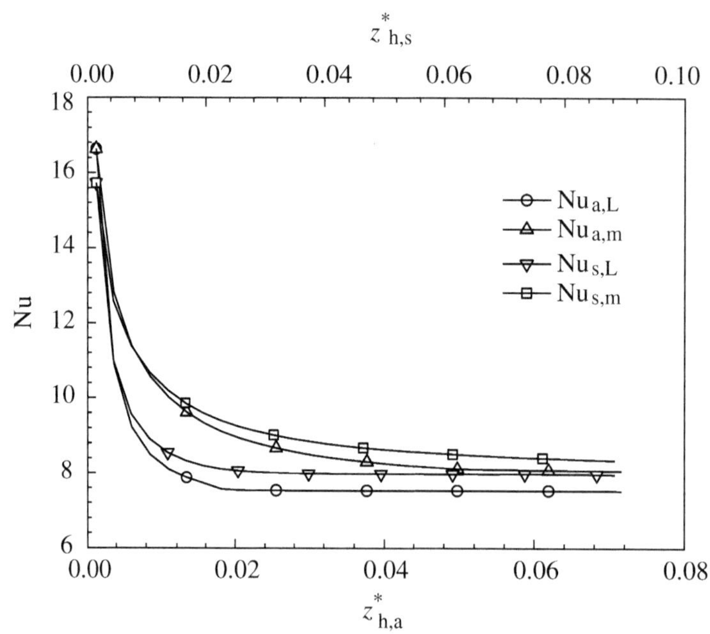

FIGURE 6.10

Local and mean Nusselt numbers for both air and solution streams along the two flow directions, $b/a=50$.

Table 6.3 Fully developed Nusselt and Sherwood numbers for various aspect ratios (*b*/*a*).

b/*a*	Nu_H	Nu_T	Nu_C^*	Sh_C^*	$Nu_{C,a}$	$Sh_{C,a}$	$Nu_{C,s}$
Sources	**Ref. [13,14]**	**Ref. [13,14]**	**Ref. [9]**	**Ref. [9]**	**(This duct)**	**(This duct)**	**(This duct)**
1.0	3.61	2.98	1.88	1.89	3.12	3.30	3.41
1.43	3.73	3.08	2.49	2.52	3.23	3.41	3.64
2.0	4.12	3.39	3.06	3.10	3.48	3.65	4.05
3.0	4.79	3.96	4.06	4.00	4.15	4.28	4.74
4.0	5.33	4.44	4.64	4.52	4.61	5.11	5.35
8.0	6.49	5.60	6.06	6.03	5.79	5.83	6.41
50.0	–	–	7.78	7.81	7.54	7.74	7.91
100.0	–	–	8.07	8.05	7.70	7.98	8.08
∞	8.23	7.54	–	–	–	–	–

Note: * *These values of* Nu_C^* *and* Sh_C^* *are for air-to-air membrane ducts.*

of variations are similar. The local Nusselt numbers decrease and approach lower limiting values. After that period, the flow is called the thermally developed region. The lower limiting value of the Nusselt number is denoted as Nu_C. The developing length for the air flow ($z_{h,a}^* = 0.02$) is a little shorter than the solution flow ($z_{h,s}^* = 0.022$). The two ducts are long enough (10 cm, $z_h^* = 0.08$) for the two flows to get thermally fully developed.

The detailed values of Nu_C, Nu_T and Nu_H under various aspect ratios are listed in Table 6.3. The values under a uniform temperature boundary condition (Nu_T) or a uniform heat flux boundary condition (Nu_H) for common ducts are also listed for comparison [13,14]. The values for air-to-air membrane duct Nu_C^* are also listed. As can be seen, for the membrane-based air-to-air heat mass exchanger mentioned in Ref. [9], which has no absorption heat transfer between the two fluids, the Nu_C values for the two fluids are the same. Also, Nu_C is generally less than Nu_H. It is larger than Nu_T at larger aspect ratios, but 37% less than Nu_T at smaller aspect ratios. However, for the membrane-formed parallel-plate channels for liquid desiccant air dehumidification, the fully developed Nusselt number on the air side ($Nu_{C,a}$) is between Nu_T and Nu_H. The air-side $Nu_{C,a}$ is closer to Nu_T. The absorption heat generated on the solution side has an important influence on the fully developed Nusselt numbers for the two fluids. The solution-side $Nu_{C,s}$ is close to Nu_H.

As mentioned, the heat flux on the solution side consists of the air-side heat flux and the latent or absorption heat flux. The average air-side heat flux and the average latent heat flux on the membrane surfaces are denoted as $q_{h,a,m}$ and $q_{Lat,m}$, respectively. In order to show the influence of the latent heat flux on the Nusselt numbers for the two streams, the fully developed Nusselt numbers for the two streams under different values of $q_{Lat,m}/q_{h,a,m}$ are listed in Table 6.4. As can be seen, $q_{Lat,m}/q_{h,a,m}$ can have positive or negative values. This is because that heat and mass transfer can be in the same or opposite directions. The greater $q_{Lat,m}/q_{h,a,m}$ is, the higher the air-side $Nu_{C,a}$ and the solution-side $Nu_{C,s}$ are. The solution-side $Nu_{C,s}$ is greater than the air-side $Nu_{C,a}$, regardless of positive or negative values of $q_{Lat,m}/q_{h,a,m}$. These results are caused by the close coupling of the heat and mass transfer between the two fluids inside the membrane.

Table 6.4 Fully developed Nusselt numbers for the two fluids under different ratios of the average latent heat flux and the average air-side heat flux ($q_{Lat,m}/q_{h,a,m}$).

b/a ↓	$Nu_{C,a}$						$Nu_{C,s}$					
$q_{Lat,m}/q_{h,a,m}$ →	**8.0**	**6.0**	**4.0**	**−2.0**	**−4.0**	**−6.0**	**8.0**	**6.0**	**4.0**	**−2.0**	**−4.0**	**−6.0**
1.0	3.22	3.12	3.09	2.82	2.72	2.62	3.43	3.41	3.39	3.50	3.44	3.40
1.43	3.33	3.25	3.17	2.91	2.85	2.71	3.66	3.65	3.62	3.73	3.68	3.64
2.0	3.53	3.48	3.41	3.19	3.06	2.95	4.07	4.05	4.03	4.14	4.09	4.05
3.0	4.27	4.15	4.12	3.85	3.75	3.65	4.76	4.73	4.72	4.84	4.78	4.73
4.0	4.75	4.61	4.58	4.38	4.23	4.14	5.37	5.34	5.33	5.45	5.38	5.34
8.0	5.91	5.79	5.75	5.51	5.41	5.28	6.43	6.41	6.39	6.49	6.44	6.41
50.0	7.63	7.54	7.48	7.21	7.10	6.96	7.92	7.91	7.89	8.06	8.00	7.97
100.0	7.81	7.70	7.66	7.38	7.25	7.14	8.10	8.08	8.05	8.16	8.11	8.06

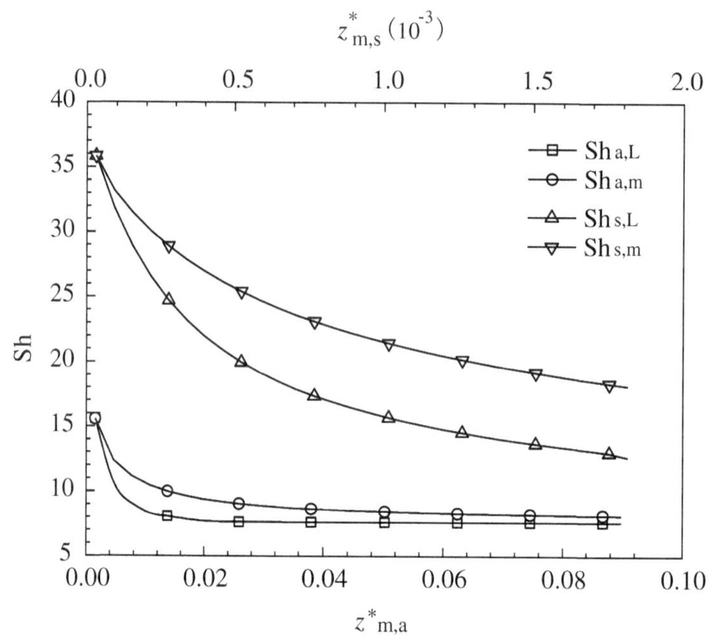

FIGURE 6.11

Local and mean Sherwood numbers for both air and solution streams along the two flow directions, $b/a = 50$.

6.7 Sherwood numbers

The peripherally local and mean Sherwood numbers for the two fluids along the two flow directions are shown in Figure 6.11. Variations of the Sherwood numbers for the air stream are similar to Nusselt numbers. In the developing region for concentration, the local Sherwood numbers decrease sharply,

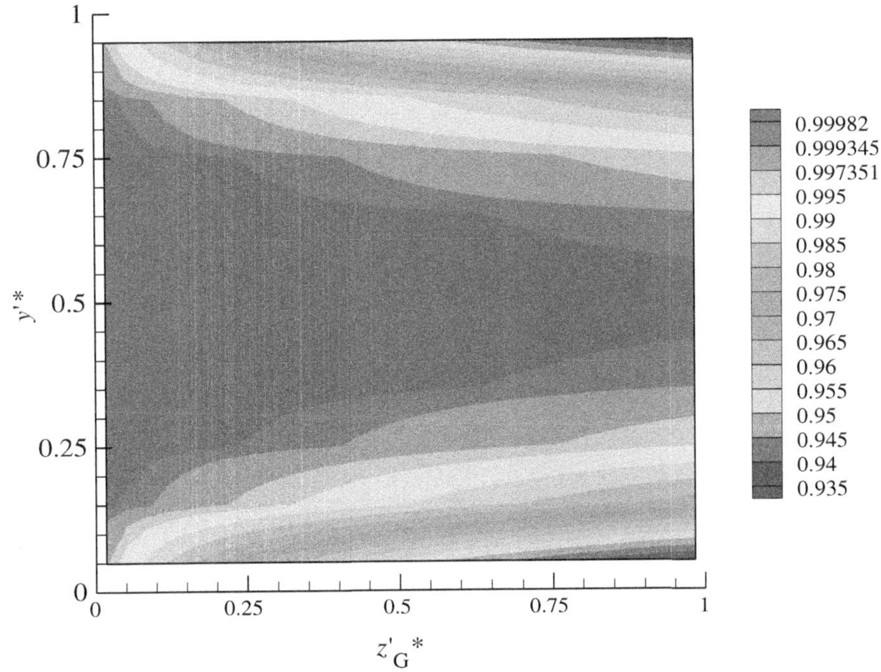

FIGURE 6.12

Dimensionless mass fraction profiles in the y'–z' plane at $x'^* = 0.5$ for the solution stream. The color represents the dimensionless mass friction of the solution (Θ_s).

from high values to lower limiting values denoted as Sh_C. The flow is called developed in concentration after that period. The detailed Sh_C values on the air side for various aspect ratios are also listed in Table 6.3. As can be seen, the fully developed Sherwood numbers on the air side ($Sh_{C,a}$) are somewhat higher than the air-side ($Nu_{C,a}$) for most aspect ratios.

For the solution flow, the situation is different. Since the Schmidt number for the solution (LiCl solution) is very large ($Sc = 1390$), the concentration boundary layer on the solution side develops much more slowly than that on the air side. As a result, the Sherwood numbers for the solution stream are much higher than for the air stream. The concentration on the solution side is not fully developed for the whole channel of $D_h = 3.92$ mm and 2b = 10 cm ($z^*_{m,s} = 2.0 \times 10^{-3} \ll 0.02$). To further illustrate this, the profiles of the dimensionless mass fraction of the solution on y'–z' plane at $x'^* = 0.5$ are plotted in Figure 6.12. As can be seen, the concentration boundary layer on the solution side is still developing at the outlet.

6.8 Temperature and humidity profiles

As observed for the air-to-air duct, temperature profiles on membrane surfaces also exhibit non-uniform value non-uniform flux behaviors [7]. There are the same trends to the humidity profiles, but more complex. Figure 6.13 shows the contours of the dimensionless humidity on the membrane surface on the air side, the arrow in the figure showing the direction of decreasing contour value. Figure 6.14 shows the

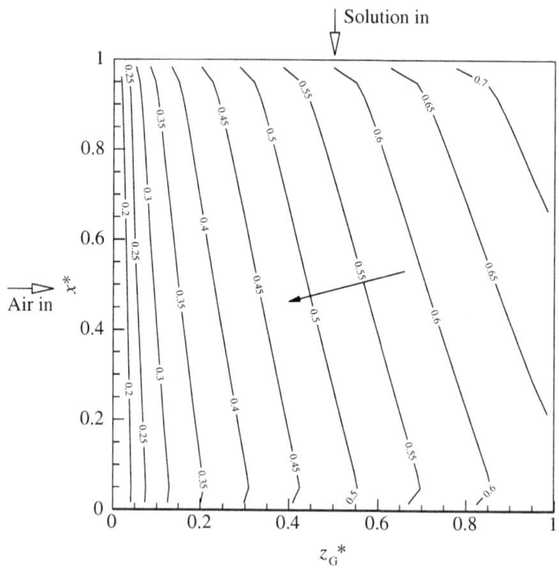

FIGURE 6.13

Contours of the dimensionless humidity on the membrane surface on the air side, the arrow in the figure showing the direction of decreasing contour value.

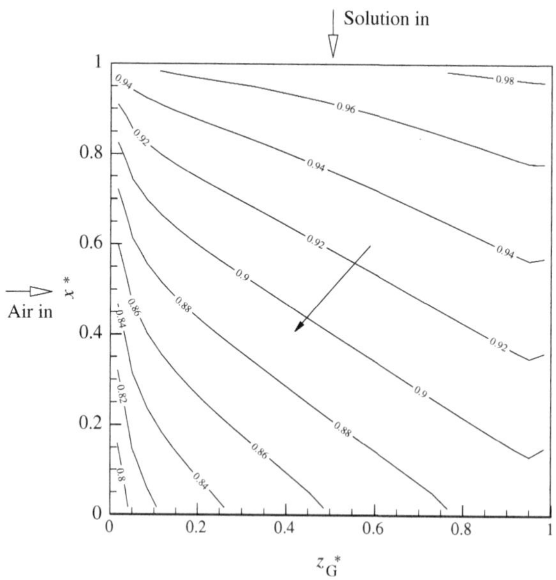

FIGURE 6.14

Contours of the dimensionless humidity on the membrane surface on the solution side, the arrow in the figure showing the direction of decreasing contour value.

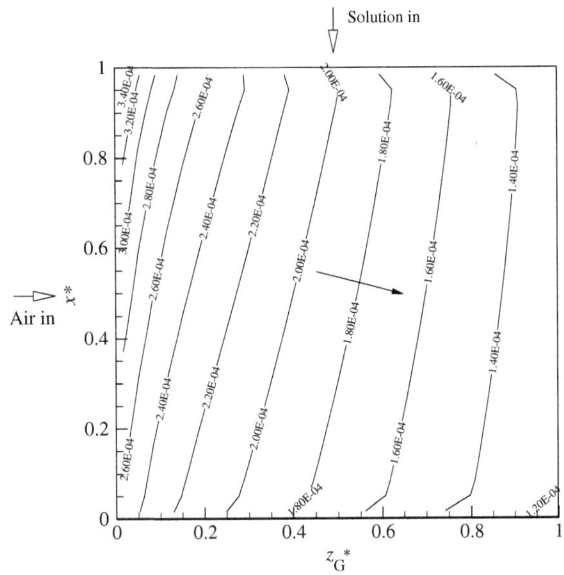

FIGURE 6.15

Contours of mass flux through the membrane (kg m^{-2}s^{-1}), the arrow in the figure showing the direction of decreasing contour value.

contours of the dimensionless humidity on the membrane surface on the solution side, the arrow in the figure showing the direction of decreasing contour value. Figure 6.15 shows the contours of mass flux through the membrane (kg m^{-2}s^{-1}), the arrow in the figure showing the direction of decreasing contour value. As can be seen, they are quite inhomogeneous.

6.9 Conclusion

The structures for liquid-to-air parallel-plate membrane ducts are the same as for air-to-air membrane ducts; however, the transport data are different, especially for the liquid side. The real boundary conditions are neither uniform value nor uniform flux boundary conditions. The flows are developing both thermally and in concentration at the entrances of the ducts. The detailed Nusselt and Sherwood numbers have been obtained for component design and optimization. For the air stream, the fully developed Nusselt numbers for air fluid ($Nu_{C,a}$) are between Nu_T and Nu_H. However, $Nu_{C,a}$ is closer to Nu_T. The Sherwood numbers are somewhat higher than the Nusselt numbers. For the solution side, heat transfer is developed shortly after the entry, but mass transfer is still developing even at the outlet. As a result, the Sherwood numbers for the solution stream are much higher than for the air side. The solution-side $Nu_{C,s}$ is about 15% higher than the air-side $Nu_{C,a}$. The traditional way of assuming the solution-side Nusselt numbers as the fully developed Nusselt numbers for common parallel-plate channels under uniform temperature leads to under-prediction. The Sherwood numbers are rather high since they are in the developing regions. Similar to air-to-air membrane ducts, the temperature values on the two membrane surfaces are the same. However, heat flux is different due to absorption heat on the solution side. The humidity flux on the two surfaces is the same; however the humidity values are different due to the high resistance to mass transfer through the membranes.

References

[1] Babakhani D, Soleymani M. An analytical solution for air dehumidification by liquid desiccant in a packed column. Int Commun Heat Mass Transfer 2009;36:969–77.
[2] Martin V, Goswami DY. Effectiveness of heat and mass transfer processes in a packed bed liquid desiccant dehumidifier/regenerator. HVAC&R Res 2000;6:21–39.
[3] Liu XH, Jiang Y, Qu KY. Heat and mass transfer model of cross-flow liquid desiccant air dehumidifier/regenerator. Energy Convers Manage 2007;48:546–54.
[4] Khan AY, Sulsona FJ. Modeling and parametric analysis of heat and mass transfer performance of refrigerant cooled liquid desiccant absorbers. Int J Energy Res 1998;22:813–32.
[5] Stevens DI, Braun JE, Klein SA. An effectiveness model of liquid desiccant system heat/mass exchangers. Sol Energy 1989;42:449–55.
[6] Vali A, Simonson CJ, Besant RW, Mahmood G. Numerical model and effectiveness correlations for a run-around heat recovery system with combined counter and cross flow exchangers. Int J Heat Mass Transfer, 2009;52:5827–40.
[7] Huang SM, Zhang LZ, Tang K, Pei LX. Fluid flow and heat mass transfer in membrane parallel-plates channels used for liquid desiccant air dehumidification. Int J Heat Mass Transfer 2012;55:2571–80.
[8] Larson MD, Simonson CJ, Besant RW. The elastic and moisture transfer properties of polyethylene and polypropylene membranes for use in liquid-to-air energy exchangers. J Membr Sci 2007;302:136–49.
[9] Zhang LZ. Heat and mass transfer in a cross-flow membrane-based enthalpy exchanger under naturally formed boundary conditions. Int J Heat Mass Transfer 2007;50:151–62.
[10] Zhang LZ, Liang CH, Pei LX. Conjugate heat and mass transfer in membrane-formed channels in all entry regions. Int J Heat Mass Transfer 2010;53:815–24.
[11] Zhang LZ. Air dehumidification technologies. Beijing: China Chemical Industry Press; 2005.
[12] Zhang LZ. Fabrication of a lithium chloride solution based composite supported liquid membrane and its moisture permeation analysis. J Membr Sci 2006;276:91–100.
[13] Kays WM, Crawford ME. Convective heat and mass transfer. 3rd ed. New York: McGraw-Hill; 1990.
[14] Shah RK, London AL. Laminar flow forced convection in ducts. New York: Academic Press; 1978.

CHAPTER 7

Heat and Mass Transfer Across a Hollow Fiber Membrane Bundle

7.1 Introduction

Parallel-plate membrane ducts and plate-fin membrane ducts are suitable structures for air-to-air heat and mass exchangers. For liquid-to-air heat and mass exchangers, shell-and-tube structures are suitable. When membranes are made into hollow fibers they can endure high pressures. Numerous fibers are packed in a shell to form a shell-and-tube structure, which makes the packing densities rather high. Liquid always flows inside the fibers, and gas flows across the bundle. The contact area between the liquid and the gas is very high, which is often used to offset the drawback of slow liquid-air transport coefficients. For such structures, air-side pressure drops are usually low and acceptable. Due to the tube structure, the liquid side can endure large pressures, which makes the driving force high. The benefits are that the pressure drops for both sides are within acceptable levels. Hollow fiber membrane bundles are used in a wide variety of applications, for instance, gas absorption, liquid distillation, gas purification, air dehumidification, humidification, etc. The structure of a hollow fiber membrane bundle for liquid air dehumidification is shown in Figure 7.1.

Figure 7.2 shows the microstructure of a single fiber. Figure 7.3 shows an amplified picture of the membrane cross section. It has finger-like macrovoids in the wall to enhance gas permeations. Due to the high packing density and microstructure of the membranes, the heat and mass transfer capabilities of the membrane contactors are very high. Further, the membranes can prevent liquid from leaking, and only permit the permeation of gases.

Fluid flow and conjugate heat and mass transfer across a hollow fiber membrane bundle are the key for performance analysis and optimization. The transport data are co-determined by the fluid flow, membrane properties, and fluid properties. In fact, the bundle can be regarded as ducts composed of fibers and fluids. The Nusselt and Sherwood numbers in these ducts are of great importance. Previous analyses of fiber bundles always used the convective heat and mass transfer coefficients for common heat transfer tube banks [1]. That's only a coarse approximation. Conjugate heat and mass transfer were not studied until recently [2–6]. For a more accurate analysis, this chapter will provide the conjugate heat and mass transfer data for these special ducts. The influences of membrane properties will be discussed.

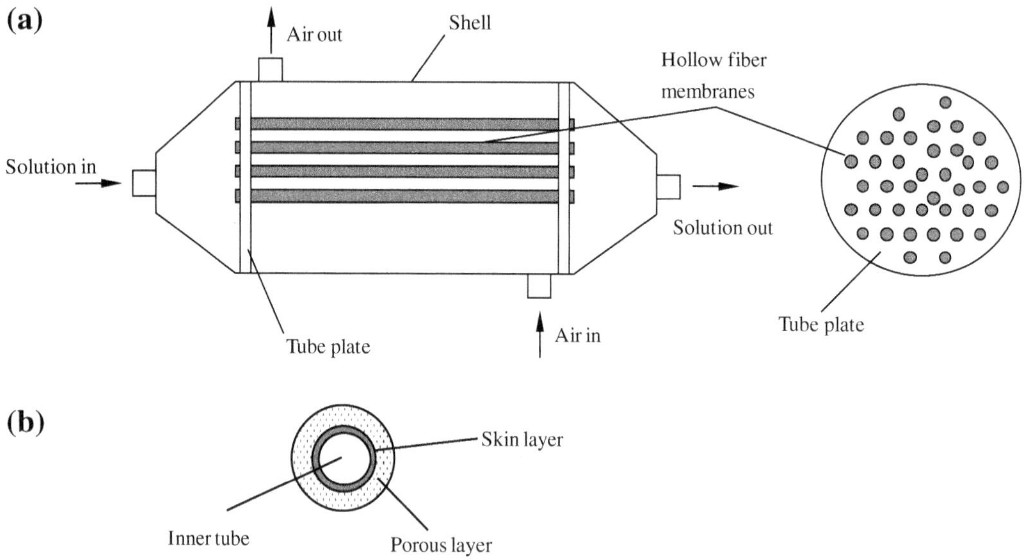

FIGURE 7.1

Schematic of a hollow fiber membrane bundle membrane contactor for liquid air dehumidification. (a) The whole structure; (b) a single fiber.

7.2 Characteristics of a bundle

Pressure drop and convective heat mass transfer coefficients

For a counter flow bundle, Reynolds number

$$Re = \frac{D_h u}{\nu} \tag{7.1}$$

$$\Delta P = f \cdot L \cdot \frac{\rho_f}{2} \frac{u^2}{D_h} \tag{7.2}$$

where Re is the Reynolds number; D_h is the hydrodynamic diameter (m); for the tube it is d_i, for the shell side it should be calculated; u is fluid mean velocity (m/s); ν is fluid kinematic viscosity (m^2/s); ΔP is the pressure drop (Pa); f is the friction factor; L is module length (m); ρ_f is fluid density (kg/m^3); for the tube side, the fluid is liquid solution (subscript "s"), and for the shell side, the fluid is air (subscript "a").

The convective mass transfer coefficient k (m/s) and the heat transfer coefficient h (kW $m^{-2}K^{-1}$) can be represented by Sherwood and Nusselt numbers respectively as

$$Sh = \frac{kD_h}{D_f} \tag{7.3}$$

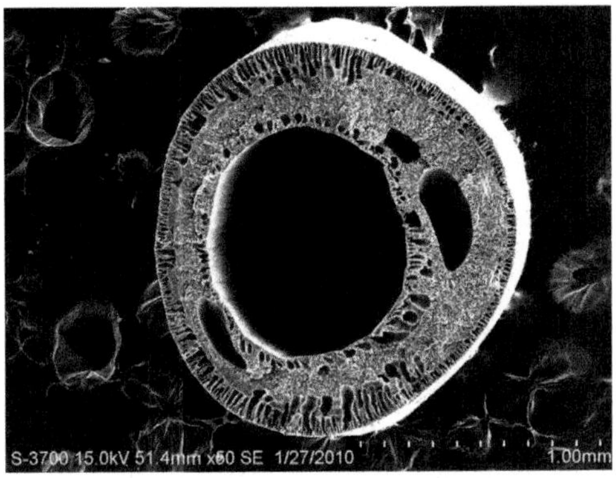

FIGURE 7.2

Scanning electron micrograph of a hollow fiber membrane.

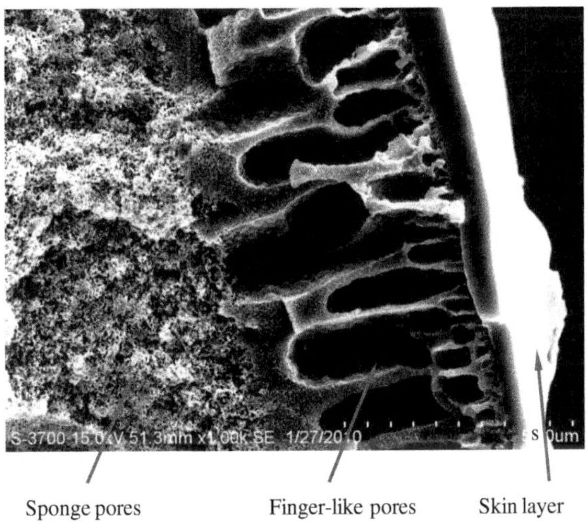

FIGURE 7.3

The microstructure of the fiber wall. It has finger-like macrovoids for permeation enhancement.

$$Nu = \frac{hD_h}{\lambda} \tag{7.4}$$

where according to the Chilton–Colburn analogy [7]

$$Nu = Sh \cdot Le^{1/3} \tag{7.5}$$

where

$$Le = \frac{Pr}{Sc} \tag{7.6}$$

$$Pr = \frac{\nu}{a_f} \tag{7.7}$$

$$Sc = \frac{\nu}{D_f} \tag{7.8}$$

where D_f is diffusivity, which represents D_{ws}, water diffusivity in solution for the tube side, and D_{va}, moisture diffusivity in air for the shell side, respectively, if taking liquid air dehumidification as an example; Le is the Lewis number of the fluid; Pr, and Sc are Prandtl number and Schmidt number for fluids respectively; a_f is thermal diffusivity (m^2/s).

The Chilton–Colburn analogy is the traditional way to calculate Nusselt and Sherwood numbers, or to calculate one from the other. The Nusselt numbers are usually adopted from heat transfer tube banks [1]. In this book, the concrete Nusselt and Sherwood numbers for the hollow fiber membrane bundle will be calculated.

The packing fraction of the module, ϕ, which is the ratio of total cross-sectional area of the fibers to the module, is defined by

$$\phi = \frac{n_f d_o^2}{D_0^2} \tag{7.9}$$

where D_0 is the diameter of the module shell (m); d_o is the outer diameter of a single fiber (m); n_f is the number of fibers in the module.

For counter flow, the air flow area on the shell side

$$A_{flow} = (1-\phi)\cdot\frac{\pi D_0^2}{4} \tag{7.10}$$

The hydrodynamic diameter on the shell side is defined as 4 times the flow area divided by the total perimeter of membrane fibers

$$D_h = \frac{4A_{flow}}{n_f \pi d_0} = \frac{(1-\phi)D_0^2}{n_f d_0} \tag{7.11}$$

The packing density A_v (m^2/m^3) is the ratio of total membrane area to module volume

$$A_v = \frac{4n_f d_0}{D_0^2} \tag{7.12}$$

Resistance through the shell side is measured. It is related to packing density and flow configurations. A correlation can be summarized to estimate the product of friction factor and Reynolds number as [6]

$$(f \cdot Re) = \text{Const} \tag{7.13}$$

For a cross-flow module, the shell-side characteristic length is the fiber outer diameter.

Overall heat and mass transfer coefficients

Resistances from solution through membranes to air can be considered as a sum of resistances in series. The overall mass and heat transfer coefficients from solution to air on the basis of the outer surface of the fibers are calculated by

$$\frac{1}{k_{\text{tot}}} = \frac{1}{k_{\text{i}}^{\text{e}}}\left(\frac{d_{\text{o}}}{d_{\text{i}}}\right) + \frac{\delta}{D_{\text{vm}}}\left(\frac{d_{\text{o}}}{\bar{d}}\right) + \frac{1}{k_{\text{o}}} \tag{7.14}$$

$$\frac{1}{h_{\text{tot}}} = \frac{1}{h_{\text{i}}}\left(\frac{d_{\text{o}}}{d_{\text{i}}}\right) + \frac{\delta}{\lambda_{\text{m}}}\left(\frac{d_{\text{o}}}{\bar{d}}\right) + \frac{1}{h_{\text{o}}} \tag{7.15}$$

where $\bar{d}$ is the arithmetic mean diameter of a fiber, i.e., $(0.5*(d_{\text{o}}+d_{\text{i}}))$; D_{vm} is the effective moisture diffusivity in membranes (m^2/s); δ is membrane thickness (m). For hydrophobic membranes, vapor diffusion is the diffusion mechanism. Estimation of diffusivity is based on humidity differences. Equations for gas diffusivity in microporous membranes can be used to estimate the transport data. This has been discussed in previous chapters.

The equivalent moisture transfer coefficient in the solution side is defined by

$$\rho_{\text{a}} k_{\text{i}}^{\text{e}} \Delta\omega_{\text{s}} = \rho_{\text{s}} k_{\text{i}} \Delta X_{\text{s}} \tag{7.16}$$

or

$$k_{\text{i}}^{\text{e}} = \frac{\rho_{\text{s}}}{\rho_{\text{a}}} k_{\text{i}} K_{\text{p}} \tag{7.17}$$

where X_{s} is the mass fraction of water in solution (kg water/kg solution); ω is the humidity ratio (kg moisture/kg dry air); ω_{s} is the equilibrium humidity with solution at a given liquid temperature and concentration; K_{p} is the partition coefficient of a solution. It is the slope of the water uptake in solution to its equilibrium humidity.

$$K_{\text{p}} = \frac{\Delta X_{\text{s}}}{\Delta\omega_{\text{s}}} \tag{7.18}$$

Fiber arrays

Fibers can be arranged in three manners: randomly, in-line and staggered. The fibers arrays are schematically shown in Figure 7.4. Strictly speaking, all the fibers are randomly distributed because they are numerous. In-line or staggered arrays are just approximations.

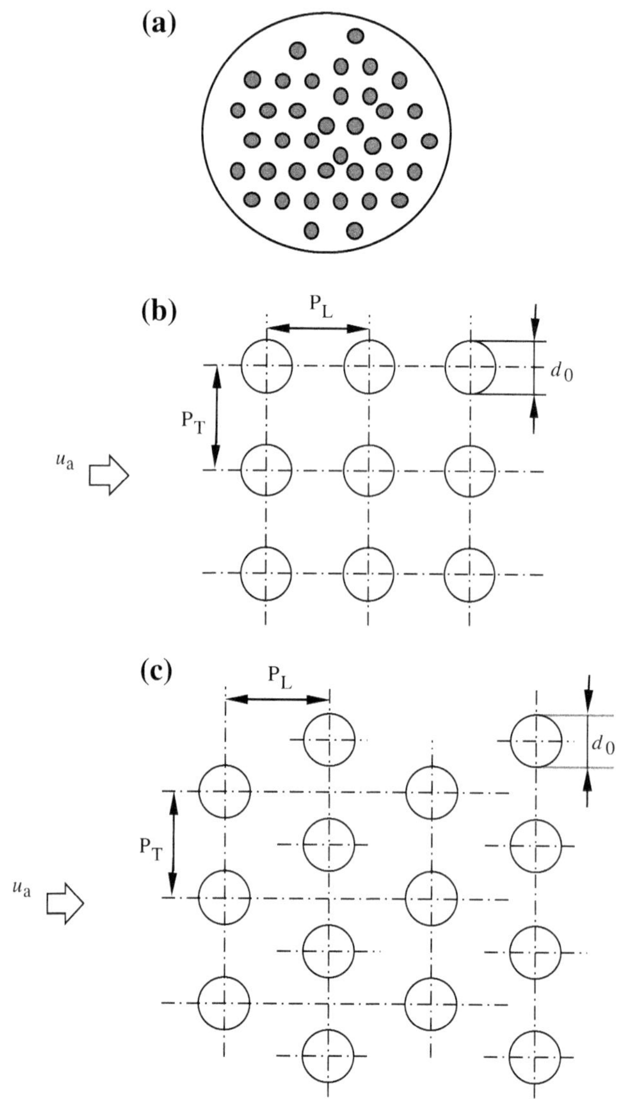

FIGURE 7.4

Fibers arrays in a hollow fiber membrane bundle. (a) Random; (b) in-line; (c) staggered.

Module effectiveness

Taking liquid air dehumidification as an example, performance is defined by the following. Sensible and latent (moisture) effectivenesses of the module are defined by

$$\varepsilon_s = \frac{T_{ao} - T_{ai}}{T_{si} - T_{ai}} \tag{7.19}$$

$$\varepsilon_L = \frac{\omega_{ao} - \omega_{ai}}{\omega_{si} - \omega_{ai}} \tag{7.20}$$

where T is temperature (K); subscripts "i" and "o" here mean inlet and outlet respectively. The latent effectiveness is also the air dehumidification effectiveness. The sensible effectiveness reflects a temperature change in the dehumidification process. Differently from sensible-only heat exchangers, absorption heat is released in dehumidification. As a result, sensible effectiveness may be negative in some cases, due to the absorption heat generated when the transfers of sensible heat and moisture are in different directions.

Enthalpy of the solution is a sum of the product of specific heat and temperature and the absorption heat

$$h_s = c_s T_s + X_s h_{abs} \tag{7.21}$$

where c_s is the specific heat of solution (kJ kg^{-1} K^{-1}); h_{abs} is absorption heat (kJ/kg); the reference states where enthalpy is equal to zero are taken to be dry at 0 °C and liquid water at 0 °C.

Equivalent enthalpy of the solution is the enthalpy of the air that is in equilibrium with the solution

$$h_s^e = c_{pa} T_s + \omega_s h_v \tag{7.22}$$

where h_v is the heat of evaporation of water at 0 °C (kJ/kg). Sometimes h_{abs} is also approximated by h_v, by ignoring the heat of mixing.

Enthalpy effectiveness of the module is defined by

$$\varepsilon_{ent} = \frac{h_{ao} - h_{ai}}{h_{si}^e - h_{ai}} \tag{7.23}$$

Substituting Eqs. (7.19), (7.20) and (7.22) into Eq. (7.23), it becomes

$$\varepsilon_{ent} = \frac{\varepsilon_s + H^* \varepsilon_L}{1 + H^*} \tag{7.24}$$

where

$$H^* = \frac{h_v\left(\omega_{si} - \omega_{ai}\right)}{c_{pa}\left(T_{si} - T_{ai}\right)} \tag{7.25}$$

As can be seen, the operating condition factor H^* is essentially a ratio of latent to sensible energy differences across the module, which is familiar to HVAC engineers.

7.3 Counter-flow arrangement

Heat and mass transfers in the hollow fiber membrane module play a key role in performance analysis and structural optimization. Regrettably the fundamental data on Nusselt and Sherwood numbers in the module are difficult to find. Several authors have investigated the transport phenomena of the absorption of CO_2, SO_2 and H_2S from flue gases by absorbents in hollow fiber membrane modules [8–10]; however, the heat transfer problem didn't attract enough attention in these studies. In contrast, the heat transfer problem in an air dehumidification process is a dominant factor. Furthermore, the well-established Nusselt correlations for traditional shell-and-tube heat exchangers are not suitable for this process because the boundary conditions are different. Moreover, the neighboring fluids are coupled together through the membranes. In this chapter, the results with laminar and counter flow are presented. To overcome the difficulties in the direct modeling of the numerous fibers, the Happel free surface model [11] is used to set up a representative cell in the module. The two streams in the cell are hydrodynamically fully developed, but developing both thermally and in concentration. The methodology is that the fluid flow is solved first. Then the Nusselt and Sherwood numbers under naturally formed boundary conditions are investigated.

7.3.1 Free surface model

The hollow fiber membrane module is used for liquid desiccant air dehumidification. The solution stream flows inside the fibers, while the air stream flows outside the fibers, as shown in Figure 7.5 in pictures of the real module. A direct modeling of the whole bundle on a fiber-to-fiber basis is difficult to perform because the fibers are numerous, usually 200 to 600 in a small shell 4 cm in diameter. To solve this problem, Happel's free surface model [11] is employed. According to this approach, each fiber in the shell space is assumed to be surrounded by a fluid envelope, and there is no momentum, heat or mass transfer on the outer surface. The fibers are distributed homogeneously and the flow is purely axial. The bundle is composed of a series of such free surface cells. Each cell has only one fiber in the

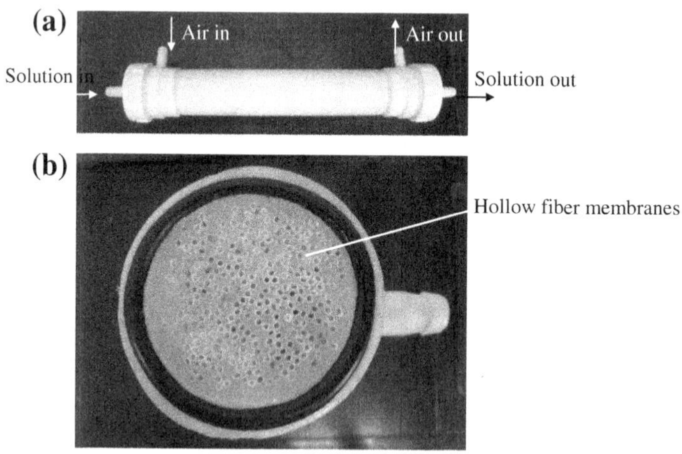

FIGURE 7.5

Structure of the hollow fiber membrane module and the cell with a free surface. (a) The shell structure; (b) the tube structure.

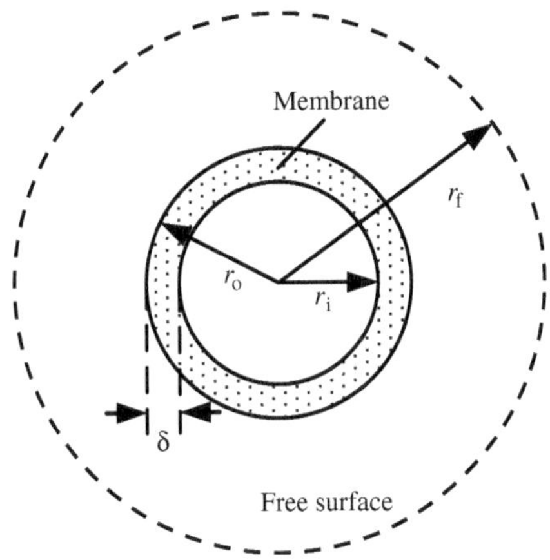

FIGURE 7.6

A free surface model. The representative cell with a free surface.

center and is surrounded by homogeneous fluid. The cross section of the tube channel is a circle, while the cross section of the shell-side channel is an annulus, as depicted in Figure 7.6. The free surface radius of a single fiber can be obtained by [11]

$$r_{\mathrm{f}} = r_{\mathrm{o}} \cdot \left(\frac{1}{\phi}\right)^{1/2} \tag{7.26}$$

where r_o is the fiber outer radius (m); ϕ is the packing fraction of the module, which has been defined by Eq. (7.9).

7.3.2 Governing equations in the cell

In the hollow fiber membrane module, the two streams flow in a counter-flow configuration. For reasons of symmetry and simplicity in calculation, one-fourth of a cell is selected as the calculation domain. The coordinate systems of the unit cell are shown in Figure 7.7. As can be seen, the air stream flows along the z axis in the annular channel, while the solution stream flows along the $-z$ axis in the circular channel. Heat and moisture can be exchanged through the membrane effectively. When water vapor is absorbed by the solution, absorption heat is released on the interface between the solution and the membrane.

In most practical applications, Reynolds numbers are usually much less than 2300, therefore the air and the solution streams are laminar. The fluids are Newtonian with constant thermal properties. The fibers are considered to be distributed evenly through the shell space, allowing the results obtained with a single fiber to be generalized to the entire module. The flows are assumed to be hydrodynamically fully developed, but developing both thermally and in concentration. Bulk transport is considered and axial heat and mass diffusion in the two flows are ignored.

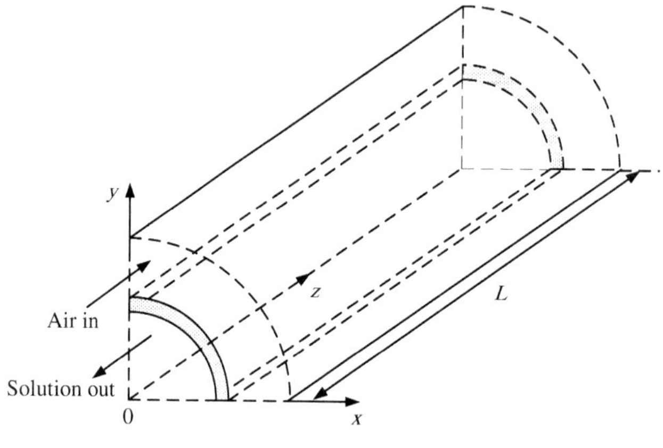

FIGURE 7.7

The coordinate system of the calculated domain, one-fourth of a free cell.

The governing momentum, energy and mass diffusion equations for the air stream are [2]

$$\mu_a \left(\frac{\partial^2 u_a}{\partial x^2} + \frac{\partial^2 u_a}{\partial y^2} \right) = \frac{dP_a}{dz} \tag{7.27}$$

$$\lambda_a \left(\frac{\partial^2 T_a}{\partial x^2} + \frac{\partial^2 T_a}{\partial y^2} \right) = \rho_a c_{pa} u_a \frac{\partial T_a}{\partial z} \tag{7.28}$$

$$D_{va} \left(\frac{\partial^2 \omega_a}{\partial x^2} + \frac{\partial^2 \omega_a}{\partial y^2} \right) = u_a \frac{\partial \omega_a}{\partial z} \tag{7.29}$$

where μ_a is the dynamic viscosity of air (Pa s); u_a is the air velocity (m/s); P_a is the pressure of the air stream (Pa); λ_a is the heat conductivity of air (kW m^{-1} K^{-1}); T_a is the temperature of air (K); c_{pa} is the specific heat of air (kJ kg^{-1} K^{-1}); D_{va} is the moisture diffusivity in air (m^2/s); ω_a is the humidity ratio of air (kg moisture/kg dry air).

The governing equations of momentum and heat transfer for the solution stream are in the same forms as those for the air stream. However, the thermo-physical properties of the LiCl solution are used to replace those of air, and coordinate z in Eqs. (7.27) and (7.28) is replaced by ($L-z$).

The mass transfer equation for the solution stream is

$$D_{ws} \left(\frac{\partial^2 X_s}{\partial x^2} + \frac{\partial^2 X_s}{\partial y^2} \right) = -u_s \frac{\partial X_s}{\partial z} \tag{7.30}$$

where the mass traction of solute is 1-X_s.

Normalization of equations

The governing equations for the air stream, Eqs. (7.27) to (7.30), can be normalized as

$$\frac{\partial^2 u_a^*}{\partial x^{*2}} + \frac{\partial^2 u_a^*}{\partial y^{*2}} = -\frac{r_f^2}{D_{h,a}^2} \tag{7.31}$$

$$\frac{\partial^2 \theta_a}{\partial x^{*2}} + \frac{\partial^2 \theta_a}{\partial y^{*2}} = U_a \frac{\partial \theta_a}{\partial z_{h,a}^*} \tag{7.32}$$

$$\frac{\partial^2 \xi_a}{\partial x^{*2}} + \frac{\partial^2 \xi_a}{\partial y^{*2}} = U_a \frac{\partial \xi_a}{\partial z_{m,a}^*} \tag{7.33}$$

$$\frac{\partial^2 \Theta}{\partial x^{*2}} + \frac{\partial^2 \Theta}{\partial y^{*2}} = -U_s \frac{\partial \Theta}{\partial z_{m,s}^*} \tag{7.34}$$

where subscripts "a" and "s" refer to the air and the solution streams, respectively.

The dimensionless velocity is defined by

$$u^* = -\frac{\mu u}{D_h^2 \frac{dP}{dz}} \tag{7.35}$$

and the dimensionless temperature is defined by

$$\theta = \frac{T - T_{ai}}{T_{si} - T_{ai}} \tag{7.36}$$

where T_{ai} is the inlet temperature of air (K), and T_{si} is the inlet temperature of solution (K).

The dimensionless humidity is defined by

$$\xi = \frac{\omega - \omega_{ai}}{\omega_{si} - \omega_{ai}} \tag{7.37}$$

where ω_{ai} is the inlet humidity of air (kg/kg), and ω_{si} is the equilibrium air humidity with the solution at the inlet temperature (T_{si}) and mass fraction (X_{si}).

The dimensionless mass fraction of solution is defined by

$$\Theta = \frac{X - X_{ei}}{X_{si} - X_{ei}} \tag{7.38}$$

where X_{si} is the inlet mass fraction of solution (kg water/kg solution), and X_{ei} is the equilibrium mass fraction of the solution with the air stream at the inlet temperature (T_{ai}) and humidity (ω_{ai}).

Dimensionless coordinates

$$x^* = \frac{x}{r_f} \tag{7.39}$$

$$y^* = \frac{y}{r_f} \tag{7.40}$$

$$z_h^* = \frac{z_1}{RePrD_h} \tag{7.41}$$

$$z_m^* = \frac{z_1}{ReScD_h} \tag{7.42}$$

where z_1 is coordinate of duct (m). It is z for the air stream and $(L-z)$ for the solution stream. Pr and Sc are the values of air or the liquid for the shell side or tube side respectively.

In Eqs. (7.32) to (7.34), dimensionless velocity coefficients U are defined by

$$U = \frac{u^*}{u_m^*}\frac{r_f^2}{D_h^2} \tag{7.43}$$

where u_m^* is the average dimensionless velocity on a cross section, which is equal to

$$u_m^* = \frac{\iint u^* \mathrm{d}A}{A_c} \tag{7.44}$$

The characteristic of the fluid flow in the channels can be represented by the product of the friction factor and the Reynolds number as

$$(fRe) = \left(\frac{-D_h \frac{\mathrm{d}P}{\mathrm{d}z}}{\rho u_m^2/2}\right)\left(\frac{\rho D_h u_m}{\mu}\right) = \frac{2}{u_m^*} \tag{7.45}$$

Dimensionless bulk temperature

$$\theta_b(z_h^*) = \frac{\iint u^* \theta \mathrm{d}A}{\iint u^* \mathrm{d}A} \tag{7.46}$$

Dimensionless bulk humidity

$$\xi_b(z_m^*) = \frac{\iint u^* \xi \mathrm{d}A}{\iint u^* \mathrm{d}A} \tag{7.47}$$

Dimensionless bulk mass fraction of solution

$$\Theta_b(z_m^*) = \frac{\iint u^* \Theta \mathrm{d}A}{\iint u^* \mathrm{d}A} \tag{7.48}$$

The peripherally local and mean Nusselt numbers can be obtained by [2]

$$Nu_L = -\frac{1}{4(\theta_w - \theta_b)}\frac{\mathrm{d}\theta_b}{\mathrm{d}z_h^*} \tag{7.49}$$

$$Nu_{\mathrm{m}} = \frac{1}{z_{\mathrm{h}}^{*}} \int_{0}^{z_{\mathrm{h}}^{*}} Nu_{\mathrm{L}} \mathrm{d}z_{\mathrm{h}}^{*} \tag{7.50}$$

where subscripts "w" and "b" mean "wall" and "bulk," respectively.

Similarly, a mass balance on a control volume for the air stream will be the following equations

$$Sh_{\mathrm{L}} = -\frac{1}{4(\xi_{\mathrm{w}} - \xi_{\mathrm{b}})} \frac{\mathrm{d}\xi_{\mathrm{b}}}{\mathrm{d}z_{\mathrm{m}}^{*}} \tag{7.51}$$

$$Sh_{\mathrm{m}} = \frac{1}{z_{\mathrm{m}}^{*}} \int_{0}^{z_{\mathrm{m}}^{*}} Sh_{\mathrm{L}} \mathrm{d}z_{\mathrm{m}}^{*} \tag{7.52}$$

Similarly, a mass balance on the control volume for the solution stream will be

$$Sh_{\mathrm{L}} = -\frac{1}{4(\Theta_{\mathrm{w}} - \Theta_{\mathrm{b}})} \frac{\mathrm{d}\Theta_{\mathrm{b}}}{\mathrm{d}z_{\mathrm{m}}^{*}} \tag{7.53}$$

$$Sh_{\mathrm{m}} = \frac{1}{z_{\mathrm{m}}^{*}} \int_{0}^{z_{\mathrm{m}}^{*}} Sh_{\mathrm{L}} \mathrm{d}z_{\mathrm{m}}^{*} \tag{7.54}$$

Under uniform temperature (concentration) boundary condition, the fully developed value is denoted as Nu_{T} (Sh_{T} for mass). Under uniform heat flux (mass flux) boundary condition, it is denoted as Nu_{H} (Sh_{H} for mass). In this conjugate case, under the naturally formed boundary conditions, the fully developed value is denoted as Nu_{C} (Sh_{C} for mass transfer).

Boundary conditions

$$\text{On membrane surface, } u^{*} = 0 \tag{7.55}$$

Inlet conditions for the air stream

$$z_{\mathrm{h,a}}^{*} = 0, \quad \theta_{\mathrm{a}} = 0 \tag{7.56}$$

$$z_{\mathrm{m,a}}^{*} = 0, \quad \xi_{\mathrm{a}} = 0 \tag{7.57}$$

Inlet conditions for the solution stream

$$z_{\mathrm{h,s}}^{*} = 0, \quad \theta_{\mathrm{s}} = 1 \tag{7.58}$$

$$z_{\mathrm{m,s}}^{*} = 0, \quad \Theta = 1 \tag{7.59}$$

Other surfaces, including the free surface, the symmetric boundary conditions

$$\frac{\partial \psi}{\partial n} = 0 \tag{7.60}$$

where ψ stands for variables such as velocity, temperature, humidity or mass frication, and n is the normal direction to the boundaries.

Conjugate boundary conditions on membrane surfaces

Since the membrane is rather thin (150 μm), temperature differences in the membrane thickness are so small that they can be ignored [2,7]. When water vapor is absorbed by the liquid desiccant, absorption heat is released on the membrane surface on the solution side. The solution-side heat flux is the sum of the air-side heat flux and the absorption (latent) heat flux. Then the normalized heat balance equation on the membrane surfaces between the air and the solution streams is

$$\lambda^* \frac{\partial \theta_a}{\partial n}\bigg|_{\text{surface, a}} + h^*_{\text{abs}} \frac{\partial \xi_a}{\partial n}\bigg|_{\text{surface, a}} = \frac{\partial \theta_s}{\partial n}\bigg|_{\text{surface, s}} \tag{7.61}$$

where the dimensionless absorption heat and the dimensionless heat conductivity are defined by

$$h^*_{\text{abs}} = \frac{\rho_a D_{va} h_{\text{abs}}}{\lambda_s}\left(\frac{\omega_{si} - \omega_{ai}}{T_{si} - T_{ai}}\right) \tag{7.62}$$

$$\lambda^* = \frac{\lambda_a}{\lambda_s} \tag{7.63}$$

Heat flux on membrane surfaces on the air and the solution sides is equal to

$$q_h = -\lambda \frac{\partial T}{\partial n}\bigg|_{\text{surface}} \tag{7.64}$$

Since temperature and humidity are strongly coupled between the two streams and the membrane, the temperature of membrane surfaces is not a fixed value across the surface.

Mass boundary conditions on membrane surfaces

$$\text{Air-side membrane surface,} \quad q_m(x_1^*, y_1^*, z_G^*) = m_v \tag{7.65}$$

$$\text{Solution-side membrane surface,} \quad q_m(x_2^*, y_2^*, z_G^*) = m_v \tag{7.66}$$

where subscripts "1" and "2" stand for the membrane surfaces on the air side and the solution side, respectively.

Parameter z_G^* is the dimensionless geometric position

$$z_G^* = \frac{z}{L} \tag{7.67}$$

where $\dot{m}_v$ is the moisture emission rate through the membrane, which is determined by the diffusion equation in membrane as

$$\dot{m}_v = \rho_a D_{vm} \frac{\omega_{m1} - \omega_{m2}}{\delta} \tag{7.68}$$

where D_{vm} is moisture diffusivity in membrane (m^2/s); δ is membrane thickness (m). Moisture diffusivity is different from water diffusivity in membranes, which has been discussed in previous chapters. Moisture diffusivity is calculated based on humidity differences.

On the other hand, moisture emission on the membrane surface on the air side is calculated by

$$q_m = -\rho_a D_{va} \left. \frac{\partial \omega_a}{\partial n} \right|_{\text{surface, a}} \tag{7.69}$$

Correspondingly, moisture emission on the membrane surface on the solution side is obtained by

$$q_m = -\rho_s D_{ws} \left. \frac{\partial X_s}{\partial n} \right|_{\text{surface, s}} \tag{7.70}$$

Moisture emission rate q is not a constant value. It is different from point to point on the membrane surfaces along the channels.

Besides the governing equations for fluid flow and membrane, the state equations for the salt solution, which has been described in Chapter 6, should also be combined in the model.

7.3.3 Numerical scheme

A boundary-fitted coordinate system is used to convert the physical plane onto a square numerical plane [2]. The method to transform the coordinates has been described in detail in Chapter 2.

After the set-up of the boundary-fitted coordinate systems, the governing dimensionless equations are transformed to the computational plane, and are then discretized by means of a finite volume method. Since the two streams and the membrane closely interact, and temperature and humidity are also related to each other, ADI (alternate direction implicit) techniques are used for solving these equations. Iterations are performed until all the governing equations are satisfied simultaneously.

To ensure the accuracy of the results calculated, numerical tests were conducted to check the effects of the grid size. They indicate that 41×41 grids in the x–y plane and 40 in the z axis are adequate, showing less than 1.0% difference compared with $51 \times 51 \times 60$ grids. The final numerical uncertainty is 1.0%.

7.3.4 Experimental work

In order to study the heat and mass transfer in the hollow fiber membrane module for liquid desiccant air dehumidification, a continuous air dehumidification system has been designed. The whole test set-up has been described in the previous chapter. Two hollow fiber membrane modules are constructed to replace the previously used plate-type modules. There are two hollow fiber membrane modules, one for air dehumidification and the other for solution regeneration. The structures of the two modules are the same. The former is called a dehumidifier, while the latter is called a regenerator. The structure is shown in Figure 7.5.

There are many components in this test system; however, we are concerned only with heat and mass transfer in the dehumidifier. The dehumidifier is like a shell-and-tube heat exchanger. Liquid desiccant flows inside the tubes, while the air stream flows on the shell side in a counter-flow arrangement. The fibers are made up of a layer of modified PVDF porous membranes which only permit the permeation of water vapor from air to the liquid solution. The tested thermodynamic properties of the membrane are summarized in Table 7.1. Some transport properties measured in the laboratory are also listed.

The test procedures and the uncertainties are the same as before. Inlet and outlet temperature, humidity, flow rate, and mass fraction for the solution are measured. The performances are calculated. Figure 7.8

Table 7.1 Membrane properties and some transport properties for the hollow fiber membrane bundle.

Symbol	Unit	Value	Symbol	Unit	Value
D_o	cm	4.0	ρ_s	kg/m^3	1245
L	cm	30.0	Re_a		160
ϕ		0.285	Re_s		3
n_f		200	Pr_a		0.71
r_i	μm	600	Pr_s		28.36
r_o	μm	750	Sc_a		0.564
r_f	μm	1414	Sc_s		1390
D_{vm}	m^2/s	1.2×10^{-7}	T_{ai}	°C	35.0
D_{va}	m^2/s	2.82×10^{-5}	T_{si}	°C	25.0
D_{ws}	m^2/s	3.0×10^{-9}	ω_{ai}	kg/kg	0.021
δ	μm	140	ω_{si}	kg/kg	0.0055
λ_a	Wm^{-1}K^{-1}	0.0263	X_{si}	kg/kg	0.65
λ_s	Wm^{-1}K^{-1}	0.5	X_{ei}	kg/kg	0.762
Porosity		0.63	Pore size	μm	0.45

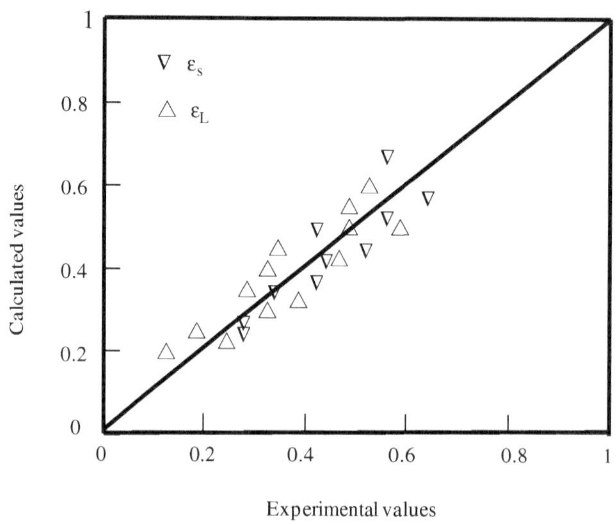

FIGURE 7.8

Comparison of the calculated and tested data for sensible and latent effectiveness.

shows the comparison of calculated and tested data for sensible and latent effectiveness. As can be seen, the predictions are satisfactory.

Numerical validation

The friction factor and the Nusselt numbers for channels of annular cross sections with a T or H boundary condition on the inner surface and an adiabatic boundary condition on the outer surface are given in Refs. [1,12]. They are compared in this study. The boundary conditions in the code are revised to be T or H conditions at the inner surface but adiabatic at the outer. For fully developed laminar flow in channels, (fRe) and Nu_T (under uniform temperature conditions) are constants. They are compared with the data in Refs. [1,12] in Table 7.2. As can be seen, the maximum difference is less than 0.8%, which means that the code is successful in predicting the flows.

7.3.5 Nusselt and Sherwood numbers

The peripherally local and mean Nusselt and Sherwood numbers for the two fluids along the axial directions are shown in Figures 7.9 and 7.10 respectively. The local Nusselt and Sherwood numbers decrease and approach the lower limiting values. For heat transfer, the flow is developed. However, for mass transfer, especially for the liquid, the boundary layers develop slowly.

The developing length for the air stream in the annular duct ($z^*_{h,a} = 0.2$) is nearly equal to that for the solution stream in the circular duct ($z^*_{h,s} = 0.21$). The two ducts are long enough ($L = 30\,\text{cm}$, $z^*_{h,a} = 2.1, z^*_{h,s} = 2.15$) for the two streams to get thermally fully developed.

Table 7.2 Comparisons of (fRe) and Nu_T of fully developed flows for two ducts [2].

Shape	r_o/r_f	Nu_T			(fRe)		
		Refs. [1,12]	(This study)	Error (%)	Refs. [1,12]	(This study)	Error (%)
Circular	–	3.66	3.64	0.55	64.00	64.11	0.17
Annular	0.05	17.46	17.51	0.29	86.27	85.99	0.33
Annular	0.10	11.56	11.54	0.17	89.37	89.16	0.24
Annular	0.25	7.37	7.40	0.41	93.21	92.95	0.28
Annular	0.40	6.15	6.19	0.65	94.73	94.47	0.28
Annular	0.45	–	5.99	–	95.01	94.62	0.41
Annular	0.50	5.74	5.77	0.52	95.25	95.05	0.21
Annular	0.53	–	5.68	–	95.36	95.18	0.19
Annular	0.60	5.45	5.48	0.55	95.59	95.39	0.21
Annular	0.65	–	5.38	–	95.71	95.40	0.33
Annular	0.70	5.23	5.26	0.57	95.80	95.61	0.20
Annular	0.75	5.15	5.19	0.77	95.87	95.72	0.16
Annular	0.80	–	5.12	–	95.92	95.74	0.19
Annular	0.83	–	5.07	–	95.95	95.78	0.18
Annular	1.00	4.86	–	–	96.00	–	–

Notes: The hydrodynamic diameter here is defined by 4 times the flow area divided by the perimeter of the fiber and the outer wall (except the circular duct).

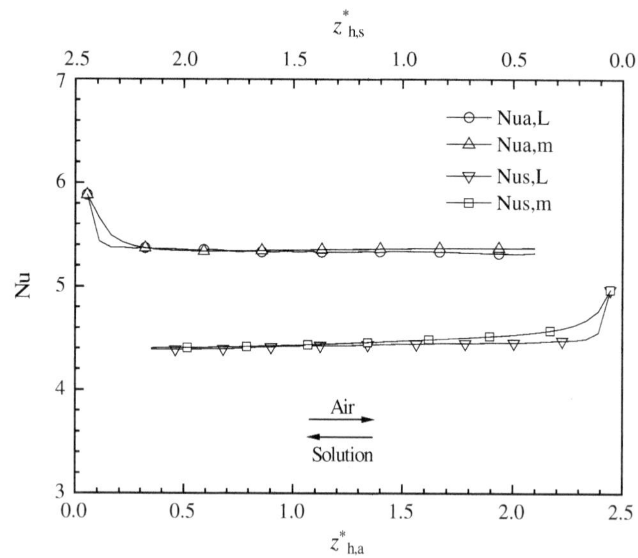

FIGURE 7.9

Local and mean Nusselt numbers for both air and solution streams along the axial directions, $r_o/r_f = 0.53$.

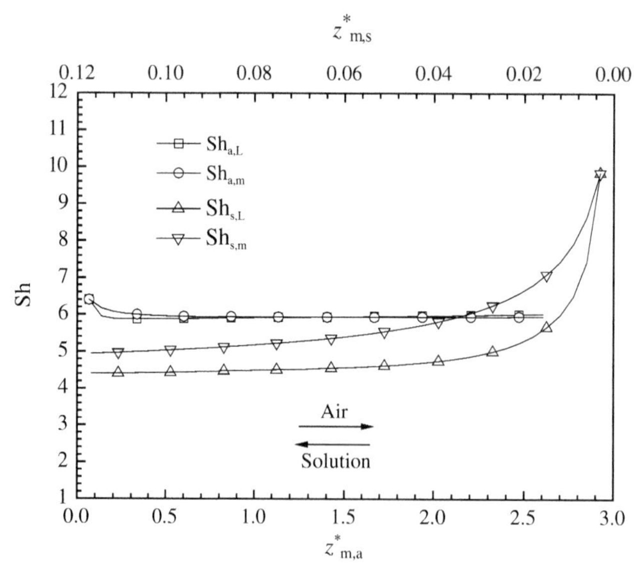

FIGURE 7.10

Local and mean Sherwood numbers for both air and solution streams along the axial directions, $r_o/r_f = 0.53$.

The fully developed Nusselt numbers for the circular channel (liquid side) under these naturally formed boundary conditions is 4.38, which is nearly equal to Nu_H (4.36, Ref. [1]), but larger than Nu_T (3.66, Ref. [1]). The values of $Nu_{C,a}$ under various radius ratios (r_o/r_f) for the annular ducts (air side) are listed in Table 7.3. The packing fractions of the module (ϕ) are commonly used values. Also listed are the calculated Nusselt numbers for the annular channel under uniform temperature boundary conditions (Nu_T) and under uniform heat flux boundary conditions (Nu_H). In these two cases, the annular tube has a free outer surface, but the inner surface is set to uniform temperature (T) or uniform heat flux (H) boundary conditions. Air-side $Nu_{C,a}$ is between Nu_H and Nu_T. $Nu_{C,a}$ is closer to Nu_H. Further, $Nu_{C,a}$ decreases with an increase in the packing fraction of the module. For the annular channel, (fRe) in this boundary condition is much less than that under the zero-velocity boundary conditions.

Variations of the Sherwood numbers are similar to the Nusselt numbers. At the entrances of the two channels, the Sherwood numbers for the solution stream are much higher than those for the air stream. This is because the entry length for the solution stream is longer than that for the air stream. The entry length for the solution stream is $z^*_{m,s} = 0.1$ ($L = 25.5$ cm). The entry length for the air stream is $z^*_{m,a} = 0.2$ (L=2.3 cm). The solution has a longer entry length because the Schmidt number of the LiCl solution (Sc=1390) is larger. On the other hand, the two channels are long enough (L=30 cm, $z^*_{m,a}$=2.6, $z^*_{m,s}$ =0.117) for the two streams to get fully developed in concentration.

The detailed Sh_C for the two channels under various radius ratios are also listed in Table 7.3. As seen, for the annular channel, the fully developed Sherwood number on the air side ($Sh_{C,a}$) is somewhat less than the air-side $Nu_{C,a}$. The higher the packing fraction is, the less the air-side $Sh_{C,a}$ is. However, for the circular channel, the fully developed Sherwood number on the solution side ($Sh_{C,s}$=4.46) is somewhat higher than the solution-side $Nu_{C,s}$. The solution-side $Sh_{C,s}$ is almost unchanged with an increase in the packing fraction.

Table 7.3 Fully developed (fRe), Nusselt and Sherwood numbers for the annular duct with a free surface, counter flow arrangement.

Packing fraction (ϕ)	r_o/r_f	Nu_T	Nu_H	$Nu_{C,a}$	$Sh_{C,a}$	(fRe)
0.063	0.25	35.138	37.331	37.082	36.384	0.556
0.162	0.40	19.200	20.803	20.698	20.350	1.377
0.203	0.45	16.870	18.254	18.158	17.933	1.691
0.253	0.50	14.911	16.225	16.136	15.956	2.060
0.281	0.53	14.115	15.356	15.298	15.067	2.258
0.360	0.60	12.427	13.600	13.520	13.280	2.791
0.422	0.65	11.478	12.570	12.494	12.341	3.187
0.491	0.70	10.704	11.674	11.650	11.529	3.607
0.563	0.75	10.054	11.011	10.987	10.824	4.026
0.641	0.80	9.536	10.413	10.346	10.233	4.460
0.689	0.83	9.260	10.120	10.054	9.921	4.715

Notes: The annular duct here has an outer free surface. The inner surface is uniform temperature (T), uniform heat flux (H) or naturally formed boundary conditions (C). The hydrodynamic diameter is defined by 4 times the free cell surface area divided by the perimeter of the fiber.

As mentioned, the heat flux on the solution side consists of the air-side heat flux and the latent or absorption heat flux. The latent heat flux has an impact on the Nusselt numbers for the two channels [2]. Heat and mass transfer can be in the same or opposite directions. The larger the $q_{Lat,m}/q_{h,a,m}$ is, the higher the air-side $Nu_{C,a}$ is. However, the solution-side $Nu_{C,s}$ is almost unchanged when $q_{Lat,m}/q_{h,a,m}$ is varied. These results are caused by the close couplings of the heat and mass transfer between the two fluids on the membrane.

The boundary conditions on the membrane surfaces are neither uniform temperature nor uniform heat flux. The boundary conditions on the membrane surfaces are naturally formed boundary conditions. Further, due to the absorption heat generated on the solution side, the heat flux is different on the two membrane surfaces. The peripherally mean heat flux ($q_{h,s}$) on the solution side is much larger than that in the air side, which implies that the latent heat transfer is much more effective than the sensible convective heat transfer between the two fluids.

The humidity on the two membrane surfaces and the mass flux change from point to point along the z axis. It is clear that the membrane surface concentration boundaries are neither uniform concentration nor uniform mass flux boundary conditions. It should be noted that the humidity values on the two surfaces are different, while the mass flux at the same point on the two surfaces is the same. This is the opposite to heat transfer.

7.3.6 Summary of transport data

Air side: The boundary conditions are neither uniform temperature/concentration nor uniform heat/mass flux boundary conditions. They are developed shortly after the inlets. The fully developed Nusselt numbers for air-stream $Nu_{C,a}$ are between Nu_H and Nu_T. $Nu_{C,a}$ is closer to Nu_H. The Sherwood numbers ($Sh_{C,a}$) are somewhat less than the air-side $Nu_{C,a}$. Both $Nu_{C,a}$ and $Sh_{C,a}$ decrease with an increase in packing fraction.

Solution side: The boundary conditions are non-uniform value and non-uniform flux boundary conditions as well. Though both heat transfer and mass transfer are fully developed at the outlets, the heat transfer boundary develops much faster than mass transfer. The solution-side $Nu_{C,s}$ is nearly equal to Nu_H. The Sherwood numbers ($Sh_{C,s}$) are a bit larger than $Nu_{C,s}$. Both $Nu_{C,s}$ and $Sh_{C,s}$ are almost unchanged when the packing fraction is varied. The obtained data can be used for future system design.

Membrane: The temperature values on the two membrane surfaces are the same. However, heat flux is different due to absorption heat on the solution side. The humidity fluxes on the two surfaces are the same; however, the humidity values are different due to high resistance in mass transfer through the membranes.

7.4 Cross-flow arrangement with laminar flow, free surface cell model

Counter flow is the most commonly used arrangement for membrane contactors. Heat and mass transfer in such contactors have been thoroughly investigated. However, this structure has a shortcoming: The air pressure drops are high, especially for air conditioning applications where air flow rates are high. To address this problem, a cross-flow membrane contactor, as depicted in Figure 7.11, has been proposed and investigated for liquid desiccant air dehumidification.

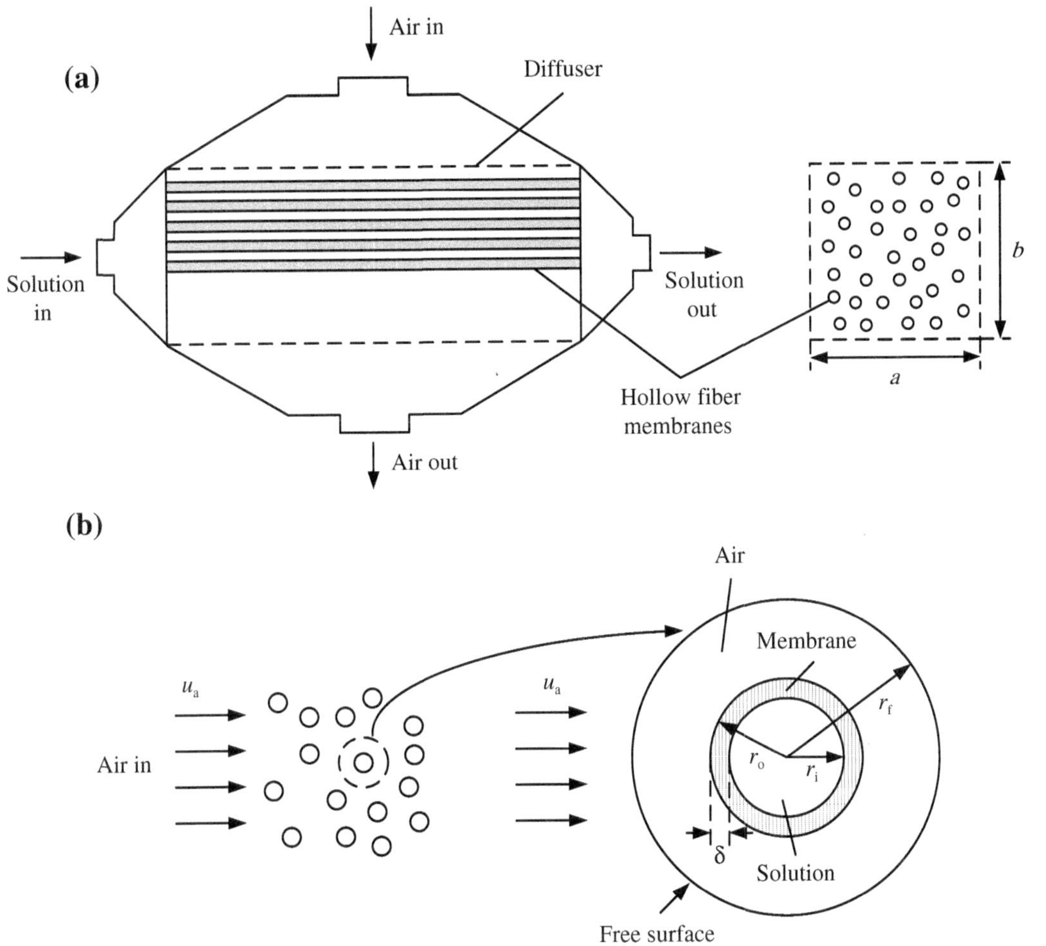

FIGURE 7.11

Schematic of a cross-flow hollow fiber membrane contactor. (a) The shell and the tube structure; (b) the free surface cell.

The structure is similar to a cross-flow shell-and-tube heat exchanger, which has been well-studied. The resistance and Nusselt numbers for a cross-flow tube bank under uniform temperature or uniform heat flux boundary conditions can be found in several well-known textbooks [1,13,14]. However, these data and the available correlations are not suitable for the membrane contactor for air dehumidification because the boundary conditions are different. The boundary conditions here are neither uniform heat flux nor uniform temperature. Rather, it's a conjugate problem. Finally, the heat and mass transfer are strongly coupled because of vapor condensation on the membrane surface.

In this section, the fluid flow and the coupled heat and mass transfer in the cross-flow hollow fiber membrane contactor are investigated. As in the previous section, Happel's free surface model is employed to simplify the analysis and to build the calculation domain.

7.4.1 Mathematical model

Happel's free surface model is employed. According to this approach, the bundle is assumed to be composed of a series of frictionless free surface cells. Each cell has only one fiber in the center and is surrounded by an air envelope. The cross sections of the conjugated tube channel and air flow channel (separated and coupled by the membrane) are a circle and an annulus respectively, as depicted in Figure 7.11.

In the hollow fiber membrane contactor, the two flows are in a cross-flow arrangement. For reasons of symmetry, only half a cell is selected as the calculation domain. The coordinate system of the unit cell is shown in Figure 7.12. Here the solution stream flows along the z axis in the inner circular channel, while the air stream flows over the fiber, which is oriented normal to the air flow. The air stream approaches the fiber with a uniform velocity u_a, at a uniform temperature T_{ai} and a uniform humidity ω_{ai}. Other boundaries are similar to a counter flow.

For the air flow across the fiber, normalized governing equations of the fluid flow, heat and mass are described as follows [3]:

$$\frac{\partial u_x^*}{\partial x^*} + \frac{\partial u_y^*}{\partial y^*} = 0 \tag{7.71}$$

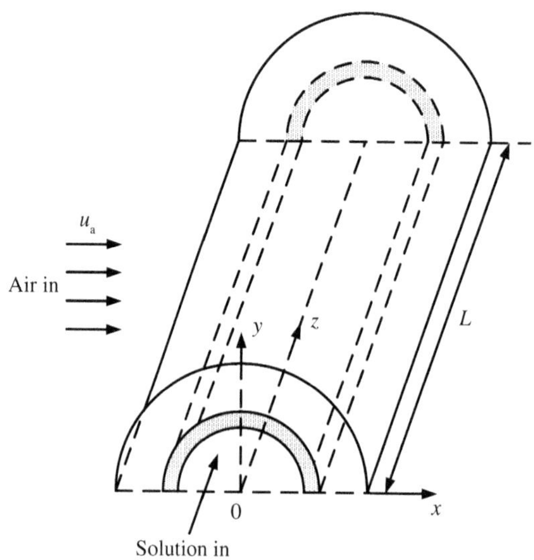

FIGURE 7.12

The coordinate system of the calculated domain of a cross-flow fiber with the free cell model.

$$u_x^* \frac{\partial u_x^*}{\partial x^*} + u_y^* \frac{\partial u_x^*}{\partial y^*} = -\frac{1}{2} \frac{\partial P_a^*}{\partial x^*} + \frac{2r_o}{r_f} \cdot \frac{1}{Re_a} \cdot \left(\frac{\partial^2 u_x^*}{\partial x^{*2}} + \frac{\partial^2 u_x^*}{\partial y^{*2}} \right) \tag{7.72}$$

$$u_x^* \frac{\partial u_y^*}{\partial x^*} + u_y^* \frac{\partial u_y^*}{\partial y^*} = -\frac{1}{2} \frac{\partial P_a^*}{\partial y^*} + \frac{2r_o}{r_f} \cdot \frac{1}{Re_a} \cdot \left(\frac{\partial^2 u_y^*}{\partial x^{*2}} + \frac{\partial^2 u_y^*}{\partial y^{*2}} \right) \tag{7.73}$$

$$u_x^* \frac{\partial \theta_a}{\partial x^*} + u_y^* \frac{\partial \theta_a}{\partial y^*} = \frac{2r_o}{r_f} \cdot \frac{1}{Re_a Pr_a} \cdot \left(\frac{\partial^2 \theta_a}{\partial x^{*2}} + \frac{\partial^2 \theta_a}{\partial y^{*2}} \right) \tag{7.74}$$

$$u_x^* \frac{\partial \xi_a}{\partial x^*} + u_y^* \frac{\partial \xi_a}{\partial y^*} = \frac{2r_o}{r_f} \cdot \frac{1}{Re_a Sc_a} \cdot \left(\frac{\partial^2 \xi_a}{\partial x^{*2}} + \frac{\partial^2 \xi_a}{\partial y^{*2}} \right) \tag{7.75}$$

where subscripts "x," "y" and "a" refer to the x axis, y axis and the air stream, respectively; superscript "*" represents the dimensionless form; u is velocity (m/s); P is pressure (Pa); θ and ξ are dimensionless temperature and humidity, respectively; Re, Pr and Sc are Reynolds, Prandtl and Schmidt numbers, respectively.

For the solution in the circular channel (fiber tube), the governing equations of the fluid flow, heat and mass are the same as in the previous section.

The dimensionless velocity for the air stream in the x axis direction is defined by

$$u_x^* = \frac{u_x}{u_{ai}} \tag{7.76}$$

and the dimensionless velocity for the air stream in the y axis direction is defined by

$$u_y^* = \frac{u_y}{u_{ai}} \tag{7.77}$$

The Reynolds number for the air stream is defined by [3]

$$Re_a = \frac{\rho_a u_{ai} D_{h,a}}{\mu_a} \tag{7.78}$$

where $D_{h,a} = 2r_o$ is the hydraulic diameter for the air flow channel (m). This is different from the previous section for counter flow.

The dimensionless pressure for the air stream is defined by

$$P_a^* = \frac{P_a}{\frac{1}{2}\rho_a u_{ai}^2} \tag{7.79}$$

The definitions of dimensionless temperature, humidity, mass fraction, and coordinates are the same as in the previous section.

The pressure drag coefficient for the air flow across the fiber can be calculated by [3]

$$C_{Dp} = \int_0^{\pi} P_a^* \bigg|_{r=r_o} \cos \beta \, d\beta \tag{7.80}$$

The friction drag coefficient for the air flow is equal to [3]

$$C_{\mathrm{Df}} = \left(\frac{2D_{\mathrm{h}}}{r_{\mathrm{f}}}\frac{1}{Re}\right)_{\mathrm{a}} \int_0^{\pi} \left.\frac{\partial u_{\beta}^{*}}{\partial n}\right|_{r=r_{\mathrm{o}}} \sin\beta \mathrm{d}\beta \tag{7.81}$$

where subscript "β" means the tangential direction; r is the radius, ranging from 0 to r_{f} (m); n is normal direction to the boundaries.

Then the total drag coefficient for the air flow can be obtained by

$$C_{\mathrm{D}} = C_{\mathrm{Dp}} + C_{\mathrm{Df}} \tag{7.82}$$

The pressure drop for the air flow across a single fiber can be calculated by

$$\Delta P = C_{\mathrm{D}} \cdot \frac{1}{2}\rho_{\mathrm{a}} u_{\mathrm{ai}}^{2} \tag{7.83}$$

The local Nusselt numbers are varied along the membrane surface on the air side. The angularly local Nusselt and Sherwood numbers for the air stream at a tangential position are [3]:

$$Nu_{\beta,\mathrm{a}} = -\frac{D_{\mathrm{h,a}}}{r_{\mathrm{f}}} \frac{\left.\frac{\partial \theta_{\mathrm{a}}}{\partial n}\right|_{r=r_{\mathrm{o}}}}{\theta_{\mathrm{w,a}} - \theta_{\mathrm{b,a}}} \tag{7.84}$$

$$Sh_{\beta,\mathrm{a}} = -\frac{D_{\mathrm{h,a}}}{r_{\mathrm{f}}} \frac{\left.\frac{\partial \xi_{\mathrm{a}}}{\partial n}\right|_{r=r_{\mathrm{o}}}}{\xi_{\mathrm{w,a}} - \xi_{\mathrm{b,a}}} \tag{7.85}$$

For the air stream, the overall mean values across a fiber are:

$$Nu_{\mathrm{ave,a}} = \frac{1}{\pi}\int_0^{\pi} Nu_{\beta,\mathrm{a}} \mathrm{d}\beta \tag{7.86}$$

$$Sh_{\mathrm{ave,a}} = \frac{1}{\pi}\int_0^{\pi} Sh_{\beta,\mathrm{a}} \mathrm{d}\beta \tag{7.87}$$

The calculation of friction factors, Nusselt and Sherwood numbers for the solution in the fibers are the same as in the previous section.

Boundary conditions

A body-fitted coordinates transformation method is used to transfer the round physical shape to a rectangular calculating shape [3]. The boundary conditions of velocity for the air stream based on Happel's free surface model are [2,3]

$$\text{On membrane surface,} \quad u_x^{*} = 0 \quad \text{and} \quad u_y^{*} = 0 \tag{7.88}$$

$$\text{On free surface,} \quad u_{\mathrm{r}}^{*} = -\cos\beta \quad \text{and} \quad \tau_{1\beta} = 0 \tag{7.89}$$

where subscript "r" represents the radial direction; τ_1 is shear stress (Pa).

Other boundary conditions, and the conjugate boundary conditions on the membrane surface, are similar to those for counter flow.

7.4.2 The module geometry

With numerical schemes and experimental validation techniques similar to those in the previous section, the code has been verified. Then the Nusselt and Sherwood numbers for the cross flow arrangements are calculated. The transport parameters of the module are similar to the previous counter-flow module. However, the counter-flow arrangements are used. Figure 7.13 shows the modeled hollow fiber membrane module for the cross flow. The geometries for the module are: $a=b=200\,\text{mm}$, $L=360\,\text{mm}$, $\phi=0.265$, $n_f=6000$.

7.4.3 Drag coefficient and Nusselt numbers

It has been found that the boundary conditions on the membrane surfaces are neither uniform temperature (concentration) nor uniform heat flux (mass flux) boundary conditions. They are naturally formed by the coupling between the two flows. So the Nusselt and Sherwood numbers calculated under uniform temperature or uniform heat flux boundary conditions are not appropriate to be adopted here. The values will be given below.

The flow in the circular channel (fiber tube) is the same as that in a common round tube [13,14]. The velocity vectors for the air flow across the fiber are plotted in Figure 7.14. As can be seen, the velocity vectors are zero on the membrane surface and they rise sharply above the membrane surface. The tangential velocity gradients are zero on the free surface. As illustrated, a vortex appears behind the fiber.

For the air flow across a fiber, the angularly local Nusselt numbers along the main flow direction under different Reynolds numbers are shown in Figure 7.15. As can be seen, the Nusselt numbers are the highest at the front stagnation point and they decrease steadily along the flow. The lowest values are in the rear of the fiber.

For the solution stream, the distributions of axially local and mean Nusselt numbers are similar to those in common round tubes [13,14]. Due to the increasing thickness in thermal boundary layers,

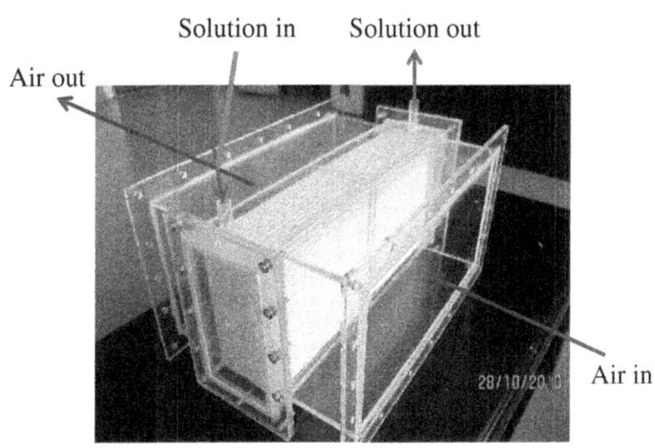

FIGURE 7.13

The cross-flow hollow fiber membrane module.

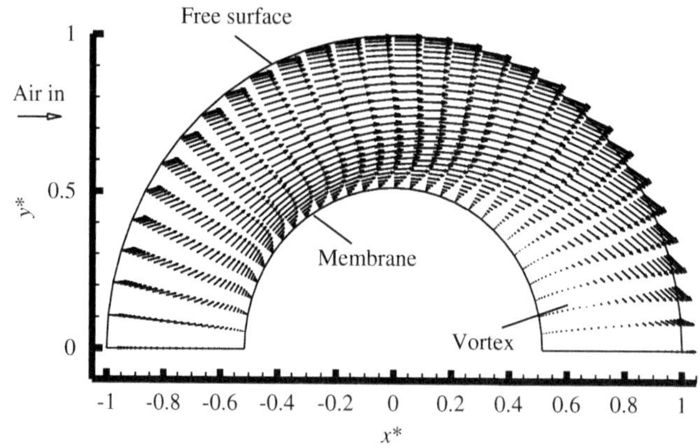

FIGURE 7.14

Velocity vectors for the air flow, $\phi = 0.265$, $Re_a = 100$.

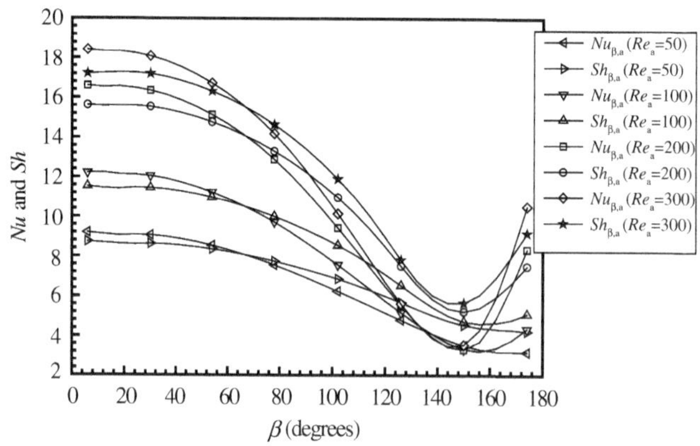

FIGURE 7.15

Variations of the angularly local Nusselt and Sherwood numbers for the air flow along the main flow direction under different Reynolds numbers, $\phi = 0.265$.

the local Nusselt number will decrease and approach asymptotically to a lower limiting value with the continuing flow. After that period, the flow is called the thermally developed region. The channel is long enough (practical length 36 cm) for the solution flow to get thermally fully developed (required length 5 mm).

For the air flow across a fiber (air side), the overall mean Nusselt number ($Nu_{ave,a}$) and the total drag coefficients (C_D) are listed in Table 7.4. The packing fraction of the contactors (ϕ) and the Reynolds numbers are those used in practical conditions. Also listed are the overall mean values under uniform

Table 7.4 The total drag coefficient and the overall mean Nusselt and Sherwood numbers for the air flow across a fiber, free surface model, laminar flow.

ϕ	r_o/r_f	$Re_a = 50$					$Re_a = 100$				
		$Nu_{ave,T}$	$Nu_{ave,H}$	$Nu_{ave,a}$	$Sh_{ave,a}$	C_D	$Nu_{ave,T}$	$Nu_{ave,H}$	$Nu_{ave,a}$	$Sh_{ave,a}$	C_D
0.091	0.30	4.28	4.96	4.92	5.04	2.92	5.77	6.63	6.46	6.63	2.39
0.160	0.40	4.65	5.39	5.54	5.72	4.03	6.23	7.18	7.11	7.36	3.25
0.200	0.45	4.87	5.58	5.90	6.13	4.81	6.51	7.51	7.49	7.82	3.85
0.265	0.51	5.25	6.01	6.51	6.85	6.35	6.98	8.05	8.15	8.59	5.01
0.362	0.60	5.91	6.74	7.58	8.13	9.86	7.79	8.99	9.26	10.00	7.59
0.490	0.70	7.14	8.05	9.41	10.49	19.09	9.19	10.58	11.10	12.54	14.21
0.562	0.75	8.21	9.12	10.76	12.42	29.72	10.26	11.77	12.41	14.58	21.52
0.642	0.80	10.01	10.89	12.71	15.57	54.33	11.93	13.57	14.21	17.70	37.23

ϕ	r_o/r_f	$Re_a = 200$					$Re_a = 250$				
		$Nu_{ave,T}$	$Nu_{ave,H}$	$Nu_{ave,a}$	$Sh_{ave,a}$	C_D	$Nu_{ave,T}$	$Nu_{ave,H}$	$Nu_{ave,a}$	$Sh_{ave,a}$	C_D
0.091	0.30	7.99	9.18	8.88	9.02	2.07	9.11	10.45	9.87	10.04	2.02
0.160	0.40	8.57	9.88	9.61	9.87	2.80	9.72	11.21	10.66	10.96	2.73
0.200	0.45	8.93	10.28	10.01	10.37	3.29	10.06	11.62	11.08	11.48	3.21
0.265	0.51	9.47	10.95	11.01	11.49	4.24	10.64	12.32	11.78	12.36	4.12
0.362	0.60	10.45	12.12	12.39	12.71	6.29	11.67	13.55	12.96	13.97	6.09
0.490	0.70	12.20	14.16	14.82	15.53	11.35	13.52	15.72	14.93	16.82	10.92
0.562	0.75	13.52	15.69	16.58	17.77	16.84	14.95	17.37	16.36	19.11	16.07
0.642	0.80	15.48	17.92	19.20	21.27	28.37	17.07	19.81	20.32	22.65	26.82

temperature ($Nu_{ave,T}$) and uniform heat flux conditions ($Nu_{ave,H}$). In these two cases, the fiber has an outer free surface, but the membrane surface is set to uniform temperature (T) or uniform heat flux (H) conditions. Air-side $Nu_{ave,a}$ and C_D rise with an increase in the packing fraction. However, C_D decreases when the Reynolds numbers rise. Air-side $Nu_{ave,a}$ is between $Nu_{ave,T}$ and $Nu_{ave,H}$ for sparsely distributed contactors ($\phi < 0.2$), but $Nu_{ave,a}$ is about 10% larger than $Nu_{ave,H}$ for densely populated contactors ($\phi > 0.2$). It is related to packing fractions and mass flow.

For the solution flow in the circular channel (liquid side), the fully developed axially local Nusselt number ($Nu_{C,s}$) and the overall mean Nusselt number ($Nu_{ave,s}$) under the naturally formed boundary conditions are 4.35 and 4.40, respectively. The solution-side $Nu_{C,s}$ is almost equal to the fully developed local value under uniform heat flux conditions ($Nu_H = 4.36$; Ref. [1]), but about 20% larger than the fully developed local value under uniform temperature conditions ($Nu_T = 3.66$; Ref. [1]). Further, both $Nu_{C,s}$ and $Nu_{ave,s}$ are almost unchanged when the packing fraction is varied.

As mentioned, the heat flux on the solution side consists of the air-side heat flux and the latent or absorption heat flux. The solution-side heat flux ($q_{h,s}$) is larger than the air-side flux ($q_{h,a}$) due to latent heat release. This implies that the latent heat transfer is more dominant than the sensible convective heat transfer between the two fluids. The values of air to solution-side heat flux ratio, $q_{Lat,ave}/q_{h,a,ave}$, can be positive or negative. This is because heat transfer and mass transfer can be in the same or opposite directions. The larger the $q_{Lat,ave}/q_{h,a,ave}$ is, the higher the air-side $Nu_{ave,a}$ and the solution-side $Nu_{C,s}$ are.

Air-side overall mean Nusselt and Sherwood numbers and the pressure drops under the conjugate heat and mass transfer boundary conditions of this cross-flow contactor are compared to the data for a counter-flow contactor of similar geometry [2]. When the Reynolds numbers are smaller than 35, both the cross-flow Nu_{ave} and Sh_{ave} are less than the counter-flow values. However, the cross-flow Nu_{ave} and Sh_{ave} become larger than the counter-flow values when the Reynolds numbers are greater than 35. This is because vortexes appear behind the fibers, which tend to enhance the heat and mass transfer [13,14]. To further illustrate, the real temperature and humidity distributions for the air flow across a single fiber are shown in Figure 7.16. The operating conditions are: $m_a = 496.3$ kg/h, $m_s = 309.4$ kg/h, $T_{ai} = 35.0$ °C, $T_{si} = 25.0$ °C, $\omega_{ai} = 0.021$ kg/kg, $\omega_{si} = 0.0055$ kg/kg, $X_{si} = 0.65$ kg/kg. As can be seen, the heat and mass

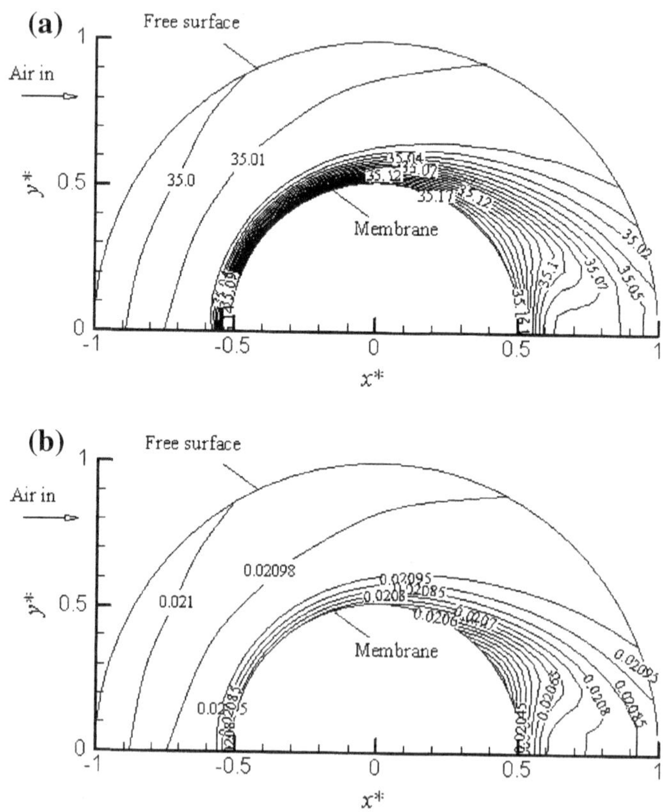

FIGURE 7.16

Temperature (°C) and humidity (kg/kg) contours for the air flow, $\phi = 0.265, z = 18$ cm $= 0.5L$. (a) Temperature contour; (b) humidity contour. Inlet conditions: $m_a = 496.3$ kg/h, $m_s = 309.4$ kg/h, $T_{ai} = 35.0$°C, $T_{si} = 25.0$°C, $\omega_{ai} = 0.021$ kg/kg, $\omega_{si} = 0.0055$ kg/kg, $X_{si} = 0.65$ kg/kg.

transfer can be enhanced by the vortexes generated behind the fiber. The overall air-side pressure drop for the cross-flow is much less than that for a counter-flow contactor. Though the counter-flow arrangement is the most commonly used contactor type, the cross-flow arrangement seems more attractive due to the higher heat mass transfer performance and lower pressure-drop losses.

The structure is similar to a shell-and-tube heat exchanger; however, the transport phenomena are totally different. The available data from classic literature are only applicable to large and sparsely populated heat transfer tubes. This research provides the specific data for densely populated small fibers. From Table 7.4, it is seen that the Nusselt and Sherwood numbers for this equipment are quite different from those for heat exchangers. There are big differences in transport data. As for the analogy, it is revealed in Table 7.4 that the analogy doesn't hold in a simple way. For instance in Table 7.4, for $Re=50$, when the packing fraction ϕ is 0.2, the ratio of $Sh/Nu=1.04$. When $\phi=0.642$, however, $Sh/Nu=1.23$. Therefore the analogy changes with the packing fraction. In contrast, for common heat exchangers, the effects of tube density are always ignored and a constant analogy is believed to be held regardless of packing fractions. There are big differences in how the analogy is held.

7.4.4 Sherwood numbers

For the air flow across the fiber, the angularly local Sherwood numbers along the main flow direction are shown in Figure 7.15. As can be seen, variations of the angularly local Sherwood numbers are similar to the Nusselt numbers. When β ranges from 0 to 80°, the Nusselt numbers are larger than the Sherwood numbers. When β ranges from 80 to 180°, the Nusselt numbers are less than the Sherwood numbers for Re_a from 50 to 100. However, when the Reynolds numbers are between 200 and 300, the Nusselt numbers become higher than the Sherwood numbers in the rear of the fiber. This is because turbulence becomes more intense here, which makes convective-side transport more dominant. Heat transfer is more sensitive to turbulence than mass transfer due to less heat transfer resistance through membranes.

For the solution flow in the circular channel, the variations of the axially local and mean Sherwood numbers along the z axis are similar to common round tubes. However, the concentration boundary layer ($z^*_{m,s}=0.06, 30$ cm) develops much more slower than the thermal boundary layer ($z^*_{h,s}=0.5, 5$ cm). This is because the Lewis number for the LiCl solution is rather small ($Le_s=0.02$) compared to that of the moist air ($Le_a=0.8$). The Schmidt number for the solution is much higher than that for the air. However, the channel is long enough ($L=36$ cm) for the solution stream to get fully developed ($z=25$ cm) finally.

For the air flow across the fiber (air side), the detailed values of the overall mean Sherwood numbers ($Sh_{ave,a}$) are also listed in Table 7.4. The air-side $Sh_{ave,a}$ values are always higher than the air-side $Nu_{ave,a}$. As can be seen from Table 7.4, for $Re=50$, when the packing fraction is less than 0.2, $Sh_{ave,a}$ is nearly equal to $Nu_{ave,a}$, if considering the uncertainty. However, when the packing fraction is 0.642, the $Sh_{ave,a}$ is 16.7% larger than $Nu_{ave,a}$, far beyond the uncertainty.

For the solution flow in the circular channel, the fully developed Sherwood number ($Sh_{C,s}$) and the axially mean Sherwood number ($Sh_{ave,s}$) under naturally formed conditions are 4.40 and 5.21, respectively. The solution-side $Sh_{C,s}$ is a little bit larger than the solution-side $Nu_{C,s}$ (= 4.35).

7.5 Cross-flow arrangement with turbulent flow, free surface cell model

Cross flow is more complex than counter flow. The fibers provide disturbances for the shell-side fluid flow. When the disturbances are strong, turbulence will be formed in the fiber bundle. Therefore in this case turbulent models should be used. In this section, a low-*Re* *k*-ε turbulence model is used to describe the air flow, while a laminar model is used for the liquid flow. The governing equations for the momentum, heat and mass transfer in the two flows and in the membrane are coupled and solved numerically. No commercial software package has the capability to solve these conjugate problems, so they are solved by a validated self-built code [4]. To overcome the difficulties in the direct modeling of the numerous fibers, still, Happel's free surface model is employed to simplify the analysis and to build the calculation domain. The Nusselt and Sherwood numbers under naturally formed boundary conditions are obtained and investigated.

7.5.1 The turbulence model

In practical applications, the Reynolds number for the solution stream in the inner circular channel is below 10 (much less than 2300), therefore laminar flow is assumed. However, for the air stream, though the Reynolds number for the air stream is still below 2300 (around 200–600), the flow would become turbulent due to the continuous disturbances from the fine fibers [15]. The laminar model is not suitable for the air stream when the Reynolds number is larger than 300 [15]. Certainly a turbulence model is required. However, the local turbulent Reynolds numbers ($Re_t = \rho k_a^2/\mu\varepsilon_a$) are less than 150. In this case, the near-wall flow cannot be accurately modeled by a standard *k*-ε turbulence model [16]. To address this problem, a low-*Re* *k*-ε model [4,16] is employed for the air stream. Other assumptions are similar to the previous section.

The computational domain and the coordinates are the same as for Figure 7.12. For the air flow over the fiber, the low-*Re* *k*-ε turbulence model is employed. In this model, the turbulent kinetic energy and the dissipation rate are calculated via transport equations. They are solved simultaneously with the conversion equations for the fluid flow. The normalized governing momentum, heat and mass equations for the air flow are described as follows [4,16]:

$$\frac{\partial u_x^*}{\partial x^*} + \frac{\partial u_y^*}{\partial y^*} = 0 \tag{7.90}$$

$$\begin{aligned}\frac{\partial (u_x^* u_x^*)}{\partial x^*} + \frac{\partial (u_y^* u_x^*)}{\partial y^*} = &\frac{\partial}{\partial x^*}\left(\Gamma_x \frac{\partial u_x^*}{\partial x^*}\right) + \frac{\partial}{\partial y^*}\left(\Gamma_x \frac{\partial u_x^*}{\partial y^*}\right) - \frac{1}{2}\frac{\partial P_a^*}{\partial x^*} \\ &+ \frac{\partial}{\partial x^*}\left(\Gamma_x \frac{\partial u_x^*}{\partial x^*}\right) + \frac{\partial}{\partial y^*}\left(\Gamma_x \frac{\partial u_y^*}{\partial x^*}\right)\end{aligned} \tag{7.91}$$

$$\begin{aligned}\frac{\partial (u_x^* u_y^*)}{\partial x^*} + \frac{\partial (u_y^* u_y^*)}{\partial y^*} = &\frac{\partial}{\partial x^*}\left(\Gamma_y \frac{\partial u_y^*}{\partial x^*}\right) + \frac{\partial}{\partial y^*}\left(\Gamma_y \frac{\partial u_y^*}{\partial y^*}\right) - \frac{1}{2}\frac{\partial P_a^*}{\partial y^*} \\ &+ \frac{\partial}{\partial x^*}\left(\Gamma_y \frac{\partial u_x^*}{\partial y^*}\right) + \frac{\partial}{\partial y^*}\left(\Gamma_y \frac{\partial u_y^*}{\partial y^*}\right)\end{aligned} \tag{7.92}$$

$$\frac{\partial\left(u_x^* k_a^*\right)}{\partial x^*}+\frac{\partial\left(u_y^* k_a^*\right)}{\partial y^*}=\frac{\partial}{\partial x^*}\left(\Gamma_k \frac{\partial k_a^*}{\partial x^*}\right)+\frac{\partial}{\partial y^*}\left(\Gamma_k \frac{\partial k_a^*}{\partial y^*}\right)+\frac{\mu_t u_{ai}}{\rho_a r_f k_{ai}} G_k^*-\frac{r_f \varepsilon_{ai}}{u_{ai} k_{ai}} \varepsilon_a^*-E^* \tag{7.93}$$

$$\frac{\partial\left(u_x^* \varepsilon_a^*\right)}{\partial x^*}+\frac{\partial\left(u_y^* \varepsilon_a^*\right)}{\partial y^*}=\frac{\partial}{\partial x^*}\left(\Gamma_\varepsilon \frac{\partial \varepsilon_a^*}{\partial x^*}\right)+\frac{\partial}{\partial y^*}\left(\Gamma_\varepsilon \frac{\partial \varepsilon_a^*}{\partial y^*}\right)+\frac{c_1 \mu_t u_{ai}}{\rho_a r_f k_{ai}} \frac{\varepsilon_a^* G_k^*}{k_a^*}-\frac{c_2 r_f \varepsilon_{ai}}{u_{ai} k_{ai}} \frac{\varepsilon_a^{*2}}{k_a^*}+F^* \tag{7.94}$$

$$\frac{\partial\left(u_x^* \theta_a\right)}{\partial x^*}+\frac{\partial\left(u_y^* \theta_a\right)}{\partial y^*}=\frac{\partial}{\partial x^*}\left(\Gamma_T \frac{\partial \theta_a}{\partial x^*}\right)+\frac{\partial}{\partial y^*}\left(\Gamma_T \frac{\partial \theta_a}{\partial y^*}\right) \tag{7.95}$$

$$\frac{\partial\left(u_x^* \xi_a\right)}{\partial x^*}+\frac{\partial\left(u_y^* \xi_a\right)}{\partial y^*}=\frac{\partial}{\partial x^*}\left(\Gamma_\omega \frac{\partial \xi_a}{\partial x^*}\right)+\frac{\partial}{\partial y^*}\left(\Gamma_\omega \frac{\partial \xi_a}{\partial y^*}\right) \tag{7.96}$$

where subscripts "x," "y" and "a" refer to the x axis, y axis and the air stream, respectively; Superscript "*" represents the dimensionless form; u here is time-averaged velocity (m/s); P is average pressure (Pa); k_a, ε_a, θ and ξ are turbulent kinetic energy, dissipation rate, dimensionless time-averaged temperature and humidity, respectively.

The generation term and two extra terms appearing in Eqs. (7.93) and (7.94) are [4,16]:

$$G_k^*=2\left[\left(\frac{\partial u_x^*}{\partial x^*}\right)^2+\left(\frac{\partial u_y^*}{\partial y^*}\right)^2\right]+\left(\frac{\partial u_x^*}{\partial y^*}+\frac{\partial u_y^*}{\partial x^*}\right)^2 \tag{7.97}$$

$$E^*=\frac{2\mu_a}{\rho_a r_f u_{ai}}\left(\frac{\partial\left(k_a^*\right)^{1/2}}{\partial y^*}\right)^2 \tag{7.98}$$

$$F^*=\frac{2\mu_t \mu_a u_{ai}}{\rho_a^2 r_f^3 \varepsilon_{ai}}\left(\frac{\partial^2 u_y^*}{\partial y^{*2}}\right)^2 \tag{7.99}$$

In Eqs. (7.91)–(7.96), the associated diffusion coefficients are [4,16]:

$$\Gamma_x=\Gamma_y=\frac{2r_o}{r_f}\frac{1+\frac{\mu_t}{\mu_a}}{Re_a} \tag{7.100}$$

$$\Gamma_k=\frac{2r_o}{r_f}\frac{1+\frac{\mu_t}{\mu_a \sigma_k}}{Re_a} \tag{7.101}$$

$$\Gamma_{\varepsilon} = \frac{2r_o}{r_f} \frac{1 + \frac{\mu_t}{\mu_a \sigma_\varepsilon}}{Re_a} \tag{7.102}$$

$$\Gamma_T = \frac{2r_o}{r_f} \frac{\frac{1}{Pr_a} + \frac{\mu_t}{\mu_a \sigma_T}}{Re_a} \tag{7.103}$$

$$\Gamma_{\omega} = \frac{2r_o}{r_f} \frac{\frac{1}{Sc_a} + \frac{\mu_t}{\mu_a \sigma_\omega}}{Re_a} \tag{7.104}$$

where μ_a represents air molecular viscosity (Pa s). The turbulent viscosity is equal to [4,16]

$$\mu_t = c_\mu c_t \rho_a \frac{k_a^2}{\varepsilon_a} \tag{7.105}$$

where

$$c_t = \exp\left(-2.5/\left(1 + Re_t/50\right)\right) \tag{7.106}$$

The turbulent Reynolds number

$$Re_t = \frac{\rho_a k_a^2}{\mu_a \varepsilon_a} \tag{7.107}$$

The Reynolds number for the air stream is defined by [17]

$$Re_a = \frac{\rho_a u_{ai} D_{h,a}}{\mu_a} \tag{7.108}$$

where$D_{h,a}$ ($= 2r_o$) is the hydraulic diameter for the air flow channel (m). This definition is the same as in the previous section.

In Eq. (7.94), c_2 is determined by [16]

$$c_2 = 1.92\left(1 - 0.3\exp\left(-Re_t^2\right)\right) \tag{7.109}$$

Other coefficients are: $c_\mu = 0.09$, $c_1 = 1.44$, $\sigma_k = 1.0$, $\sigma_\varepsilon = 1.3$, $\sigma_T = 0.95$, $\sigma_\omega = 1.0$.

The dimensionless time-averaged velocity for the air stream in the x axis direction is defined by

$$u_x^* = \frac{u_x}{u_{ai}} \tag{7.110}$$

and that for the air stream in the y axis direction is defined by

$$u_y^* = \frac{u_y}{u_{ai}} \tag{7.111}$$

The dimensionless turbulent kinetic energy for the air stream is defined by

$$k_{\mathrm{a}}^{*} = \frac{k_{\mathrm{a}}}{k_{\mathrm{ai}}} \tag{7.112}$$

where k_{ai} is the turbulent kinetic energy for the approaching air stream ($\mathrm{m^2/s^2}$), which is obtained by [18]

$$k_{\mathrm{ai}} = 0.01\left(u_{\mathrm{ai}}^{2}/2\right) \tag{7.113}$$

The dimensionless turbulent dissipation rate for the air stream is defined by

$$\varepsilon_{\mathrm{a}}^{*} = \frac{\varepsilon_{\mathrm{a}}}{\varepsilon_{\mathrm{ai}}} \tag{7.114}$$

where $\varepsilon_{\mathrm{ai}}$ is the turbulence dissipation rate for the approaching air stream ($\mathrm{m^2/s^3}$), which is equal to [18]

$$\varepsilon_{\mathrm{ai}} = \frac{500 c_{\mu} k_{\mathrm{ai}}^{2}}{u_{\mathrm{ai}} D_{\mathrm{h,a}}} \tag{7.115}$$

The definitions of dimensionless pressure, temperature, humidity, the pressure drag, and the governing equations for the solution inside the fibers are the same as in the previous section for a laminar flow. However, here the air-side values are the time-averaged values.

7.5.2 Boundary conditions

A boundary-fitted coordinate technology is used to transfer the round physical shape to a rectangular calculating shape [4]. The boundary conditions of velocity, turbulent kinetic energy and turbulence dissipation rate for the air stream based on Happel's free surface model are [3,4]

$$\text{On membrane surface: } u_x^{*} = 0, \quad u_y^{*} = 0, \quad k_{\mathrm{a}}^{*} = 0 \quad \text{and} \quad \varepsilon_{\mathrm{a}}^{*} = 0 \tag{7.116}$$

$$\text{On the free surface: } u_{\mathrm{r}}^{*} = -\cos\beta, \quad \tau_{1\beta} = 0, \quad k_{\mathrm{a}}^{*} = \frac{0.01(u_{\mathrm{f}}^{2}/2)}{k_{\mathrm{ai}}} \quad \text{and} \quad \varepsilon_{\mathrm{a}}^{*} = \frac{500 c_{\mu} \left(k_{\mathrm{a}}^{*}\right)^{2}}{u_{\mathrm{ai}} D_{\mathrm{h,a}} \varepsilon_{\mathrm{ai}}} \tag{7.117}$$

where subscript "r" represents the radial direction; τ_1 is shear stress (Pa); u_{f} is the resultant velocity on the free surface. The turbulence kinetic energy and the dissipation rate on the free surface are updated by the local velocity. Besides the inlet conditions mentioned in the previous section, a 5% turbulence intensity is assumed for the inlet velocity.

On the symmetric boundary conditions

$$\frac{\partial \psi}{\partial n} = 0 \tag{7.118}$$

where ψ stands for variables such as pressure, velocity, turbulent kinetic energy, dissipation rate, temperature, humidity or mass fraction. Concentrations and temperature on the membrane surfaces are conjugated by the method mentioned previously.

The turbulence model and the whole procedure are solved and validated both experimentally and numerically. The studied module is the same as in the previous section. The methodology is the same as before. Besides the low-*Re* k-ε model, a laminar model for the air stream is also used for comparison [4]. It was found that the two models fit the test data well when Re_a is less than 300. The maximum difference is less than 6.0% for the results. However the low-*Re* k-ε model is better when Re_a is higher than 300. Generally the low-*Re* k-ε model is successful in modeling the fluid flow, heat and mass transfer across the bundle. After the model validations, numerical work will be performed for packing fractions (ϕ) from 0.160 to 0.642 and Re_a from 300 to 600.

7.5.3 Transport data for turbulent flow

Nusselt numbers

For the air flow across a fiber, the angularly local Nusselt numbers along the main flow direction under different Reynolds numbers are shown in Figure 7.17. As can be seen, the Nusselt numbers are the highest at the front stagnation point and they decrease steadily along the flow. The lowest values are in the rear of the fiber. The values rise with increasing Reynolds numbers.

As observed in the previous section, for the solution stream, the distributions of axially local and mean Nusselt numbers are similar to those in common round tubes. It has been shown that the axially local Nusselt numbers decrease sharply from large values at the inlet to fully developed values after the entry length. The channel is long enough ($L = 36$ cm) for the solution flow to get thermally fully developed ($z = 5$ cm).

Air-side overall mean Nusselt numbers ($Nu_{\text{ave,a}}$) and the total drag coefficients (C_D) for the turbulent flow are listed in Table 7.5. Also listed are the overall mean values under uniform temperature ($Nu_{\text{ave,T}}$) and uniform heat flux conditions ($Nu_{\text{ave,H}}$). In these two ideal cases, the fiber has an outer free

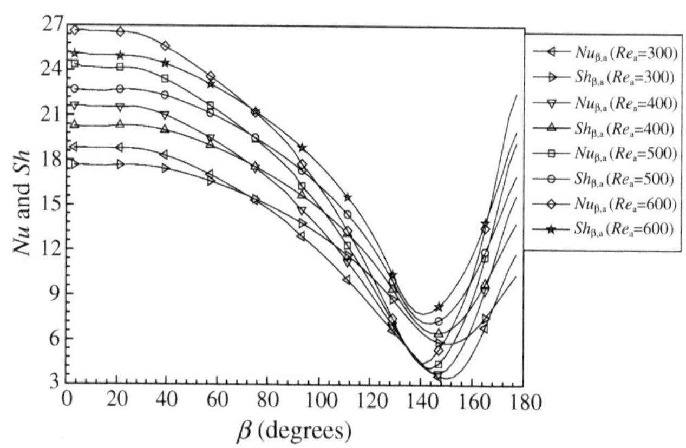

FIGURE 7.17

Variations of the angularly local Nusselt and Sherwood numbers for the air flow along the main flow direction under different Reynolds numbers, $\phi = 0.265$.

Table 7.5 The total drag coefficient and the overall mean Nusselt and Sherwood numbers for the air flow across a fiber, free surface model, turbulent flow.

ϕ	r_o/r_f	$Re_a = 300$					$Re_a = 400$				
		$Nu_{ave,T}$	$Nu_{ave,H}$	$Nu_{ave,a}$	$Sh_{ave,a}$	C_D	$Nu_{ave,T}$	$Nu_{ave,H}$	$Nu_{ave,a}$	$Sh_{ave,a}$	C_D
0.160	0.40	10.29	11.94	11.38	11.59	1.82	11.98	13.87	13.22	13.40	1.70
0.200	0.45	10.67	12.39	11.88	12.10	2.16	12.39	14.38	13.54	13.91	2.02
0.265	0.51	11.32	13.17	12.79	13.04	2.81	13.10	15.23	14.51	14.91	2.63
0.362	0.60	12.49	14.54	14.37	14.67	4.24	14.38	16.75	16.21	16.70	3.94
0.490	0.70	14.60	16.95	17.27	17.66	7.80	16.69	19.42	19.28	19.89	7.17
0.562	0.75	16.19	18.77	19.54	20.03	11.67	18.43	21.44	21.66	22.39	10.65
0.642	0.80	18.54	21.45	23.00	23.65	19.80	21.03	24.42	25.68	26.26	17.89
ϕ	r_o/r_f	$Re_a = 500$					$Re_a = 600$				
		$Nu_{ave,T}$	$Nu_{ave,H}$	$Nu_{ave,a}$	$Sh_{ave,a}$	C_D	$Nu_{ave,T}$	$Nu_{ave,H}$	$Nu_{ave,a}$	$Sh_{ave,a}$	C_D
0.160	0.40	13.52	15.61	14.90	15.04	1.63	14.55	16.77	16.01	16.10	1.41
0.200	0.45	13.97	16.17	15.22	15.60	1.93	15.42	17.84	16.77	17.16	1.83
0.265	0.51	14.73	17.11	16.26	16.70	2.51	16.27	18.88	17.90	18.35	2.42
0.362	0.60	16.11	18.78	18.11	18.61	3.73	17.73	20.66	19.87	20.41	3.59
0.490	0.70	18.60	21.65	21.79	22.09	6.76	20.39	23.82	23.81	24.11	6.47
0.562	0.75	20.48	23.88	23.89	24.66	9.98	22.40	26.20	26.02	26.83	9.51
0.642	0.80	23.28	27.15	28.22	28.74	16.64	25.39	29.71	30.65	31.13	15.78

surface, but the membrane surface is set to uniform temperature (T) or uniform heat flux (H) conditions. The friction drag coefficients (C_{Df}) are less than the pressure drag coefficients (C_{Dp}). Air-side $Nu_{ave,a}$, C_{Df}, C_{Dp} and C_D rise with increasing packing fraction. Further, C_{Df}, C_{Dp} and C_D decrease when the Reynolds numbers rise. Air-side $Nu_{ave,a}$ is between $Nu_{ave,T}$ and $Nu_{ave,H}$ for sparsely distributed contactors (low ϕ), but $Nu_{ave,a}$ is larger than $Nu_{ave,H}$ for densely populated contactors (high ϕ). For the solution flow in the circular channel (liquid side), the fully developed axially local Nusselt number ($Nu_{C,s}$) and the overall mean Nusselt number ($Nu_{ave,s}$) under the naturally formed boundary conditions are 4.50 and 4.60, respectively. The solution-side $Nu_{C,s}$ is almost equal to the fully developed local value under uniform heat flux conditions ($Nu_H = 4.36$; Refs. [1,13,14]), but larger than that under uniform temperature conditions ($Nu_T = 3.66$; Refs. [1,13,14]). Further, both $Nu_{C,s}$ and $Nu_{ave,s}$ are almost unchanged when the packing fraction is varied.

Sherwood numbers

For the air flow across the fiber, the angularly local Sherwood numbers along the main flow direction are shown in Figure7.17. As can be seen, variations of the angularly local Sherwood numbers are similar to the Nusselt numbers. When β ranges from 0 to 80°, the Nusselt numbers are larger than the Sherwood numbers. When β ranges from 80 to 165°, the Nusselt numbers are lower than the

Sherwood numbers. However, the Nusselt numbers become higher than the Sherwood numbers in the rear of the fiber.

For the solution flow in the circular channel, the variations of the axially local and mean Sherwood numbers along the z axis are similar to those in common round tubes [1,13,14]. However, the concentration boundary layer ($z^*_{m,s} = 0.06, 30$ cm) develops much more slowly than the thermal boundary layer ($z^*_{h,s} = 0.5, 5$ cm). The reason is that the Schmidt number for the LiCl solution is rather large ($Sc_s = 1390$). Nevertheless, the channel is long enough ($L = 36$ cm) for the solution stream to get fully developed.

For the air flow across the fiber, the detailed values of the overall mean Sherwood numbers ($Sh_{ave,a}$) are listed in Table 7.5. The air-side $Sh_{ave,a}$ are somewhat higher than the air-side $Nu_{ave,a}$. The air-side $Sh_{ave,a}$ increases with increasing packing fraction (ϕ) and/or Reynolds numbers. For the solution flow in the circular channel, the fully developed Sherwood number ($Sh_{C,s}$) and the axially mean Sherwood number ($Sh_{ave,s}$) under naturally formed condition are 4.64 and 5.33, respectively. The solution-side $Sh_{C,s}$ is a bit larger than the solution-side $Nu_{C,s}$ (4.50).

7.6 Cross flow arrangement with laminar flow, periodic cell model

Free surface models have been successfully applied in the analysis of fluid flow and heat and mass transfer in a fiber bundle. However, the free surface model is only a hypothetic cell around a fiber. According to this method, the inter-fiber effects are negligible. In fact, this is only an approximation. The hypothesis is only valid when the fibers are sparsely distributed, for instance $\phi \leq 0.25$ [19]. When the fiber packing density is high, for instance, $\phi > 0.25$, the effects of neighboring fibers should be considered. In this section, a periodic cell including neighboring fibers and the air duct surrounded by the fibers is selected as the computational domain. The transport data are obtained. This section assumes a laminar fluid flow across a fiber.

7.6.1 The computational cell

The fiber arrangement can be either in-line or staggered, as shown in Figure 7.4b and c. The process air flows across the fiber bank. Fluids are affected by the neighboring fibers. To account for these interactions between the neighboring fibers, this section takes a new approach: The fluid in the fibers is modeled together with the neighboring fibers. The calculating domains are selected as the periodic areas surrounded by the neighboring fibers, as shown in Figures 7.18 and 7.19 by the shaded areas for the in-line and staggered arrays respectively. In these representative cells, the fluid flow and the conjugate heat and mass transfer between the fluid and these surrounding fibers are investigated. The Nusselt and Sherwood numbers accounting for these interactions are obtained and analyzed. They are compared to those data calculated with a free surface model.

7.6.2 Mathematical model

Governing equations

In the hollow fiber membrane bundle, the two streams flow in a cross-flow arrangement. The air flow and the heat and mass transfer across the bundle show periodic behaviors [20–22]. For reasons of

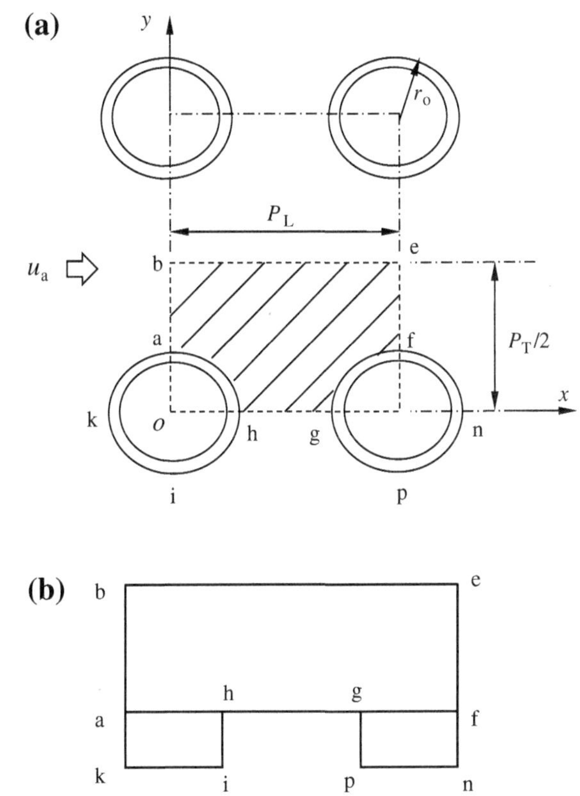

FIGURE 7.18

Coordinates and the computational cell in a fiber bundle with in-line array. The shaded areas are the cell. (a) Physical plane; (b) computational plane.

symmetry and simplicity in calculation, two unit cells, as shown in Figures 7.18 and 7.19 for the in-line and the staggered respectively, are selected as the calculation domains. The packing fraction of the whole bundle is equal to that of the unit cell, which can be calculated by

$$\phi = \frac{\pi r_o^2}{P_T P_L} \tag{7.119}$$

where r_o is the fiber outer radius (m); P_L and P_T are longitudinal and transverse pitches (m), respectively.

Due to the complex geometry of the unit cell, a boundary-fitted coordinate transformation technology is used in calculations. The physical planes shown in Figures 7.18 a and 7.19 a are transformed to the computational planes as shown in Figures 7.18 b and 7.19 b respectively. The fibers are oriented normal to the air flow. The solution stream flows along the z axis in the circular fiber tubes (two round channels for the in-line, three round channels for the staggered), while the air stream flows over the fibers. On the

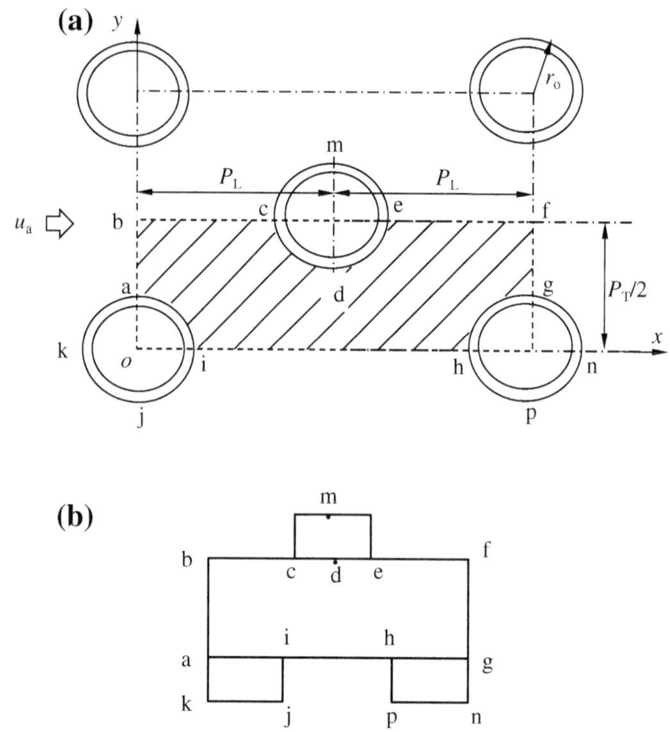

FIGURE 7.19

Coordinates and the computational cell in a fiber bundle with staggered array. The shaded areas are the cell. (a) Physical plane; (b) computational plane.

computational planes, the air duct is "a-b-c-d-e-f-g-h-a" for the in-line array and "a-b-c-d-e-f-g-h-i-a" for the staggered array respectively.

Laminar flow is considered first. Reynolds numbers for both the air and the solution streams are less than 400, therefore steady laminar flows are assumed. The air flow is two-dimensional with constant physical properties, meaning the velocities are functions of x and y only. The equations for the liquid solution are the same as before.

For the air flow across the fiber bank, the normalized governing equations of the fluid flow, heat and mass are described as follows:

$$\frac{\partial u_x^*}{\partial x^*} + \frac{\partial u_y^*}{\partial y^*} = 0 \tag{7.120}$$

$$u_x^* \frac{\partial u_x^*}{\partial x^*} + u_y^* \frac{\partial u_x^*}{\partial y^*} = -\frac{1}{2}\frac{\partial P_a^*}{\partial x^*} + \frac{D_{h,a}}{P_L} \cdot \frac{1}{Re_a} \cdot \left(\frac{\partial^2 u_x^*}{\partial x^{*2}} + \frac{\partial^2 u_x^*}{\partial y^{*2}} \right) \tag{7.121}$$

$$u_x^* \frac{\partial u_y^*}{\partial x^*} + u_y^* \frac{\partial u_y^*}{\partial y^*} = -\frac{1}{2} \frac{\partial P_a^*}{\partial y^*} + \frac{D_{h,a}}{P_L} \cdot \frac{1}{Re_a} \cdot \left(\frac{\partial^2 u_y^*}{\partial x^{*2}} + \frac{\partial^2 u_y^*}{\partial y^{*2}} \right) \quad (7.122)$$

$$u_x^* \frac{\partial \theta_a}{\partial x^*} + u_y^* \frac{\partial \theta_a}{\partial y^*} = \frac{D_{h,a}}{P_L} \cdot \frac{1}{Re_a Pr_a} \cdot \left(\frac{\partial^2 \theta_a}{\partial x^{*2}} + \frac{\partial^2 \theta_a}{\partial y^{*2}} \right) \quad (7.123)$$

$$u_x^* \frac{\partial \xi_a}{\partial x^*} + u_y^* \frac{\partial \xi_a}{\partial y^*} = \frac{D_{h,a}}{P_L} \cdot \frac{1}{Re_a Sc_a} \cdot \left(\frac{\partial^2 \xi_a}{\partial x^{*2}} + \frac{\partial^2 \xi_a}{\partial y^{*2}} \right) \quad (7.124)$$

For the solution stream in the circular inner tubes, the normalized governing equations of the fluid flow, heat and mass are in the same form, which can be described as follows:

$$\frac{\partial^2 u_s^*}{\partial x^{*2}} + \frac{\partial^2 u_s^*}{\partial y^{*2}} = -\frac{P_L^2}{D_{h,s}^2} \quad (7.125)$$

$$\frac{\partial^2 \theta_s}{\partial x^{*2}} + \frac{\partial^2 \theta_s}{\partial y^{*2}} = U_s \frac{\partial \theta_s}{\partial z_{h,s}^*} \quad (7.126)$$

$$\frac{\partial^2 \Theta_s}{\partial x^{*2}} + \frac{\partial^2 \Theta_s}{\partial y^{*2}} = U_s \frac{\partial \Theta_s}{\partial z_{m,s}^*} \quad (7.127)$$

where the dimensionless parameters are defined as those in Section 7.4 if not mentioned here. $D_{h,a}(= 2r_o)$ is the hydraulic diameter for the air flow channel (m), and $D_{h,s}(= 2r_i)$ is the hydraulic diameter for the solution (m).

Under the new coordinate system, dimensionless coordinates are defined by

$$x^* = \frac{x}{P_L} \quad (7.128)$$

$$y^* = \frac{y}{P_L} \quad (7.129)$$

The dimensionless velocity coefficient U_s is defined by

$$U_s = \frac{u_s^*}{u_m^*} \frac{P_L^2}{D_{h,s}^2} \quad (7.130)$$

where u_m^* is the average dimensionless velocity on a cross section.

The characteristic of the solution fluid flow in the round channels is represented by the product of the friction factor and the Reynolds number as

$$(fRe)_s = \left(\frac{-D_h \frac{dP}{dz}}{\rho u_m^2/2} \right)_s \left(\frac{\rho D_h u_m}{\mu} \right)_s = \left(\frac{2}{u_m^*} \right)_s \quad (7.131)$$

The friction factor for the air flow across the bundle can be calculated by [20–22]

$$f_{\mathrm{a}} = \frac{(P_{\mathrm{a}})_{\mathrm{i}} - (P_{\mathrm{a}})_{\mathrm{o}}}{\frac{1}{2} N \rho_{\mathrm{a}} u_{\mathrm{a,max}}^2} \tag{7.132}$$

where subscripts "i" and "o" mean the inlet and the outlet plane of the calculation unit cell, respectively; N is the number of rows in the flow direction; $u_{\mathrm{a,max}}$ is equal to [1,13,14]

$$u_{\mathrm{a,max}} = \frac{P_{\mathrm{T}}}{P_{\mathrm{T}} - 2r_{\mathrm{o}}} u_{\mathrm{ai}} \tag{7.133}$$

The transversely mean Nusselt and Sherwood values for the air stream are useful for energy and mass conservation analysis along the air flow. The mean Nusselt and Sherwood numbers for the air stream are:

$$Nu_{\mathrm{ave,a}} = \frac{P_{\mathrm{T}} L}{2A_{\mathrm{m}}} Re_{\mathrm{a}} Pr_{\mathrm{a}} \frac{(\theta_{\mathrm{b,a}})_{\mathrm{i}} - (\theta_{\mathrm{b,a}})_{\mathrm{o}}}{\Delta\theta_{\mathrm{a}}} \tag{7.134}$$

$$Sh_{\mathrm{ave,a}} = \frac{P_{\mathrm{T}} L}{2A_{\mathrm{m}}} Re_{\mathrm{a}} Sc_{\mathrm{a}} \frac{(\xi_{\mathrm{b,a}})_{\mathrm{i}} - (\xi_{\mathrm{b,a}})_{\mathrm{o}}}{\Delta\xi_{\mathrm{a}}} \tag{7.135}$$

where A_{m} means the membrane fibers area in the air-stream side, which is $\pi r_{\mathrm{o}} L$ and $2\pi r_{\mathrm{o}} L$ for the in-line and the staggered arrangements, respectively; $\Delta\theta_{\mathrm{a}}$ and $\Delta\xi_{\mathrm{a}}$ are the log mean temperature and humidity differences between membrane surface and the air stream, respectively. $\Delta\theta_{\mathrm{a}}$ and $\Delta\xi_{\mathrm{a}}$ are equal to

$$\Delta\theta_{\mathrm{a}} = \frac{(\theta_{\mathrm{w,a}} - \theta_{\mathrm{b,a}})_{\mathrm{i}} - (\theta_{\mathrm{w,a}} - \theta_{\mathrm{b,a}})_{\mathrm{o}}}{\ln[(\theta_{\mathrm{w,a}} - \theta_{\mathrm{b,a}})_{\mathrm{i}}/(\theta_{\mathrm{w,a}} - \theta_{\mathrm{b,a}})_{\mathrm{o}}]} \tag{7.136}$$

$$\Delta\xi_{\mathrm{a}} = \frac{(\xi_{\mathrm{w,a}} - \xi_{\mathrm{b,a}})_{\mathrm{i}} - (\xi_{\mathrm{w,a}} - \xi_{\mathrm{b,a}})_{\mathrm{o}}}{\ln[(\xi_{\mathrm{w,a}} - \xi_{\mathrm{b,a}})_{\mathrm{i}}/(\xi_{\mathrm{w,a}} - \xi_{\mathrm{b,a}})_{\mathrm{o}}]} \tag{7.137}$$

The Nusselt and Sherwood numbers for the liquid side are defined as in Section 7.4.

Boundary conditions

As mentioned, a boundary-fitted coordinate transformation method is employed to transfer the physical domains to rectangular calculating domains [19]. They are shown in Figures 7.18 and 7.19. For the in-line array, air duct is "a-b-c-d-e-f-g-h-a." Two surrounding hollow fibers are "a-k-i-h" and "g-p-n-f." The air duct and the liquid solution are conjugated by membrane surfaces "a-h" and "g-f." For the staggered array, the air duct is "a-b-c-d-e-f-g-h-i-a." The three neighboring fibers are "a-i-j-k," "g-n-p-h," and "c-d-e-m." The air duct and the liquid are conjugated by membrane surfaces "a-i," "h-g," and "c-d-e."

$$\text{On membrane surfaces,} \quad u_x^* = 0 \text{ and } u_y^* = 0 \tag{7.138}$$

The periodic boundary conditions of velocity for the air stream

$$\left(u_x^*\right)_i = \left(u_x^*\right)_o \text{ and } \left(u_y^*\right)_i = \left(u_y^*\right)_o \tag{7.139}$$

Due to the periodicities of the heat and mass transfer for the air flow across the fiber bundle, the boundary conditions of temperature and humidity for the air stream can be described as

$$\left(\Phi_a\right)_i = \left(\Phi_a\right)_o \text{ and } \left(\Omega_a\right)_i = \left(\Omega_a\right)_o \tag{7.140}$$

where Φ and Ω are dimensionless temperature and humidity used for the periodic boundary conditions, which are defined as [13,14]:

$$\Phi = (\theta - \theta_w)/(\theta_b - \theta_w) \tag{7.141}$$

$$\Omega = (\xi - \xi_w)/(\xi_b - \xi_w) \tag{7.142}$$

The boundary conditions for the liquid are the same as in Section 7.4. Further, fluid and air are conjugated by heat flux and mass flux on membrane surfaces, as treated in previous sections.

7.6.3 Solution procedure

The above equations are then discretized by means of a finite volume method [19] on the computational domain. The convective terms in Eqs. (7.121)–(7.124) are discretized by a power law scheme [14–16]. Since the air and the solution streams and the membrane interact, and temperature and concentration are also related to each other, ADI techniques are used to solve these equations. The Navier-Stokes equations for the air stream (Eqs. (7.120)–(7.122)) are solved using the SIMPLE algorithm [20–22]. The periodic velocity boundary conditions are realized using the mutual replacements method [20,21]. The velocity at the inlet plane is updated by the velocity at the outlet plane after certain iterations. Normal outlet velocity should be revised to satisfy the overall mass conservation at each iteration [20,21].

After these procedures, the couplings between the two streams and the periodicity are satisfied simultaneously. The temperature and concentration profiles on membrane surfaces are the naturally formed boundary conditions. It is a conjugate problem.

To ensure the accuracy of the results calculated, grid independence tests are conducted. Then the code is experimentally validated with the test rig built before. Two modules are made with these membranes and are experimented upon for air dehumidification. Module A: in-line arrangement; module length (L=30.0 cm); width (a=20.0 cm); height (b=20.0 cm); number of fibers (n_f=4500) and packing fraction (ϕ=0.196), transfer area (A_{tot}=6.4 m^2), longitudinal pitch (P_L=3 mm), transverse pitch (P_T=3 mm). Module B: staggered arrangement; module length (L=30.0 cm); width (a=20.0 cm); height (b=20.0 cm); number of fibers (n_f=4433) and packing fraction (ϕ=0.194), transfer area (A_{tot}=6.3 m^2), longitudinal pitch (P_L=3 mm), transverse pitch (P_T=3 mm).

Comparisons are made between the calculated and the experimentally obtained mean Nusselt and Sherwood numbers ($Nu_{ave,a}$ and $Sh_{ave,a}$) and the friction factors (f_a) for the air stream, which are shown

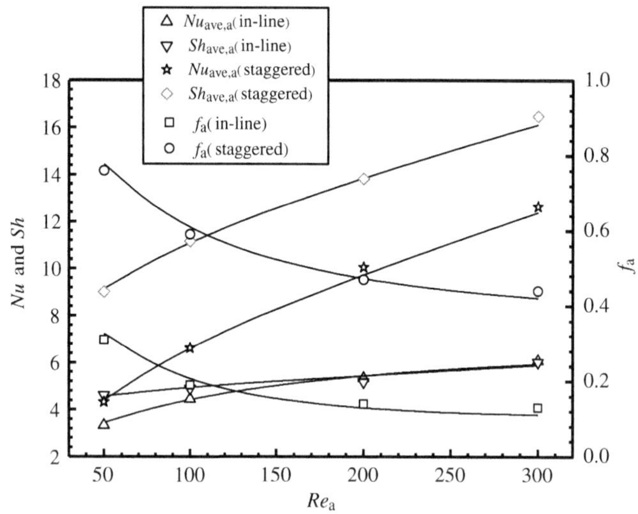

FIGURE 7.20

The mean Nusselt and Sherwood numbers and the frication factors for the air stream, $P_L/(2r_o) = P_T/(2r_o) = 2.0$, $\phi = 0.196$. The solid lines represent the simulated data. The discrete symbols are the measured values.

in Figure 7.20. As can be seen, discrepancies between the numerical and tested data are below 7.5%. For the solution flow in the round tubes, the mean Nusselt and Sherwood numbers ($Nu_{ave,s}$ and $Sh_{ave,s}$) and $(fRe)_s$ measured are 4.44, 5.45 and 65.02, respectively. They are close to the calculated values: 4.56, 5.32 and 63.89 respectively. Finally, the code is verified with the transport data for a tube bank under T and H boundary conditions from Refs. [1,13,14].

7.6.4 Nusselt numbers

The fluid flow in the circular tube (liquid side) is the same as that in a common round tube [13,14]. Air-side is the focus now. The velocity vectors for the air flow across the bundle are plotted in Figure 7.21. As can be seen, the velocity vectors are zero on the membrane surfaces and they rise sharply above the membrane surfaces. A vortex appears between the two neighboring fiber tubes for the in-line arrangement as shown in Figure 7.21a. For the staggered arrangement as shown in Figure 7.21b, two vortexes, one clockwise and the other counter-clockwise, appear between the fiber tubes. Velocities at the inlet and the outlet show a periodic feature for both the in-line and the staggered arrangements.

Air-side mean Nusselt numbers ($Nu_{ave,a}$) and the friction factors (f_a) are listed in Tables 7.6 and 7.7 for the in-line and the staggered arrays respectively. The ranges of the packing fraction (ϕ) and the Reynolds numbers are those typical in engineering. Also listed are the mean values under uniform temperature ($Nu_{ave,T}$) and uniform heat flux boundary conditions ($Nu_{ave,H}$). In such cases, the fiber membrane surfaces are set either to uniform temperature (T) or to uniform heat flux (H) conditions. As can be seen, the air-side $Nu_{ave,a}$ and f_a rise with increasing packing fractions (ϕ), but f_a decreases with rising Reynolds numbers. Generally the air-side $Nu_{ave,a}$ is between the ($Nu_{ave,T}$) and the ($Nu_{ave,H}$).

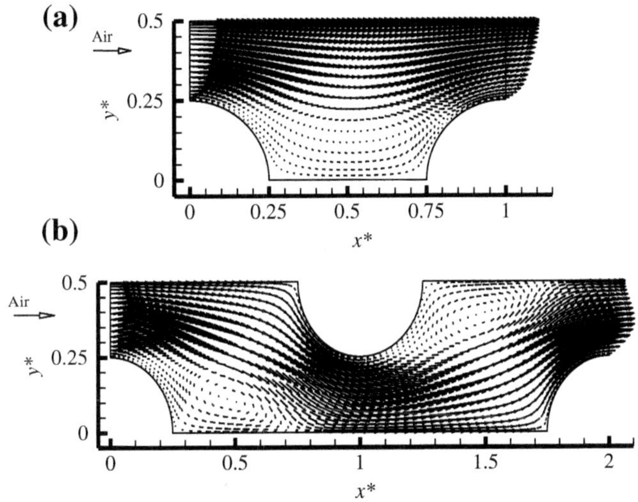

FIGURE 7.21

Velocity vectors for the air flow, $P_L/(2r_o) = P_T/(2r_o) = 2.0, Re_a = 100$. (a) In-line arrangement; (b) staggered arrangement.

For the in-line arrangement, when Re_a is higher than 100 and ϕ is less than 0.16, $Nu_{ave,a}$ is higher than $Nu_{ave,H}$. For the staggered arrangement, when the packing fractions are less than 0.16, the air-side $Nu_{ave,a}$ is somewhat less than $Nu_{ave,T}$. The values of $Nu_{ave,T}$, $Nu_{ave,H}$, $Nu_{ave,a}$ and f_a for the staggered arrangement are higher than those for the in-line arrangement.

For the solution flow in the round tubes (liquid side), the channel is long enough ($L = 30$ cm) for the solution flow to get thermally fully developed ($z = 5$ mm). The fully developed local Nusselt number ($Nu_{C,s}$) and the overall mean Nusselt number ($Nu_{ave,s}$) under the naturally formed conditions are invariant with different arrangements and different packing fractions. They are 4.48 and 4.56, respectively.

7.6.5 Sherwood numbers

Air-side mean Sherwood numbers ($Sh_{ave,a}$) for the in-line and the staggered arrays are listed in Tables 7.6 and 7.7, respectively. For the in-line array, air-side $Sh_{ave,a}$ is mostly between$Nu_{ave,T}$ and $Nu_{ave,H}$. However, it becomes higher than $Nu_{ave,H}$ when Re_a is higher than 100 and ϕ is less than 0.16. For the staggered arrangement, air-side $Sh_{ave,a}$ is between $Nu_{ave,T}$ and $Nu_{ave,H}$ all the time. Further, it is closer to $Nu_{ave,H}$. For both the in-line and the staggered arrangements, the air-side $Sh_{ave,a}$ increases with increasing packing fraction (ϕ) and/or Reynolds numbers.

For the solution flow in the circular channel (liquid side), the variations of the local and mean Sherwood numbers along the z axis are similar to common round tubes [13,14]. However, the concentration boundary layer ($z^*_{m,s} = 0.05, 25$ cm) develops much more slowly than the thermal boundary layer ($z^*_{h,s} = 0.3, 3$ cm). This is because that the Schmidt number of the LiCl solution is rather large ($Sc_s = 1390$). However, the duct is still long enough ($L = 30$ cm) for the solution flow to get

Table 7.6 The friction factor and the mean Nusselt and Sherwood numbers for the air stream across the hollow fiber membrane bundle, with in-line arrangement and $P_T = P_L$, laminar flow, considering interactions between neighboring fibers.

$P_L/(2r_o)$	ϕ	$Re_a = 50$					$Re_a = 100$				
		$Nu_{ave,T}$	$Nu_{ave,H}$	$Nu_{ave,a}$	$Sh_{ave,a}$	f_a	$Nu_{ave,T}$	$Nu_{ave,H}$	$Nu_{ave,a}$	$Sh_{ave,a}$	f_a
1.25	0.502	7.25	19.56	7.68	18.85	0.880	7.86	20.25	8.46	19.81	0.517
1.5	0.349	4.90	9.11	4.75	8.78	0.554	5.31	9.29	5.99	9.02	0.318
1.75	0.256	3.96	6.07	4.01	5.85	0.411	4.30	6.46	5.09	6.27	0.232
2.0	0.196	3.43	4.71	3.44	4.56	0.327	3.74	5.09	4.56	4.97	0.180
2.25	0.155	3.07	3.92	3.04	3.81	0.270	3.33	4.24	4.07	4.18	0.144
2.5	0.125	2.81	3.39	2.69	3.31	0.229	3.02	3.62	3.59	3.62	0.115
$P_L/(2r_o)$	ϕ	$Re_a = 200$					$Re_a = 300$				
		$Nu_{ave,T}$	$Nu_{ave,H}$	$Nu_{ave,a}$	$Sh_{ave,a}$	f_a	$Nu_{ave,T}$	$Nu_{ave,H}$	$Nu_{ave,a}$	$Sh_{ave,a}$	f_a
1.25	0.502	9.25	21.32	9.44	20.21	0.346	10.74	22.58	10.69	21.26	0.293
1.5	0.349	6.27	9.68	6.62	9.34	0.213	7.36	10.40	7.11	9.88	0.183
1.75	0.256	5.13	6.94	5.82	6.72	0.157	6.13	7.53	6.15	7.23	0.140
2.0	0.196	4.46	5.52	5.52	5.43	0.121	5.41	5.98	5.93	5.88	0.111
2.25	0.155	3.93	4.54	5.24	4.60	0.0899	4.50	4.83	5.78	4.98	0.0828
2.5	0.125	3.31	3.78	4.96	4.02	0.0632	3.58	3.83	5.58	4.29	0.0488

Table 7.7 The friction factor and the mean Nusselt and Sherwood numbers for the air stream across the hollow fiber membrane bundle, with staggered arrangement and $P_T = P_L$, laminar flow, considering interactions between neighboring fibers.

$P_L/(2r_o)$	ϕ	$Re_a = 50$					$Re_a = 100$				
		$Nu_{ave,T}$	$Nu_{ave,H}$	$Nu_{ave,a}$	$Sh_{ave,a}$	f_a	$Nu_{ave,T}$	$Nu_{ave,H}$	$Nu_{ave,a}$	$Sh_{ave,a}$	f_a
1.25	0.502	12.84	18.19	12.73	17.62	1.343	16.45	21.37	15.15	20.29	1.004
1.5	0.349	9.72	12.53	6.85	11.88	1.059	12.09	15.23	10.64	14.71	0.786
1.75	0.256	7.93	10.50	5.41	10.04	0.903	9.93	12.66	8.17	12.41	0.670
2.0	0.196	7.01	9.42	4.40	9.14	0.777	8.63	11.37	6.81	11.26	0.577
2.25	0.155	6.41	8.74	3.69	8.60	0.691	7.81	10.59	5.87	10.57	0.515
2.5	0.125	5.91	8.25	3.18	8.21	0.614	7.23	10.05	5.18	10.10	0.456
$P_L/(2r_o)$	ϕ	$Re_a = 200$					$Re_a = 300$				
		$Nu_{ave,T}$	$Nu_{ave,H}$	$Nu_{ave,a}$	$Sh_{ave,a}$	f_a	$Nu_{ave,T}$	$Nu_{ave,H}$	$Nu_{ave,a}$	$Sh_{ave,a}$	f_a
1.25	0.502	22.04	27.17	20.16	25.92	0.773	26.46	31.62	26.28	30.26	0.653
1.5	0.349	15.63	18.79	14.38	18.29	0.615	18.61	21.71	18.57	21.10	0.538
1.75	0.256	12.75	15.63	11.51	15.32	0.533	15.28	18.16	14.27	17.69	0.473
2.0	0.196	11.14	14.15	9.79	13.92	0.460	13.46	16.54	12.38	16.12	0.421
2.25	0.155	10.12	13.28	8.28	13.09	0.413	12.31	15.61	11.11	15.24	0.364
2.5	0.125	9.41	12.71	7.31	12.56	0.344	11.55	15.02	10.21	14.67	0.320

fully developed. The fully developed Sherwood numbers ($Sh_{C,s}$) and the overall mean Sherwood numbers ($Sh_{ave,s}$) are the same for different fiber tubes. This feature is similar to the solution-side $Nu_{C,s}$ and $Nu_{ave,s}$. The solution-side $Sh_{C,s}$ and $Sh_{ave,s}$ under the naturally formed boundary conditions are 4.35 and 5.32, respectively. The solution-side $Sh_{C,s}$ is a bit less than the solution-side $Nu_{C,s}$ (4.48).

Comparison with free surface model

The f_a estimated by the free surface model is lower than a staggered arrangement but higher than an in-line arrangement. The values calculated by a free surface model may deviate 20% higher than a regularly packed bundle when the packing fractions are higher than 0.25.

Similarly, the mean Nusselt numbers calculated by a free surface model are between the in-line and staggered arrangements. The higher the packing fractions are, the larger the deviations between the free surface models and the current predictions are. A 30% deviation from a free surface model may occur for a packed bundle when the packing fractions are larger than 0.25. The free surface model is reasonable when the fibers are sparsely populated, say, the packing fraction less than 0.25.

7.7 Cross-flow arrangement and turbulent flow, periodic cell model

Now let's go to turbulent flow. A hollow fiber has been investigated with periodic cells. The fluid flow and the conjugate heat and mass transfer between the fluid and these surrounding fibers were investigated. The shell-side flow was assumed to be laminar. In reality, when the air velocities are high, turbulence would be generated in the fiber bundles. To account for the turbulence on the shell side generated by the impinging fibers, a low-*Re* k-ε turbulence model is used to describe the air flow on the shell side in this section. The Nusselt and Sherwood numbers in the bundle are obtained and analyzed. The transport data for this case are presented.

7.7.1 Mathematical model

Governing equations

The computational cells are the same as those in the previous section for laminar flow. In practical applications, the Reynolds number for the solution stream in the inner circular channel is below 10 (much less than 2300), so laminar flow is assumed. However, for the shell-side air stream, it has been found that the flow tends to become turbulent due to the continuous disturbances from the numerous fine fibers. The laminar model is not suitable for the air stream when the Reynolds number is larger than 300 [5]. Further, the local turbulent Reynolds numbers ($Re_t = \rho k_a^2/\mu\varepsilon_a$) are less than 150. In this case, the near-wall flow cannot be accurately modeled by a standard k-ε turbulence model. To address this problem, a low-*Re* k-ε model is used for the air stream. The normalized governing momentum, heat and mass equations for the air stream are described as follows [5,20,21]:

$$\frac{\partial u_x^*}{\partial x^*} + \frac{\partial u_y^*}{\partial y^*} = 0 \tag{7.143}$$

$$\frac{\partial\left(u_x^* u_x^*\right)}{\partial x^*}+\frac{\partial\left(u_y^* u_x^*\right)}{\partial y^*}=\frac{\partial}{\partial x^*}\left(\Gamma_x \frac{\partial u_x^*}{\partial x^*}\right)+\frac{\partial}{\partial y^*}\left(\Gamma_x \frac{\partial u_x^*}{\partial y^*}\right)-\frac{1}{2} \frac{\partial P_{\mathrm{a}}^*}{\partial x^*}$$
$$+\frac{\partial}{\partial x^*}\left(\Gamma_x \frac{\partial u_x^*}{\partial x^*}\right)+\frac{\partial}{\partial y^*}\left(\Gamma_x \frac{\partial u_y^*}{\partial x^*}\right) \tag{7.144}$$

$$\frac{\partial\left(u_x^* u_y^*\right)}{\partial x^*}+\frac{\partial\left(u_y^* u_y^*\right)}{\partial y^*}=\frac{\partial}{\partial x^*}\left(\Gamma_y \frac{\partial u_y^*}{\partial x^*}\right)+\frac{\partial}{\partial y^*}\left(\Gamma_y \frac{\partial u_y^*}{\partial y^*}\right)-\frac{1}{2} \frac{\partial P_{\mathrm{a}}^*}{\partial y^*}$$
$$+\frac{\partial}{\partial x^*}\left(\Gamma_y \frac{\partial u_x^*}{\partial y^*}\right)+\frac{\partial}{\partial y^*}\left(\Gamma_y \frac{\partial u_y^*}{\partial y^*}\right) \tag{7.145}$$

$$\frac{\partial\left(u_x^* k_{\mathrm{a}}^*\right)}{\partial x^*}+\frac{\partial\left(u_y^* k_{\mathrm{a}}^*\right)}{\partial y^*}=\frac{\partial}{\partial x^*}\left(\Gamma_{\mathrm{k}} \frac{\partial k_{\mathrm{a}}^*}{\partial x^*}\right)+\frac{\partial}{\partial y^*}\left(\Gamma_{\mathrm{k}} \frac{\partial k_{\mathrm{a}}^*}{\partial y^*}\right)$$
$$+\frac{\mu_{\mathrm{t}} u_{\mathrm{ai}}}{\rho_{\mathrm{a}} P_{\mathrm{L}} k_{\mathrm{ai}}} G_{\mathrm{k}}^*-\frac{P_{\mathrm{L}} \varepsilon_{\mathrm{ai}}}{u_{\mathrm{ai}} k_{\mathrm{ai}}} \varepsilon_{\mathrm{a}}^*-E^* \tag{7.146}$$

$$\frac{\partial\left(u_x^* \varepsilon_{\mathrm{a}}^*\right)}{\partial x^*}+\frac{\partial\left(u_y^* \varepsilon_{\mathrm{a}}^*\right)}{\partial y^*}=\frac{\partial}{\partial x^*}\left(\Gamma_{\varepsilon} \frac{\partial \varepsilon_{\mathrm{a}}^*}{\partial x^*}\right)+\frac{\partial}{\partial y^*}\left(\Gamma_{\varepsilon} \frac{\partial \varepsilon_{\mathrm{a}}^*}{\partial y^*}\right)$$
$$+\frac{c_1 \mu_{\mathrm{t}} u_{\mathrm{ai}}}{\rho_{\mathrm{a}} P_{\mathrm{L}} k_{\mathrm{ai}}} \frac{\varepsilon_{\mathrm{a}}^* G_{\mathrm{k}}^*}{k_{\mathrm{a}}^*}-\frac{c_2 P_{\mathrm{L}} \varepsilon_{\mathrm{ai}}}{u_{\mathrm{ai}} k_{\mathrm{ai}}} \frac{\varepsilon_{\mathrm{a}}^{*2}}{k_{\mathrm{a}}^*}+F^* \tag{7.147}$$

$$\frac{\partial\left(u_x^* \theta_{\mathrm{a}}\right)}{\partial x^*}+\frac{\partial\left(u_y^* \theta_{\mathrm{a}}\right)}{\partial y^*}=\frac{\partial}{\partial x^*}\left(\Gamma_{\mathrm{T}} \frac{\partial \theta_{\mathrm{a}}}{\partial x^*}\right)+\frac{\partial}{\partial y^*}\left(\Gamma_{\mathrm{T}} \frac{\partial \theta_{\mathrm{a}}}{\partial y^*}\right) \tag{7.148}$$

$$\frac{\partial\left(u_x^* \xi_{\mathrm{a}}\right)}{\partial x^*}+\frac{\partial\left(u_y^* \xi_{\mathrm{a}}\right)}{\partial y^*}=\frac{\partial}{\partial x^*}\left(\Gamma_{\omega} \frac{\partial \xi_{\mathrm{a}}}{\partial x^*}\right)+\frac{\partial}{\partial y^*}\left(\Gamma_{\omega} \frac{\partial \xi_{\mathrm{a}}}{\partial y^*}\right) \tag{7.149}$$

The equations for the solution side are the same as in the previous section. The definitions for dimensionless coordinates are the same as in the previous section as well, where $D_{\mathrm{h,s}}$ ($= 2r_{\mathrm{i}}$) is the hydraulic diameter for the solution flow channel (m). *Re*, *Pr* and *Sc* are Reynolds, Prandtl and Schmidt numbers, respectively.

The generation term and two extra terms appear in the above equations:

$$G_{\mathrm{k}}^*=2\left[\left(\frac{\partial u_x^*}{\partial x^*}\right)^2+\left(\frac{\partial u_y^*}{\partial y^*}\right)^2\right]+\left(\frac{\partial u_x^*}{\partial y^*}+\frac{\partial u_y^*}{\partial x^*}\right)^2 \tag{7.150}$$

$$E^*=\frac{2 \mu_{\mathrm{a}}}{\rho_{\mathrm{a}} P_{\mathrm{L}} u_{\mathrm{ai}}}\left(\frac{\partial\left(k_{\mathrm{a}}^*\right)^{1/2}}{\partial y^*}\right)^2 \tag{7.151}$$

$$F^* = \frac{2\mu_t \mu_a u_{ai}}{\rho_a^2 P_L^3 \varepsilon_{ai}} \left(\frac{\partial^2 u_y^*}{\partial y^{*2}} \right)^2 \tag{7.152}$$

where term E^* is added due to the anisotropy of the kinetic energy in the viscous layer.

In Eqs. (7.144)–(7.149), the associated diffusion coefficients are [5,20,21]:

$$\Gamma_x = \Gamma_y = \frac{2r_o}{P_L} \frac{1 + \frac{\mu_t}{\mu_a}}{Re_a} \tag{7.153}$$

$$\Gamma_k = \frac{2r_o}{P_L} \frac{1 + \frac{\mu_t}{\mu_a \sigma_k}}{Re_a} \tag{7.154}$$

$$\Gamma_\varepsilon = \frac{2r_o}{P_L} \frac{1 + \frac{\mu_t}{\mu_a \sigma_\varepsilon}}{Re_a} \tag{7.155}$$

$$\Gamma_T = \frac{2r_o}{P_L} \frac{\frac{1}{Pr_a} + \frac{\mu_t}{\mu_a \sigma_T}}{Re_a} \tag{7.156}$$

$$\Gamma_\omega = \frac{2r_o}{P_L} \frac{\frac{1}{Sc_a} + \frac{\mu_t}{\mu_a \sigma_\omega}}{Re_a} \tag{7.157}$$

Other coefficients in the turbulence model are the same as in Section 7.5 for the free surface model. The definitions of friction factor, maximum velocity, Nusselt and Sherwood numbers in the cell are the same as in the previous section for laminar flow, except that the values are time-averaged values.

Boundary conditions

Similar to the solution with a laminar model, the boundary-fitted coordinate transformation technique is used.

The inlet boundary conditions of the turbulent kinetic energy and dissipation rate for the air stream [18]

$$k_a^* = \frac{0.01(u_f^2/2)}{k_{ai}} \quad \text{and} \quad \varepsilon_a^* = \frac{500 c_\mu \left(k_a^*\right)^2}{V_{ai} D_{h,a} \varepsilon_{ai}} \tag{7.158}$$

where u_f is the resultant velocity. The inlet turbulence kinetic energy and dissipation rate for the air stream are updated by the new velocity at each iteration. Other periodic boundary conditions at inlets and outlets, on membrane surfaces, and symmetric boundaries are the same as those for laminar flow.

Heat and mass fluxes on membrane surfaces are the conjugate conditions for the air stream and the liquid solution.

The numerical scheme is similar to laminar flow. The code is validated both numerically and experimentally with the modules constructed with the membrane fibers fabricated in the laboratory [5].

7.7.2 Turbulence for the air flow

Turbulence intensities (u_a'/u_a) in the air flow for the in-line and the staggered arrangements are shown in Figure 7.22a and b, respectively. As can be seen, the turbulence intensities are higher where velocities are higher. It has been found that vortexes appear behind the fibers. The turbulence intensities behind the fibers where vortexes appear are higher than other regions. For both the in-line and the staggered arrangements, the turbulence intensities are higher than 0.07, meaning that the fluctuations of velocities for the air flow in the bundle are relatively large [13,14]. In other words, the turbulent behavior of the air flow across the bundle should be taken into account. Therefore the turbulence model employed here is reasonable and necessary to account for the disturbances from the fibers.

7.7.3 Nusselt and Sherwood numbers

Air-side mean Nusselt numbers ($Nu_{ave,a}$) and the friction factors (f_a) are listed in Tables 7.8 and 7.9 for the in-line and the staggered arrangements, respectively. The packing fraction (ϕ) and the Reynolds numbers are those commonly used in engineering applications. Also listed are the mean values under uniform temperature ($Nu_{ave,T}$) and uniform heat flux boundary conditions ($Nu_{ave,H}$). In such cases, the

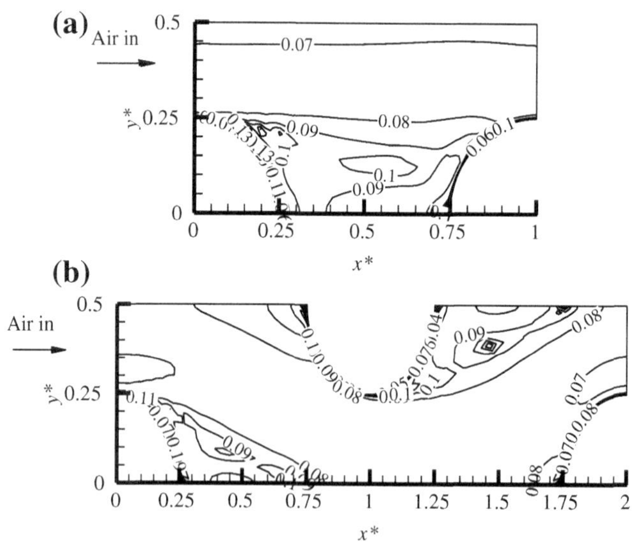

FIGURE 7.22

Contours of the turbulence intensities for the air stream, $P_L/(2r_o) = P_T/(2r_o) = 2.0$, $\phi = 0.196$, $Re_a = 300$. (a) In-line arrangement; (b) Staggered arrangement.

Table 7.8 The friction factor and the mean Nusselt and Sherwood numbers for the air stream across the hollow fiber membrane bundle, with in-line arrangement and $P_T = P_L$, turbulent flow, considering interactions between neighboring fibers.

$P_L/(2r_o)$	ϕ	$Re_a = 300$					$Re_a = 400$				
		$Nu_{ave,T}$	$Nu_{ave,H}$	$Nu_{ave,a}$	$Sh_{ave,a}$	f_a	$Nu_{ave,T}$	$Nu_{ave,H}$	$Nu_{ave,a}$	$Sh_{ave,a}$	f_a
1.25	0.502	13.86	19.12	11.75	18.88	0.362	15.23	21.00	15.49	20.21	0.328
1.5	0.349	8.85	10.96	7.92	10.56	0.227	10.07	12.52	8.98	11.72	0.211
1.75	0.256	7.18	8.52	6.84	7.90	0.172	8.34	10.20	7.67	8.89	0.164
2.0	0.196	6.30	7.32	6.51	6.50	0.135	7.45	8.93	7.03	7.36	0.131
2.25	0.155	5.62	6.22	6.30	5.53	0.106	6.76	7.71	6.95	6.27	0.104
2.5	0.125	4.89	5.68	6.03	4.75	0.0884	5.84	6.72	6.62	5.32	0.0875

$P_L/(2r_o)$	ϕ	$Re_a = 500$					$Re_a = 600$				
		$Nu_{ave,T}$	$Nu_{ave,H}$	$Nu_{ave,a}$	$Sh_{ave,a}$	f_a	$Nu_{ave,T}$	$Nu_{ave,H}$	$Nu_{ave,a}$	$Sh_{ave,a}$	f_a
1.25	0.502	16.70	23.06	17.98	21.97	0.310	18.19	25.31	20.66	23.91	0.299
1.5	0.349	11.33	14.44	12.66	12.98	0.203	12.57	16.48	14.93	14.31	0.198
1.75	0.256	9.52	12.03	8.48	9.92	0.160	10.70	13.86	12.28	11.02	0.156
2.0	0.196	8.62	10.60	7.78	8.26	0.129	9.78	12.19	8.59	9.15	0.128
2.25	0.155	7.95	9.24	7.55	7.02	0.103	9.14	10.70	8.30	7.71	0.102
2.5	0.125	6.53	7.82	7.37	5.95	0.0868	8.50	9.91	8.12	6.38	0.0852

Table 7.9 The friction factor and the mean Nusselt and Sherwood numbers for the air stream across the hollow fiber membrane bundle, with staggered arrangement and $P_T = P_L$, turbulent flow, considering interactions between neighboring fibers.

$P_L/(2r_o)$	ϕ	$Re_a = 300$					$Re_a = 400$				
		$Nu_{ave,T}$	$Nu_{ave,H}$	$Nu_{ave,a}$	$Sh_{ave,a}$	f_a	$Nu_{ave,T}$	$Nu_{ave,H}$	$Nu_{ave,a}$	$Sh_{ave,a}$	f_a
1.25	0.502	26.60	31.86	25.14	30.46	0.690	30.68	35.72	29.46	34.19	0.645
1.5	0.349	18.75	21.88	17.78	21.22	0.518	21.62	24.57	20.88	23.74	0.487
1.75	0.256	15.39	18.33	14.44	17.82	0.431	17.84	21.69	17.12	20.01	0.408
2.0	0.196	13.31	17.62	12.29	17.07	0.362	15.21	20.42	14.53	19.63	0.340
2.25	0.155	12.38	15.82	11.22	15.40	0.317	14.51	18.07	13.57	17.47	0.300
2.5	0.125	11.58	15.23	10.31	14.85	0.276	13.62	17.51	12.56	16.94	0.262

$P_L/(2r_o)$	ϕ	$Re_a = 500$					$Re_a = 600$				
		$Nu_{ave,T}$	$Nu_{ave,H}$	$Nu_{ave,a}$	$Sh_{ave,a}$	f_a	$Nu_{ave,T}$	$Nu_{ave,H}$	$Nu_{ave,a}$	$Sh_{ave,a}$	f_a
1.25	0.502	34.58	40.02	33.52	37.52	0.611	38.37	43.92	37.33	40.58	0.592
1.5	0.349	24.39	27.06	23.80	26.06	0.471	27.10	29.40	26.59	28.23	0.460
1.75	0.256	20.22	22.88	19.67	22.03	0.395	22.53	24.94	22.13	23.95	0.386
2.0	0.196	16.91	23.11	16.57	22.10	0.325	18.42	25.65	18.43	24.45	0.313
2.25	0.155	16.55	20.20	15.87	19.43	0.291	18.51	22.26	18.05	21.31	0.284
2.5	0.125	15.55	19.72	14.74	18.86	0.253	17.39	21.85	16.89	20.92	0.247

fiber membrane surfaces are set manually either to uniform temperature (T) or to uniform heat flux (H) conditions. As can be seen, the air-side $Nu_{ave,a}$ and f_a rise with increasing packing fractions (ϕ), but f_a decreases with rising Reynolds numbers. For the in-line arrangement, generally the air-side $Nu_{ave,a}$ is between the ($Nu_{ave,T}$) and the ($Nu_{ave,H}$). There are some exceptions. When Re_a is smaller than 400 and ϕ is less than 0.16, $Nu_{ave,a}$ is higher than $Nu_{ave,H}$. For the staggered arrangement, the air-side $Nu_{ave,a}$ is somewhat less than $Nu_{ave,T}$. The values of $Nu_{ave,T}$, $Nu_{ave,H}$, $Nu_{ave,a}$ and f_a for the staggered arrangement are higher than those for the in-line arrangement.

Local Nusselt numbers for the liquid side decrease sharply from high values at the inlet to fully developed values after the entry length. The duct is long enough ($L=30\,\text{cm}$) for the solution flow to get thermally fully developed ($z=5\,\text{mm}$). The fully developed local Nusselt number ($Nu_{C,s}$) and the overall mean Nusselt number ($Nu_{ave,s}$) under the naturally formed conditions are invariant with different arrangements and different packing fractions. They are 4.48 and 4.56, respectively.

Air-side mean Sherwood numbers ($Sh_{ave,a}$) for the in-line and the staggered arrangements are listed in Tables 7.8 and 7.9, respectively. For the in-line arrangement, the air-side $Sh_{ave,a}$ is mostly between $Nu_{ave,T}$ and $Nu_{ave,H}$. However, it becomes less than $Nu_{ave,T}$when ϕ is less than 0.2. For the staggered arrangement, the air-side $Sh_{ave,a}$ is between $Nu_{ave,T}$ and $Nu_{ave,H}$ all the time. Further, it is closer to $Nu_{ave,H}$. For both the in-line and the staggered arrangements, the air-side $Sh_{ave,a}$ increases with increasing packing fraction and/or Reynolds numbers.

For the solution flow in the circular channel (liquid side), the variations of the local and mean Sherwood numbers along the z axis are similar to common round tubes [13,14]. However, the concentration boundary layer ($z^*_{m,s} = 0.05, 25$ cm) develops much more slowly than the thermal boundary layer ($z^*_{h,s} = 0.3, 3$ cm). This is because that the Schmidt number of the LiCl solution is rather large ($Sc_s = 1390$). The liquid-side $Sh_{C,s}$ and $Sh_{ave,s}$ under the naturally formed boundary conditions are 4.34 and 5.31, respectively. The solution-side $Sh_{C,s}$ is a bit less than the solution-side $Nu_{C,s}$ (= 4.47).

7.8 Conclusion

This chapter provides a systematic review of the fluid flow and conjugate heat and mass transfer in a hollow fiber membrane bundle. Two methods are used in analysis: the free surface cell model and the periodic cell model. When the packing densities are below 0.25, free surface models are suitable. When the packing densities are higher than 0.25, periodic cells, which consider the interactions between neighboring fibers, should be used. The fiber bundles can be simplified to in-line arrays and staggered arrays.

Two flow arrangements between the shell side and the tube side are considered: counter flow and cross flow. Air pressure drops for the cross flow are lower, and the heat and mass transfer coefficients are higher. For liquid-to-air processes, cross flow is more suitable. For liquid-to-liquid processes, a counter-flow arrangement is most common. For the cross-flow arrangement, both the laminar flow and turbulent flow on the shell side are considered. When the Reynolds numbers are below 300, a laminar flow assumption is acceptable. When the shell-side Reynolds numbers are higher than 300, turbulent models should be used. A low-*Re* k-ε model is successfully used in the prediction of air-side flows. The transport data for these different operating conditions are presented, and can provide design guidelines and tools for performance analysis.

For convenience in selecting appropriate transport data, the following guidelines are recommended:

(1) Counter flow, Table 7.3.
(2) Cross flow, packing fraction $\leq$0.25, laminar flow, Table 7.4.
(3) Cross flow, packing fraction $\leq$0. 25, turbulent flow, Table 7.5.
(4) Cross flow, packing fraction >0.25, laminar flow, in-line and staggered arrays, Table 7.6 and Table 7.7 respectively.
(5) Cross flow, packing fraction >0.25, turbulent flow, in-line and staggered arrays, Table 7.8 and Table 7.9 respectively.

References

[1] Incropera FP, Dewitt DP. Introduction to heat transfer. 3rd edn. New York: John Wiley & Sons; 1996.

[2] Zhang LZ, Huang SM, Chi JH, Pei LX. Conjugate heat and mass transfer in a hollow fiber membrane module for liquid desiccant air dehumidification: A free surface model approach. Int J Heat Mass Transfer 2012;55:3789–99.

[3] Zhang LZ, Huang SM, Pei LX. Conjugate heat and mass transfer in a cross-flow hollow fiber membrane contactor for liquid desiccant air dehumidification. Int J Heat Mass Transfer 2012;55(25–26):8065–76.

[4] Huang SM, Zhang LZ, Tang K, Pei LX. Turbulent heat and mass transfer across a hollow fiber membrane tube bank in liquid desiccant air dehumidification. ASME J Heat Transfer 2012;134:082001-1–082001-10.

[5] Zhang LZ, Huang SM, Zhang WB. Turbulent heat and mass transfer across a hollow fiber membrane bundle considering interactions between neighboring fibers. Int J Heat Mass Transfer (in press).

[6] Zhang LZ. Heat and mass transfer in a randomly packed hollow fiber membrane module: a fractal model approach. Int J Heat Mass Transfer 2011;54:2921–31.

[7] Zhang LZ. Total heat recovery, heat and moisture recovery from ventilation air. New York: Nova Science; 2009.

[8] Porcheron F, Drozdz S. Hollow fiber membrane contactor transient experiments for the characterization of gas/liquid thermodynamics and mass transfer properties. Chem Eng Sci 2009;64:265–75.

[9] Park HH, Deshwal BR, Kimb IW, Lee HK. Absorption of SO_2 from flue gas using PVDF hollow fiber membranes in a gas/liquid contactor. J Membr Sci 2008;319:29–37.

[10] Luis P, Garea A, Irabien A. Modelling of a hollow fibre ceramic contactor for SO_2 absorption. Sep Purif Technol 2010;72:174–9.

[11] Happel J. Viscous flow relative to arrays of cylinders. AIChE J 1959;5:174–8.

[12] Manglik RM, Fang PP. Effect of eccentricity and thermal boundary conditions on laminar fully developed flow in annular ducts. Int J Heat Fluid Flow 1995;16:298–306.

[13] Kays WM, London AL. Compact heat exchangers. New York: McGraw-Hill; 1984.

[14] Kays WM, Crawford ME. Convective heat and mass transfer. New York: McGraw-Hill, Inc.; 1993.

[15] Frank K, Mark SB. Principles of heat transfer. 4th edn.. New York: Harper & Row; 1986.

[16] Jones WP, Launder BE. The prediction of low-Reynolds-number phenomena with two-equation model of turbulence. Int J Heat Mass Transfer 1973;16:1119–30.

[17] Shibu S, Chhabra RP, Eswaran V. Power law fluid flow over a bundle of cylinders at intermediate Reynolds numbers. Chem Eng Sci 2001;56:5545–54.

[18] Patankar SV, Sparrow EM, Ivanovic M. Thermal interaction among the confining walls of a turbulent recirculating flow. Int J Heat Mass Transfer 1978;24:269–74.

[19] Huang SM, Zhang LZ, Pei LX. Transport phenomena in a cross-flow hollow fiber membrane bundle used for liquid desiccant air dehumidification. Indoor and Built Environment, doi: http://dx.doi.org/10.1177/1420326X12452881 (in press).

[20] Antonopoulos KA. Heat transfer in tube banks under conditions of turbulent inclined flow. Int J Heat Mass Transfer 1985;28:1645–56.

[21] Antonopoulos KA. Pressure drop during laminar oblique flow through in-line tube assemblies. Int J Heat Mass Transfer 1987;30:673–81.

[22] Mandhani VK, Chhabra RP, Eswaran V. Forced convection heat transfer in tube banks in cross-flow. Chem Eng Sci 2002;57:379–91.

CHAPTER

8 Heat and Mass Transfer in Hollow Fiber Membrane Bundles with Randomly Distributed Fibers

8.1 Introduction

Hollow fiber membrane modules have gained increasing attention in many industries. The processes include gas separation, waste heat recovery, water treatment, product purification, liquid distillation, pervaporation [1–4], etc. The reasons for the widespread use are: (1) The packing density is high; there are large contact areas between tube side and shell side; (2) the tube side can endure high pressure and the driving force can be very high; (3) convenience in sealing of flows: Shell-side sealing of fibers can be performed with epoxy resins at the both ends of the shell.

A module has numerous fibers. They are grouped together to form a hollow fiber bundle. The bundle is then packed into a larger tube shell, forming a shell-and-tube heat exchanger structure, as shown in Figure 8.1. The number of fibers is much higher than the number of tubes in a heat exchanger [5,6].

The previous chapter investigated the fluid flow and conjugate heat and mass transfer in a hollow fiber membrane bundle. The fibers are assumed to be uniformly distributed. This is only an ideal approximation, at least in some cases. Though a uniformly distributed fiber bundle is possible by a strict arrangement of fibers, in many cases the fiber distribution is rather random, as shown in Figure 8.2. There are several reasons. (1) Correct control of fiber position is difficult, because fibers are very fine, with outer diameters from 0.5 mm to 2.0 mm. The number of fibers is very large. For example, there are 4000 fibers in a shell 10 cm in diameter. For so many fine and soft fibers, precise control of positions is rather difficult. (2) Correct control of fiber position is expensive. It needs laser drilling and machining processes on plates. In most cases, the fibers are grouped and packed together by hand unmethodically to reduce costs. As a result, in most applications, fiber distribution in a bundle is quite random.

For such modules, shell-side heat and mass transfer is a key parameter influencing system performance. Many studies have been conducted to investigate the shell-side mass transfer in a randomly packed membrane module [7–18]. Some studies have employed the Voronoi Tessellation theory [14] to model the randomly packed hollow fiber modules [15–18]. According to this theory, the voids in the module are classified into a series of cells which have a Voronoi Tessellation function distribution. It was found that the irregularity in packing tends to depress the performance seriously. The theory provides a possibility to analyze the randomly packed module quantitatively. Regretably, however, the

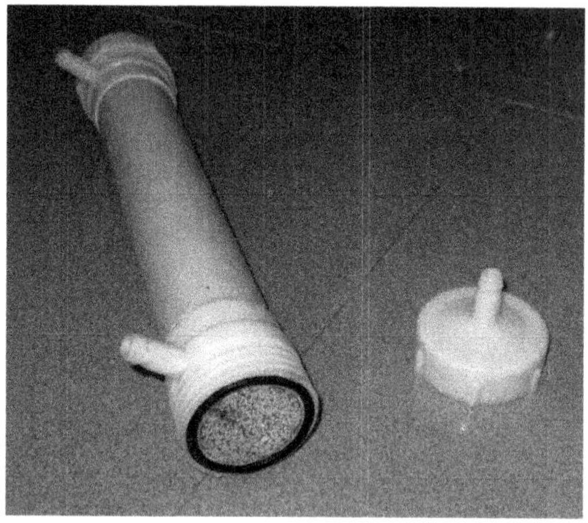

FIGURE 8.1

Structure of a hollow fiber membrane module.

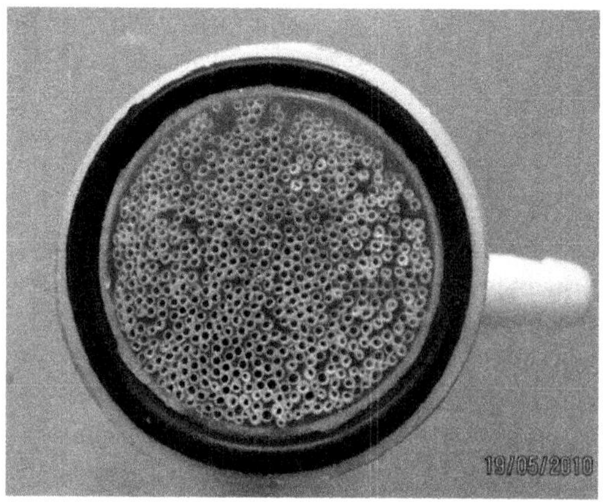

FIGURE 8.2

Picture of the head showing the randomly packed fibers in a shell.

degree of irregularity is not truly reflected in these studies. The reason is that this theory assumes a uniform probability function for all the modules, while in reality the degree of irregularity may be different from one module to another, even under the same packing fractions. Consequently it is still not known how exactly the irregularity affects the flow and transport phenomenon in a module. Recently, fractal theory has been used in the estimation of the degree of irregularity in fiber distribution [19]. Correlations were proposed for the calculation of shell-side convective mass transfer coefficients, by considering the degree of fiber irregularity. However, the transport data used for the basic cells were originally deduced for heat transfer in tube arrays under uniform temperature boundary conditions [5,6], not for fiber bundles.

This chapter is a step forward; similar to Ref. [19], fractal theory is employed to address the problem of non-uniformity in the packing. The transport data deduced in the previous chapter, which were specifically calculated for the conjugate heat and mass transfer in fiber bundles, are used as the properties for the unit cells.

8.2 Fractal theory

Fractal theory is an efficient tool for analyzing irregular but self-similar structures such as the fiber modules. The key issue of this theory is to use the fractal dimension of the packing to reflect the degree of irregularity in the module.

Exact fractal geometries are shown in Figure 8.3. They show exact self-similarities on different scales. In nature, many objects are non-exact fractal geometries. They seem irregular, but they have some degree of self-similarities. Euclidean geometry describes ordered objects using integer dimensions 0 (a dot), 1 (a line), 2 (a surface), and 3 (a cube), respectively. However, numerous objects in nature, such as rough surfaces, coastlines, mountains, and islands, are disordered and irregular, and they do not follow the Euclidean description since their length, area and volume are scale-dependent. Such objects are called fractal, mostly non-exact fractals, and are described using a non-integral dimension called the fractal dimension.

In recent years, fractal theory has been used extensively in the analysis of heat and mass transfer in porous media. The measure of a fractal structure, $M(L)$ is related to the length scale L, through a power law in the form of:

$$M(L) \sim L^{\mathrm{Df}} \tag{8.1}$$

where "~" means "scale as," and D_f is the fractal dimension of the structure. Associated with Eq. (8.1), the property of self-similarity implies that the value of D_f calculated from the relationship in Eq. (8.1) remains constant over a range of length scales, L. Geometric constructs such as the Sieroinsky gasket and the Koch curve are examples of exact fractals which exhibit identical self-similarity over an infinite range of length scales. However, self-similarity in a global sense is seldom observed in nature and the fractal description is based on a statistical self-similarity, i.e., the objects exhibit self-similarity in some average sense, over a certain local range of length scale, L, relevant to the problem. The concept of statistical self-similarity introduced above is used in the following section to develop a geometric description of the random distribution of hollow fiber membranes.

Fractal characteristics of fiber distributions

Orderly distributed arrays are the basic cells for heat and mass transfer. There are typically two kinds of arrays: in-line (or square array, SA) and staggered arrays (or triangular array, TA). For an orderly distributed array, the pitch to diameter ratio (σ) of the array is an important parameter and it can be calculated from the packing fraction as:

For [SA]

$$\sigma = \left(\frac{\pi}{4\phi}\right)^{0.5} \tag{8.2}$$

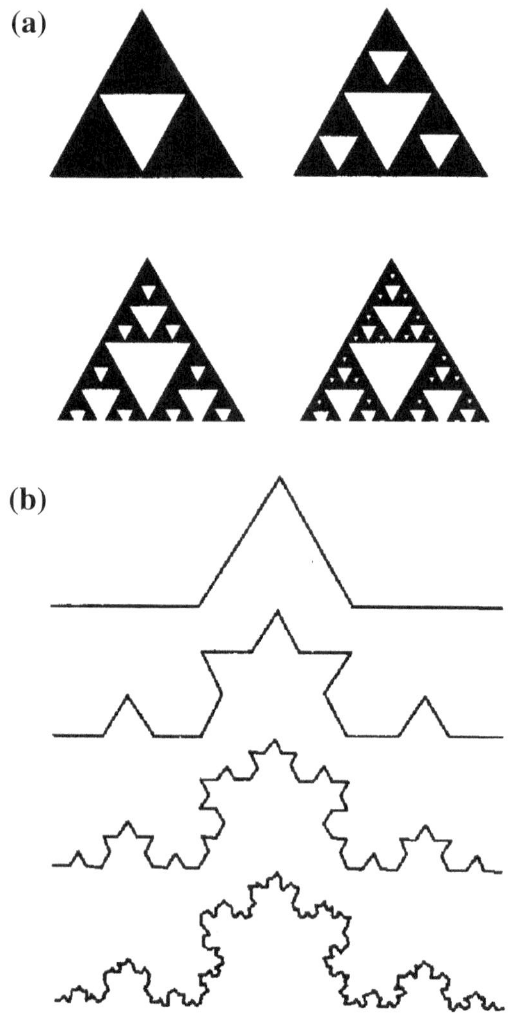

FIGURE 8.3

Classic exact fractal geometries. (a) Sieroinsky gasket; (b) Koch curve.

For [TA]

$$\sigma = \left(\frac{\pi}{2\sqrt{3}\phi}\right)^{0.5} \tag{8.3}$$

The fluid flow and heat mass transfer in the shell side are considered. Fluid flows in the cavities formed between the fibers. The fibers are randomly distributed in a shell, as shown in Figure 8.2. To show this clearly, Figure 8.4 gives an amplified view of a section in the sealing head in the module.

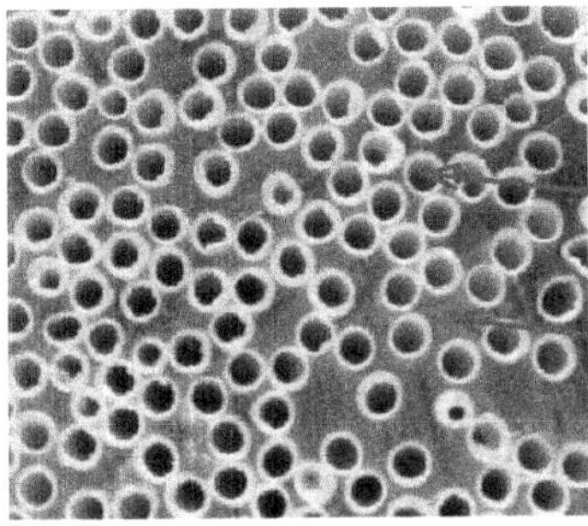

FIGURE 8.4

A representative section of the randomly distributed fiber bundle.

The tube-side flow is simple and can be calculated with well-developed correlations [12,13], as well as the data from previous sections. This chapter will focus on the shell-side flow. As can be seen, the cavities formed between the neighboring fibers exhibit a fractal nature. They are self-similar over a certain range of length scales.

This study employs an “imaginary free-surface cells” technique as used in Ref. [19] to set up the model. According to this methodology, the empty volume on the shell side is divided by the number of the fibers. Each volume, also called a cell, has a fiber in the center. The annular flow around each fiber is considered to be equivalent to the flow through an annular space formed by two concentric cylinders. One cylinder is the outer diameter of the fiber, and the other is a hypothetic free surface. All the flow is concentrated inside the imaginary cells. The concept is schematically shown in Figures 8.5 and 8.6.

For each cell, the local packing density is

$$\phi = \frac{d_o^2}{d_e^2} \tag{8.4}$$

where d_o is the outer diameter of a fiber, d_e is the diameter of the hypothetic cell having the fiber.

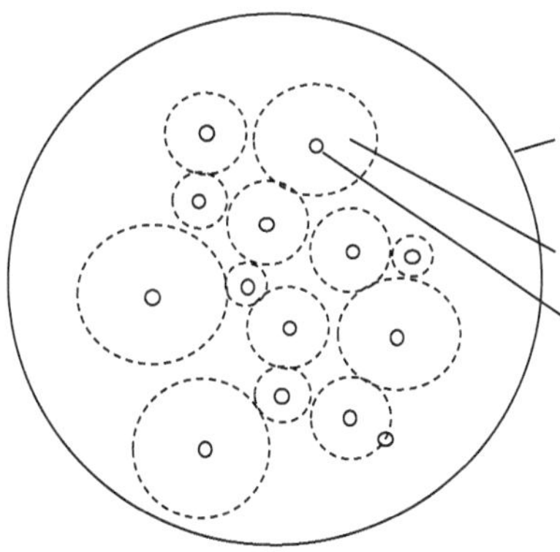

FIGURE 8.5

A bundle of equivalent hypothetic cells in the shell.

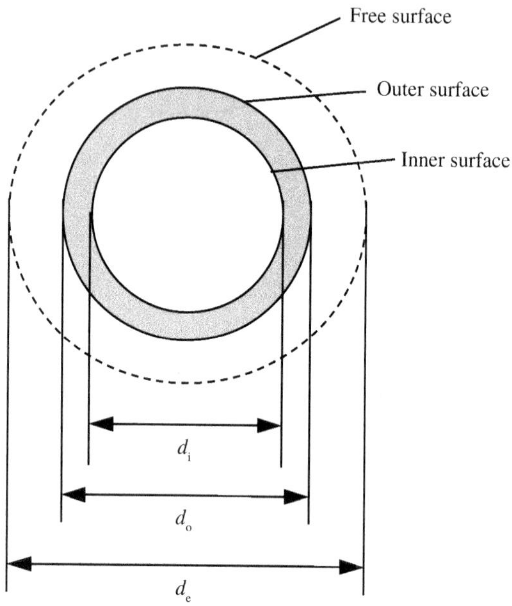

FIGURE 8.6

A hypothetic cell in the shell.

Local porosity

$$\varepsilon_t = 1 - \phi \tag{8.5}$$

The equivalent void diameter of each cell

$$d_v = \varepsilon_t^{1/2} d_e = \left(\frac{\varepsilon_t}{1 - \varepsilon_t}\right)^{1/2} d_o \tag{8.6}$$

Hydrodynamic diameter of the cell

$$D_h = \frac{4A_c}{P_f} = \frac{\varepsilon_t}{1 - \varepsilon_t} d_o = \frac{d_v^2}{d_o} \tag{8.7}$$

The shell can be considered as a bundle of cells with variable diameters. Due to the irregularity of the cavities formed by the fibers, cell diameters are also irregular. However, they exhibit a self-similar fractal nature. Each cell has a void space. Let the void diameter be d_v, and the relationship between the number of voids that are larger than d_v and the void size d_v is an important property of fractals.

The voids in a shell are analogous to the islands or lakes on Earth; they are self-similar. Their cumulative size distribution follows the power law relation [19]:

$$N(d_v) = \left(\frac{d_{v,max}}{d_v}\right)^{D_f} \tag{8.8}$$

where $N(d_v)$ represents the total number of voids with diameter greater than d_v on a unit area; $d_{v,max}$ is the maximum void diameter. From Eq. (8.8), the total void number can be expressed by

$$N(d_{v,min}) = \left(\frac{d_{v,max}}{d_{v,min}}\right)^{D_f} = N_f \tag{8.9}$$

and

$$-dN = D_f d_{v,max}^{D_f} d_v^{-(D_f+1)} d(d_v) \tag{8.10}$$

The negative sign in Eq. (8.10) implies that the void population decreases with the increase of void size. Equations (8.8) to (8.10) hold true for both exactly and statistically self-similar fractal geometries. For two-dimensional space, $1 \leq D_f \leq 2$. By combining Eqs. (8.8) and (8.10) we have

$$-\frac{dN}{N_f} = D_f d_{v,max}^{D_f} d_{v,max}^{-(D_f+1)} d(d_v) = f(d_v) d(d_v) \tag{8.11}$$

where $f(d_v)$ is the probability density function, and

$$\int_{-\infty}^{+\infty} f(d_v) d(d_v) = \int_{d_{v,min}}^{d_{v,max}} f(d_v) d(d_v) = 1 - \left(\frac{d_{v,min}}{d_{v,max}}\right)^{D_f} \tag{8.12}$$

The above equation equals unity if and only if

$$\frac{d_{v,min}}{d_{v,max}} = 0 \tag{8.13}$$

This implies that $d_{v,min} \ll d_{v,max}$ must be satisfied for fractal analysis of a shell. If not, the shell is a non-fractal medium. For those shells that have relatively uniform or orderly distributed fibers, the fractal theory is not appropriate to use. Fortunately, most hollow fiber membrane contactors are packed randomly. Even if it is intentionally packed in an orderly manner, a small discrepancy in fiber positioning would lead to irregularity in fiber distributions, given the large quantities of fibers (200–500 in a shell) and their fine diameters (around 1 mm).

On the other hand, from Eq. (8.9) the relation between $d_{v,min}$ and $d_{v,max}$ can be obtained as

$$d_{v,max} = d_{v,min} (N_f)^{1/D_f} \tag{8.14}$$

The geometric mean diameter of the voids is $d_{v,m}$, then

$$d_{v,m}^2 = -\frac{1}{N_f}\int_{d_{v,min}}^{d_{v,max}} d_{vi}^2 dN \tag{8.15}$$

Substituting Eq. (8.11) into Eq. (8.15), an integration gives

$$d_{v,m}^2 = \left[\frac{D_f d_{v,min}^{D_f} d_v^{-D_f+2}}{(-D_f+2)}\right]_{d_{v,min}}^{d_{v,max}} \tag{8.16}$$

The equation is further simplified to

$$d_{v,m}^2 = \frac{D_f d_{v,min}^{D_f} d_{v,max}^{-D_f+2} - D_f d_{v,min}^2}{(-D_f+2)} \tag{8.17}$$

For a specific hollow fiber membrane module, the number of fibers, the fiber diameters, and the shell diameter are always known. The mean porosity ε_{tm} is determined by the module parameter as

$$\varepsilon_{tm} = 1 - \frac{N_f\, d_o^2}{D_s^2} = \frac{N_f d_{v,m}^2}{D_s^2} \tag{8.18}$$

As can be seen, when the module parameters are known, the values of $d_{v,min}$ and $d_{v,max}$ can be calculated from Eqs. (8.1), (8.17), and (8.18). There are many parameters for a module. Sometimes the producer does not specify them clearly. In such cases, when the module parameters are unknown, the fiber diameter, number of fibers, and mean porosity can be calculated on a unit area of shell cross section, as shown in Figure 8.4. The representative cross section is observed and the cross-sectional area is calculated. Then the number of fibers and the diameter of a single fiber are determined, and the mean porosity can be calculated.

$$d_{v,min} = d_{v,m}\sqrt{\frac{-D_f+2}{D_f\left(N_f^{(-D_f+2)/D_f} - 1\right)}} \tag{8.19}$$

It should also be noted that if the tubes are arranged uniformly in a triangular array, the maximum packing fraction is 0.907. If the tubes are arranged in a square array, the maximum packing fraction is 0.785. They are the limits for the minimum void diameter in a module.

8.3 Flow and heat mass transfer in the bundle

The cells are classified into M categories according to their void diameters. For each category $i(I = 1\ldots M, \sum N_i = n_f)$, the fluid flow

$$V_i = \frac{\pi d_{v,i}^2}{4} u_i N_i \tag{8.20}$$

$$V_t = n_f u_m \frac{\pi d_{v,m}^2}{4} \tag{8.21}$$

where V_i and V_t are the volumetric flow rates of fluid through the *i*th category of cells and the whole module respectively (m^3/s), u_m is the flow area weighted mean velocity, N_i is the number of cells in the *i*th category ($N_i = -dN_i$). It can be calculated by

$$u_m = \frac{\sum_{i=1}^{M} d_{v,i}^2 u_i N_i}{n_f d_{v,m}^2} \tag{8.22}$$

The bulk Reynolds number of flow through the module is defined by

$$Re_m = \frac{\rho u_m D_{hm}}{\mu} \tag{8.23}$$

where μ is hydrodynamic viscosity (Pa s). The bulk Darcy friction factor is

$$f_m = \frac{\Delta P_b}{L} \cdot \frac{D_{hm}}{\frac{1}{2}\rho u_m^2} \tag{8.24}$$

where L is the fiber length (m), D_{hm} is the hydrodynamic diameter of the whole module. The product of the Reynolds number and the friction factor is

$$(f\mathrm{Re})_m = \frac{\Delta P_b}{L} \cdot \frac{2D_{hm}^2}{\mu u_m} \tag{8.25}$$

Since each cell has the same pressure drop as the module, for the *ith* cells

$$(f\mathrm{Re})_i = \frac{\Delta P_b}{L} \cdot \frac{2D_{hi}^2}{\mu u_i} \tag{8.26}$$

where $(fRe)_i$ is the product of the local friction factor and the Reynolds number. For laminar flow in ducts, the product of friction factor and Reynolds number satisfies

$$(f \cdot \mathrm{Re}) = const \tag{8.27}$$

The values of the constants are from the calculated values for fiber bundles in the previous section. That is, $(fRe)_i$ is a function of the local packing fraction. $(fRe)_i$ is first calculated. Then from Eqs. (8.25) and (8.26), it has

$$\frac{u_i}{u_m} = \left(\frac{D_{hi}}{D_{hm}}\right)^2 \frac{(f\mathrm{Re})_m}{(f\mathrm{Re})_i} \tag{8.28}$$

Substituting Eqs. (8.25) and (8.26) into (8.20) to (8.22), we have

$$\frac{1}{(f\mathrm{Re})_m} = \sum_{i=1}^{M} \left(\frac{n_i d_{v,i}^2}{n_f d_{v,m}^2}\right) \left(\frac{D_{hi}}{D_{hm}}\right)^2 \cdot \frac{1}{(f\mathrm{Re})_i} \tag{8.29}$$

The fractional flow through the ith category

$$W_i = \frac{V_i}{V_t} = \left(\frac{N_i d_{v,i}^2}{n_f d_{v,m}^2}\right)\left(\frac{D_{hi}}{D_{hm}}\right)^2 \cdot \frac{(f\mathrm{Re})_m}{(f\mathrm{Re})_i} \tag{8.30}$$

It should be noted that a significant amount of flow occurs through a small number of large cells. It is found that cells in the largest 20% of the total number of cells carry 50%. Further, the following equation should be satisfied,

$$\sum_{i=1}^{M} W_i = 1 \tag{8.31}$$

The pressure drop of the bundle can be calculated by Eq. (8.25) with known mean velocity u_m and $(fRe)_m$ calculated by Eq. (8.29).

Cells with larger diameters have a great influence on the overall fluid flow and pressure drop. Note that when $N_i < 0.5$, the probability of cells in the ith category is negligible. So no cells are placed in this category. Without this treatment, errors may be generated from these numerically reasonable but physically meaningless cells.

8.4 Heat and mass transfer in the bundle

Shell-side convective heat and mass transfer coefficients can be expressed in terms of the Nusselt number and the Sherwood number respectively.

$$Nu = \frac{hD_h}{\lambda} \tag{8.32}$$

$$Sh = \frac{kD_h}{D_{va}} \tag{8.33}$$

where h and k are the heat transfer coefficient (kW m^{-2} K^{-1}) and the mass transfer coefficient (m/s) respectively; λ is the heat conductivity of fluid (kW m^{-1} K^{-1}); D_{va} is diffusivity in shell-side fluids (m^2/s).

Up until now, many correlations have been proposed for shell-side convective mass transfer coefficient predictions. They are summarized in Table 8.1, from Eqs. (8.34)–(8.45).

In this investigation, the whole module is divided into a series of cells. The local heat and mass transfer coefficient in each category of cells is required. In the past, the local heat and mass transfer coefficients outside one single hollow fiber in each cell were adopted from Miyatake and Iwashita [20,21], which were proposed for the heat transfer coefficient expression of shell-and-tube heat exchangers with tubes arrayed regularly, with a laminar flow outside the tubes under constant temperature and heat flux boundary conditions. In this study, the local heat and mass transfer coefficients calculated from the previous chapter for fiber bundles are used. Both the TA array and the SA array are considered.

Table 8.1 Correlations for predicting shell-side mass transfer in a randomly packed hollow fiber membrane module.

Correlations	Eqs. No.	Packing Densities (ϕ, %)	Reynolds Number Range	References
$Sh = 1.25\left(\frac{ReD_h}{L}\right)^{0.93} Sc^{0.33}$	(8.34)	2.5, 26	0–500	Yang and Cussler [7]
$Sh = 5.85\,(1-\phi)\left(\frac{D_h}{L}\right) Re^{0.6} Sc^{0.33}$	(8.35)	4, 8.7, 19.7, 40	0–500	Prasad and Sirkar [8]
$Sh = (0.53 - 0.58\phi)\, Re^{0.53} Sc^{0.33}$	(8.36)	31.9–75.8	25–300	Costello et al. [9]
$Sh = 0.41\left(\frac{1}{N_f}\right)\left(\frac{D_s}{d_o}\right)^2\left(\frac{D_h}{L}\right)^{0.25} Re^{0.33} Sc^{0.33}$	(8.37)	5.06–15.7	10–300	Nii and Takeuchi [10]
$Sh = \left(Sh_1^3 + Sh_2^3 + Sh_3^3\right)^{1/3}$	(8.38)	10–75	Laminar	Lipnizki et al. [11]
$Sh_1 = 3.66 + 1.2\left(\sqrt{\phi}\right)^{-0.8}$	(8.39)			
$Sh_2 = 1.615\left(1 + 0.14\left(\sqrt{\phi}\right)^{-0.5}\right)^3\left(\frac{ReScD_h}{L}\right)^{1/2}$	(8.40)			
$Sh_3 = \left(\frac{2}{1+22Sc}\right)^{1/6}\left(\frac{ReScD_h}{L}\right)^{1/2}$	(8.41)			
$Sh = 1.62\left(ReSc\frac{D_h}{L}\right)^{1/3}$	(8.42)	Tube flow	Laminar, $Gz \geq 25$	Leveque equation [12,13]
$Sh = \left(0.31\phi^2 - 0.34\phi + 0.1\right) Re^{0.9} Sc^{0.33}$	(8.43)	29.78, 49.78, 69.78, 91.6	32–1287	Wu and Chen [15]
$Sh = (0.163 + 0.27\phi)\left(\frac{u}{D_v L}\right)^{0.6}$	(8.44)	20, 30, 40, 50	178–1194	Zheng et al. [18]
$(fRe) = \begin{pmatrix} -428.62\phi^3 + 620.49\phi^2 \\ -304.24\phi + 107.9 \end{pmatrix} \cdot \psi_f$ $\psi_f = 1.149D_f - 1.0$ $Sh = \begin{pmatrix} 14.06\phi^4 - 29.21\phi^3 \\ +22.59\phi^2 - 7.71\phi + 1.03 \end{pmatrix} Re^{0.33} Sc^{0.33} \psi_k$ $\psi_k = 0.882D_f - 0.535$	(8.45)	10–80	Laminar	Zhang [19]

Considering the fluid in the *i*th category of cells, heat transfer in the *i*th category of cells satisfies the following equation

$$\rho c_p V_i \frac{dT_i}{dx} = h_i N_i \pi d_o \left(T_w - T_{in}\right) \tag{8.46}$$

The solution of Eq. (8.46) is

$$T_{i,out} = T_w - \left(T_w - T_{in}\right) e^{-NTU_i} \tag{8.47}$$

The number of transfer units for sensible heat

$$NTU_{si} = \frac{h_i N_i \pi d_o}{\rho c_p V_i} \tag{8.48}$$

The total number of transfer units for heat transfer

$$NTU_{st} = \frac{h_m n_f \pi d_o}{\rho c_p V_t} \tag{8.49}$$

The bulk temperature

$$T_{b,out} = \frac{\sum_{i=1}^{M} V_i T_{i,out}}{V_t} \tag{8.50}$$

The total number of transfer units is

$$NTU_{st} = -\ln\left(\sum_{i=1}^{M} W_i e^{-NTU_i}\right) \tag{8.51}$$

After calculation of NTU_{st}, the mean heat transfer coefficient and the Nusselt number of the whole module can be obtained by Eq. (8.49).

Mass transport in a single cell can be expressed by the following equation

$$V_i \frac{dC_i}{dx} = k_i \pi d_o N_i \left(C_w - C_{in}\right) \tag{8.52}$$

where C is the solute concentration in the fluid (kg/m^3), for a more general analysis, C_w is the concentration at fiber wall surfaces, which is a constant under uniform wall concentration boundary conditions, C_{in} is the bulk inlet concentration, and x is the axial coordinate (m). The solution of Eq. (8.52) gives the concentration at outlet as

$$C_{i,out} = C_w - \left(C_w - C_{in}\right) e^{-NTU_{Li}} \tag{8.53}$$

$$NTU_{Li} = \frac{k_i \pi d_o N_i}{V_i} \tag{8.54}$$

where NTU_L is the number of transfer units for mass transfer. The bulk concentration at the outlet is defined by

$$C_{b,out} = \frac{\sum_{i=1}^{M} V_i C_{i,out}}{V_t} \tag{8.55}$$

If the mean convective mass transfer coefficient for the module is k_m, the overall mass transfer for the module is

$$V_t \frac{dC_b}{dx} = n_f k_m \pi d_o (C_w - C_{in}) \tag{8.56}$$

The solution is

$$C_{b,out} = C_w - (C_w - C_{in}) e^{-NTU_{Lt}} \tag{8.57}$$

where the total number of mass transfer units is

$$NTU_{Lt} = \frac{k_m n_f \pi d_o}{V_t} \tag{8.58}$$

Substituting Eqs. (8.55) into (8.57), and combining with (8.30), then

$$NTU_{Lt} = -\ln\left(\sum_{i=1}^{M} W_i e^{-NTU_{Li}}\right) \tag{8.59}$$

Then the mean convective mass transfer coefficients and the Sherwood number for the module can be calculated from Eq. (8.58).

The performance indices of a randomly packed module are defined by

$$\eta_f = \frac{(f\mathrm{Re})_m}{(f\mathrm{Re})_0} \tag{8.60}$$

$$\eta_h = \frac{Nu_m}{Nu_0} \tag{8.61}$$

$$\eta_m = \frac{Sh_m}{Sh_0} \tag{8.62}$$

where the subscript "0" refers to the values for a regularly distributed bundle with the same packing fraction as the fiber module, where TA arrays are assumed for the orderly bundles. They are the ratios of performance of a real random module to an ideal orderly module.

8.5 Results and discussion

Determination of D_f

The area dimension D_f can be determined by the box-counting method [19]. This method is based on the image analysis of a sufficiently large section on the module cross section. According to this method, the module cross-sectional sample under consideration is discretized using square boxes of size L_c. Then the number, $N(L_c \geq d_v)$, of boxes required to completely cover the fibers is counted. The number of fibers is equal to the number of cells. The cell area fractal dimension, D_f, can be determined by the value of the slope of a linear fit through data on a logarithmic plot of the cumulative number of cells $N(L_c \geq d_v)$ verses the square box size L_c.

Two modules, Module A and Module B, with properties listed in Table 8.2 are investigated. The representative sections of module cross sections are taken with a digital camera. The numbers of cells which are completely covered by boxes of size L_c are counted. Then the logarithmic plots of the cumulative numbers of cells versus box size L_c for the module are plotted in Figure 8.7. The two modules are both illustrated in the figure. As can be seen, the numbers of cumulative cells increase as the boxes sizes increase. The data follow a relatively ideal linear relationship on the logarithmic scale, and this confirms the statistical fractal nature of the distributions of fibers in the modules. From the slopes, the fractal dimension for Module A is $D_f = 1.744$ and for Module B it is $D_f = 1.882$. As can be seen, generally, the higher the packing density of the module is, the larger the fractal dimension is. This is reasonable because the more the fibers are packed, the fewer are the larger voids.

Table 8.2 Physical and transport properties of the two hollow fiber membrane modules.

Name of Property	Symbol	Unit	Module A	Module B
Shell diameter	D_s	mm	42	42
Module fiber length	L	mm	290	210
Number of fibers in the module	n_f		300	494
Fiber outer diameter	d_o	mm	1.4	1.4
Fiber inner diameter	d_i	mm	1.1	1.1
Packing fraction	ϕ		0.333	0.549
Packing density	A_v	m^2/m^3	953	1568
Fractal dimension	D_f		1.74	1.88
Moisture diffusivity in air	D_{va}	m^2/s	2.82×10^{-5}	2.82×10^{-5}
Effective moisture diffusivity in membrane	D_{vm}	m^2/s	3.7×10^{-10}	3.7×10^{-10}
Heat conductivity of membrane	λ_m	$Wm^{-1} K^{-1}$	0.43	0.43
Reynolds number	Re		23–300	14–200

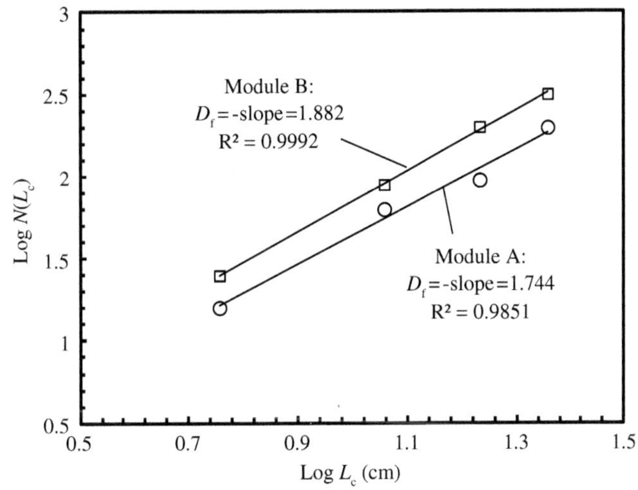

FIGURE 8.7

Schematic illustration of the procedure for evaluation of fractal area dimensions for fiber distribution with Modules A and B.

Distribution of cells

When the fractal dimensions are obtained, the cell distributions can be calculated. The numbers of voids with void size d_V for Modules A and B, $N(d_V)$, are calculated and plotted in Figure 8.8. The solid lines represent the cells that really exist. The dashed lines represent the fake cells which only have numerical meanings under fractal theory assumptions. As can be seen, the number of cells decreases sharply with an increase in cell size. The voids of these cells are less than 5 mm in diameter, which is in agreement with picture observations. Both modules have similar trends.

The modules are randomly packed, so there are no two modules that are identical. The degrees of irregularity are related to the fractal dimensions. To illustrate this point quantitatively, the standard deviation of cell diameters can be used to reflect the degree of non-uniformity. It is an index of cell non-uniformity. The higher the d_V is, the larger the irregularity is. For two-dimensional geometries, the fractal dimensions are from 1.0 to 2.0. When D_f is 2.0, the packing is strictly orderly. The higher the fractal dimensions are, the less the non-uniformity is.

The packing fraction also has an influence on irregularity. Module B is more homogeneous than Module A. The larger the packing fraction is, the less the non-uniformity is. The reason is that the more densely packed the module is, the more homogeneously the fibers are placed in the module. From the packing densities listed in Table 8.1, most of the membrane contactors have a packing fraction from 0.2 to 0.6, with a fractal dimension from 1.6 to 1.9.

Model validation

The modules are tested on the humidification test rig introduced in Chapter 7. The construction of the test rig has been described. While the inlet conditions are set to the design points, the test rig is operated under different mean air flow velocities. The pressure drop through the shell side is deduced.

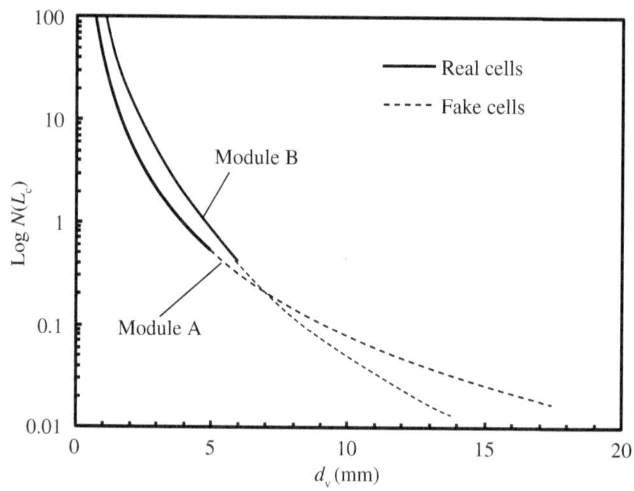

FIGURE 8.8

Cell size distributions for fibers in Modules A and B.

After the inlet and outlet temperature and humidity are measured, the overall mass transfer coefficients are obtained. Then the air-side mass transfer coefficients are deduced. These are the experimentally obtained values. On the other hand, the pressure drop and mass transfer coefficients can be directly calculated from the fractal model set up in this chapter. These are called the model results. The two modules are similar in geometries, but have different packing fractions and fiber lengths.

Experimental results are used to validate the calculated values. Comparisons are made between the calculated and experimentally obtained pressure drop, and heat and mass transfer coefficients. They are demonstrated in Figures 8.9–8.11 respectively. As can be seen, the fractal model usually predicts the results acceptably. In other words, the predictions of the pressure drop and heat mass transfer processes are satisfactory with the fractal models. The calculated performances are in good agreement with the experimentally obtained values. The discrepancies between the model prediction and the test data are below 10.4% for pressure drop and 10.6% for heat and mass transfer coefficients. The mass transfer is more accurately predicted than the pressure drop and heat transfer coefficients because resistance in the inlet/outlet tubes is difficult to predict very accurately. Heat dissipation from the module walls is much higher than mass losses. After the model validation, further numerical work can be performed to evaluate performance.

Performance deterioration

With the fractal model, the friction factor and the Nusselt and Sherwood numbers can be calculated, taking into account the fractal dimensions in the modules. The packing fraction has a dominant influence on pressure drop: The pressure drop increases with an increase in packing fraction. The product of the friction factor and the Reynolds number is a key parameter for pressure drop analysis and module design. The ratio of (*fRe*) of the membrane module to an orderly tube bundle with staggered array

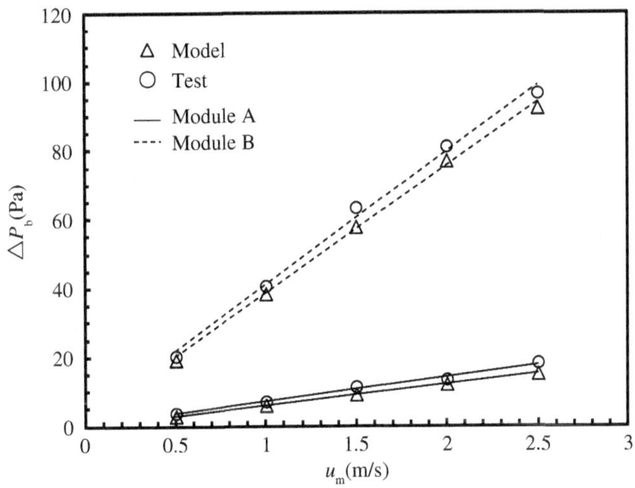

FIGURE 8.9

Comparisons of the pressure drops between the experimental data and the fractal models for the two modules.

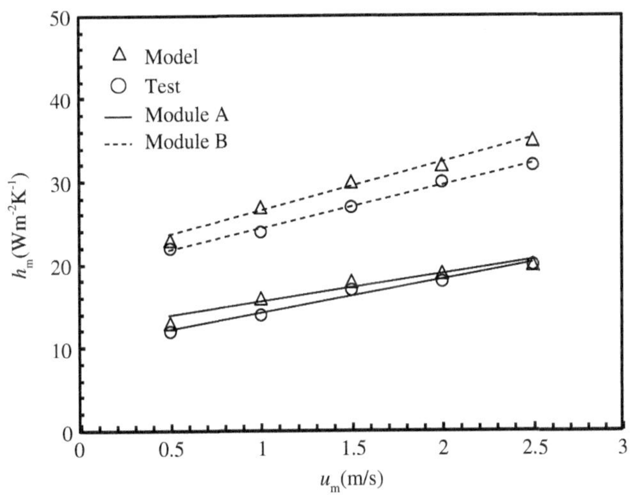

FIGURE 8.10

Comparisons of the mean convective heat transfer coefficients between the experimental data and the fractal models for the two modules.

is shown in Figure 8.12. It can be observed that the (fRe) ratio, η_f, increases with packing fraction. All the η_f values are less than 1.0. This means that the larger the packing fraction is, the larger the (fRe) ratio is, and the higher the resistance is. However, the irregularity tends to depress the pressure drop.

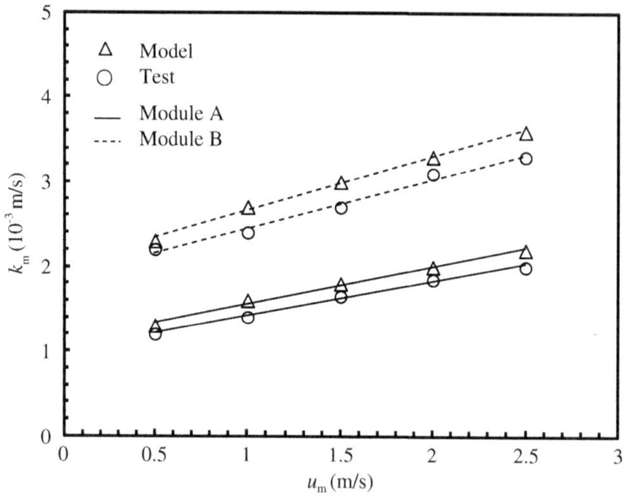

FIGURE 8.11

Comparisons of the mean mass transfer coefficients between the experimental data and the fractal models for the two modules.

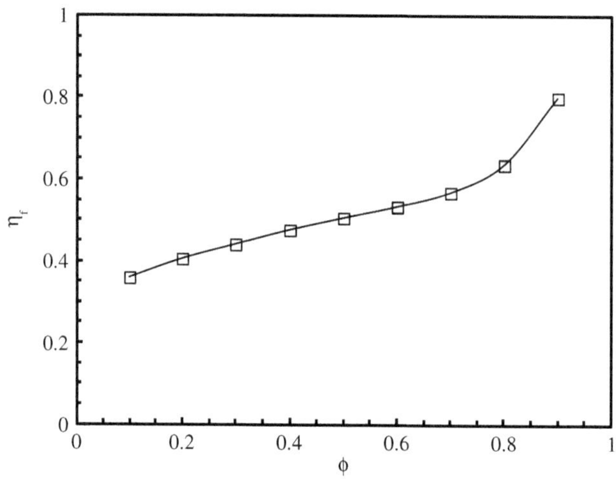

FIGURE 8.12

Variations of the ratios of the product of friction factor and Reynolds number to that of regularly packed bundles with packing fractions.

The reduction of resistance is due to the larger holes, gaps, and channelings in the fiber bundles, which makes the fluid short-cut. The cells with larger voids, though not many in number, have a great influence on fluid flow. The largest portion of fluid flows in these large cells, making the overall

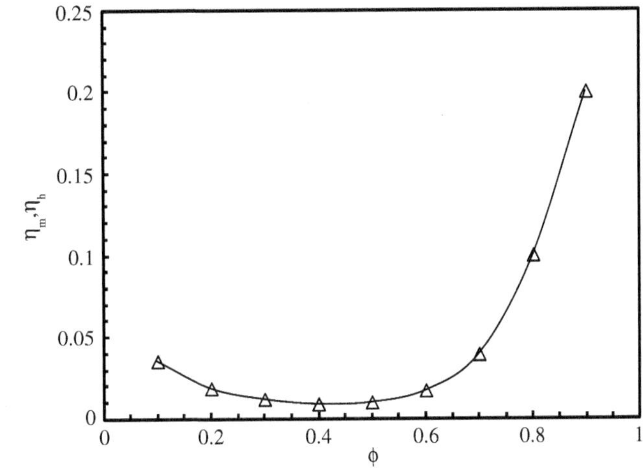

FIGURE 8.13

Variations of the ratios of Nusselt and Sherwood numbers of randomly packed bundles to those of regularly packed bundles with packing fractions.

pressure drop depressed. In other words, the fluids are short-cut by these large cells, diminishing the pressure drop.

The heat and mass transfer performance deteriorates with non-uniformity as well. The ratios of the Nusselt and Sherwood numbers of membrane modules to those of orderly distributed bundles are plotted in Figure 8.13. The higher the packing fractions are, the lower the Nusselt and Sherwood numbers are for the membrane modules. When modules are packed randomly, the Nusselt and Sherwood numbers are lower than in orderly distributed staggered arrays. Most of the fluids flows in the large pinholes and channelings in the bundle, while much less fluid flows in the locations densely populated by fibers. A large portion of the membranes is not used effectively. As a consequence, the heat and mass transfer performance deteriorates seriously. Further, as can be seen in Figure 8.13, the Sherwood or Nusselt number ratio, η_m or η_h, increases with increase in packing fraction when the packing fraction is above 0.4. This indicates that when the packing fraction is increased, the module becomes more homogeneous and the flow becomes more uniform. The ratios decrease a little bit when the packing fraction is from 0.2 to 0.4. This decrease is due to the much more rapid decrease of Sh_0. It doesn't mean that the irregularity increases. Certainly this decrease is very small. When the packing fraction is further increased above 0.7, the membrane module tends to become a common regularly spaced shell-and-tube heat exchanger with pure axial flow. As observed in Figure 8.13, for most membrane contactors the packing fraction is from 0.2 to 0.6, and the resulting Nusselt and Sherwood number ratios are from 1% to 5%. These results demonstrate that compared to shell-and-tube heat exchangers, the performance deteriorates considerably for randomly packed hollow fiber membrane contactors. However, thanks to the high packing density of the hollow fibers, the membrane area is large enough (1000 m^2/m^3) to offset the deterioration in Nusselt and Sherwood numbers. The final exchange performance is determined by (A_t h_m) or (A_t k_m), which is still several times higher than a common shell-and-tube heat exchanger, especially so for a high packing density module which allows high pressure drops.

Correlations

Correlations are summarized to estimate the mean friction factor and the mean heat and mass transfer coefficients on the shell side. Based on the numerical data with the above experimentally validated fractal model, correlations are proposed for the overall friction factor and mean Sherwood number estimations.

$$(f\mathrm{Re}) = \left(-426.62\phi^3 + 611.49\phi^2 - 257.24\phi + 92.9\right)\psi_f \tag{8.63}$$

$$Nu = \left(13.86\phi^4 - 28.21\phi^3 + 21.29\phi^2 - 7.21\phi + 1.08\right)\mathrm{Re}^{0.33}\mathrm{Sc}^{0.33}\psi_h \tag{8.64}$$

$$Sh = \left(11.25\phi^4 - 23.36\phi^3 + 18.07\phi^2 - 6.17\phi + 1.13\right)\mathrm{Re}^{0.33}\mathrm{Sc}^{0.33}\psi_m \tag{8.65}$$

$$\psi_f = 1.149D_f - 1.0 \tag{8.66}$$

$$\psi_h = 0.872D_f - 0.534 \tag{8.67}$$

$$\psi_m = 0.862D_f - 0.489 \tag{8.68}$$

where ψ_f, ψ_h and ψ_m are the correction factors for the friction factor, heat and mass transfer respectively, which takes into account the irregularity in fiber distribution. For highly irregular packing, $D_f = 1$; for highly regular packing, $D_f = 2$. Generally, for practical modules used in engineering, D_f has a value from 1.6 to 1.9. Other conditions for use are: $\phi = 0.1$ to 0.8; laminar flow.

The established correlations for the Sherwood number are compared to those in the references listed in Table 8.1. The comparisons are illustrated in Figure 8.14. The assumed D_f is 1.74 for this study.

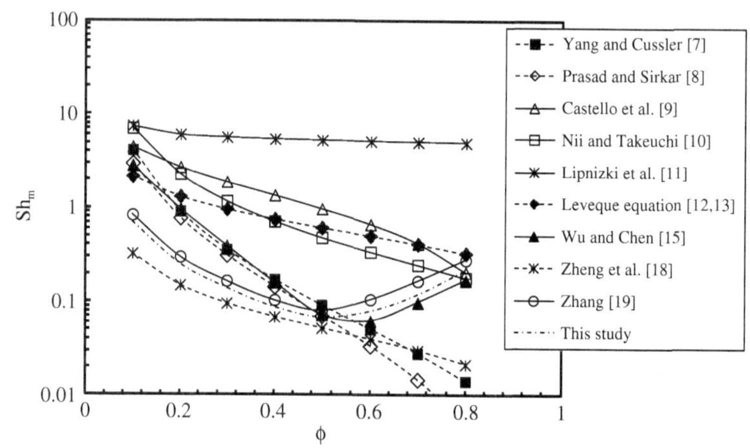

FIGURE 8.14

Variations of the mean Sherwood numbers of randomly packed bundles with various packing fractions, calculated with various correlations. $D_f = 1.74$.

The correlations proposed with fractal models can reflect the effects of the degree of irregularity on performance; however, others cannot.

8.6 Conclusion

Though membrane contactors have been used extensively, the earlier correlations didn't consider the random distribution at all. Recent correlations have indeed considered the irregularities; however, they didn't consider the degree of irregularity. This study is a step forward from previous studies in that it considers not only the maldistribution and the degree of maldistribution, but also the conjugate heat and mass transfer in a hollow fiber membrane bundle. The new approach is proposed just for such randomly spaced objects in nature. Fractal dimensions are obtained to represent the self-similar irregularities in fiber distribution. The friction factor and the Sherwood and the Nusselt number depressions from the non-uniformity in fiber packing are investigated. The results are that the higher the packing fraction is, the larger the fractal dimension is, and the more homogeneous the fiber distributions are. The flow maldistribution seriously deteriorates the heat and mass transfer performance. The randomly packed module has Nusselt and Sherwood numbers only 1–5% of those for a regularly spaced tube array. High packing density should be used to offset the flow maldistribution.

With the newly available data, the proposed correlations take into account the degree of irregularity in terms of fractal dimensions, as well the conjugate heat and mass transfer in the bundle. The values of the Sherwood number exhibit a first-decrease-then-increase behavior with increasing packing fraction. They can be used to estimate the performance.

References

[1] Delgado JA, Uguina MA, Sotelo JL, Agueda VI, Sanz A. Simulation of CO_2 absorption into aqueous DEA using a hollow fiber membrane contactor: evaluation of contactor performance. Chemical Engineering Journal 2009;152:396–405.

[2] Zhang LZ. Total Heat Recovery: Heat and Moisture Recovery in Ventilation Air. New York: Nova Science; 2009.

[3] Liu LY, Ding ZW, Lu Y, Ma RY. Modeling the bubbling enhanced microfiltration for submerged hollow fiber membrane module. Desalination 2010;256:77–83.

[4] Zhang GJ, Wang NX, Song X, Ji SL, Liu ZZ. Preparation of pilot-scale inner skin hollow fiber pervaporation membrane module: effects of dynamic assembly conditions. Journal of Membrane Science 2009;338:43–50.

[5] Incropera FP, Dewitt DP. Fundamentals of Heat and Mass Transfer. 6th edn. New York: John Wiley & Sons; 2007.

[6] Kays WM, London AL. Compact Heat Exchangers. New York: McGraw-Hill; 1980.

[7] Yang MC, Cussler EL. Designing hollow-fiber contactors. AIChE Journal 1986;32:1910–6.

[8] Prasad P, Sirkar KK. Dispersion-free solvent extraction with microporous hollow-fiber modules. American Institute of Chemical Engineers Journal 1987;33:1057–66.

[9] Costello MJ, Fane AG, Hogan PA, Schofield RW. The effect of shell side hydrodynamics on the performance of axial flow hollow fibre modules. Journal of Membrane Science 1993;80:1–11.

[10] Nii S, Takeuchi H. Removal of CO_2 and/or SO_2 from gas streams by a membrane absorption method. Gas Separation & Purification 1994;8:107–13.

[11] Lipnizki F, Field RW. Mass transfer performance for hollow fiber modules with shell-side axial feed flow using engineering approach to develop a framework. Journal of Membrane Science 2001;193:195–208.
[12] Viegas RMC, Rodriguez M, Luque S, Alvarez JR, Coelhoso IM, Crespo JPSG. Mass transfer correlations in membrane extraction: analysis of Wilson-plot methodology. Journal of Membrane Science 1998;145:129–42.
[13] Mengual JI, Khayet M, Godino MP. Heat and mass transfer in vacuum membrane distillation. International Journal of Heat and Mass Transfer 2004;47:865–75.
[14] Chen V, Hlavacek M. Application of Voronoi Tessellation for modeling randomly packed hollow fiber bundles. AIChE Journal 1994;40:606–12.
[15] Wu J, Chen V. Shell-side mass transfer performance of randomly packed hollow fiber modules. Journal of Membrane Science 2000;172:59–74.
[16] Bao LH, Lipscomb GG. Well-developed mass transfer in axial flows through randomly packed fiber bundles with constant wall flux. Chemical Engineering Science 2002;57:125–32.
[17] Zheng JM, Xu YY, Xu ZK. Flow distribution in a randomly packed hollow fiber membrane module. Journal of Membrane Science 2003;211:263–9.
[18] Zheng JM, Xu YY, Xu ZK. Shell side mass transfer characteristics in a parallel flow hollow fiber membrane module. Separation Science and Technology 2003;38:1247–67.
[19] Zhang LZ. Heat and mass transfer in a randomly packed hollow fiber membrane module: a fractal model approach. International Journal of Heat and Mass Transfer 2011;54:2921–31.
[20] Miyatake O, Iwashita H. Laminar flow heat transfer to a fluid flowing axially between cylinders with a uniform surface temperature. International Journal of Heat and Mass Transfer 1990;33:417–25.
[21] Miyatake O, Iwashita H. Laminar flow heat transfer to a fluid flowing axially between cylinders with a uniform wall heat flux. International Journal of Heat and Mass Transfer 1991;34:322–7.

CHAPTER

Conjugate Heat Transfer in Plate-Fin and Tube Heat Exchangers

9

9.1 Introduction

When liquid exchanges heat with air through tubes, the air-side convective heat transfer coefficients are several magnitudes lower than those on the liquid side. Therefore air-side heat transfer should be intensified. Extended surfaces are recommended for heat transfer tubes. Due to the limited air-side heat transfer coefficients, dense parallel plates are assembled perpendicular to a tube bank to provide an extended air-side heat transfer surface. As a consequence, the overall heat transfer rates can be greatly increased. Due to their high heat transfer capabilities, as well as their compactness and large packing density, they are widely used in areas such as air-conditioning evaporators, condensers, process gas heater and coolers, etc. There are several choices to achieve extended surfaces: finned circular tubes, plate-fins and tubes. Among these two alternatives, plate-fin and tube heat exchangers are widely used in various industries due to their compactness and ease of construction. A plate-fin and tube heat exchanger is pictured in Figure 9.1. The fins can be further modified into various types: plain plate fins, corrugated fins (wavy fins, serrated fins), and interrupted fins (louvered fins, perforate fins, slit fins). Figure 9.2 shows two fins of the slit louvered type.

The application of plate-fin and tube heat exchangers usually involves heat exchange between a liquid fluid and an air stream, such as evaporators and condensers. Fins of various types are designed to further air-side heat transfer coefficients. Hydrophilic coatings are often added to the fin surfaces. Though these enhanced fin surfaces can significantly improve air-side heat transfer coefficients in comparison with their plain fin counterparts, the plain fins are still by far the most popular pattern used in air-cooled heat exchangers. The reason behind this is that plain plate fins have superior reliability for long-term operation. They have lower friction coefficients and, most importantly, they are cheaper to manufacture and easy to clean up.

The performance of plate-fin and tube heat exchangers is influenced by many factors such as operating conditions and geometric parameters. The fins, tubes and fluids are coupled together. The complexities have drawn much attention in this area. Fluid flow and heat transfer in plate-fin and tube heat exchangers have been a focus in the past decades. Experimental investigations on the heat transfer and pressure drop performance can be dated back to the 1970s [1–4], and have been continuously conducted until recently [5–7]. Empirical correlations to predict the Colburn (j) and friction factor (f) vs Reynolds numbers for plain fins were developed by McQuiston [8], Gray and Webb [9], and Wang et al. [7], for various geometric and operating conditions. Though many experimental data are available for

FIGURE 9.1

Picture of a plate-fin and tube heat exchanger with copper tubes and aluminum fins.

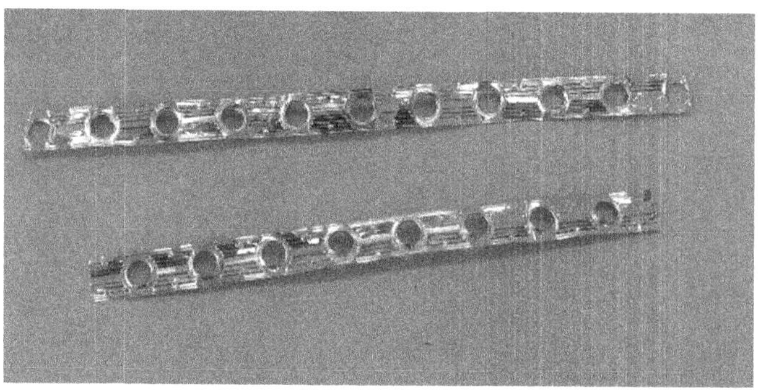

FIGURE 9.2

Picture of two slit louvered fins.

heat exchanger design, the data are still very limited because they are case sensitive. They were tested only for specific tube diameters, tube distances, and fin pitches.

Use of computational fluid dynamics (CFD) in modeling the fluid flow and heat transfer in plate-fin and tube heat exchangers is a new trend with fast-developing computing techniques. Besides the aforementioned experimental work, due to the ability to provide insight into the local fluid flow and heat transport dynamics in the fine heat exchanger channels, there have been many numerical investigations in this area as well. Jang and Wu [10] conducted a three-dimensional numerical study of fluid flow and heat transfer in plate-fin and tube heat exchangers. Using a laminar model and assuming uniform fin temperature, the effects of different geometrical parameters were investigated in detail. Both the local and mean values of heat transfer coefficients and pressure drop characteristics were predicted. In addition, Tsai and Sheu Tony [11] conducted a 3-D numerical simulation for a two-row staggered finned tube heat exchanger. Mendez et al. [12] analyzed the influence of fin spacing on heat transfer coefficients of a single-row finned tube heat exchanger. He et al. [13] conducted a three-dimensional numerical study of heat exchangers from the viewpoint of the field synergy principle. It was found that the effects of the five parameters on the heat transfer performance of the finned tube banks can be well described by the field synergy principle. Besides plain fins, there are also many numerical studies of non-plain plate fins such as wavy fins [13–15], slotted fins [16], louver fins [17,18], and so on. In these researches, either laminar or turbulent models were used.

Flows around tube banks are usually subcritical (mixed, transition to turbulence occurs after boundary separation), or critical (predominantly turbulent, only part of the boundary layer developing on the tube surface is laminar) [19]. Vortex shedding and the consequent anisotropic velocity oscillations are the dominant mechanisms for momentum and heat transfer in bare tube banks [20–23]. The combination of the flow instabilities and the transitional phenomena present in the boundary layers makes this type of flow difficult to model with a simple laminar model. It is thus reasonable to guess that plate-fin and tube heat exchangers, which are composed of parallel-plate fins and tube banks, may have similar unsteady behaviors in the flow. As a result, the effects of unsteady flow on momentum and heat transfer in plate-fin and tube heat exchangers need to be addressed. Regrettably, though there have been many studies of the plate-fin and tube heat exchangers, the problems of unsteady flow in the heat exchanger are seldom mentioned. Previous researches only considered steady-state laminar flows. Furthermore, it should be noted that to capture the transient behavior in the channel, the computational time-step should be at least of the order of 1 ms, because velocity oscillations in tube banks are of the order of 100 Hz [21]. Otherwise, the temporal information in the flow may be lost due to a large time-step. Knowledge of the unsteady behavior in this relatively well-studied heat exchanger has not accumulated sufficiently yet. For the limited number of studies involving plain plate-fin and tube heat exchangers, heat transfer in fins and fin efficiencies were ignored by assuming uniform temperature for both fins and tubes. The problem of conjugate heat transfer has not been fully studied before. Recently, there were several studies investigating the transient fluid flow and heat transfer in plate-fin and tube, and cross-corrugated ducts [24,25]. Insights into fluid flow and heat transfer were revealed to some extent.

In this chapter, unsteady fluid flow and conjugate heat transfer in a plain plate-fin and tube heat exchanger with a staggered arrangement are investigated with a new representative duct. The representative duct surrounded by the tubes and fins is selected as the computational domain. The effects of unsteady velocity oscillations on the momentum and heat transfer characteristics in the exchanger are discussed. To ease the computational efforts with direct numerical simulation, a large eddy simulation model (LES) is used to calculate the temporal behavior of velocity, momentum, and temperature fields in the flow channel.

9.2 Numerical models

9.2.1 Physical model

A schematic diagram of a plain plate-fin and tube heat exchanger is shown in Figure 9.3. There are in total four rows of tubes in the flow direction, with a staggered arrangement. The geometric parameters of the heat exchanger are listed in Table 9.1. They are identical to the parameters of the heat exchanger

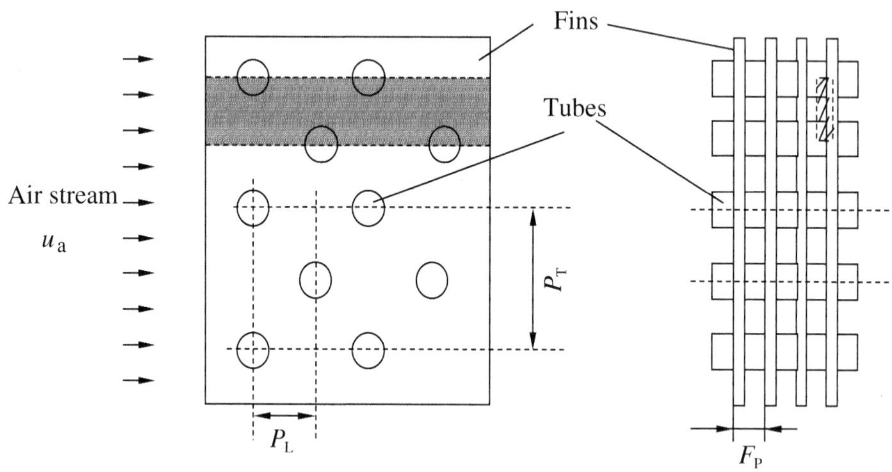

FIGURE 9.3

Geometries of the plate-fin and tube heat exchanger with staggered arrangement. The shaded area is the computational domain.

Table 9.1 Geometric parameters of the plain plate-fin and tube heat exchanger investigated.

Parameters	Unit	Value
Tube outer diameter (d_o)	mm	15.9
Tube thickness	mm	1.05
Coil length	mm	400
Coil width	mm	132
Coil height	mm	247
Fin pitch (F_p)	mm	3.175
Fin thickness	mm	0.25
Fin spacing (H_z)	mm	2.925
Transverse tube pitch (P_T)	mm	33
Longitudinal tube pitch (P_L)	mm	38
Tube number		24
Row number		4
Pass number		4
Tube material		Copper
Fin material		Aluminum

studied in Ref. [10], for the sake of model validation and comparison of results. In Figure 9.3, the shaded area between two neighboring spanwise tubes is selected as the computational domain. Mid-planes between two neighboring columns of tubes and half the duct between two neighboring fins are selected for symmetry.

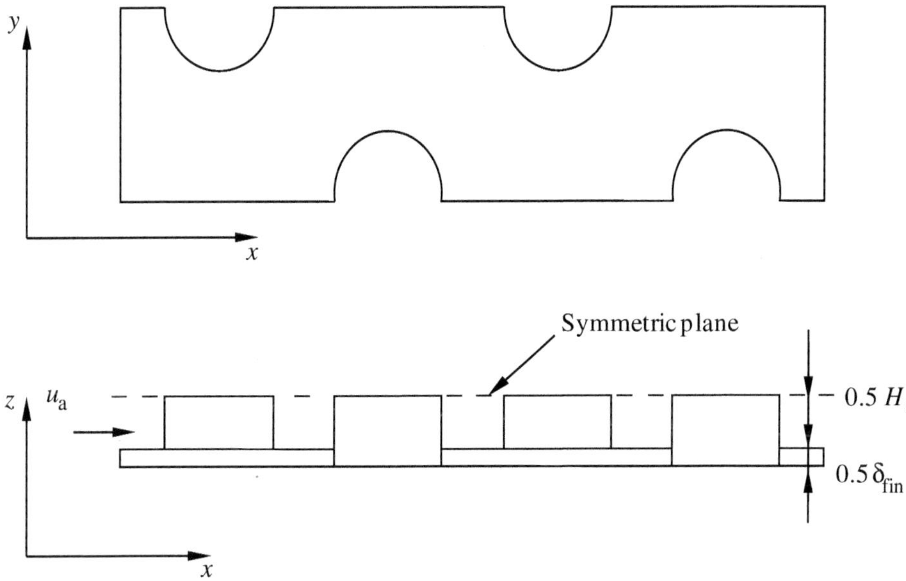

FIGURE 9.4

The computational domain and the coordinate system for the plate-fin and tube duct.

The three-dimensional computational domain is shown separately in Figure 9.4. It is like a parallel-plate duct flow, where the fin is the lower boundary, and the symmetric mid-plane between two fins is the upper boundary. The half tubes, which connect with the fin, constitute part of the left and right boundaries. The coordinate system is also illustrated in the figure. A uniform flow enters the computational domain along the streamwise (x) direction. The fin spanwise direction is the y axis and the fin normal direction is the z axis.

9.2.2 Large eddy simulation

Modeling of fluid flow has been an interesting topic for a long period. At the Reynolds numbers considered here (1000–5000, which is typical for air-cooled heat exchangers) transitional flow is expected. As is well known, such flows are difficult to simulate with conventional steady-state models, either laminar or turbulent ones [24,25]. The reason is that turbulence consists of chaotic motions, and often persistent organized motions as well, at a range of scales that increases rapidly with Reynolds numbers. This restricts a complete numerical solution to very low Reynolds numbers. A direct numerical simulation is hard to perform with limited computer capacity within a reasonable time-span. Numerical calculation of turbulence has evolved into three major categories: direct numerical simulation (DMS), large eddy simulation (LES), and Reynolds-averaged Navier-Stokes (RANS) equation simulation. In DNS all scales, up to their viscous scales (Kolomogorov eddies), are resolved with accuracy by the use of three-dimensional Navier-Stokes equations. However, the computation power required is always too high. In RANS, all the eddies smaller than those of the flow domain are averaged so there is no information on the small scales of motion. LES of turbulence may be considered as a method that lies between the two extremes, where all the length scales that are greater than the size of a mesh are computed directly, while the eddies of sizes smaller than the mesh size are modeled through transport techniques [20,21]. Therefore, in this study, a large eddy simulation technique is employed to capture the unsteady turbulence in the channel without leading to too much computational burden.

The governing equations of continuity, momentum (Navier-Stokes) and temperature conservation for Newtonian incompressible fluids are [24]:

$$\frac{\partial u_j}{\partial x_j} = 0 \tag{9.1}$$

$$\frac{\partial u_i}{\partial t} + \frac{\partial u_i u_j}{\partial x_j} = -\frac{1}{\rho}\frac{\partial P}{\partial x_i} + \frac{\partial}{\partial x_j}[\nu + \nu_s]\frac{\partial u_i}{\partial x_j} \tag{9.2}$$

$$\frac{\partial T}{\partial t} + \frac{\partial u_j T}{\partial x_j} = \frac{\partial}{\partial x_j}\left[\frac{\nu}{Pr} + \frac{\nu_s}{Pr_s}\right]\frac{\partial T}{\partial x_j} \tag{9.3}$$

where u_j is the resolved (or filtered) instantaneous velocity components of large-scale motion (m/s), ($j=1$, 2, 3 for x, y, z respectively), x_j is the coordinate component (m), t is time (s), ρ is density (kg/m^3), T is temperature (K), ν is kinematic viscosity (m^2/s), Pr is the Prandtl number, subscript "s" refers to subgrid, and P is modified pressure (Pa),which is expressed by

$$P = P_0 + \frac{2}{3}\rho k \tag{9.4}$$

where P_0 is static pressure (Pa), k is subgrid (or unresolved) turbulence energy (m^2/s^2).

The subgrid viscosity is computed by the Smagorinsky subgrid-scale model as

$$\nu_s = (C_s\Delta)^2|\bar{S}| \tag{9.5}$$

where the local strain rate is defined by

$$|\bar{S}| = \left(2\bar{S}_{ij}\bar{S}_{ij}\right)^{1/2} \tag{9.6}$$

$$\bar{S}_{ij} = \frac{1}{2}\left(\frac{\partial u_i}{\partial x_j} + \frac{\partial u_j}{\partial x_i}\right) \tag{9.7}$$

where in Eq. (9.5) C_s is the Smagorinsky constant with a value from 0.065 to 0.25. It is commonly selected as 0.1 [20]. Δ is the length scale used in the definition of the filter, and it is selected as the average size of the generic grid cell (cubic root of its volume). The value for the subgrid Prandtl number Pr_s in Eq. (9.3) is from 0.5 to 1.0 [24]. In this chapter, it is chosen as 0.85.

The hydrodynamic diameter of the flow channel ($\approx 2H_z$) is used for the definition of the similarity numbers in accordance to

$$Re = \frac{2u_{\max}H_z}{\nu} \tag{9.8}$$

$$St_r = \frac{2fH_z}{u_a} \tag{9.9}$$

$$Pr = \frac{\nu}{a} \tag{9.10}$$

$$Nu = \frac{2hH_z}{\lambda} \tag{9.11}$$

where u_{max} is the maximum velocity at the narrowest channel cross section (m/s), St_r is the Strouhal number, f is the oscillation frequency (Hz), a is thermal diffusivity (m^2/s), λ is heat conductivity (kW m^{-1} K^{-1}), and h is convective heat transfer coefficient between tubes, fin and air stream (kW m^{-2} K^{-1}). Here, $Pr = 0.707$ is assumed for air.

The dimensionless axial coordinate is defined by

$$x^* = \frac{x}{P_L} \tag{9.12}$$

As can be seen, $x^* = 1, 2, 3, 4$, corresponds to the first cycle (tube), the second cycle, the third cycle, and the last cycle (the outlet), respectively. In this regard, the channel can be considered as a cyclic duct flow. The fluid flow and heat transfer in the exchanger can be regarded as phenomena happening in the representative duct.

Dimensionless spanwise and normal coordinates are defined respectively as

$$y^* = \frac{y}{0.5P_T} \tag{9.13}$$

$$z^* = \frac{z}{0.5H_z} \tag{9.14}$$

The mean Nusselt number from the channel inlet to position x^* is defined by

$$Nu_m = \frac{2h_m H_z}{\lambda} \tag{9.15}$$

where h_m is the mean heat transfer coefficient from channel inlet to position x^*. It can be numerically calculated by the heat balance equation as

$$h_m = \frac{mc_p(T_i - T)}{A\Delta T} \tag{9.16}$$

where m is mass flow rate (kg/s), c_p is the specific heat of fluid, kJ/(kg K), T_i is the fluid temperature at the inlet, A is the tube and fin surface area from inlet to position x^* (m^2), and ΔT is the logarithmic temperature difference between the tube and the fluid, which is calculated by

$$\Delta T = \frac{(T_i - T_{tube}) - (T - T_{tube})}{\ln \frac{T_i - T_{tube}}{T - T_{tube}}} \tag{9.17}$$

where T_{tube} is the tube temperature (K).

The pressure drop from the inlet to axial position x^* is

$$\Delta P_0 = P_0 - P_0 \tag{9.18}$$

In calculating mean heat transfer coefficients and pressure drop, the temperature and pressure are chosen as the time-averaged values.

9.2.3 Conjugate heat transfer in fins

Due to the small thickness of the fins, two-dimensional conductive heat transfer is assumed in fins. The conjugate heat transfer between fin and air flow is written as

$$\lambda_{\text{fin}}\delta_{\text{fin}}\left(\frac{\partial^2 T_{\text{fin}}}{\partial x^2} + \frac{\partial^2 T_{\text{fin}}}{\partial y^2}\right) = -2\lambda\left(\frac{\partial T}{\partial z}\right)_{z^*=0} \tag{9.19}$$

9.2.4 Initial and boundary conditions

The initial conditions are: $t=0$, $u_j=0$; $T=T_i=300\,\text{K}$.

The boundary conditions at the inlet are:

$$x^* = 0, u_i = u_a, u_j = u_k = 0; T = T_i \tag{9.20}$$

In addition, the flow at the inlet is assumed to be laminar with no perturbations. At the outlet, streamwise gradients for all the variables are set to zero. At the symmetric plane $z^*=1$, normal gradients are set to zero. At the solid surfaces, no-slip conditions are set. At the tube walls, a uniform temperature condition of 333 K is assumed. At fin locations where they connect with the tubes, fin temperature is equal to the tube temperature. At the channel right and left boundaries ($y^*=0$, 1), symmetric boundary conditions are assumed. The boundaries on fin-air interfaces are the conjugate boundary conditions.

9.3 Numerical methods

The three-dimensional computational domain is created and meshed first with unstructured grids. The generated mesh structures on the duct surface are shown in Figure 9.5. To give an enlarged illustration, the figure only shows two cycles (two tubes). Average mesh sizes are: $\Delta x=1\,\text{mm}$, $\Delta y=1.0\,\text{mm}$, $\Delta z=0.15\,\text{mm}$. In total 50,314 grids are generated. Then a grid independence test is conducted. It is found that a finer mesh of 96,346 grids will have little merit in further increasing the accuracy of the average Nusselt numbers and pressure drops. Therefore it is considered that the current mesh number is enough. Further, in the case of LES, the computational mesh requirements are typically high. Too fine grids would lead to grids that are not manageable by ordinary personal computers. This is another reason why a finer mesh is not encouraged.

The numerical scheme for solving the governing equations is based on the finite volume approach. Discretization of the equations is applied using second-order central differencing, except for the

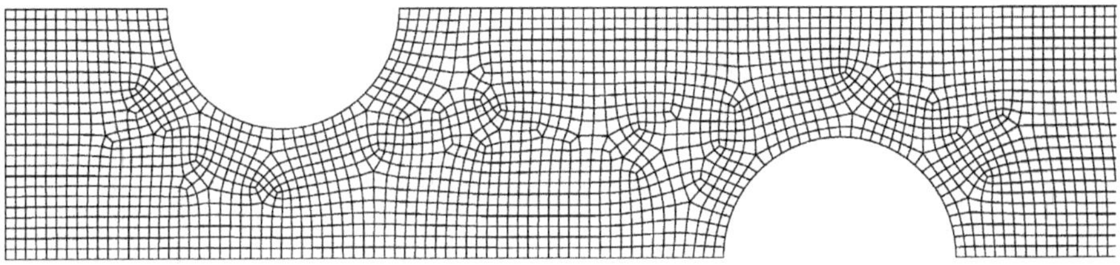

FIGURE 9.5

The mesh structure generated for the duct, showing only two tubes for clarity.

convection terms, which are discretized using the bounded second order upwind scheme (BSOU). This upwind differencing scheme allows second-order accuracy, thus overcoming the problems of numerical diffusion associated with first-order upwind differencing, while being constantly bounded as opposed to some third-order schemes (such as QUICK). Discretization in time is expressed as a first-order, fully implicit Euler scheme. Although the scheme is of first-order accuracy, it is fully implicit, which allows stability even for large time-steps. Solution of the equations is based on the SIMPLE algorithm for pressure correction. Typical time-steps vary from 50 μs to 1 ms, depending on Reynolds numbers. According to the Nyquist criterion [19], in order to calculate a frequency of *f*, one must use a sampling rate of at least 2*f* for the micro-time-scale frequency.

Time-marching calculations are stopped when either a steady or a periodic flow is found, or in the case of aperiodic solutions when average values of heat transfer coefficients and pressure drop values become approximately constant.

After the solution of temperature fields in the fluid, the temperature fields in the fins are calculated with the just calculated temperature fields of the fluid. The temperatures on the fluid side are taken as the boundary conditions on fin surfaces. Then the temperature fields in the fluid are re-solved with the fin temperatures as the boundary conditions for the fluids. The values are outputted until the old and new values for all the nodes both in fluids and in fins are converged. At this time, temperature and pressure values are outputted. The conjugate heat transfer problem is solved.

Model validation

The numerical results for the plate-fin and tube heat exchanger depicted in Table 9.1 with inlet frontal velocity ranging from 2.6 m/s ($Re=1788$) to 6.0 m/s ($Re=4125$) are shown in Figure 9.6 for mean heat transfer coefficients and Figure 9.7 for the pressure drop of the whole exchanger. In total, four models are used: laminar (LAM), standard k-ε turbulence (SKE), shear stress transport k-ω turbulence (SSTKW) and large eddy simulation (LES). Experimental results from Ref. [10] are also plotted in the two figures. The data are calculated from correlations for large tube diameters. From Figure 9.6, it is seen that the SSTKW predicts the heat transfer coefficients well for a wide range of *Re* numbers. The laminar model predicts the heat transfer coefficients well when the frontal velocity is below 4.0 m/s ($Re<2330$), but it underestimates the heat transfer coefficients when the frontal velocity is beyond 4.0 m/s. In contrast, the SKE model overestimates the heat transfer coefficients by at least 50% for all the *Re* numbers studied. From this analysis, the SKE model is phased out at first. The laminar model

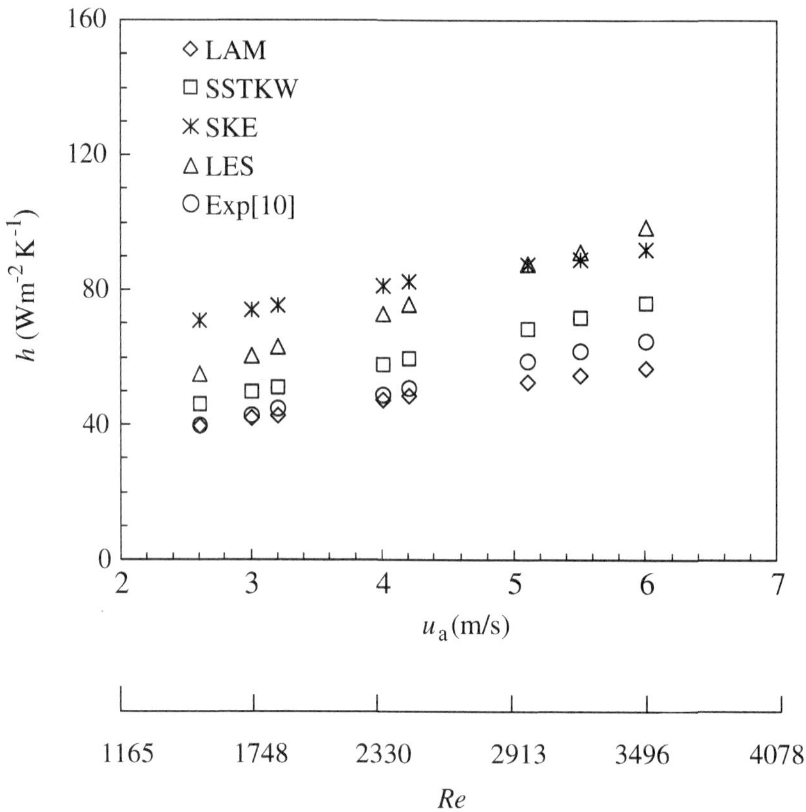

FIGURE 9.6

Comparison of the calculated mean heat transfer coefficients with the experimental data.

can be used when *Re* numbers are below 2330. The LES model is next to the SSTKW. It is also acceptable. It is not as accurate as the SSTKW model, but it can predict the transitional behaviors of fluid flow in the ducts. This is the benefit with LES simulations.

From Figure 9.7, it is seen that the laminar model predicts the pressure drop correctly only when *Re* is below 2330. Surprisingly, the SKE model fails to predict the pressure drop even at higher *Re* numbers. All the laminar, the SSTKW and the SKE models fail to estimate the pressure drop correctly when *Re* exceeds 2330. The greater the *Re* numbers, the greater are the disparities between the calculated values with the three models and the experimental data. The trends are in agreement with the numerical results in Ref. [10]. They underestimate the pressure drop by 40% at $Re=3496$. In contrast, the LES model is the only model that can predict the pressure drop quite satisfactorily in the transitional flow regime. The predictions with this model at lower *Re* numbers are also acceptable. These two figures provide validations for the methodology adopted in this chapter. The LES model is acceptable in the prediction of performance in the ducts.

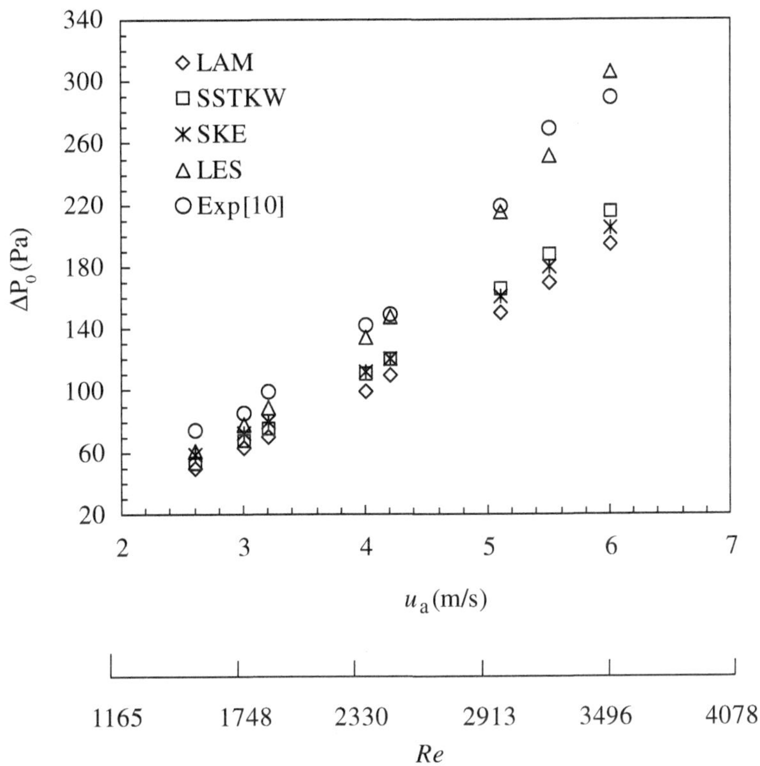

FIGURE 9.7

Comparison of the calculated pressure drop with the experimental literature data.

9.4 Results and discussion

9.4.1 Local mean performances

The locally mean Nusselt numbers calculated by Eq. (9.15) along the flow direction x are shown in Figure 9.8. The effects of row numbers on performance can be observed. As can be seen, generally, the mean Nusselt numbers decrease with row numbers, but only to a small extent. A higher frontal velocity usually leads to higher Nusselt numbers. The trends are similar to the numerical and experimental results in Refs. [5,10]. However, it was found that it is the second row (not the first row) that has the highest Nusselt numbers. The reason lies in the coherent components of the exchanger: parallel plates and tube banks. Two factors influence the performance simultaneously: (1) Flow is developing along the channel length and the heat transfer coefficients tend to decrease along flow; (2) at the same time, turbulence is developing with tube rows, which would increase heat transfer coefficients. The final results are a balance between these two contradictory effects and heat transfer coefficients reach maximum at the second row.

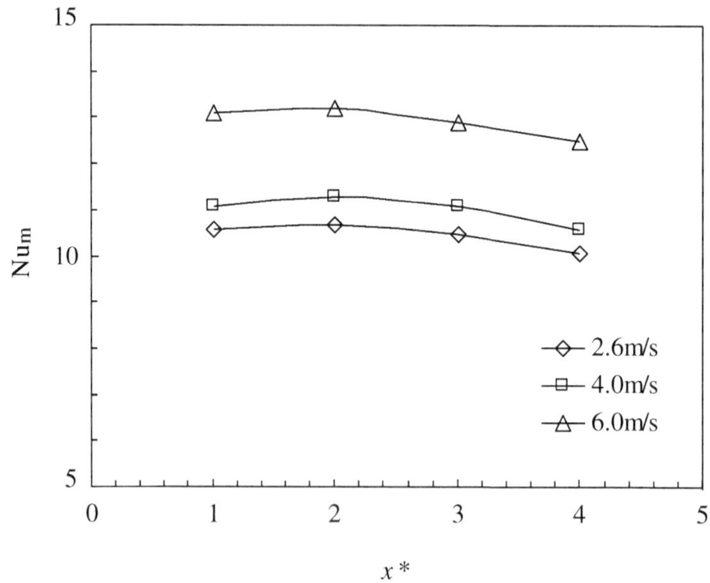

FIGURE 9.8

Variations of locally mean Nusselt numbers along the axial direction.

The variations of pressure drop along the flow defined by Eq. (9.18) are plotted in Figure 9.9. The pressure drops continuously with tube rows, but the first row has the highest pressure drops. The next three rows have somewhat equal pressure drops. This phenomenon indicates that the fluid flow becomes nearly full developed after the second row.

9.4.2 Unsteady behavior

Now, the oscillation behavior is analyzed. The oscillations of velocity components (streamwise, spanwise, and normal) at a monitoring point (center on face $x^*=3$) under three frontal velocities are shown in Figure 9.10 ($u_a=2.6\,\text{m/s}$), Figure 9.11 ($u_a=4.0\,\text{m/s}$) and Figure 9.12 ($u_a=6.0\,\text{m/s}$). As can be seen, the higher the frontal velocities are, the larger the scale of amplitudes of oscillations is. In Figure 9.10, the velocity oscillations are large at the beginning, then the oscillation amplitudes decrease steadily with iteration time-steps. After about 0.1 s, the flow is considered to be quasi-steady state, because constant oscillation amplitudes are reached. The amplitude of oscillation under this frontal velocity is rather low, of the order of 0.03 m/s, therefore flow in this condition can be approximated by a laminar flow. The oscillation amplitudes of streamwise and spanwise velocity are almost equal. The normal velocity oscillations are negligible. The streamwise and spanwise velocity oscillations under 4.0 m/s frontal velocity shown in Figure 9.11 exhibit similar trends to those in Figure 9.10. However, the amplitudes are much larger: 0.15 m/s. Nevertheless the fluctuations are still regular. Flow in this condition can be considered transitional. With increasing frontal velocities, the oscillations become larger and larger, and it become more turbulent, as shown in Figure 9.12 for 6.0 m/s. The oscillations can be

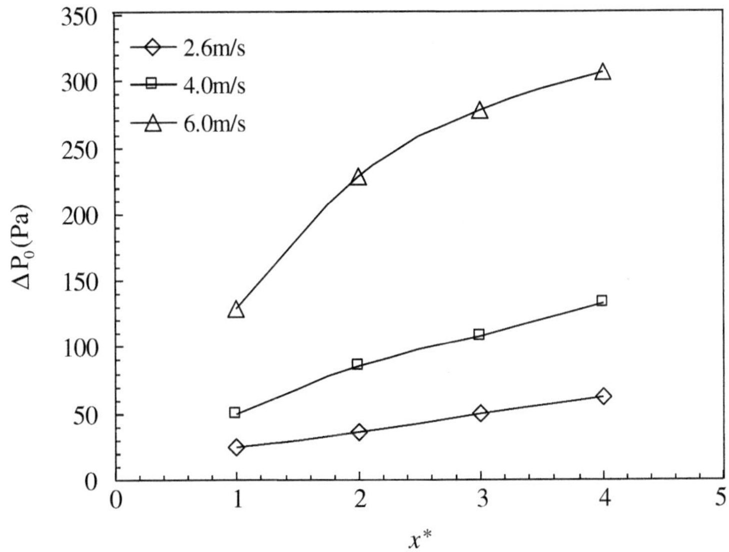

FIGURE 9.9

Local pressure drop along the axial direction.

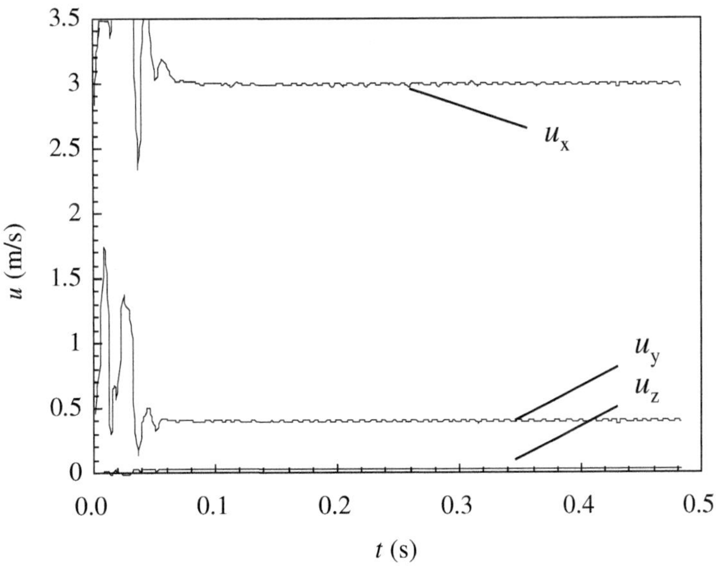

FIGURE 9.10

Time dependence of three velocity components at a monitoring point (center on face $x^*=3$) for frontal air velocity 2.6 m/s.

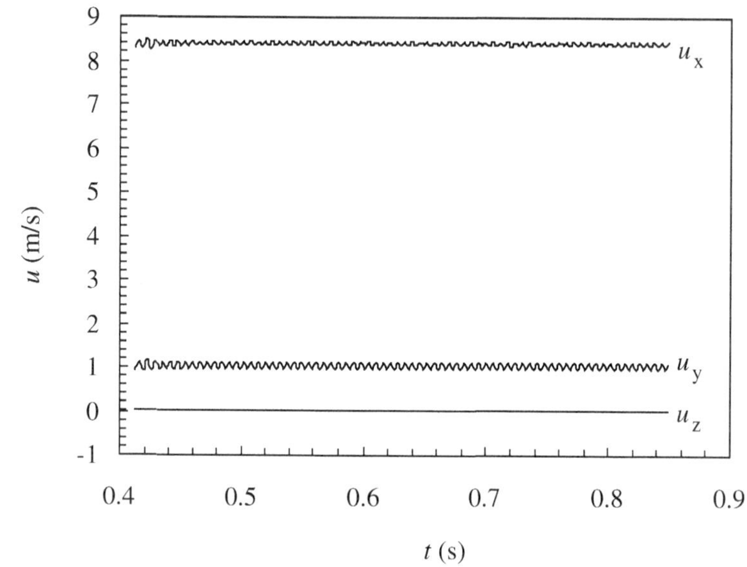

FIGURE 9.11

Time dependence of three velocity components at a monitoring point (center on face $x^*=3$) for frontal air velocity 4.0 m/s.

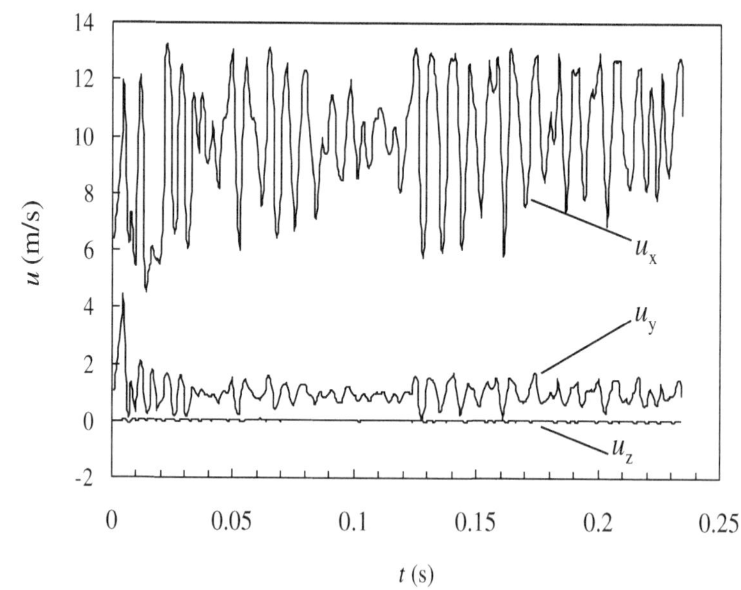

FIGURE 9.12

Time dependence of three velocity components at a monitoring point (center on face $x^*=3$) for frontal air velocity 6.0 m/s.

as large as 6 m/s for streamwise velocity and 2 m/s for spanwise velocity. The cyclic manner is totally irregular. Flow in this condition can be considered nearly turbulent. As for the normal velocity, the oscillations are still very small. The reason may be that the flow is mainly two-dimensional, as a result of flow confinement by two parallel plates of very small pitch. This is the demonstrable difference between a flow in a plate-fin tube and that in a bare tube bank. That is the reason why the hydrodynamic diameter (double the fin spacing) is selected as the characteristic length of flow in this research. The flow is more like a duct flow than a flow through bare tubes.

The Fourier transforms of the time series of the streamwise velocity oscillations are shown in Figures 9.13–9.15 for frontal velocities 2.6 m/s, 6.0 m/s, and 10.0 m/s, respectively. In the figures, the x axis represents dimensionless frequency, the Strouhal number. The y axis represents energy level or series amplitudes. The dominant St_r numbers for 2.6 m/s and 6.0 m/s frontal velocities are 0.23 (102 Hz) and 0.12 (123 Hz) respectively. The data set around the dominant frequency for 6.0 m/s is more sparsely populated than that for 2.6 m/s, meaning the dominant frequency is "less dominant" with increasing frontal velocity, and the flow is more turbulent. As for the frequency spectrum for a high frontal velocity of 10 m/s shown in Figure 9.15, there is no dominant frequency observed at all, meaning it is totally aperiodic and fully turbulent now.

In a bare tube bank, the dominant frequency of vortex shedding is around $St_r = 0.366$, where the Strouhal number is defined by the tube outer diameter [19]. The calculated corresponding vortex shedding frequencies are about 59.8 Hz for 2.6 m/s and 138 Hz for 6.0 m/s frontal velocity. As can be seen, the dominant oscillation frequencies are not equal to the vortex shedding frequencies. This indicates that the turbulence in this plate-fin and tube channel is not a direct result of vortex shedding. Rather, it is due to the zigzag nature of the channel structure.

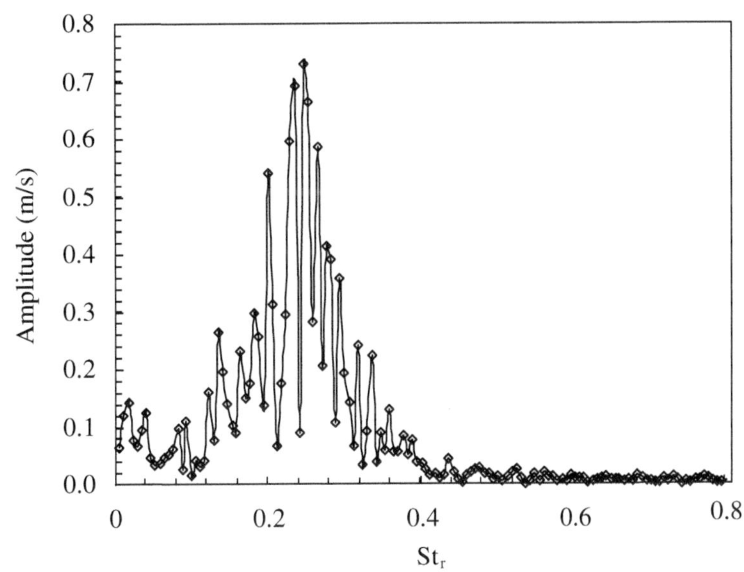

FIGURE 9.13

Frequency spectrum of streamwise velocity oscillation under frontal velocity 2.6 m/s.

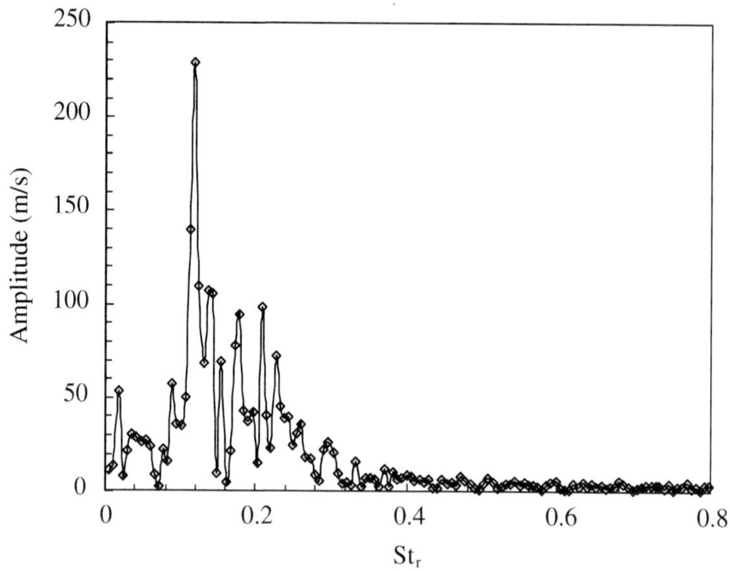

FIGURE 9.14

Frequency spectrum of streamwise velocity oscillation under frontal velocity 6.0 m/s.

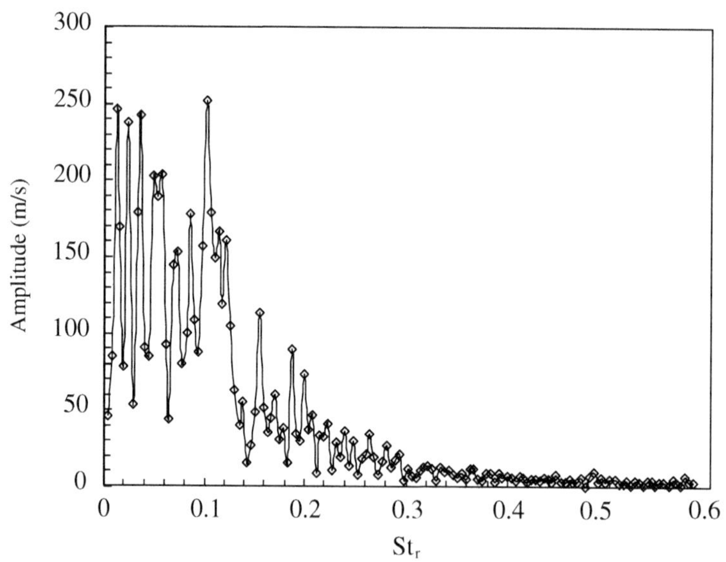

FIGURE 9.15

Frequency spectrum of streamwise velocity oscillation under frontal velocity 10.0 m/s.

9.4.3 Flow field and heat transfer

The velocity vectors on the horizontal plane $z^* = 1$ under frontal velocity of 4 m/s are shown in Figure 9.16. To show the vectors clearly, the figure only demonstrates three cycles. As can be seen, steady vortexes are formed behind the tubes. The vortexes have a mixing effect that would intensify momentum and heat transfer. Velocity vectors at different time-steps show that there is no demonstrable vortex shedding. The reason may be that the parallel plates with small spacing have a large confinement effect on the flow. The energy of the vortex that would lead it to shed itself from the tubes in a bare tube bundle is not large enough to overcome the viscous force that makes the flow adhere to the fins.

When the velocity is high enough, vortex shedding is observed [24]. The contours of the mean temperature on the x–y plane $z^* = 1$ are illustrated in Figure 9.17 for frontal velocity 4 m/s. The contours are concentric circles around the tubes locally, but are similar to a zigzag duct flow in general. Therefore, the heat transfer and fluid flow in a plate-fin and tube heat exchanger is more like a duct flow.

9.4.4 Fin efficiency

Conjugate heat transfer between fins and air stream is taken into account for the exchanger. The mean temperature distributions on the fin surface were obtained and are plotted in Figure 9.18. A 1–5 °C temperature non-uniformity is observed on the surface. The predominant temperature contours are concentric circles, indicating that the heat transfer from the tubes is the dominant heat transfer mechanisms in fins.

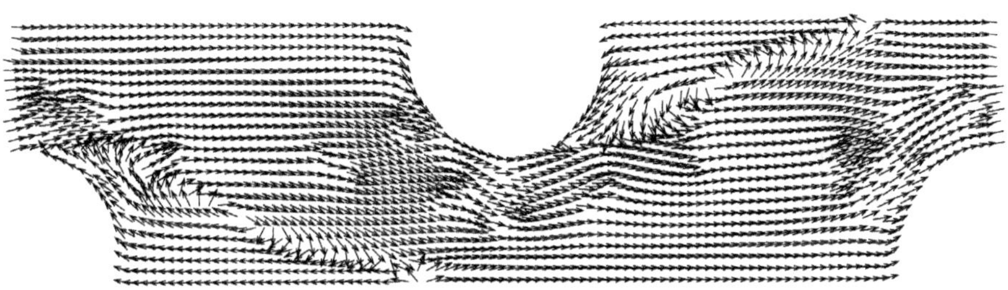

FIGURE 9.16

Velocity vectors on horizontal plane $z^*=1$, showing middle three cycles, 4.0 m/s, $t=0.2$ s.

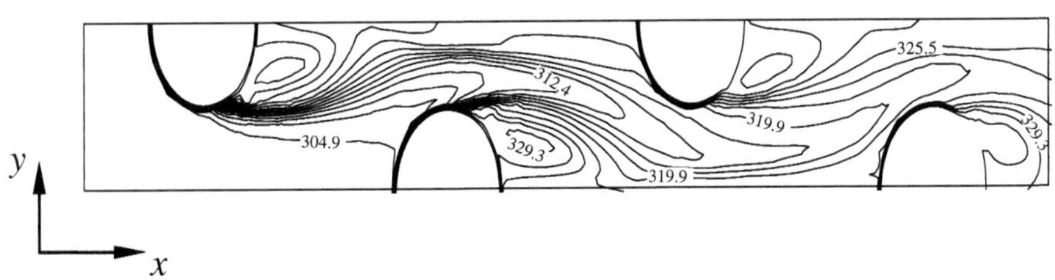

FIGURE 9.17

Contours of temperature (K) on plane $z^*=1$ under frontal velocity 4.0 m/s.

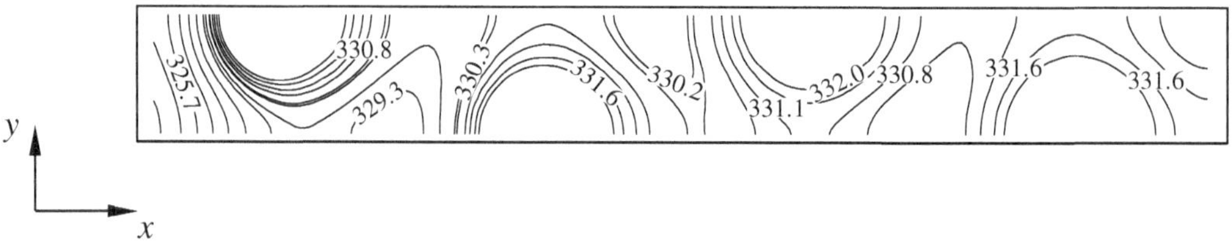

FIGURE 9.18

Contours of fin temperature (K) under frontal velocity of 4.0 m/s.

Table 9.2 Fin efficiencies of the plate-fin and tube heat exchanger.

u_a (m/s)	2.6	3.0	3.2	4.0	4.2	5.1	5.5	6.0
Re	1787	2054	2186	2710	2842	3429	3693	4089
η_{fin} (this study)	0.910	0.902	0.901	0.893	0.885	0.878	0.871	0.869
η_{fin} (Schmidt approximation)	0.876	0.869	0.867	0.855	0.853	0.840	0.836	0.829

Table 9.2 lists the calculated fin efficiencies under various frontal velocities. Fin efficiency is defined as the ratio of the mean Nusselt numbers under real fin conductivity to those under ideal infinite fin conductivity (or uniform fin temperature assumptions). To make a comparison, the fin efficiencies calculated by Schmidt approximation [26] are also listed in the table. As can be seen, the Schmidt approximation underestimates the prediction by 3.7% to 4.6%. The higher the frontal velocities are, the greater the fin temperature non-uniformity is, and the larger the differences between the current prediction with the Schmidt approximation are.

9.5 Conclusion

The plate-fin and tube heat exchanger is a special duct flow. This chapter has shown the characteristics of the fluid flow and conjugate heat transfer in the duct. The unsteady behavior of the fluid flow and heat transfer in plate-fin and tube heat exchangers was investigated. The oscillations of the velocity components show a transition from laminar to fully turbulent with increasing frontal velocities. Though the velocities oscillate seriously, steady vortexes are formed behind each tube, due to the viscous drag force from the closely spaced parallel plates. A frequency spectrum analysis and velocity vector plot prove that vortex shedding is not the reason behind the unsteady flow. In this respect, the flow is more like a duct flow with a zigzag structure. The strong mixing effects of the vortexes cause an improvement of heat transfer and an increase in pressure drop.

Even under very low frontal velocities, streamwise and spanwise velocities exhibit a certain degree of periodic and regular oscillation. The flow becomes aperiodic and chaotic with increasing Reynolds numbers. The large eddy simulation technique provides a reliable tool to analyze the temporal fluid flow and heat transfer in this heat exchanger, which usually works in a transitional flow regime.

References

[1] Saboya FEM, Sparrow EM. Transfer characteristics of two row plate fin and tube heat exchanger configurations. Int J Heat Mass Transfer 1976;19:41–9.

[2] Saboya FEM, Sparrow EM. Experiments on a three-row fin and tube heat exchanger. ASME J Heat Transfer 1976;98:26–34.

[3] Rich DG. The effect of fin spacing on the heat transfer and friction performance of multi-row plate fin-and-tube heat exchangers. ASHRAE Trans 1973;17:137–45.

[4] Rich DG. The effect of the number of tube rows on the heat transfer performance of smooth plate and fin-and-tube heat exchangers. ASHRAE Trans 1975;81:307–17.

[5] Wang CC, Chi KY. Heat transfer and friction characteristics of plain fin-and-tube heat exchangers. Part I: New experimental data. Int J Heat Mass Transfer 2000;43:2681–91.

[6] Wang CC, Lin YT, Lee CJ. An air side correlation for plain fin-and-tube heat exchangers in wet conditions. Int J Heat Mass Transfer 2000;43:1869–72.

[7] Wang CC, Chi KY, Chang CJ. Heat transfer and friction characteristics of plain fin-and-tube heat exchangers. Part II: Correlation. Int J Heat Mass Transfer 2000;43:2693–700.

[8] McQuiston FC. Correlation for heat, mass and momentum transport coefficients for plate-fin-tube heat transfer surfaces with staggered tube. ASHRAE Trans 1978;84:194–309.

[9] Gray DL, Webb RL. Heat transfer and friction correlations for plate fin-and-tube heat exchangers having plain fins. San Francisco: Proceedings of the ninth international heat transfer conference; 1986.

[10] Jang JY, Wu MC. Numerical and experimental studies of three-dimensional plate-fin and tube heat exchangers. Int J Heat Mass Transfer 1996;39:3057–66.

[11] Tsai SF, Sheu Tony WH. Some physical insight into a two-row finned tube heat transfer. Comput Fluids 1998;27:29–46.

[12] Mendez RR, Sen M, Yang KT, McClain R. Effect of fin spacing on convection in a plate fin and tube heat exchanger. Int J Heat Mass Transfer 2000;43:39–51.

[13] He YL, Tao WQ, Song FQ, Zhang W. Three-dimensional numerical study of heat transfer characteristics of plain plate fin-and-tube heat exchangers from view point of field synergy principle. Int J Heat Fluid Flow 2005;26:459–73.

[14] Tao YB, He YL, Wu ZG, Tao WQ. Three-dimensional numerical study and field synergy principle analysis of wavy fin heat exchangers with elliptic tubes. Int J Heat Fluid Flow 2007;28:1531–44.

[15] Tao YB, He YL, Huang J, Wu ZG, Tao WQ. Three-dimensional numerical study of wavy fin-and-tube heat exchangers and field synergy principle analysis. Int J Heat Mass Transfer 2007;50:1163–75.

[16] Tao WQ, Chen YP, Lee TS. The influence of strip location on the pressure drop and heat transfer performance of a slotted fin. Numer Heat Transfer A: Appl 2007;52:463–80.

[17] Perrotin T, Clodic D. Thermal-hydraulic CFD study in louvered fin-and-flat-tube heat exchangers. Int Journal of Refrig 2004;27:422–32.

[18] Wu HL, Gong Y, Zhu X. Air flow and heat transfer in louver-fin round-tube heat exchangers. ASME J Heat Transfer 2007;129:200–10.

[19] Bouris D, Bergeles G. Two dimensional time dependent simulation of the subcritical flow in a staggered tube bundle using a subgrid scale model. Int J Heat Fluid Flow 1999;20:105–14.

[20] Barsamian HR, Hassan YA. Large eddy simulation of turbulent crossflow in tube bundles. Nucl Eng Des 1997;172:103–22.

[21] Hassan YA, Barsamian HR. Tube bundle flows with the large eddy simulation technique in curvilinear coordinates. Int J Heat Mass Transfer 2004;47:3057–71.

[22] aus der Wiesche S. Large-eddy simulation study of an air flow past a heated square cylinder. Heat Mass Transfer 2007;43:515–25.

[23] Rollet-Miet P, Laurence D, Ferziger J. LES and RANS of turbulent flow in tube bundles. Int J Heat Fluid Flow 1999;20:241–54.

[24] Zhang LZ, Zhong WC, Chen JM, Zhou JR. Fluid flow and heat transfer in plate-fin and tube heat exchangers in transitional flow regime. Numer Heat Transfer A: Appl 2011;60:766–84.

[25] Zhang LZ, Chen ZY. Convective heat transfer in cross-corrugated triangular ducts under uniform heat flux boundary conditions. Int J Heat Mass Transfer 2011;54(1–3):597–605.

[26] Tao YB, He YL, Huang J, Wu ZG, Tao WQ. Numerical study of local heat transfer coefficients and fin efficiency of wavy fin-and-tube heat exchangers. Int J Therm Sci 2007;46:768–78.

CHAPTER

Exchanger Structure-Induced Flow Maldistribution and Performance Deterioration

10

10.1 Introduction

Heat and mass exchangers have been widely used in various industries such as gas separation, air conditioning, and pollution control. Heat and mass transfer ducts can be regarded as the representative elements in the exchanger cores. The conjugate heat and mass transfers in ducts have been investigated thoroughly in previous chapters. However, cores alone cannot work. A complete exchanger must have shells and ducts which are used to hold the core and to separate the flows. Figure 10.1 shows the schematic of a typical commercialized air-to-air heat mass exchanger/total heat exchanger. As can be seen, the whole exchanger is composed of inlets, outlets, exchanger shells, separating plates, fans, and, most importantly, the core. When installed, the exchanger core is rotated 45° and inserted into the exchanger shell horizontally, forming two air ducts with the separating plates in the rectangular shell box. Fresh air is sucked into the exchanger by a fan (as shown in the figure) through the fresh air inlet. Exhaust air is sucked into the exchanger by another fan through the exhaust air inlet. The two air streams are drawn through the exchanger core in a cross-flow arrangement. The two fans themselves are packed and placed in the air ducts. They suck air axially and eject air radially. Such placements make the total heat exchanger compact and easy to install in rooms. However, the arrangements also make the flow complicated. To see the structure more clearly, the front view, the side view and the top view of the total heat exchanger is shown in Figure 10.2.

The core is independently shown in Figure 10.3. It is a parallel-plate structure. The structure of the exchanger makes it easy to install, clean, and replace.

Even though a single-core duct has been investigated in detail, the fluid flow and heat transfer in the whole exchanger are of equal importance. Considering the complex ducting work inside a practical exchanger shell, it is anticipated that the flow would undergo turnarounds, expansions, and contractions, which would finally lead to flow maldistribution across the core face. The final flows in the ducts are inhomogeneous. However, almost all previous studies in this field concentrated only on a representative single duct in the exchanger core, by ignoring the real flow distributions in the whole exchanger. In other words, a uniform flow distribution across the core face was assumed. This is certainly a very ideal assumption which needs further clarification.

The effects of flow maldistribution have been well recognized and presented for heat exchangers and cooling towers. There have been several pieces of research on this topic, mostly on plate heat

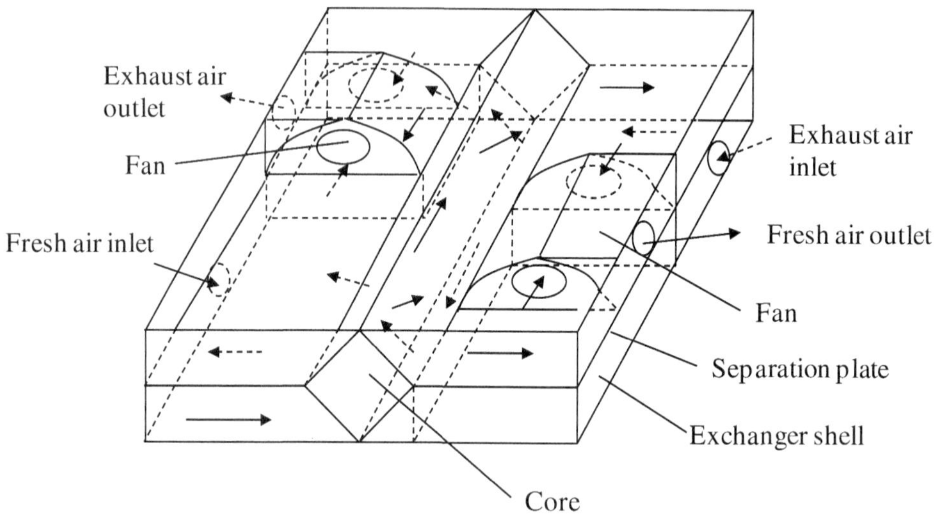

FIGURE 10.1

Schematic of a practical total heat exchanger with parallel-plate or plate-fin membrane cores.

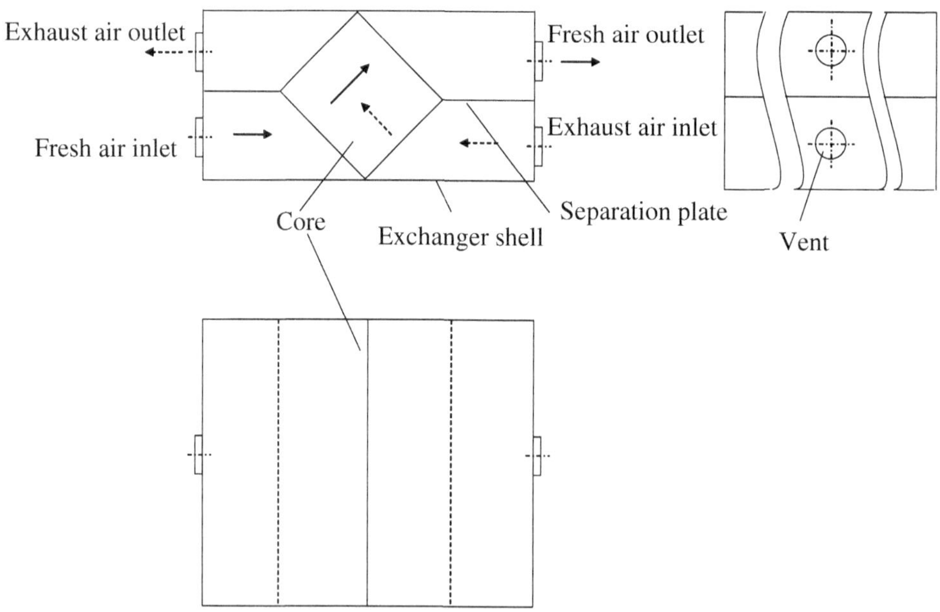

FIGURE 10.2

Front view, side view, and top view of the total heat exchanger.

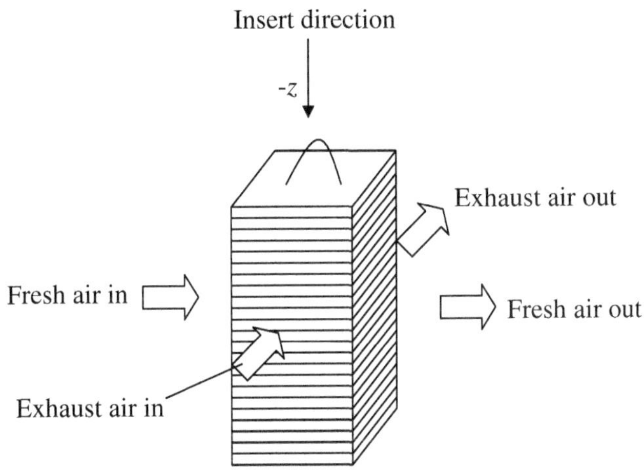

FIGURE 10.3

Schematic of a membrane-based parallel-plate total heat exchanger core.

exchangers of *U*-type or *Z*-type [1–10]. Regrettably these researches did not study the influence of the cores on the flow distribution. They only considered the effects of exchanger structure on flows. A heat mass exchanger as shown in Figures 10.1 and 10.2 is much more complex than an ideal rectangular pipe as assumed in previous works. The incoming fluids impinge on the core surface at an angle, not in normal directions. So the arrangement here is neither *U*-type nor *Z*-type, but a three-dimensional skewed *Z*-type. Previous studies solved the velocity distribution in the inlet duct without the coupling with the core. They separated the inlet duct from the core and didn't consider the effects of the core when solving the fluid fields in the inlet duct. In other words, the inverse effects of the core on flow distribution in the conduits were not reflected. This may lead to unrealistic (or too ideal) flow distribution in the exchanger, considering the strong interactions between the inlet conduits, the core and the outlet conduits. To address this problem, recently there have been several investigations on heat and mass exchangers [11–14] in which the interactions between the exchanger structure and the core are considered.

This chapter is a step forward. Besides considering the interactions between the core and exchanger shells, the basic transport data obtained in previous chapters are used in the estimation of heat and mass transfer performance and deterioration. The flow maldistribution is calculated by selecting the whole exchanger as the domain. The influences of the inlet duct, the core and the outlet duct will be considered simultaneously. After the maldistribution of flow fields are determined, its effects on the thermal performance of the heat exchanger are discussed with the new transport data.

10.2 Parallel-plate core

10.2.1 Flow distribution

The core depicted in Figure 10.3 is inserted into the exchanger shell. The fluid flow in the exchanger is numerically calculated. The fresh air duct and the exhaust air duct are in symmetry. So only the fresh

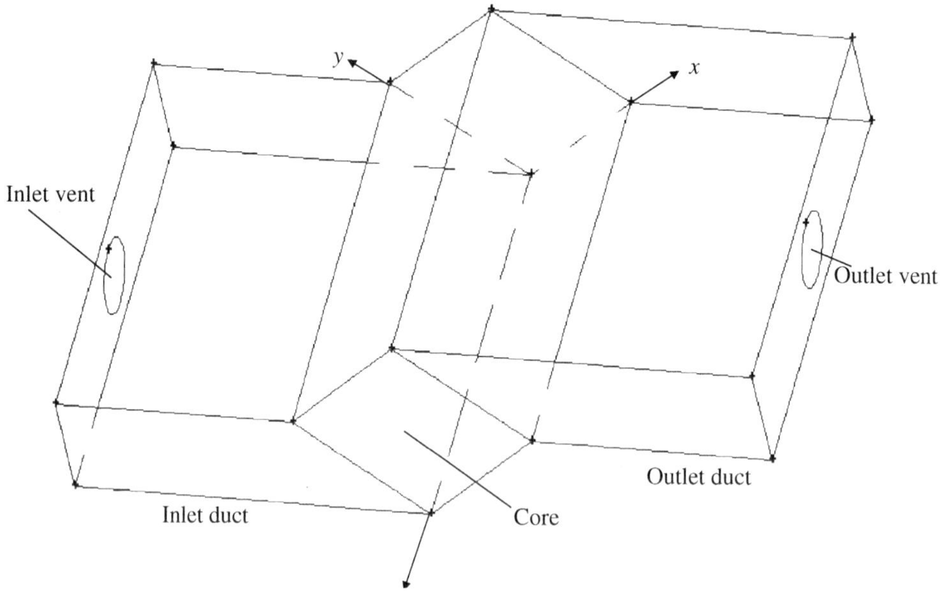

FIGURE 10.4

The modeling domain for flow distribution of the total heat exchanger fresh air duct.

air duct as shown in Figure 10.4 is selected as the calculation domain. The duct and the core should be modeled together. Modeling the flows in the core is a difficulty. There have been several numerical studies concerning the modeling of flow maldistribution in plate heat exchangers, on a duct-to-duct basis [8–10]. However, due to the limits on computer capacity and speed, the number of channels that can be modeled directly is quite limited. Duct-to-duct modeling is detailed enough to disclose local variations in a channel. However, a real core usually has hundreds, even thousands, of channels. The large numbers of channels prohibit the direct simulation of flow on a duct-to-duct basis. To solve this problem, this chapter simulates the core as a porous medium which only permits two-dimensional air flow in the duct. When modeling the fluid flow, heat and mass transfer are not considered. In other words, only the flow in one stream is considered.

The computational grids are generated and the problem is numerically solved using the finite volume method [11]. The whole geometry consists of three parts: the inlet duct, the core, and the outlet duct. The geometrical properties are listed in Table 10.1. Air passes through inlet vent, inlet duct, the core, outlet duct, and the outlet vent, consecutively. The inlet duct and the outlet duct are air flow in a "volume." The core is treated as a porous medium. The fluid domains are modeled with air at room temperature. The boundary condition at the outlet vent is defined as the volumetric flow rate leaving the outlet vent surface. The boundary condition at the inlet vent is "outflow," which means a zero axial variable gradient across the outlet vent surface. This is because air is sucked inside from the inlet vent. Other surfaces are defined as wall conditions, except the two core surfaces facing and backing the air flow (surface $x^*=0$ and 1). A porous medium is used for the core. A standard k-ε turbulence model is

Table 10.1 Structural and physical parameters of the two parallel-plates cores tested.

Symbol	Unit	Core A	Core B
n_{ch}		92	127
x_F, y_F	mm	185	185
z_F	mm	460	460
δ_m	μm	100	100
λ_m	$Wm^{-1}K^{-1}$	0.43	0.43
$2a$	mm	2.5	1.8
D_h	mm	2.15	1.55
A_f, A_e	m^2	6.30	8.69
D_{wm}	m^2/s	5.15×10^{-10}	5.15×10^{-10}
K_p		10	10
Inlet/outlet duct length	mm	262	262
Inlet/outlet vent diameter	mm	60	60

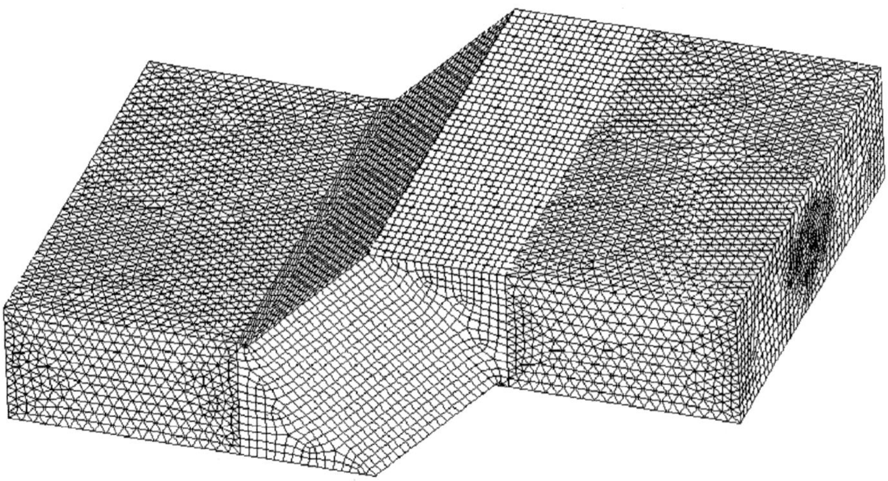

FIGURE 10.5

Mesh structure of the fresh air duct for flow calculation.

used for the inlet and outlet ducts because air flow in the duct (not inside the core) is quite turbulent due to the large characteristic sizes. The simulation is considered converged when the residual of all variables remains below 10^{-5} and the mean inlet velocities remain constant.

A view of the mesh structure can be seen in Figure 10.5. They are mixed-structure meshes including both structured and unstructured meshes. The grid sensitivity of the results is checked by comparing the results of two grids with different refinement levels. The coarse grid configurations have approximately

160,489 cells (97,851 tetrahedral cells and 62,638 hexahedral cells), while the fine grids have approximately 230,564 cells (102,251 tetrahedral cells and 128,313 hexahedral cells). The mean absolute difference between the results with the coarse and fine grids is below 1%. As the results show negligible grid sensitivity, the coarse grid is used in the rest of this work to speed up convergence.

Modeling with a porous medium is simpler than duct-to-duct modeling. The porous media model can be used for a wide variety of problems, including flows through packed beds, filter papers, perforated plates, flow distributors, and tube banks. The porous media model incorporates an empirically determined flow resistance in a region of the model defined as "porous." In reality, the porous media model is nothing more than an added momentum sink in the governing momentum equations. Porous media are modeled by the addition of a momentum source term to the standard fluid flow equations. The source term is composed of two parts, a viscous loss term (Darcy), and an inertial loss term, expressed as [11]

$$S_{\mathrm{i}} = \sum_{j=1}^{3} D_{\mathrm{ij}} \mu u_{\mathrm{j}} + \sum_{j=1}^{3} C_{\mathrm{ij}} \frac{1}{2} \rho \left| u_{\mathrm{j}} \right| u_{\mathrm{j}} \tag{10.1}$$

where S_i is the source term for the ith (x, y, or z) momentum equation, and D and C are prescribed matrices. This momentum sink contributes to the pressure gradient in the porous cell, creating a pressure drop that is proportional to the fluid velocity (or velocity squared) in the cell.

To recover the case of simple homogeneous porous media

$$S_{\mathrm{i}} = \frac{\mu}{\alpha_{\mathrm{p}}} u_{\mathrm{i}} + C_2 \frac{1}{2} \rho \left| u_{\mathrm{i}} \right| u_{\mathrm{i}} \tag{10.2}$$

where α_{p} is the permeability in Darcy's Law and C_2 is the inertial resistance factor, simply specify D and C as diagonal matrices with $1/\alpha$ and C_2, respectively, on the diagonals (and zero for the other elements).

In flows through porous media, the pressure drop is typically proportional to velocity and the constant C_2 can be considered to be zero. Ignoring convective acceleration and diffusion, the porous media model then reduces to Darcy's Law. The pressure drop across the channel can be calculated by

$$\Delta P = \frac{\mu}{\alpha_{\mathrm{p}}} u_{\mathrm{i}} \Delta x \tag{10.3}$$

where u_{i} is the face velocity in the core (in x direction).

Based on the definition of friction factor for ducts,

$$\Delta P = \frac{\Delta x}{D_{\mathrm{h}}} f \frac{1}{2} \rho u_{\mathrm{a}}^2 \tag{10.4}$$

$$Re = \frac{\rho u_{\mathrm{a}} D_{\mathrm{h}}}{\mu} \tag{10.5}$$

where u_{a} is velocity in ducts. It is 2 times the face velocity, because half the core face is blocked and sealed for the other passage.

$$u_{\mathrm{a}} = 2u_{\mathrm{i}} \tag{10.6}$$

For fully developed laminar flow in ducts

$$(f \cdot Re) = C_3 \tag{10.7}$$

where C_3 is a constant for a given cross section. The values of C_3 for ducts of various cross sections can be found in previous chapters. It should be noted that the flows in the ducts are two-dimensional, so friction factors for both axial flow and transverse flow should be used.

Substituting Eqs. (10.4) through (10.7) into (10.3), the permeability in Darcy's Law can be calculated by

$$\alpha_p = \frac{D_h^2}{C_3} \tag{10.8}$$

As can be seen, when the core structure is known, its equivalent Darcy permeability can be evaluated by the above equation. Its effects on flow distribution can then be calculated. The value of $1/\alpha_p$ can be input into the numerical program directly. Since the fluid can only flow in the duct length direction (x) and transverse direction (y), the permeability in the third direction (normal, z) is set to infinitely small. Flow in this direction is not permitted by numerous membrane sheets (impeding walls).

With the method described above, the fluid flow in the whole exchanger can be calculated. The two-dimensional velocity distributions on the core face ($x^*=0$) can be predicted. The maldistribution can be analyzed with a definition of local flow nonuniformity parameter as

$$\beta = \frac{\dot{m}}{\dot{m}_m} \tag{10.9}$$

where $\dot{m}$ is the actual local mass flux (kg m^{-2}s^{-1}), and $\dot{m}_m$ is the average mass flux across the core face. When the air is considered as an incompressible fluid, the nonuniformity can be simply calculated as the local to mean velocity ratio.

10.2.2 Heat mass transfer in cores

After the flow distribution on the core face is known, the convective heat and mass transfer between the fresh air and the exhaust air can be studied. A mesoscopic model is set up. Each duct cross section is represented by one temperature and/or humidity. The temperature or humidity varies along flow directions (x for fresh air and y for exhaust air) and the corresponding perpendicular directions (y for fresh air and x for exhaust air) simultaneously. On each duct cross section, temperature or humidity is represented by a lumped parameter. It can be considered as a semi-lumped parameter model. The two air streams, one hot and humid (fresh air), and the other cool and dry (exhaust air), exchange heat and moisture in the exchanger in a cross flow arrangement. A two-dimensional slug flow heat and mass transfer model can be set up to govern the energy and mass conservations in the two air streams:

$$\frac{\partial T_f^*}{\partial x^*} = NTU_{sf}(T_{mf}^* - T_f^*) \tag{10.10}$$

$$\frac{\partial T_e^*}{\partial y^*} = NTU_{se}(T_{me}^* - T_e^*) \tag{10.11}$$

$$\frac{\partial \omega_f^*}{\partial x^*} = NTU_{Lf}(\omega_{mf}^* - \omega_f^*) \tag{10.12}$$

$$\frac{\partial \omega_e^*}{\partial y^*} = NTU_{Le}(\omega_{me}^* - \omega_e^*) \tag{10.13}$$

where x is the flow direction for the fresh stream and y is the flow direction for the exhaust stream. Certainly since the local mass flux is variant with both core height (y direction) and core depth (z direction), the local numbers of transfer units are different from point to point on the core surface. That's the difference between models considering and not considering flow maldistribution. Subscript "f" refers to the fresh side and "e" refers to the exhaust side; "s" refers to sensible and "L" refers to latent; "mf" refers to membrane surface on the fresh side, and "me" refers to membrane surface on the exhaust side.

The dimensionless temperature and humidity are defined by

$$T^* = \frac{T - T_{ei}}{T_{fi} - T_{ei}} \tag{10.14}$$

$$\omega^* = \frac{\omega - \omega_{ei}}{\omega_{fi} - \omega_{ei}} \tag{10.15}$$

The dimensionless coordinates are defined by

$$x^* = \frac{x}{x_F} \tag{10.16}$$

$$y^* = \frac{y}{y_F} \tag{10.17}$$

$$z^* = \frac{z}{z_F} \tag{10.18}$$

where x_F and y_F are channel lengths for fresh air and exhaust air (m). Here $x_F = y_F$. Core depth, which determines the total number of channels, is z_F. The air-side local number of transfer units is defined by

$$NTU_{sf} = \frac{1}{\beta}\frac{h_f A}{\left(m_t c_p\right)_f} \tag{10.19}$$

$$NTU_{se} = \frac{1}{\beta}\frac{h_e A}{\left(m_t c_p\right)_e} \tag{10.20}$$

$$NTU_{Lf} = \frac{1}{\beta}\frac{\left(\rho_a k\right)_f A}{m_{tf}} \tag{10.21}$$

$$NTU_{Le} = \frac{1}{\beta}\frac{\left(\rho_a k\right)_e A}{m_{te}} \tag{10.22}$$

where k and h are air-side convective mass transfer coefficient (m/s) and convective heat transfer coefficient ($\text{kW m}^{-2}\text{s}^{-1}$) respectively; m_t is total air mass flow rate (kg/s); A is total transfer area of membranes for **each** stream; c_p is specific heat ($\text{kJ kg}^{-1}\text{K}^{-1}$).

As can be seen in Eqs. (10.19) to (10.22), the last term in the right-hand side is the overall number of transfer units for a stream. The local number of transfer units is equal to the overall number of transfer units divided by the local to overall mass flux. Therefore the local number of transfer units is inversely proportional to mass flux. The local number of transfer units increases with decreased mass flux.

The convective heat transfer coefficient and mass transfer coefficient can be calculated by

$$Nu = \frac{hD_h}{\lambda_a} \tag{10.23}$$

$$Sh = \frac{kD_h}{D_{va}} \tag{10.24}$$

where D_{va} is vapor diffusivity in air (m^2/s), D_h is the hydrodynamic diameter (m). For parallel-plate channels, the fully developed Nusselt numbers and Sherwood numbers can be obtained from previous chapters.

Heat conduction through the plate is in equilibrium with the convective heat transfer on both sides. The equilibrium can be expressed by

$$h_f\left(T_f^* - T_{mf}^*\right) = \frac{\lambda_m}{\delta_m}\left(T_{mf}^* - T_{me}^*\right) \tag{10.25}$$

$$h_e\left(T_e^* - T_{me}^*\right) = -\frac{\lambda_m}{\delta_m}\left(T_{mf}^* - T_{me}^*\right) \tag{10.26}$$

where λ_m is the heat conductivity of the membrane (kW m^{-1} K^{-1}); δ_m is the thickness of the membrane (m). In the above two equations, the right-hand sides are heat conduction flow through plates. They have the same value, but different signs. On the left-hand sides, the first equation represents convective heat transfer flow on the fresh-air side, and the second equation represents convective heat transfer on the exhaust-air side.

Moisture in air streams must be coupled with moisture uptake in membranes. The relation between water uptake in membrane and equilibrium humidity can be measured by adsorption isotherms as

$$w = K_p\omega \tag{10.27}$$

where w is water uptake in materials (kg water/kg material), K_p is the partition coefficient. The more hydrophilic the materials are, the higher the values of K_p are.

The relation between humidity ratio and RH is

$$RH = \frac{\exp(5294/T)}{10^6}\omega \tag{10.28}$$

Moisture diffusion through the membrane is in equilibrium with the convective mass transfer on the two surfaces. The equations can be expressed by

$$\rho_a k_f\left(\omega_f - \omega_{mf}\right) = \frac{\rho_m D_{wm}}{\delta_m}\left(W_{mf} - W_{me}\right) \tag{10.29}$$

$$\rho_a k_e \left(\omega_e - \omega_{me}\right) = -\frac{\rho_m D_{wm}}{\delta_m}\left(W_{mf} - W_{me}\right) \tag{10.30}$$

where D_{wm} is water diffusivity in the membrane material (m²/s), ρ_m is the plate density (kg/m³). The relation between the equivalent moisture diffusivity and water diffusivity in membranes is

$$D_{vm} = \frac{\rho_m K_p}{\rho_a} D_{wm} \tag{10.31}$$

The total number of transfer units for sensible heat transfer and moisture transfer is

$$NTU_{s,tot} = \frac{h_{tot} A}{\left(m_t c_p\right)_{min}} = \frac{h_{tot} A}{\left(m_t c_p\right)_{min}} \tag{10.32}$$

$$NTU_{L,tot} = \frac{k_{tot} A}{\left(m_t c_p\right)_{min}} = \frac{k_{tot} A}{\left(m_t c_p\right)_{min}} \tag{10.33}$$

respectively, where

$$(h)_{tot}^{-1} = (h)_f^{-1} + \left(\frac{\lambda_m}{\delta_m}\right)^{-1} + (h)_e^{-1} \tag{10.34}$$

$$(k)_{tot}^{-1} = (k)_f^{-1} + \left(\frac{\rho_m D_{wm} K_p}{\rho_a \delta_m}\right)^{-1} + (k)_e^{-1} \tag{10.35}$$

Sensible effectiveness

$$\varepsilon_s = \frac{\left(m_t c_p\right)_f \left(T_{fi} - T_{fo}\right)}{\left(m_t c_p\right)_{min}\left(T_{fi} - T_{ei}\right)} = \frac{\left(m_t c_p\right)_e \left(T_{eo} - T_{ei}\right)}{\left(m_t c_p\right)_{min}\left(T_{fi} - T_{ei}\right)} \tag{10.36}$$

Latent effectiveness

$$\varepsilon_L = \frac{\left(m_t\right)_f \left(\omega_{fi} - \omega_{fo}\right)}{\left(m_t\right)_{min}\left(\omega_{fi} - \omega_{ei}\right)} = \frac{\left(m_t\right)_e \left(\omega_{eo} - \omega_{ei}\right)}{\left(m_t\right)_{min}\left(\omega_{fi} - \omega_{ei}\right)} \tag{10.37}$$

The sensible effectiveness and the latent effectiveness are the key performance indexes to evaluate a total heat exchanger. When the outlet values are substituted with calculated values, the effectiveness is the numerical data.

In fact, the final sensible effectiveness and the latent effectiveness can be estimated from the total number of transfer units with established correlations [15], if a uniform flow distribution is realized. However, to know the details of heat and moisture transfer in the exchanger, detailed finite difference equations should be solved.

The exchanger thermal and latent performance deterioration factors are defined by

$$\tau_s = \frac{\varepsilon_{s,\text{uniform}} - \varepsilon_s}{\varepsilon_{s,\text{uniform}}} \tag{10.38}$$

$$\tau_L = \frac{\varepsilon_{L,\text{uniform}} - \varepsilon_L}{\varepsilon_{L,\text{uniform}}} \tag{10.39}$$

where $\varepsilon_{s,\text{uniform}}$ and $\varepsilon_{L,\text{uniform}}$ are exchanger sensible and latent effectiveness at uniform flow distribution, respectively.

Boundary conditions

Fresh:

$$T_f^*\big|_{x^*=0} = 1 \tag{10.40}$$

$$\omega_f^*\big|_{x^*=0} = 1 \tag{10.41}$$

Exhaust:

$$T_e^*\big|_{y^*=0} = 0 \tag{10.42}$$

$$\omega_e^*\big|_{y^*=0} = 0 \tag{10.43}$$

When the two streams have the same mass flow rate:

$$\varepsilon_s = \frac{(T_{fi} - T_{fo})}{(T_{fi} - T_{ei})} = \frac{(T_{eo} - T_{ei})}{(T_{fi} - T_{ei})} = T_{eo}^* \tag{10.44}$$

$$\varepsilon_L = \frac{(\omega_{fi} - \omega_{fo})}{(\omega_{fi} - \omega_{ei})} = \frac{(\omega_{eo} - \omega_{ei})}{(\omega_{fi} - \omega_{ei})} = \omega_{eo}^* \tag{10.45}$$

10.2.3 Solution procedure and validation

A finite difference technique is used to discretize the partial differential equations developed for the air streams. The calculating domain of the core is divided into a number of discrete nodes. Each node represents a control volume. The number of calculating nodes is 50 in both the x direction and the y direction. Besides the discretization of energy and mass equations in x and y directions, the whole core depth is divided into 100 cells (z direction). Under this scheme, it is as though the whole core is divided into 100 mini-cross-flow total heat exchangers in core depth. For each mini-cross-flow total heat exchanger, the inlet velocity varies with y direction for fresh air and x direction for exhaust air.

An upstream differencing scheme is used for the two air streams. The two air streams and the plate are closely coupled in heat and moisture transfer. Therefore iterative techniques are needed to solve these equations. A description of the iterative procedure is as follows:

a. Calculate the local air stream mass flux distribution on the core inlet faces ($x^*=0$, and $y^*=0$), based on the resistance of the core.
b. Assume initial temperature and humidity fields in the two streams.
c. Calculate the temperature and humidity values on the plate surfaces.
d. Taking the current values of temperature and humidity on the plate surfaces as the default values, get the temperature and humidity profiles in the two air streams by solving Eqs. (10.10) through (10.13). Substitute the local mass flux distribution obtained in (a) into the equations for both the fresh air and the exhaust air local number of transfer units.
e. Go to (c), until the old values and the newly calculated values of temperature and humidity at all calculating nodes are converged.

After these procedures, all the governing equations are solved simultaneously. To ensure the accuracy of the results presented, numerical tests were performed for the core to determine the effects of the grid size. They indicated that 50 grids are adequate (less than 0.1% difference compared with 80 grids) for the z and y directions, and 100 grids are adequate for the z direction. The final numerical uncertainty is 0.1%.

When the temperature fields in the exchanger core are calculated, the heat exchange effectiveness is calculated using mean outlet values on the outlet face $x^*=1$ or $y^*=1$. These are numerically obtained data.

Experimental validation

Experimental validation of both fluid flow and heat and mass transfer performance is performed. To measure the heat and moisture exchange effectiveness, several experiments are done. A schematic of the test-rig is shown in Figure 10.6. Two air ducts are assembled. One is for fresh air, and the other is for exhaust air. Each duct consists of a variable-speed blower downstream from the core, a wind tunnel, a set of nozzles, wind straighteners, and temperature and humidity sensors before and after the exchanger. The exchanger is connected to the two ducts with flanges. The whole test rig is established in a constant

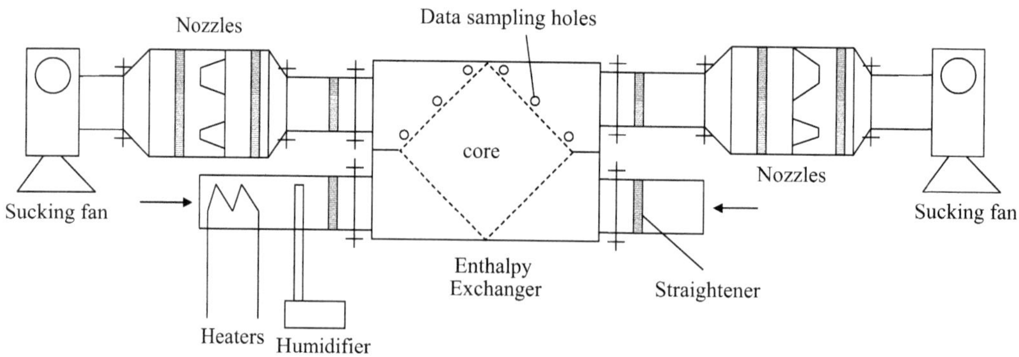

FIGURE 10.6

Experimental set-up of the total heat exchanger with a membrane core.

temperature and constant humidity room, so the inlet temperature and humidity of the exhaust air, which equals the room air, can be controlled and maintained very well even under very hot and humid ambient weather conditions. Additional electric heating circuits and humidifiers are installed for the fresh air duct. The heating and humidification power currents can be adjusted according to the fresh air set-point temperature. A 10 mm thick plastic foam insulation layer is pasted on the outer surfaces of the ducts and the exchanger shells to prevent heat dissipation from the system to the surroundings. The heat or moisture losses from the system are below 0.5%. Air leakage is prevented. Inlet and outlet temperature and humidities are measured to calculate the sensible and latent effectiveness.

Two parallel-plate membrane cores as listed in Table 10.1 are built and tested in the test rig built in the laboratory [11]. The core depths are the same: 460 mm. The geometrical and transport properties are listed in Table 10.1. As can be seen, the outer dimensions of the two cores are the same, so they can be replaced and inserted into the same exchanger cell conveniently. However, the channel pitches are different. Core A has 92 channels for each flow. Core B has 127 channels.

The sensible effectiveness of the two cores under various volumetric air flow rates is plotted in Figure 10.7. The latent effectiveness of the two cores under various air flow rates is plotted in Figure 10.8. The measured data are demonstrated by discrete symbols. The calculated data are plotted by solid lines. The maximum deviation between the calculated value and the measured data is below 6%, indicating the performance is predicted satisfactorily. Due to the compact structure and the high packing density, the heat mass transfer area is very large for such exchangers. Therefore the exchanger effectiveness is rather high, especially for core B. The outside dimensions are the same, but core A has 92 passages and core B has 127 passages. The resulting channel pitches are: core A 2.5 mm, core B

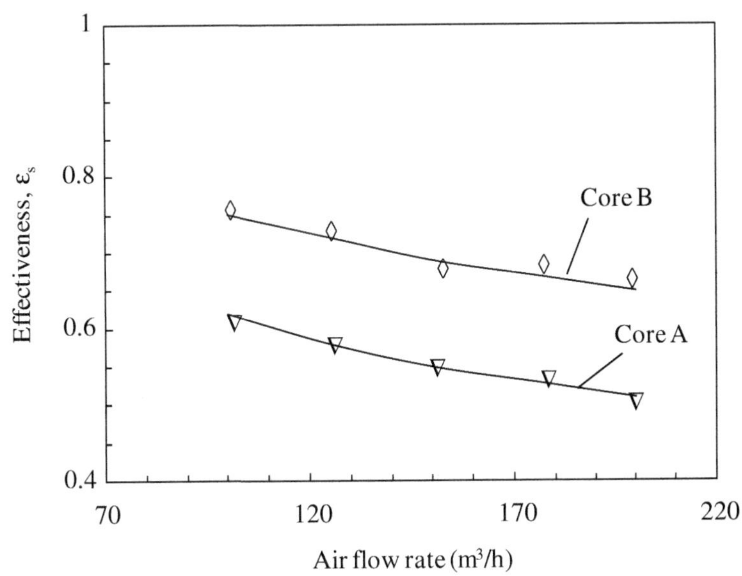

FIGURE 10.7

Sensible effectiveness of the two parallel-plate cores under various air flow rates. The solid line is the calculated values, and the discrete symbols are the measured data.

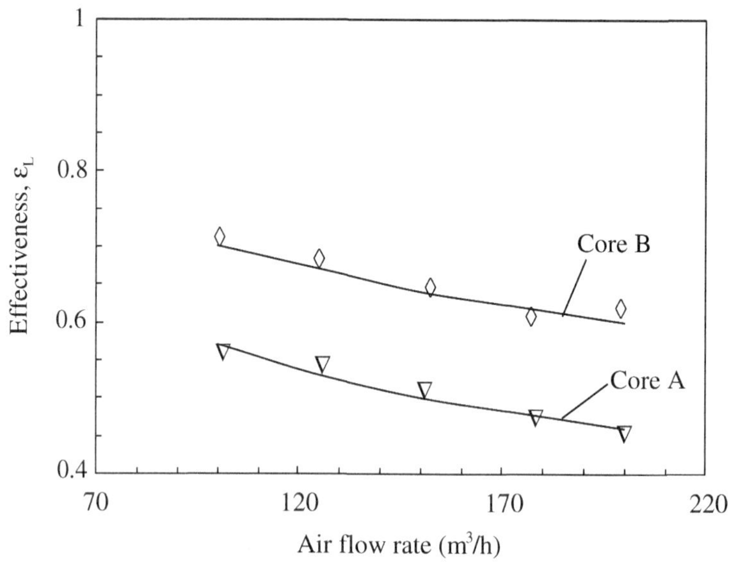

FIGURE 10.8

Latent effectiveness of the two parallel-plates cores under various air flow rates. The solid line is the calculated values, and the discrete symbols are the measured data.

1.8 mm. Core B is more compact, therefore its effectiveness is higher. The sensible effectiveness is usually higher than the latent effectiveness, because moisture resistance through the membranes is much higher than thermal resistance through the membranes.

10.2.4 Performance deterioration

Flow maldistribution

The effects of the core channel pitch on flow maldistribution are modeled. Two cores as listed in Table 10.1 with different channel pitches are considered: core A, 2.5 mm; and B 1.8 mm. The profiles of velocity nonuniformity (local to mean velocity ratio) on the core outlet face are plotted in Figures 10.9 and 10.10 for the two cores respectively. The total volumetric air flow rate for each stream is 150 m^3/h, which equals a mean face velocity of 0.49 m/s or mean duct velocity of 0.98 m/s.

As can be seen from these figures, the larger the channel pitch is, the more serious the flow maldistribution is. The maximum flow nonuniformity for the 2.5 mm pitch core A is 3.4, while it is only 1.9 for the 1.8 mm pitch core B. For both cores, the highest velocity exists at the center of the core face where the incoming fluids impinge on the core face directly. The lowest velocity exists on the corners of the core surface, because here the pressure is the least. As expected, the nonuniformity is symmetric to the plane $z^* = 0.5$ due to geometric symmetry. However, due to the skew angle between the core face and the impinging flow, the nonuniformity is asymmetric to plane $y^* = 0.5$. These results indicate that the flow maldistribution is co-determined by the inlet duct, the core and the outlet duct. The core itself plays a determinant role in the flow maldistribution. These components should be coupled together

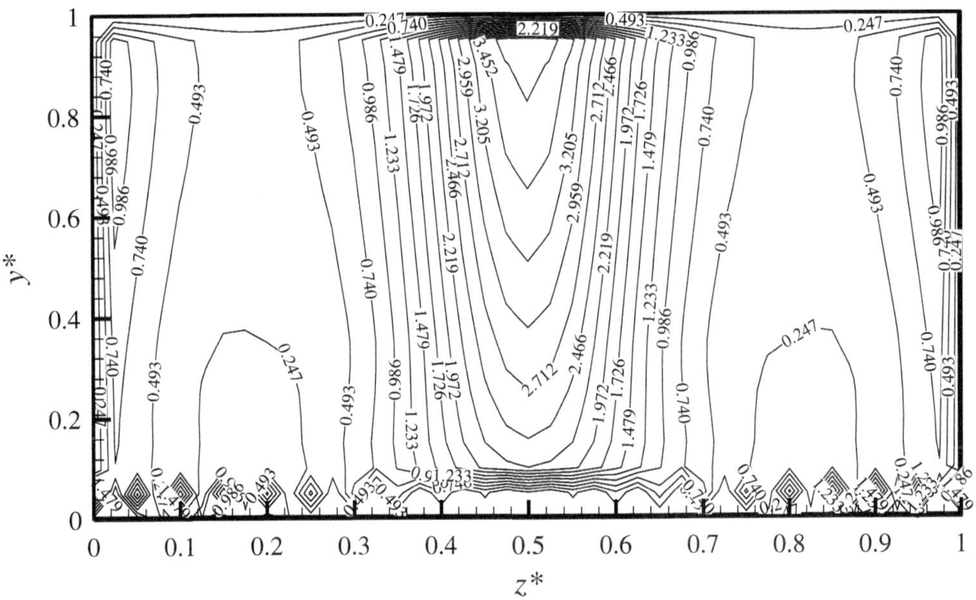

FIGURE 10.9

Velocity nonuniformity (β) on the core face: core A with channel pitch 2.5 mm.

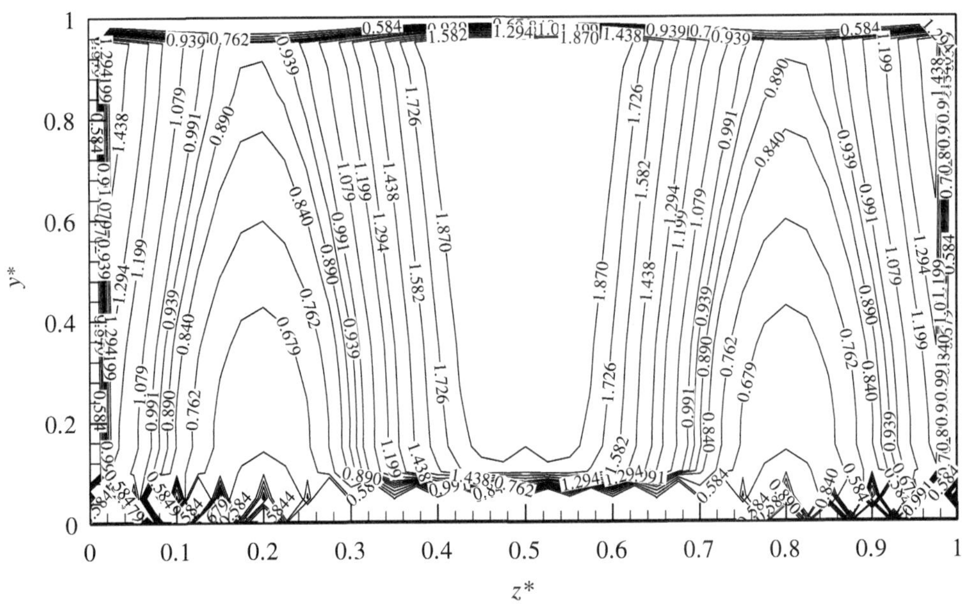

FIGURE 10.10

Velocity nonuniformity (β) on the core face: core B with channel pitch 1.8 mm.

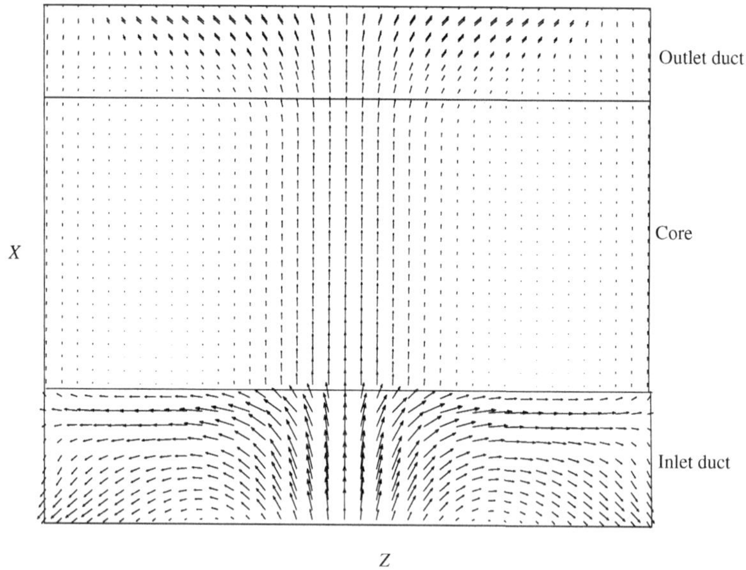

FIGURE 10.11

Velocity vectors on plane $y^*=0.5$ (center plane in core) for core A with channel pitch 2.5 mm.

to predict the flow maldistribution. The methodology of separating the core from the inlet duct is not appropriate. Further, when the core channel pitch is below 2 mm, a semi-uniform flow distribution can be assumed. On the other hand, flow maldistributions on cores with channel pitches larger than 2 mm are very serious. They should be taken into account.

Velocity vectors on plane $y^*=0.5$ for core A are plotted in Figure 10.11. The plane is cut at the center of the core. The channel pitch is 2.5 mm and flow rate is 150 m^3/h. As can be seen, the incoming air impinges on the core face and some flow is driven by the core face to the remote areas at the corners. There are swirls generated before the core face, which lead to flow redistribution. The flow in the core becomes somewhat uniform, due to this redistribution. The higher the core resistance is, the more serious the swirls are, and the more homogeneously the flow redistributes. In each parallel-plate channel, the flow is two-dimensional, though the axial velocity is higher than the transverse flow.

Thermal and latent performance deteriorations

With the heat and mass transfer model, the local temperature and humidity on the outlet face of the core can be calculated. The whole exchanger is divided into a number of mini-exchangers in the z direction. Each mini total heat exchanger is an x–y cross flow total heat exchanger. The outlet temperatures for each mini-exchanger vary with x and/or y directions. Therefore, the outlet temperature and humidity for fresh air are two-dimensional on the z–y face, while the outlet temperature and humidity for exhaust air are two-dimensional on the z–x face.

Figures 10.12 and 10.13 show the local dimensionless temperature and humidity on the fresh air outlet face ($x^*=1$) for core B. As can be seen, for the relatively homogeneous flow distribution on core B,

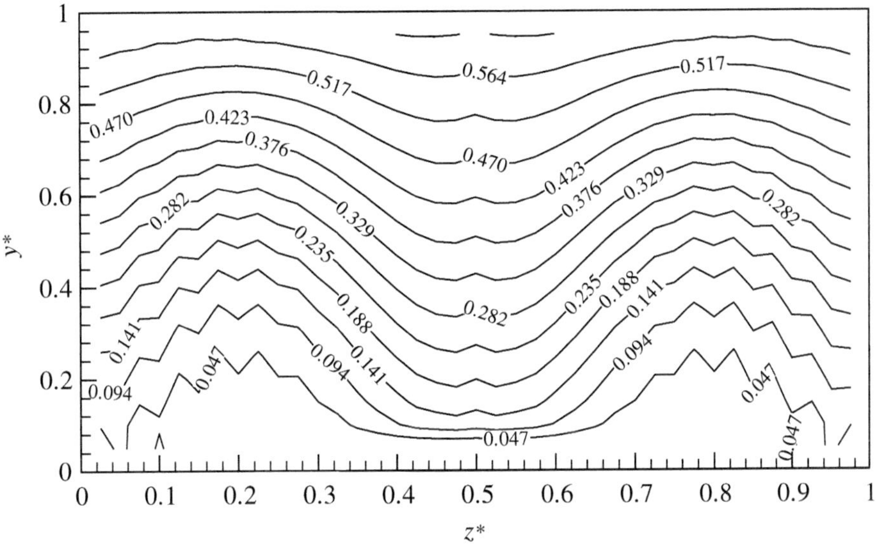

FIGURE 10.12

Local dimensionless temperature of fresh air on the core outlet face ($x^*=1$): core B with channel pitch 1.8 mm.

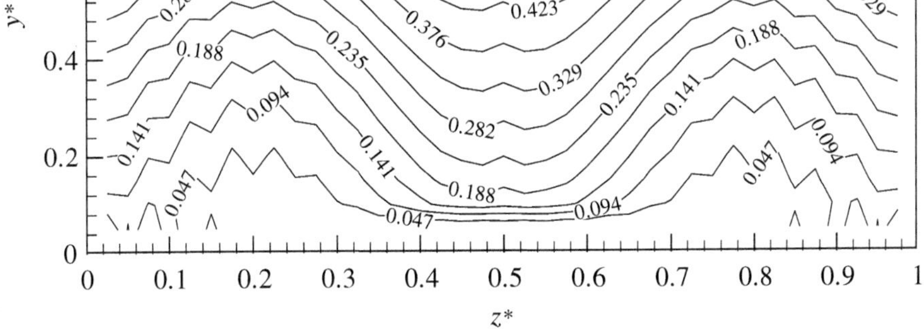

FIGURE 10.13

Local dimensionless humidity of fresh air on the core outlet face ($x^*=1$): core B with channel pitch 1.8 mm.

Table 10.2 Effectiveness and deterioration factors for the two cores.

Core A, $2a=2.5$ mm						
	Sensible effectiveness			Latent effectiveness		
V (m^3/h)	$\varepsilon_{S,uniform}$	ε_S	τ_S	$\varepsilon_{L,uniform}$	ε_L	τ_L
100	0.70	0.62	0.130	0.65	0.57	0.134
125	0.67	0.58	0.152	0.61	0.53	0.156
150	0.61	0.55	0.159	0.58	0.50	0.164
175	0.63	0.53	0.180	0.56	0.48	0.172
200	0.60	0.51	0.186	0.54	0.46	0.179
Core B, $2a=1.8$ mm						
	Sensible effectiveness			Latent effectiveness		
V (m^3/h)	$\varepsilon_{S,uniform}$	ε_S	τ_S	$\varepsilon_{L,uniform}$	ε_L	τ_L
100	0.79	0.75	0.053	0.74	0.70	0.055
125	0.77	0.72	0.068	0.72	0.67	0.070
150	0.74	0.69	0.070	0.69	0.64	0.073
175	0.72	0.67	0.071	0.67	0.62	0.075
200	0.70	0.65	0.074	0.65	0.60	0.092

the dimensionless temperature and humidity iso-lines are nearly parallel to the plates, but have waves. The waves are generated by flow maldistribution. The valleys are positions where flow rates are lower, while the crests are positions where the flow rates are higher. If the flow is uniform across the core face, the shapes become straight parallel lines. The shapes of moisture transfer are similar to heat transfer.

For engineering applications, the overall heat and moisture exchange effectiveness of the whole exchanger is more significant than the local values. Table 10.2 lists the exchanger thermal and latent performance deterioration factors under various air flow rates for the two cores. As can be seen, both the thermal and the latent performance deterioration factors for the core B are within 9% and the performance deterioration can be ignored. However, for the core A with larger channel pitches, the performance deterioration is substantial. The deterioration factors for core A can be as high as 19%. For the two cores, both the sensible and the latent effectiveness are equally influenced. The deterioration factors are as high as 20%. In contrast, for plate-fin cores, which will be discussed in next section, the sensible and latent effectiveness deteriorations are smaller with similar channel pitches [13]. Therefore compared to plate-fin cores, parallel-plate cores should have more attention paid to the problem of flow maldistribution. That's no surprise because this structure has more freedom in air flows.

10.3 Plate-fin cores

10.3.1 General considerations

Plate-fin cores, as shown in Figure 10.14, are used instead of parallel-plate cores. For plate-fin cores, flows inside the cores are considered one-dimensional. Transverse mixing is prohibited by the walls of the small ducts. Equations (10.3) to (10.8) can still be used to calculate the permeability, α_p, which is

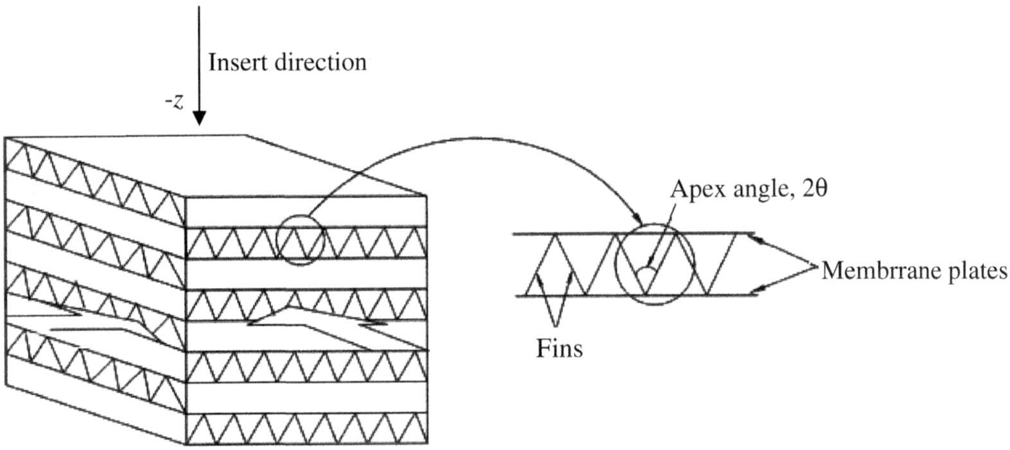

FIGURE 10.14

Schematic of a membrane exchanger core, with triangular plate-fin structure.

then used to calculate flow in the porous media. The friction factors were obtained in previous chapters for the plate-fin ducts.

Similarly, a mesoscopic model can be set up to summarize the heat and mass transfer in the core. It can be considered as a semi-lumped parameter model. The two air streams, one hot and humid (fresh air), and the other cool and dry (exhaust air), exchange heat and moisture in the exchanger in a cross-flow arrangement.

The local number of transfer units is equal to the overall number of transfer units divided by the local to overall mass flux ratio. Therefore the local number of transfer units is inversely proportional to the mass flux ratio. The local number of transfer units increases with decreased mass flux.

Heat and mass transport in the air ducts can still be calculated by Eqs. (10.10) to (10.13). However the convective heat transfer and mass transfer coefficients should be calculated by the fully developed Nusselt and Sherwood numbers for plate-fin ducts. For plate-fin ducts of finite fin conductance, the fully developed Nusselt and Sherwood numbers are influenced by aspect ratios (a/b) or apex angles for triangular ducts, and fin conductance parameters [12]. This is quite different from the simple classic data of a sensible-only heat exchanger with infinite fin conductance [13]. The areas are the overall areas including plate and fins.

Membranes are used for both the plates and membranes, then the fin conductance parameter for sensible heat transfer

$$\Omega_s = \frac{\lambda_m \delta_m}{\lambda_a (2a)} \tag{10.46}$$

The fin conductance parameter for moisture transfer

$$\Omega_L = \frac{\rho_m K_p D_{wm} \delta_m}{\rho_a D_{va} (2a)} \tag{10.47}$$

Table 10.3 Structural and transport parameters of the two plate-fin cores.

Symbol	Unit	Core A	Core B
n_{ch}		57	127
x_F, y_F	mm	185	185
z_F	mm	460	460
δ_m	μm	100	100
λ_m	$Wm^{-1}K^{-1}$	0.44	0.44
θ	degrees	60	60
$2a$	mm	4.0	1.8
D_h	mm	3.72	1.67
A_f, A_e	m^2	6.14	13.69
Ω_S		0.42	0.93
Ω_L		0.0019	0.0043
D_{wm}	m^2/s	5.15×10^{-10}	5.15×10^{-10}
K_p		10	10
Inlet/outlet duct length	mm	262	262
Inlet/outlet vent diameter	mm	60	60

10.3.2 Fluid flow maldistribution

Two plate-fin cores whose parameters are listed in Table 10.3 are modeled. The numerical scheme and the experimental validation procedures are the same as with the parallel-plate cores.

The effects of the core duct pitch on flow maldistribution are modeled. Two cores with different duct pitches are considered: A 4.0 mm and B 1.8 mm. The core dimensions and other structural parameters are described in Table 10.3. The profiles of velocity nonuniformity (local to mean velocity ratio) on the core outlet face for the two pitches can be observed in Ref. [13]. The larger the channel pitch is, the more serious the flow maldistribution. The flow maldistribution is co-determined by the inlet duct, the core and the outlet duct. The core itself plays a determinant role in the flow maldistribution. These components should be coupled together to predict the flow maldistribution. Compared to parallel-plate cores with the same duct pitches, the flow nonuniformity is smaller because plate-fins can only permit one-dimensional flows while parallel plates permits two-dimensional flows. When the core duct pitch is below 2 mm, a uniform flow distribution can be assumed. This is easy to understand. A core with such small channels is more like a flow straightener. The flow after such structures becomes uniform. On the other hand, flow maldistributions in cores with duct pitches larger than 2 mm are very serious. They should be taken into account.

10.3.3 Performance deterioration

The overall heat and moisture transfer effectiveness of the whole exchanger is more significant than the local values. Figure 10.15 depicts the exchanger thermal and latent performance deterioration factors under various air flow rates for the two cores. As can be seen, both the thermal and the latent performance deterioration factors for core B are very small and the performance deterioration can be ignored. However, for core A with larger duct pitches, the performance deterioration is substantial. For core A,

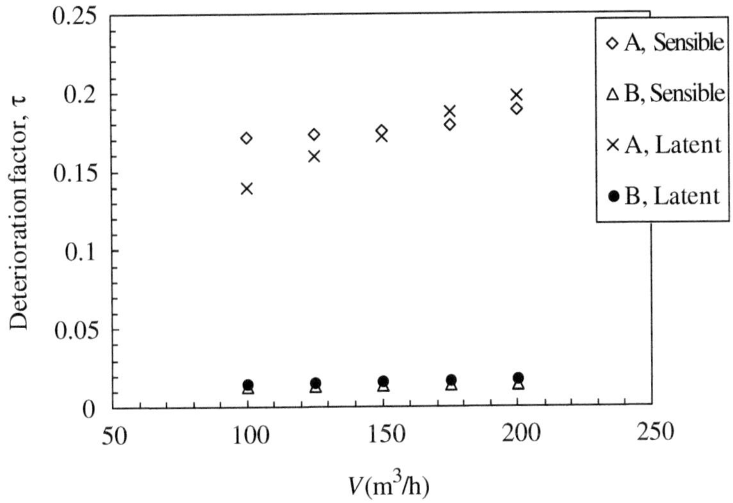

FIGURE 10.15

Performance deterioration factors for the two plate-fin cores under various air flow rates.

the thermal effectiveness is deteriorated by 15–20%, and the latent effectiveness is deteriorated by 20–22%. The higher deterioration factors for latent effectiveness are due to the higher moisture resistance through the membranes. The latent effectiveness of a heat and mass exchanger is usually low. A uniform flow distribution becomes more important because all the exchanger area should be equally and sufficiently employed to take part in moisture exchange. Therefore, flow maldistribution is more important for heat and mass exchangers than for sensible heat exchangers.

10.4 Hollow fiber membrane cores

Hollow fiber membrane contactors have been used extensively in various industries, mostly for liquid gas treatment. They are called membrane contactors. In Chapter 8, the flow maldistribution induced by the fibers' nonuniform distribution was considered. This chapter will consider flow maldistribution induced by the exchanger structure [14].

A hollow fiber membrane contactor, as shown in Figure 10.16, is considered. The shell is a rectangular box. Numerous hollow fibers are packed inside to form a shell-and-tube structure. The whole membrane contactor, the module, is composed of the fiber bundle, the inlet header, and the outlet header. The inlet and outlet headers are further connected to air ducts. Water flows inside the fibers, and air flows across the fibers. The membranes permit the permeation of moisture, but prevent the leaking of water droplets. In this way, air is humidified. The system is shown in Figure 10.17. The same apparatus can be used for air dehumidification if the pure water is replaced by concentrated hygroscopic salt solution. In this chapter, the fibers are assumed to be uniformly distributed but the flow in the bundle is maldistributed by the exchanger structures [14]. Previous research assumed a uniform air distribution

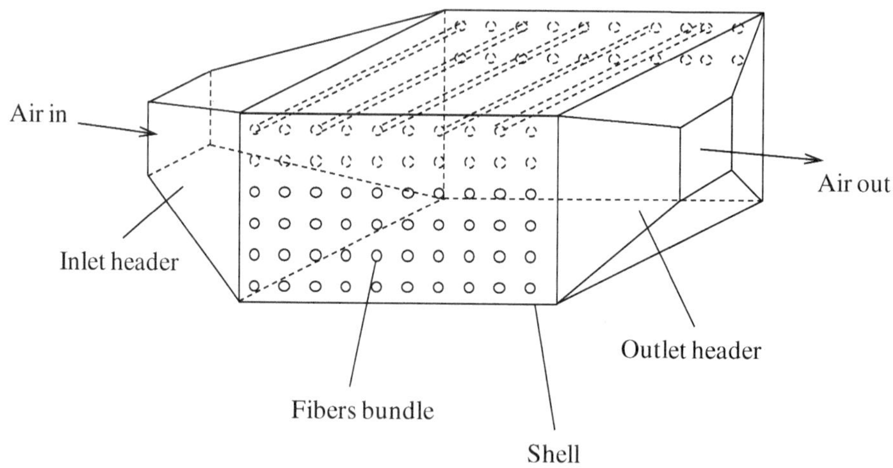

FIGURE 10.16

Configurations of a hollow fiber membrane contactor for air humidification, showing only five complete fibers for simplicity.

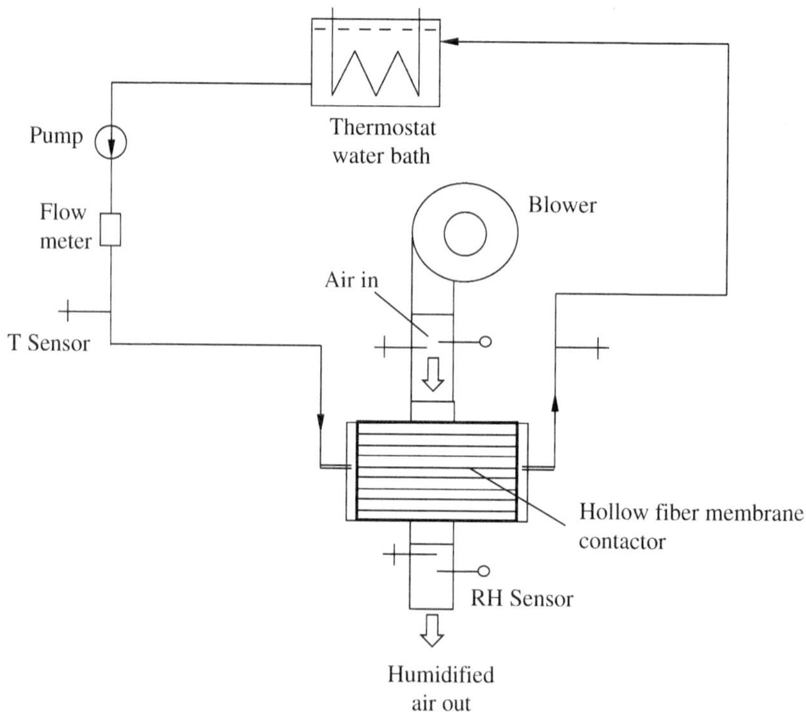

FIGURE 10.17

Schematic of the air humidification system with a hollow fiber membrane contactor.

on the inlet surface to the bundle. However, as observed, the expansion and contraction structures of the headers, as well as the influence of the air ducts, make the flow complicated in the bundle. The maldistribution of fluid flow will have adverse effects on humidification performance.

10.4.1 Theory and equations

Compared to plate-type exchangers, the flow in a bundle is more complicated. Three-dimensional flow distributions are considered. The α_p in Eq. (10.2) for the three directions is different due to the different resistance equations. The coordinate system is shown in Figure 10.18. For the transverse flows (y and z directions) across the bundle

$$\Delta P_i = N_i \chi_i \frac{\rho_a u_{max,i}^2}{2} f_i \tag{10.48}$$

$$u_{max,i} = \frac{\sigma_i}{\sigma_i - 1} u_i \tag{10.49}$$

where in the ith direction ($i = y, z$), N is the row number of fibers along the flow direction, χ is a correction factor [16], f_i is the friction factor across tube banks listed in Ref. [16], σ is the pitch to diameter

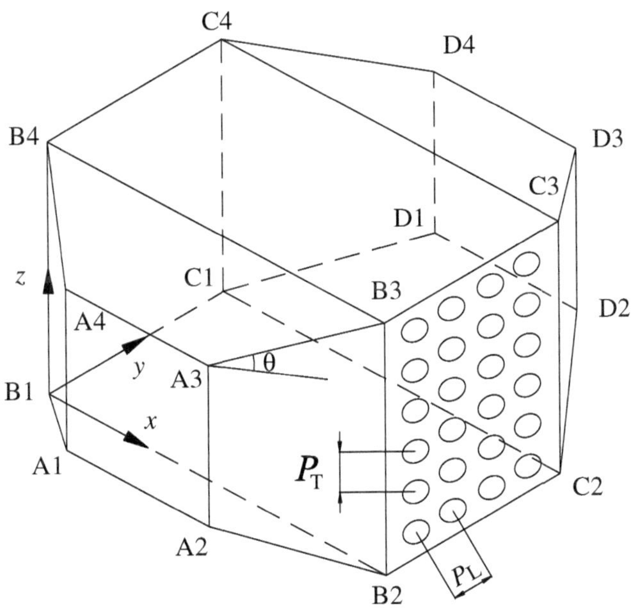

FIGURE 10.18

Schematic of the membrane module consisting of an inlet header, a membrane core and an outlet header. (A1-A2-A3-A4 is the inlet face of the module; B1-B2-B3-B4 is the inlet face of the core; C1-C2-C3-C4 is the outlet face of the core; D1-D2-D3-D4 is the outlet face of the module.)

ratio ($\sigma = P_T/d_o$), and u_i is face velocity. The values of the friction factor can also be found in previous chapters for hollow fibers.

For axial inside flow along the fibers (i.e., $i = x$ direction)

$$\Delta P_i = f_i \frac{x_F}{D_h} \frac{\rho_a u_i^2}{2} \tag{10.50}$$

where the friction factor f_i for x direction is calculated by

$$f_i \cdot Re_i = const \tag{10.51}$$

The constants can be found in previous chapters.

Reynolds number:

$$Re_i = \frac{D_h u_i \rho_a}{\mu_a} \tag{10.52}$$

The hydraulic diameter D_h for the hollow fiber bundle can be estimated by

$$D_h = 4\frac{A_c}{P_f} \tag{10.53}$$

where A_c is the cross-sectional area of the hollow fiber membrane core, which can be calculated by

$$A_c = y_F z_F - \frac{n_f \pi d_o^2}{4} \tag{10.54}$$

where n_f is the total number of fibers in the membrane core.

The wetted perimeter P_f can be obtained from

$$P_f = n_f \pi d_o \tag{10.55}$$

As can be seen, when the core structure is known, the value of $1/\alpha_p$ for each direction can be evaluated by Eqs. (10.48) to (10.55). Its effects on flow distribution can then be calculated. The value of $1/\alpha_p$ for each direction can be inputted into the software directly.

The packing fraction of the core, which is the ratio of total cross-sectional area of the fibers to the face area of the core, is

$$\phi = \frac{n_f \pi d_o^2}{4 y_F z_F} \tag{10.56}$$

The bundle has equal longitudinal and transverse pitches. Then $P_L = P_T$. The pitch to diameter ratio of the array is an important parameter and it can be calculated from the packing fraction as

$$\sigma = \left(\frac{\pi}{4\phi}\right)^{0.5} \tag{10.57}$$

Shell-side heat and mass transfer coefficients can be evaluated from the Nusselt and Sherwood numbers provided in previous chapters.

After the flow distribution on the core face is known, the convective heat and mass transfer between the air and the water streams can be studied. The fluid is considered to flow in the voids among the fibers. These voids are regarded as special ducts.

The flow maldistribution here is for the whole exchanger. It is assumed that as long as the flow enters into the ducts, it becomes stable and can be treated as a steady flow. The assumption of a fully developed laminar flow is for the flow inside the small ducts formed by the fibers. Locally it is treated as fully develop laminar flow for heat and mass transfer analysis.

A fiber-to-fiber modeling means to discretize and to model the whole bundle. Because the fibers are numerous, and each fiber needs thousands of grids, the overall calculation load is far beyond computer capacity. In contrast, with an equivalent parallel-plate model as described in the above section, the calculation load is reduced substantially. The method is valid, because the transfer capability (NTU) of the parallel-plate ducts is the same as the bundle. The resulting effectiveness is the same. The performance is the same.

As previously, a mesoscopic scale model is set up. The air stream and the water stream flow in a cross-flow arrangement, while exchanging heat and moisture through the membranes. Detailed fiber-by-fiber modeling is not likely considering that there are numerous fibers. So a conversion approach is taken. The hollow fiber membrane core is equivalently converted to a set of parallel-plate channels in the x–y plane. The channels are then stacked in the z direction. The channel length is x_{F}, channel width is y_{F}, and channel pitch is $z_{\mathrm{F'}}$. The channel length and width are identical to the length and width of the module. The channel pitch $z_{\mathrm{F'}}$ is only a hypothetic parameter. The channels have the same exchange area as the module, so the number of parallel channels is [14]

$$n_{\mathrm{ch}} = \frac{A_{\mathrm{tot}}}{x_{\mathrm{F}} y_{\mathrm{F}}} \tag{10.58}$$

where A_{tot} is the total area of the outer surface of the fibers (m^2). In this manner, the core is equivalently converted to a set of parallel-plate heat mass exchangers, which is easy to solve with a control volume-based finite difference numerical scheme.

A slug flow is assumed. Air flows along the y direction. Normalized equations for heat and moisture conservation in the air stream are

$$\frac{\partial T_{\mathrm{a}}^*}{\partial y^*} = NTU_{\mathrm{s}} \left(T_{\mathrm{w}}^* - T_{\mathrm{a}}^* \right) \tag{10.59}$$

$$\frac{\partial \omega_{\mathrm{a}}^*}{\partial y^*} = NTU_{\mathrm{L}} \left(\omega_{\mathrm{w}}^* - \omega_{\mathrm{a}}^* \right) \tag{10.60}$$

where

$$NTU_{\mathrm{s}} = \frac{h_{\mathrm{tot}} A_{\mathrm{tot}}}{\beta \rho_{\mathrm{a}} V_{\mathrm{a}} c_{\mathrm{pa}}} \tag{10.61}$$

$$NTU_{\mathrm{Lat}} = \frac{k_{\mathrm{tot}} A_{\mathrm{tot}}}{\beta V_{\mathrm{a}}} \tag{10.62}$$

In the equations β represents flow nonuniformity, which is defined by

$$\beta = \frac{u}{u_m} \tag{10.63}$$

where u_m is the average velocity on the inlet face.

Similarly, normalized equations for heat and moisture conservation in liquid water are

$$\frac{\partial T_w^*}{\partial x^*} = -m_s^* NTU_s \left(T_w^* - T_a^*\right) - m_L^* NTU_L \left(\omega_w^* - \omega_a^*\right) \tag{10.64}$$

$$\frac{\partial \omega_w^*}{\partial x^*} = E_T R_1 \frac{\partial T_w^*}{\partial x^*} \tag{10.65}$$

In the equations, the two dimensionless heat capacity ratios are defined by

$$m_s^* = \frac{\rho_a V_a c_{pa}}{\rho_w V_w c_w} \tag{10.66}$$

and

$$m_L^* = \frac{\rho_a V_a h_v \left(\omega_{w,i} - \omega_{a,i}\right)}{\rho_w V_w c_w \left(T_{w,i} - T_{a,i}\right)} \tag{10.67}$$

Inlets temperature to humidity ratio

$$R_1 = \frac{T_{w,i} - T_{a,i}}{\omega_{w,i} - \omega_{a,i}} \tag{10.68}$$

As can be seen, the first ratio defined by Eq. (10.67) is the air to water sensible heat capacity ratio, and the second one defined by Eq. (10.68) is the inlet temperature to humidity ratio of the water stream, which actually reflects the effect of evaporation heat of moisture on the water temperature.

Dimensionless temperature, humidity and coordinates are defined by

$$T^* = \frac{T - T_{a,i}}{T_{w,i} - T_{a,i}}, \quad \text{and} \quad \omega^* = \frac{\omega - \omega_{a,i}}{\omega_{w,i} - \omega_{a,i}} \tag{10.69}$$

$$x^* = \frac{x}{x_F}, \quad y^* = \frac{y}{y_F}, \quad \text{and} \quad z^* = \frac{z}{z_F} \tag{10.70}$$

Sensible cooling and humidification efficiencies of the module are defined by

$$\varepsilon_s = \frac{T_{a,o} - T_{a,i}}{T_{w,i} - T_{a,i}} \tag{10.71}$$

$$\varepsilon_h = \frac{\omega_{a,o} - \omega_{a,i}}{\omega_{w,i} - \omega_{a,i}} \tag{10.72}$$

where ω_w is the equivalent humidity of saturated air on the inner surface of the fiber. The sensible cooling efficiency reflects a temperature change in the humidification process. Different from sensible-only heat exchangers, absorption heat is extracted in humidification. As a result, the sensible cooling efficiency may be negative or greater than 1.0 in some cases due to the evaporation heat extracted. When it is negative, it means that the transfers of sensible heat and moisture are in different directions.

The deterioration factors in module sensible cooling and humidification performance are defined by

$$\tau_s = \frac{\varepsilon_{s,uniform} - \varepsilon_s}{\varepsilon_{s,uniform}} \tag{10.73}$$

$$\tau_h = \frac{\varepsilon_{h,uniform} - \varepsilon_h}{\varepsilon_{h,uniform}} \tag{10.74}$$

where $\varepsilon_{s,uniform}$ and $\varepsilon_{h,uniform}$ are exchanger sensible cooling and humidification efficiencies at uniform flow distribution, respectively. The uniform flow corresponds to the case without considering flow maldistributions. In other words, in this ideal case, all the local inlet velocities are equal to the inlet value. The zero transition angle header is such a case.

10.4.2 Numerical solutions and validation

The velocity profiles in the module are simulated by a commercial CFD package while the membrane core is assumed to be a porous medium [14]. A sample of the computational mesh is shown in Figure 10.19. In the simulation of the air flow field in the module, only half of the symmetric domain

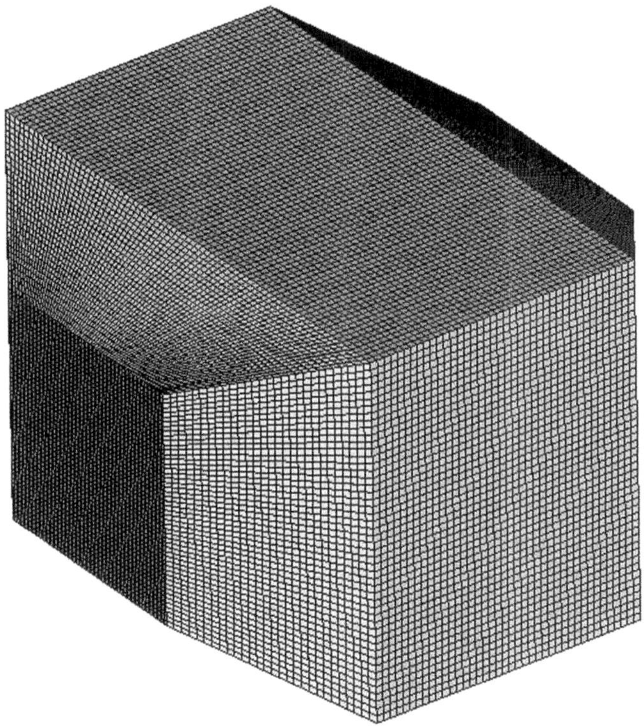

FIGURE 10.19

Mesh structure of the computational domain.

is calculated for simplification. So in following analysis, only half of the symmetric core is shown for the velocity distributions. Velocity inlet and pressure outlet boundary conditions are set at the inlet and outlet faces respectively. The velocity distributions on the inlet surface of the core, A1-A2-A3-A4 in Figure 10.18, are obtained. After the flow distribution is determined, heat and moisture transfer in the module can be calculated by the above Eqs. (10.59)–(10.68). The equations are discretized to partial difference equations by heat and mass balances on a control volume. They are then solved by the alternate direction implicit (ADI) finite difference technique. A grid independence test is conducted to optimize the grids. It is found that 50×60 grids are enough for numerical accuracy (compared to a denser mesh of 60×70 grids, the differences in outlet temperature and relative humidity are less than 0.1%). The dimensionless temperature and humidity at the water inlet are set to 0 and 1 respectively. The dimensionless temperature and humidity at the air inlet are 1 and 0, respectively.

Experimental work is used to validate the model and the methodology. Two cores, as listed in Table 10.4, are tested and investigated. The angle of the inlet duct of the tested module is 12°. By setting other parameters to their design conditions, volumetric air flow rates are changed. Performance under various air flow rates is measured and calculated. The test data are then compared with the calculated data. The average temperatures at the outlet of the core under various volumetric air flow rates are plotted in Figure 10.20 for core A and core B. As can be seen, the tested temperatures are usually a little bit higher than the numerical values. This is because the ambient temperature in the test chamber

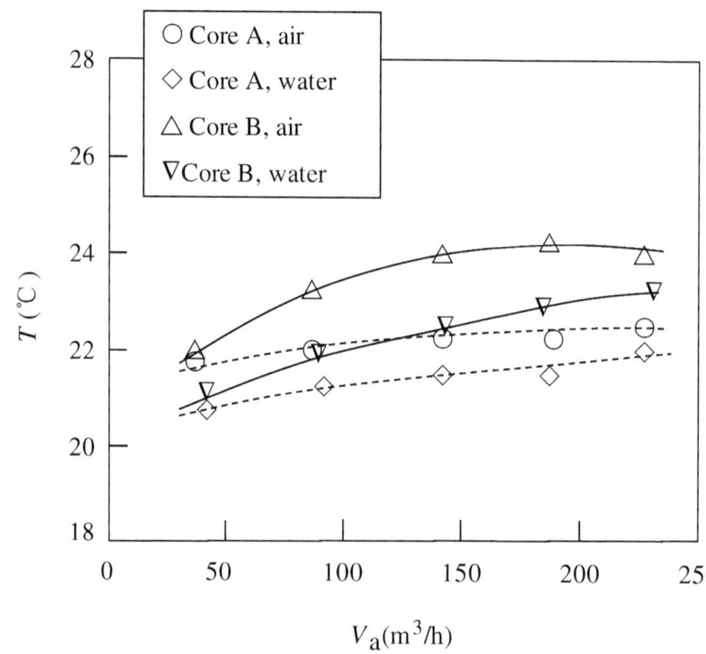

FIGURE 10.20

Comparison of outlet temperatures of air and water streams with the two cores. Discrete symbols are the measured values and the lines are calculated values. $T_{a,i} = 25\,°C$, $T_{w,i} = 20\,°C$.

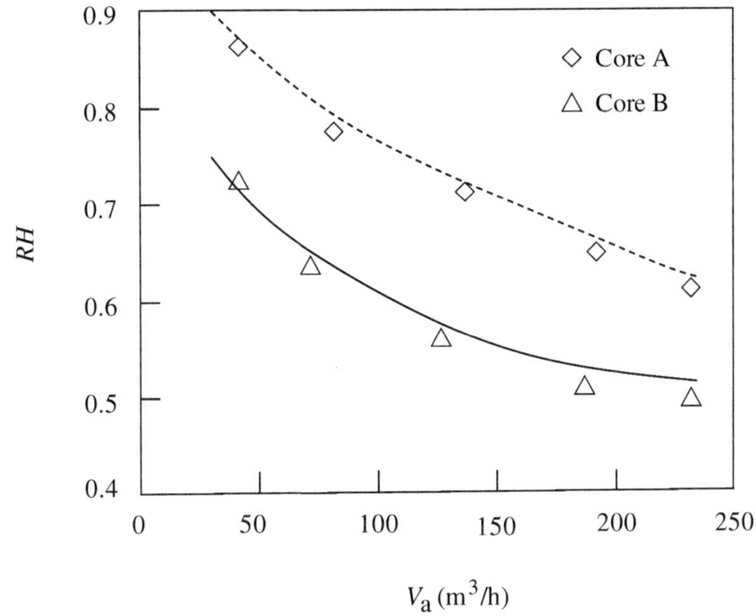

FIGURE 10.21

Comparison of outlet humidity of the air stream for the two cores. The discrete symbols are the tested data and the lines are the numerical data. $T_{a,i}=25\,°C$, $T_{w,i}=20\,°C$.

(298 K) is higher than the module temperature (293 K). Heat is transferred from the chamber to the test rig during the experiment. The relative humidity at the outlet of the core under various air flow rates is shown in Figure 10.21. The tested relative humidity is a little bit lower than the calculated, because the tested temperatures are higher. However, generally, the maximum differences between the numerical and tested temperature are below 5%. The maximum differences for humidity are below 7%. The maximum differences for pressure drops are 14%. From this analysis the model and the methodology are verified. Heat and moisture balances are satisfied. The system temperature is lower than ambient. A smaller air temperature change and a greater water temperature change are due to heat transfer from the surroundings into the system. Core A has higher sensible and latent effectiveness than core B because its packing density is higher. The pressure drops across the core are plotted in Figure 10.22. They are the pressure differences between the inlet and outlet faces of the fiber bundles. The pressure drop for core A is higher than for core B because it is more compact.

10.4.3 Flow maldistribution and performance deterioration

With the verified model, the effects of the packing fractions of cores on flow maldistribution in the module are modeled. Two cores in Table 10.4 with different packing fractions are considered: core A, 0.20; core B, 0.10. The number of fibers packed is different. All the other parameters such as core dimensions and membrane properties are the same. The flow maldistributions can be observed [14].

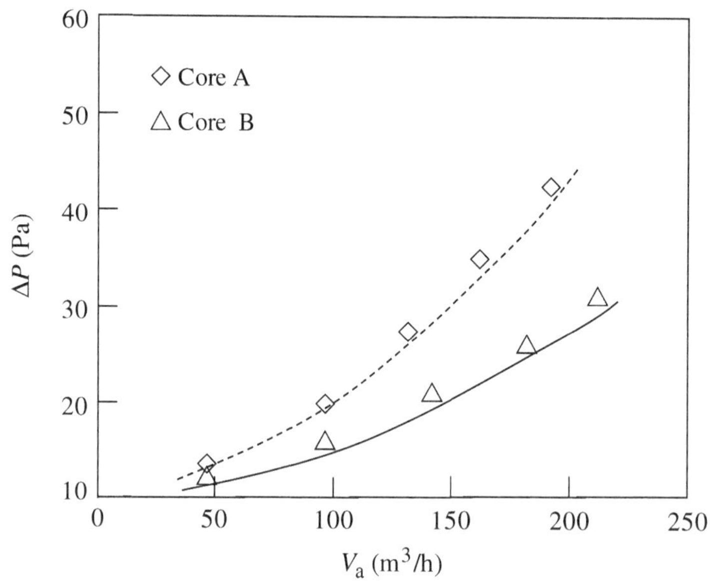

FIGURE 10.22

Comparison of pressure drops across the two cores. The discrete symbols are the test data and the lines are the numerical data. $T_{a,i}=25\,°C$, $T_{w,i}=20\,°C$.

Table 10.4 Physical and transport properties of the two hollow fiber membrane cores.

Name of Property	Symbol	Unit	Module A	Module B
Module dimensions	$x_F \times y_F \times z_F$	mm^3	$380 \times 200 \times 250$	$380 \times 200 \times 250$
Inlet/outlet duct angle	θ	deg	12	12
Transverse pitch	P_T	mm	3.2	4.8
Longitudinal pitch	P_L	mm	3.2	4.8
Effective fiber length	x_F	mm	380	380
Number of fibers in the module	n_f		4836	2132
Packing fraction	ϕ		0.20	0.10
Fiber outer diameter	d_o	mm	1.6	1.6
Membrane thickness	δ_m	mm	0.14	0.14
Moisture diffusivity in air	D_{va}	m^2/s	2.82×10^{-5}	2.82×10^{-5}
Pore diameter	d_p	μm	0.41	0.41
Water flow rate	V_w	L/h	85	85
Water temperature	$T_{w,i}$	K	293	293
Inlet air relative humidity	RH		0.31	0.32

(Continued)

Table 10.4 Continued.

Name of Property	Symbol	Unit	Module A	Module B
Inlet air temperature	$T_{a,i}$	K	298	298
Membrane porosity			0.76	0.76
Water diffusivity in membrane	D_{wm}	m^2/s	5.15×10^{-10}	5.15×10^{-10}
Partition coefficient	K_p		10	10
Thermal conductivity of membrane	λ_m	$Wm^{-1}K^{-1}$	0.43	0.43

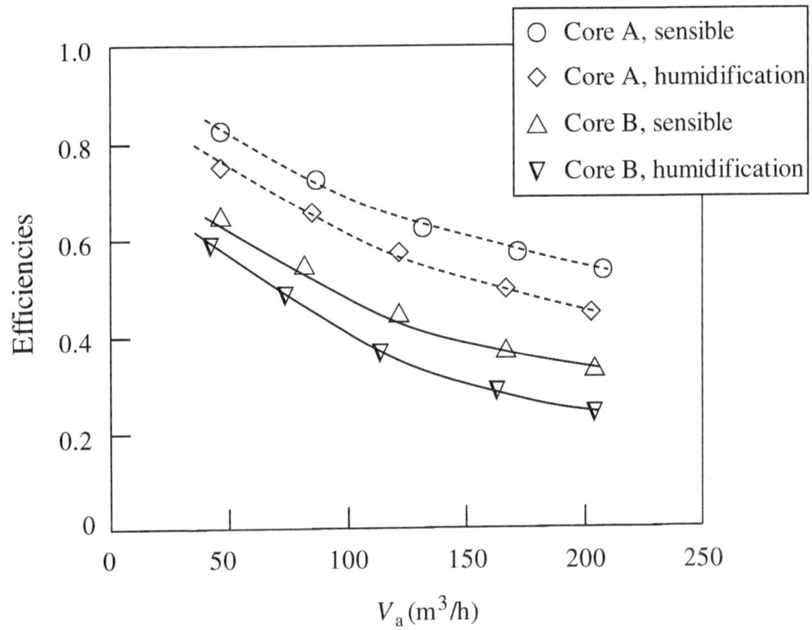

FIGURE 10.23

Sensible cooling and humidification efficiencies of the two cores.

The sensible cooling and humidification efficiencies of the whole module are more significant than the local values. Figure 10.23 depicts the sensible cooling and humidification efficiencies under various air flow rates with the two cores mentioned. Figure 10.24 shows the deterioration factors in the sensible cooling and humidification efficiencies. As can be seen, both the thermal and the humidification performance deteriorate more seriously with module B than with module A. Densely populated fibers have fewer deterioration factors. The sensible cooling efficiencies are decreased by 3–30%, and the humidification efficiencies are decreased by 30–60%. The higher deterioration factors for humidification

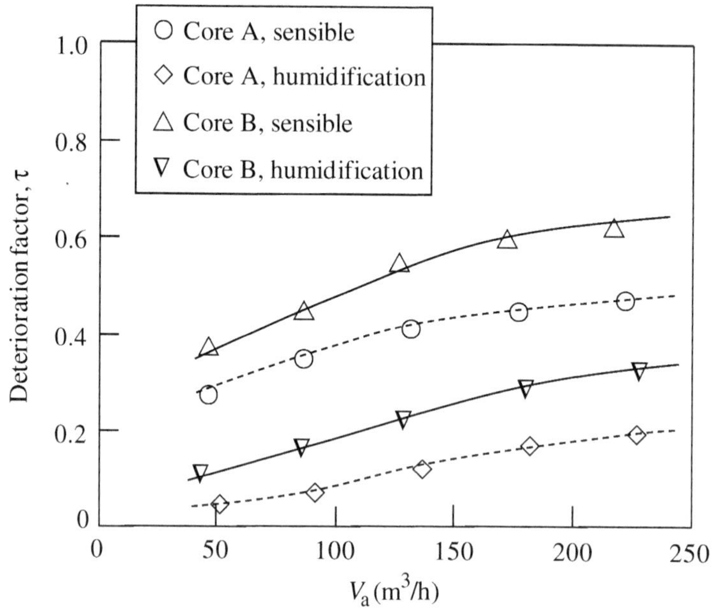

FIGURE 10.24

Deterioration factors of sensible cooling and humidification efficiencies for the two cores.

efficiencies are due to the higher moisture resistance through the membranes. Mass transfer resistance is usually much higher than heat transfer resistance. Compared to heat exchangers, flow maldistribution should be considered more seriously for membrane modules. A uniform flow distribution becomes more important for membrane contactors. This is because all the contactor areas should be equally and sufficiently involved in moisture exchange. That's in fact why "contactors" are so called. For the densely populated bundles, the deterioration factors are less than 20%.

10.5 Conclusion

Heat and mass exchangers are important, but cores cannot work alone. They should be combined together with inlet and outlet headers and ducts. The exchanger structures are usually complex, which leads to flow expansion and contraction impingement. These factors make the flow on the core surface rather nonuniform. This flow nonuniformity usually produces a 20–40% performance deterioration in heat and mass transfer performance with duct pitches higher than 2 mm. The adverse effects are more serious for mass transfer than for sensible heat transfer.

Flow maldistribution is a result of interactions between inlet duct flow, flow in the cores and flow in the outlet duct, as well as the connecting heads. The core itself plays a determining role in flow distribution. For cores with parallel-plate, plate-fin and hollow fiber structures, the problems of flow maldistribution are very serious. Among them, the flows in a plate-fin structure are more uniform because this structure only has a one-dimensional flow in the ducts. The flows in parallel-plate ducts are two-dimensional and those in hollow fibers are three-dimensional. They are more susceptible to maldistribution problems. For plate-fin and parallel-plate cores, the flow can be regarded as uniform if the duct pitches are less than 2 mm. For hollow fiber bundles, the flow is relatively uniform when the packing fraction is higher than 0.2.

References

[1] Bassiouny MK, Martin H. Flow distribution and pressure drop in plate heat exchangers-I, U-type arrangement. Chem Eng Sci 1984;39:693–700.

[2] Bassiouny MK, Martin H. Flow distribution and pressure drop in plate heat exchangers-II, Z-type arrangement. Chem Eng Sci 1984;39:701–4.

[3] Rao BP, Kumar PK, Das SK. Effect of flow distribution to the channels on the thermal performance of a plate heat exchanger. Chem Eng Process 2002;41:49–58.

[4] Dwivedi AK, Das SK. Dynamics of plate heat exchangers subject to flow variations. Int J Heat Mass Transfer 2007;50:2733–43.

[5] Shah RK. Flow maldistribution-compact heat exchangers. In: Rohsenow WM, Hartnett PJ, Ganic EN, editors. Handbook of heat transfer applications. New York: McGraw-Hill; 1985. p. 266–80.

[6] Ranganayakulu C, Seetharamu KN, Sreevatsan KV. The effects of inlet fluid flow nonuniformity on thermal performance and pressure drops in crossflow plate-fin compact heat exchangers. Int J Heat Mass Transfer 1997;40:27–37.

[7] Sparrow EM, Tong JCK, Abraham JP. Fluid flow in a system with separate laminar and turbulent zones. Numer Heat Transfer A: Appl 2008;53:341–53.

[8] Pingaud H, Le Lann JM, Koehret B. Steady-state and dynamic simulation of plate fin heat exchangers. Comput Chem Eng 1989;13:577–85.

[9] Nakoniec K. Numerical modeling of cross-flow plate-fin air-to-air heat exchanger under unsteady flow conditions. Numer Heat Transfer A: Appl 2006;49:1–24.

[10] Yuan P. Effect of inlet flow maldistribution on the thermal performance of a three-fluid cross-flow heat exchanger. Int J Heat Mass Transfer 2003;46:3777–87.

[11] Zhang LZ. Performance deteriorations from flow maldistribution in air-to-air heat exchangers: a parallel-plates membrane core case. Numer Heat Transfer A: Appl 2009;56:746–63.

[12] Zhang LZ. Flow maldistribution and performance deteriorations in a membrane-based heat and mass exchanger. ASME J Heat Transfer 2009;131:111801-1–7.

[13] Zhang LZ. Flow maldistribution and thermal performance deterioration in a cross-flow air to air heat exchanger with plate-fin cores. Int J Heat Mass Transfer 2009;52:4500–9.

[14] Zhang LZ, Li ZX, Zhong TS, Pei LX. Flow maldistribution and performance deteriorations in a cross flow hollow fiber membrane module for air humidification. J Membr Sci 2013;427:1–9.

[15] Zhang LZ. Total heat recovery: heat and moisture recovery from ventilation air. New York: Nova Science; 2009.

[16] Incropera FP, Dewitt DP. Fundamentals of heat and mass transfer. New York: John Wiley & Sons; 2007.

CHAPTER

11 Effectiveness-NTU Methods for Heat and Mass Transfer Processes

11.1 Introduction

Effectiveness-NTU (number of transfer units) methods are popular in heat exchanger design [1,2]. They are simple in form. The performances of a heat exchanger can be readily evaluated if the number of transfer units is known. The number of transfer units ($\mathrm{NTU} = UA/(mc_\mathrm{p})$) itself is a combination of overall heat transfer coefficients, transfer area, fluid flow rate and heat capacity. It summarizes these dimensional parameters into one dimensionless parameter. The performance becomes a monotone function of this dimensionless parameter. Geometric and design parameters can be estimated with these dimensionless correlations.

Similar to sensible-only heat exchangers, effectiveness-NTU (ε-NTU) methods are also convenient for analysis of heat and mass exchangers. They provide a simple way to build the relations between heat and mass transfer performance and the geometric, physical and operating conditions of heat and mass exchangers. Figure 11.1 shows an ε-NTU curve for cross-flow heat exchangers [3]. With known *NTU* values, effectiveness can be found easily from the plot. For example, if the *NTU* is 3.0, the consequent heat exchanger effectiveness is 0.68. Design of a heat exchanger will be easier. Going a step forward, if the plot can be used to estimate the mass transfer performance as well, design and optimization of heat and mass exchangers will be much simplified. However, because mass transfer is more complicated than heat transfer, and heat transfer is coupled with mass transfer, ε-NTU methods for heat and mass exchangers are more complicated. How to emulate this methodology from sensible-only heat exchangers for heat and mass exchangers is a great challenge.

Over the past 10 years, great efforts have been made in this area. The objectives have been to build dimensionless correlations between performance and the number of transfer units for both heat and mass transfer. The key issue is to set up and to clarify the overall mass transfer resistance. Various correlations have been successfully proposed for parallel-plate and plate-fin membrane total heat exchangers [3,4], hollow fiber membrane total heat exchangers [5], and hollow fiber membrane liquid desiccant air dehumidifiers [6]. The studies are useful in the design of such equipment, especially from the perspectives of engineering applications. This chapter will give a review of this research.

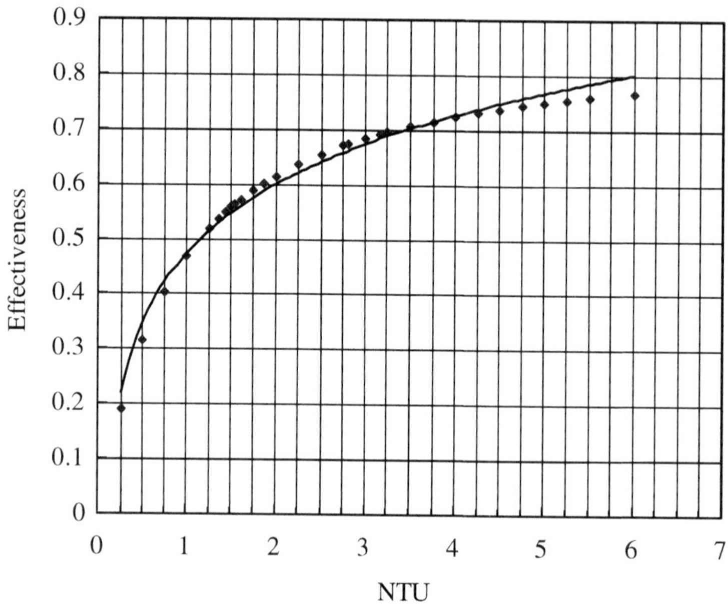

FIGURE 11.1

Effectiveness-*NTU* curve for a cross-flow heat exchanger.

11.2 Air-to-air heat and mass exchanger with a parallel-plate core

Consider a simple plain parallel-plate membrane total heat exchanger. It is in a cross-flow arrangement. The device is shown in Figure 11.2. Two air streams—the fresh and the exhaust—flow in thin, parallel, alternating membrane or paper layers, in order to transfer heat and moisture from one air stream to the other. In air conditioning, the fresh air is usually the outside air and the exhaust is the room air that needs to be discharged to the outside.

Analogous to the heat transfer effectiveness commonly used in heat exchanger analysis, the concept of effectiveness can be applied to the heat and moisture transfer processes in a membrane-based enthalpy exchanger. For a constant specific heat and heat of phase change, the effectiveness is defined as:

Sensible effectiveness

$$\varepsilon_s = \frac{(mc_{pa})_f (T_{fi} - T_{fo})}{(mc_{pa})_{min}(T_{fi} - T_{ei})} = \frac{(mc_{pa})_e (T_{ei} - T_{eo})}{(mc_{pa})_{min}(T_{fi} - T_{ei})} \tag{11.1}$$

Latent effectiveness

$$\varepsilon_L = \frac{(m)_f(\omega_{fi} - \omega_{fo})}{(m)_{min}(\omega_{fi} - \omega_{ei})} = \frac{(m)_e(\omega_{ei} - \omega_{eo})}{(m)_{min}(\omega_{fi} - \omega_{ei})} \tag{11.2}$$

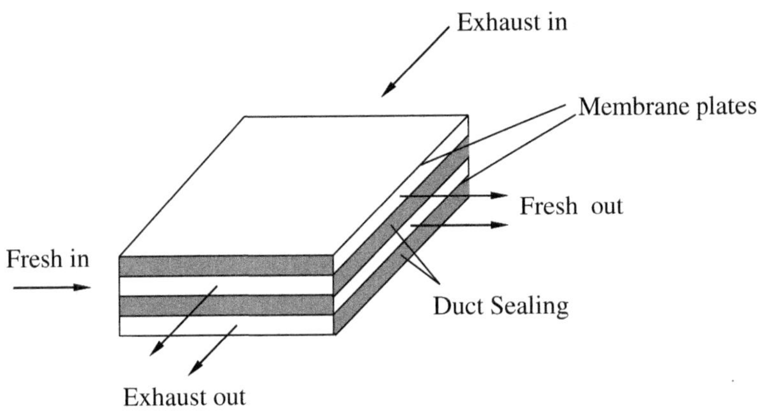

FIGURE 11.2

Schematic of a cross-flow parallel-plate total heat exchanger.

Enthalpy transfer effectiveness, i.e., total energy transfer effectiveness

$$\varepsilon_{en} = \frac{m_f(H_{fi} - H_{fo})}{m_{min}(H_{fi} - H_{ei})} = \frac{m_e(H_{ei} - H_{eo})}{m_{min}(H_{fi} - H_{ei})} \tag{11.3}$$

where m is mass flow rate (kg/s), H is the specific enthalpy of air, and it is calculated by

$$H = c_{pa}T + \omega(2501 + 1.86\,T) \tag{11.4}$$

where T is in °C.

The third term in Eq. (11.4) usually has a less than 3% effect, thus it can be ignored. Then the enthalpy effectiveness can be further simplified as

$$\varepsilon_{en} = \frac{\varepsilon_s + \varepsilon_L H_1^*}{1 + H_1^*} \tag{11.5}$$

where

$$H_1^* = \frac{2501(\omega_{fi} - \omega_{ei})}{c_{pa}(T_{fi} - T_{ei})} \approx 2501\frac{\Delta\omega}{\Delta T} \tag{11.6}$$

where H_1^* is essentially a ratio of latent to sensible energy differences between the inlets of two air streams flowing through the total heat exchanger. H_1^* can in theory vary from $-\infty$ to $+\infty$, but varies typically from -6 to $+6$ for enthalpy recovery in HVAC applications. From the above equation, it is clear that the total enthalpy effectiveness is not a simple algebraic average of sensible and latent effectiveness. When $H_1^* = 1$, $\varepsilon_{en} = (\varepsilon_s + \varepsilon_L)/2$. As $H_1^* \to \infty$, $\varepsilon_{en} \to \varepsilon_L$; as $H_1^* \to 0$, $\varepsilon_{en} \to \varepsilon_s$; as $H_1^* \to -1$, $\varepsilon_{en} \to \pm\infty$.

Correlations for sensible heat transfer and mass transfer for parallel-plate-type total heat exchangers are: For a cross-flow arrangement [4]

$$\varepsilon_s = 1 - \exp\left[\frac{\exp(-R_1 NTU_s^{0.78}) - 1}{R_1 NTU_s^{-0.22}}\right] \tag{11.7}$$

$$\varepsilon_L = 1 - \exp\left[\frac{\exp(-NTU_L^{0.78} R_2) - 1}{NTU_L^{-0.22} R_2}\right] \tag{11.8}$$

A cocurrent-flow arrangement [3]

$$\varepsilon_s = \frac{1 - \exp[-NTU_s(1 + R_1)]}{1 + R_1} \tag{11.9}$$

$$\varepsilon_L = \frac{1 - \exp[-NTU_L(1 + R_2)]}{1 + R_2} \tag{11.10}$$

For a counter-flow arrangement [3]

$$\varepsilon_s = \frac{1 - \exp[-NTU_s(1 - R_1)]}{1 - R_1 \exp[-NTU_s(1 - R_1)]}, \quad (R_1 < 1) \tag{11.11}$$

$$\varepsilon_L = \frac{1 - \exp[-NTU_L(1 - R_2)]}{1 - R_2 \exp[-NTU_L(1 - R_2)]}, \quad (R_2 < 1) \tag{11.12}$$

$$\varepsilon_s = \frac{NTU_s}{1 + NTU_s} \quad (R_1 = 1) \tag{11.13}$$

$$\varepsilon_L = \frac{NTU_L}{1 + NTU_L} \quad (R_2 = 1) \tag{11.14}$$

where in the correlations,

$$NTU_L = \beta_r \cdot NTU_s \tag{11.15}$$

$$R_1 = (mc_{pa})_{min}/(mc_{pa})_{max} \tag{11.16}$$

$$R_2 = m_{min}/m_{max} \tag{11.17}$$

Overall numbers of transfer units for both sensible heat and latent heat are used to reflect the sensible heat and mass transfer in exchangers. For the total heat exchanger that has equal area on both sides, the total numbers of transfer units for sensible and latent heat are

$$NTU_s = \frac{A_{tot} U_s}{(mc_{pa})_{min}} \tag{11.18}$$

$$NTU_L = \frac{\rho_a A_{tot} U_L}{(m)_{min}} \tag{11.19}$$

where U_s is the total sensible heat transfer coefficient. The general form is

$$U_s = \left[\frac{1}{h_f} + \frac{\delta_p}{\lambda_p} + \frac{1}{h_e}\right]^{-1} \tag{11.20}$$

The term in the middle is the thermal resistance of the plate, which value is around 0.005 (m^2K)/kW. The other two terms are convective thermal resistance. Their values are of the order of 40 (m^2K)/kW, or 8000 times larger than the plate resistance. Therefore, plate resistance for heat transfer can be ignored.

For total heat exchangers, the membranes are hydrophilic. The mechanism for moisture transfer is solution-diffusion [7]. The total moisture transfer coefficient is

$$U_L = \left(\frac{1}{k_f} + r_p + \frac{1}{k_e}\right)^{-1} \tag{11.21}$$

where the first and the third term in the equation are the convective mass transfer resistance on the air sides, and the middle term is the plate mass transfer resistance, which is expressed by

$$r_p = \frac{\rho_a \delta_p}{\rho_p D_{wp}} \frac{1}{K_{pm}} \tag{11.22}$$

where D_{wp} is water diffusivity in the plate (m^2/s). Water uptake w in the plate (membrane in this case) material is governed by an adsorption isotherm as

$$w = K_{pm}\omega \tag{11.23}$$

The comparison of total transfer units for moisture and sensible heat, assuming equal specific heats for two air streams, gives

$$\beta_r = \frac{NTU_L}{NTU_s} = \frac{U_L \rho_a c_{pa}}{U_s} \tag{11.24}$$

It can be further simplified to

$$\beta_r = \frac{1}{1+\alpha_r} \tag{11.25}$$

$$\alpha_r = \frac{r_m}{r_c} \tag{11.26}$$

$$r_c = \frac{2}{k_f} \tag{11.27}$$

where r_c is the convective moisture transfer resistance, and α_r is the ratio of diffusive resistance to convective resistance for the plate. As can be seen, the total number of transfer units for moisture can be estimated from the total number of transfer units for sensible heat, and the ratio of diffusive to convective moisture resistance. As $\alpha_r \to \infty$, $NTU_L \to 0$, no moisture can be permeated through the plate. In this case, the "plate" is like a metal plate. On the other hand, as $\alpha_r \to 0$, $NTU_L \to NTU$, $\varepsilon_L = \varepsilon_S$. If $\alpha_r = 1$, $NTU_L = NTU/2$. Under the common operating conditions, the values of α_r vary from 2 to 10 [7], which implies that plate resistance for moisture transfer cannot be ignored.

Both the sensible and the latent effectiveness can be properly represented by the correlations given. The largest discrepancies between the predictions by the correlation and the experimental data result from the cases with the smallest air flow rate, where the uncertainties of the experimental data are the biggest. The average discrepancies between the predicted and experimental results are 7.3% and 8.6% for sensible and latent effectiveness respectively.

For a given exchanger, the sensible effectiveness is a fixed value at the specified flow rates. However, the latent effectiveness will be affected by the two important dimensionless factors: the ratio of diffusive to convective resistance (α_r), and the ratio of total number of transfer units of moisture to that of sensible (β_r). The values of α and β are in turn affected by the membrane material types and operating inlet conditions.

As mentioned, for total heat recovery, hydrophilic membranes are used. The value of NTU_L decreases and increases for first-type ($C_S < 1$; C_s is constant in the sorption curve; see previous Chapter 3, Figure 3.5) and third-type ($C_S > 1$) membrane materials with increasing inlet humidity, respectively [4]. The trends of resulting latent effectiveness are the same as those of NTU_L. For the linear-type material ($C_S = 1$), the NTU_L and the latent effectiveness will not change with the outside conditions. The number of transfer units for moisture would stay at 0.45 times of that for sensible heat for this material.

11.3 Air-to-air heat and mass exchanger with a plate-fin core

For plate-fin total heat exchangers, as shown in Figure 11.3, the heat transfer area on both sides of a plate is extended by fins. Therefore the extended surface and changed heat transfer coefficients should

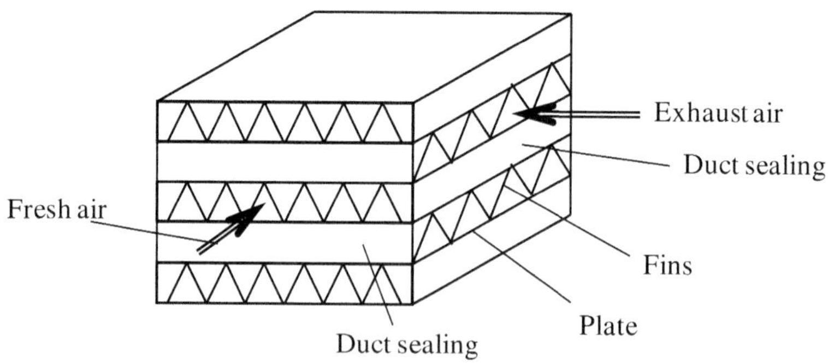

FIGURE 11.3

Schematic of a cross-flow plate-fin total heat exchanger.

be considered. Heat conduction through the plate is in equilibrium with the convective heat transfer on both sides. The equilibrium on a whole exchanger basis can be expressed by [3]

$$(hA)_f\left(T_f - T_{pf}\right) = \frac{A_p\lambda_p}{\delta_p}\left(T_{pf} - T_{pe}\right) \tag{11.28}$$

$$(hA)_e\left(T_e - T_{pe}\right) = -\frac{A_p\lambda_p}{\delta_p}\left(T_{pf} - T_{pe}\right) \tag{11.29}$$

where A_p is the total transfer area (m^2) of the plates; λ_p is the heat conductivity of the plate ($kW\,m^{-1}K^{-1}$); δ_p is the thickness of the plate (m); $(hA)_f$ is the fresh-side convective heat transfer coefficient multiplied by total fresh-air-side transfer area. The convective heat transfer coefficients in plate-fin ducts can be obtained from previous sections. Total transfer area includes plate and fin areas. If the plate is a membrane, subscript "p" can be replaced by "m."

Moisture diffusion through the plate is in equilibrium with the convective mass transfer on the two surfaces. The equations can be expressed by

$$\rho_a (kA)_f\left(\omega_f - \omega_{pf}\right) = \frac{\rho_p A_p D_{wp}}{\delta_p}\left(w_{pf} - w_{pe}\right) \tag{11.30}$$

$$\rho_a (kA)_e\left(\omega_e - \omega_{pe}\right) = -\frac{\rho_p A_p D_{wp}}{\delta_p}\left(w_{pf} - w_{pe}\right) \tag{11.31}$$

Substituting a partition coefficient into the above equations, we have

$$(kA)_f\left(\omega_f - \omega_{pf}\right) = \frac{\rho_p A_p D_{wp} K_{pm}}{\rho_a \delta_p}\left(\omega_{pf} - \omega_{pe}\right) \tag{11.32}$$

$$(kA)_e\left(\omega_e - \omega_{pe}\right) = -\frac{\rho_p A_p D_{wp} K_{pm}}{\rho_a \delta_p}\left(\omega_{pf} - \omega_{pe}\right) \tag{11.33}$$

The overall numbers of transfer units for sensible heat transfer and moisture transfer are

$$NTU_s = \frac{\left(U_s A\right)_{tot}}{\left(mc_{pa}\right)_{min}} \tag{11.34}$$

$$NTU_L = \frac{\rho_a \left(U_L A\right)_{tot}}{(m)_{min}} \tag{11.35}$$

respectively, where

$$\left(U_s A\right)_{tot}^{-1} = (hA)_f^{-1} + \left(\frac{A_p\lambda_p}{\delta_p}\right)^{-1} + (hA)_e^{-1} \tag{11.36}$$

$$(U_L A)_{tot}^{-1} = (kA)_f^{-1} + \left(\frac{\rho_p A_p D_{wp} K_{pm}}{\rho_a \delta_p}\right)^{-1} + (kA)_e^{-1} \tag{11.37}$$

Generally, the two streams have the same structure and exchanger area, therefore

$$A_f = A_e = A_p + A_{fin} \tag{11.38}$$

At this step, the effectiveness can be calculated from ε-NTU correlations developed for parallel-plate and plate-fin exchangers. The correlations deduced for parallel-plate cores are also applicable to plate-fin cores. However, the overall heat and mass transfer coefficients, the total area for the plate-fin structures, and the overall number of transfer units deduced above should be used in the correlations Eqs. (11.7)–(11.17). Examples for applications with these correlations can be found in Ref. [3].

11.4 Air-to-air heat and mass exchanger with a hollow fiber membrane bundle

11.4.1 Geometric considerations

Besides parallel-plate and plate-fin types, hollow fiber membrane bundles are also a choice for air-to-air heat and mass exchangers. In this concept, a bundle of hollow fiber membranes is packed in a shell to form a module. The structure is like a shell-and-tube heat exchanger, as show in Figure 11.4. The fresh air and the exhaust air flow on the tube side and the shell side respectively. They exchange heat and moisture through the membrane wall. The membrane-made fibers have a diameter of 1–3 mm, so the packing density is rather high. The contact area between the two air streams can be as high as 1000 m^2/m^3. As a result, the heat and moisture exchange effectiveness can be high enough to attract commercial interest.

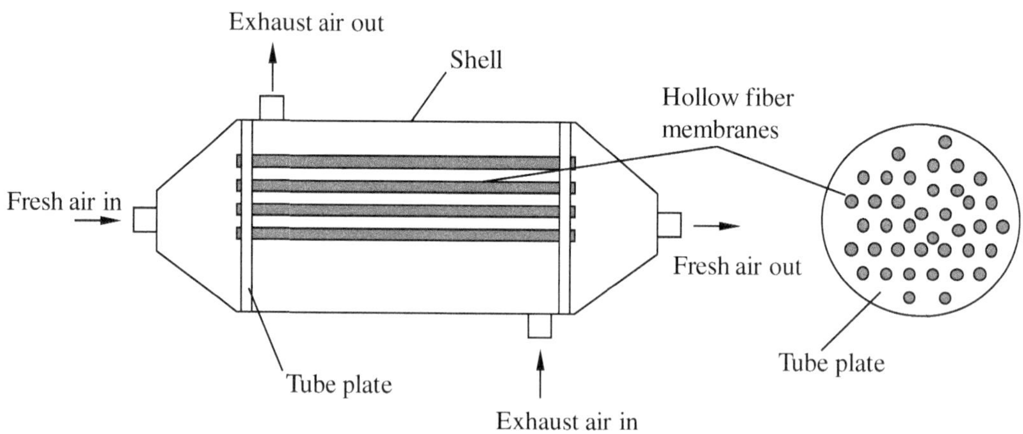

FIGURE 11.4

A counter-flow air-to-air heat and mass exchanger with hollow fiber membrane bundle.

Heat and mass transfer in the hollow fiber membrane module is the key issue for performance analysis and system optimization. Traditionally a detailed analysis of heat and mass transfer relied on numerical schemes, which is not convenient for engineering applications. Recently, an analytical solution was proposed [5] for the heat and mass transfer processes in this module. The model can be used conveniently for the estimation of sensible and latent effectiveness. Experiments were conducted to validate the solution [5]. Compared to finite difference iterations, the analytical solution is simple for engineers to use.

The packing fraction of the module, which is the ratio of the total cross-sectional area of the fibers to the module, is expressed by

$$\phi = \frac{n_f d_o^2}{D_0^2} \tag{11.39}$$

where D_0 is the diameter of the module shell (m), d_o is the outer diameter of a single fiber (m), n_f is the number of fibers in the module. In practice, the diameters and thickness of each fiber may be somewhat different; the mean values are used.

For a counter-flow arrangement, the air flow area on the shell side

$$A_{flow} = (1-\phi)\cdot\frac{\pi D_0^2}{4} \tag{11.40}$$

The total transfer area based on the outer surface of the fibers is

$$A_{tot} = n_f \pi d_o L \tag{11.41}$$

The hydrodynamic diameter on the shell side is defined as 4 times the flow area divided by the total perimeter of the membrane fibers

$$D_h = \frac{4A_{flow}}{n_f \pi d_o} = \frac{(1-\phi)D_o^2}{n_f d_o} \tag{11.42}$$

The packing density A_v(m^2/m^3) is the ratio of the total membrane area to the module volume

$$A_v = \frac{4 n_f d_o}{D_o^2} \tag{11.43}$$

For a cross-flow arrangement, the characteristic length is the outer diameter of the fibers.

11.4.2 Number of transfer units

The resistance from fresh air to exhaust air can be considered as the sum of resistances in series. The overall mass and heat transfer coefficients from fresh to exhaust air on the basis of the outer surface of the fibers are calculated by [5]

$$\frac{1}{U_s} = \frac{1}{h_i}\left(\frac{d_0}{d_i}\right) + \frac{\delta_m}{\lambda_m}\left(\frac{d_0}{\bar{d}}\right) + \frac{1}{h_o} \tag{11.44}$$

$$\frac{1}{U_L}=\frac{1}{k_i}\left(\frac{d_0}{d_i}\right)+\frac{\rho_a\delta_m}{\rho_m D_{wm}K_{pm}}\left(\frac{d_0}{\bar{d}}\right)+\frac{1}{k_o} \tag{11.45}$$

where $\bar{d}$ is the arithmetic mean diameter of a fiber, i.e., $(0.5\,(d_o+d_i))$; D_{wm} is the water (not moisture) diffusivity in membranes (m^2/s); δ_m is membrane thickness (m).

Tube-side and shell-side resistance and transport data can be obtained from friction factors, Nusselt and Sherwood numbers, calculated in previous chapters with corresponding flow arrangements and fiber array structures.

With calculated total heat and mass transfer coefficients, the total number of transfer units (NTU_s and NTU_L) can be then calculated from Eqs. (11.18) and (11.19). Finally, the correlations developed for the parallel-plate type, Eqs. (11.7) to (11.17), can be adopted for hollow fiber membrane bundles [5]. The correct flow arrangements should also be selected.

11.4.3 Performance analysis

Take a counter-flow hollow fiber membrane bundle as an example, which is used for air-to-air heat and mass exchange, or as a total heat exchanger. The geometric and transport parameters are listed in Table 11.1.

As seen from Eqs. (11.7) to (11.17), the operating and design conditions can be summarized in the non-dimensional parameters of NTU (NTU_s or NTU_L) and R (R_1 *or* R_2). The sensible and latent effectivenesses

Table 11.1 Physical and transport properties of the hollow fiber membrane module for total heat recovery under design operating conditions with counter-flow arrangement.

Name of Property	Symbol	Unit	Values
Shell diameter	D_o	mm	38
Module length	L	mm	300
Membrane type	Hydrophilic		
Number of fibers in the module	n_f		200
Fiber outer diameter	d_o	mm	1.5
Fiber inner diameter	d_i	mm	1.2
Packing fraction	ϕ		0.28
Packing density	A_v	m^2/m^3	750
Moisture diffusivity in air	D_{va}	m^2/s	2.82×10^{-5}
Water diffusivity in membrane	D_{wm}	m^2/s	3.7×10^{-10}
Partition coefficient of membrane	K_{pm}		10
Heat conductivity in membrane	λ_m	$Wm^{-1}K^{-1}$	0.47
Fresh air Temperature	T_{fi}	°C	35.0
Fresh air humidity ratio	ω_{fi}	kg/kg	0.021
Freah air mass flow rate	m_f	kg/h	5.0
Exhaust air mass flow rate	m_e	kg/h	5.0
Exhaust air temperature	T_{ei}	°C	25.0
Exhaust air humidity ratio	ω_{ei}	kg/kg	0.009

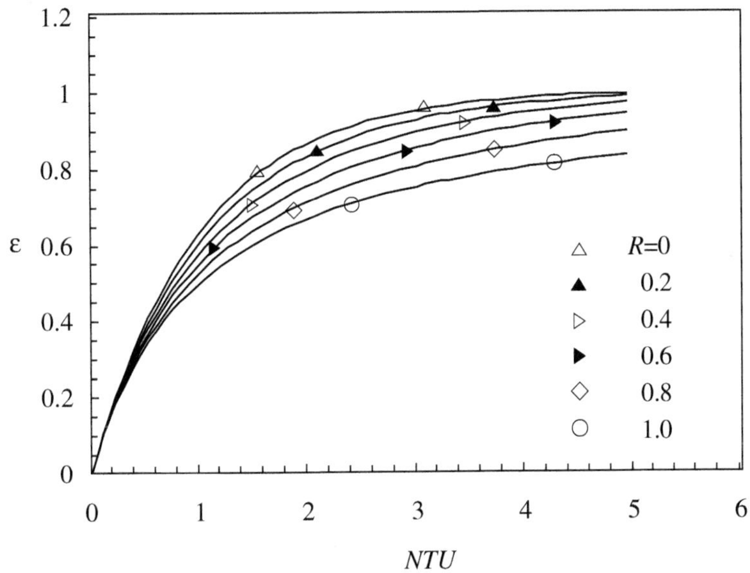

FIGURE 11.5

Effects of NTU (NTU_s or NTU_L) and R (R_1 or R_2) on sensible or latent effectiveness with a counter-flow arrangement.

are solely determined by these two dimensionless parameters. The effects of *NTU* and *R* on the effectiveness are calculated with Eqs. (11.11) to (11.14) for a counter-flow arrangement, and are plotted in Figure 11.5. As seen, the higher the *NTU*, the greater is the effectiveness. The *NTU* has greater influence when it is from 0 to 3. When the *NTU* is greater than 3, its effects become less. Therefore too many fibers will have no great benefit in further enhancing performance. The optimum *NTU* should be near 3 to 4.

The lower the *R* is, the higher the effectiveness is. When $R = 1$, or for a balanced flow, the effectiveness is the lowest. In ventilation, balanced flow is common for pressure balance purposes.

Everything has merits and demerits, but how to use the merits and overcome the demerits is the key to success. For this study, the effectiveness and pressure drop correlations would help to overcome these pressure penalties, by properly controlling the packing density and fiber diameter. The pressure drop balance between the shell and the tube sides should be considered for hollow fiber modules. When the geometries are correctly designed, balanced flow can be ensured.

The above analysis is classic for a sensible-only heat exchanger. However, from this analysis it is found that the same rules also hold for mass exchange. The established correlations provide a simple design tool for a hollow fiber membrane module, which is more complicated due to the complex structure and simultaneous heat mass exchange.

By setting other parameters unchanged under the design operating conditions, the effects of the number of fibers in the module can be examined. Figure 11.6 shows the effects on NTU_s and NTU_L, and on the consequent sensible and latent effectiveness. The inner diameter of the fiber is fixed to 1.2 mm. The number of fibers, n_f, influences the exchange area directly. As can be seen, when n_f is increased from 50 to 250, the NTU_s increases from 1.06 to 5.93, while the NTU_L increases from 0.64 to 3.4. Correspondingly, the sensible effectiveness increases from 0.52 to 0.86, while the latent effectiveness increases from 0.39 to 0.77. The rate of increase becomes lower when n_f is larger than 150.

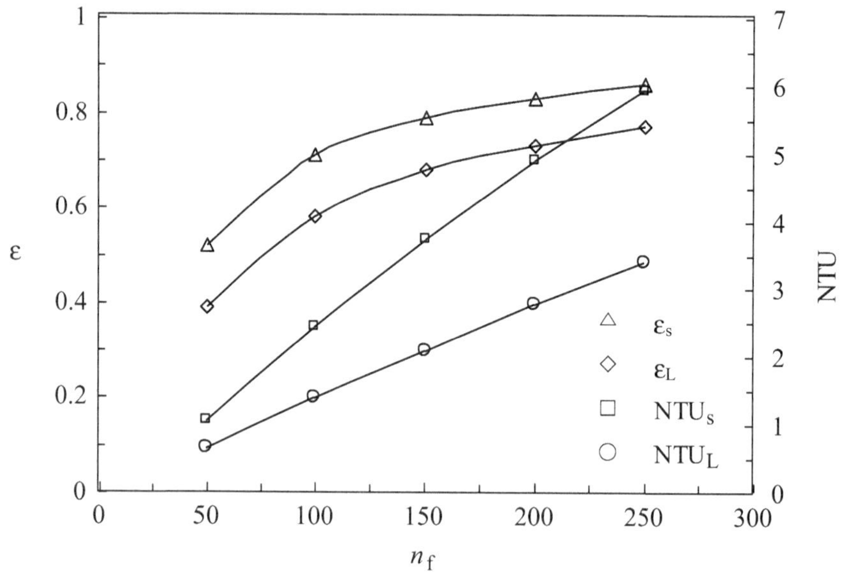

FIGURE 11.6

Effects of the number of fibers on effectiveness and *NTU* for the hollow fiber membrane bundle with a counter-flow arrangement.

On the other hand, the value of NTU_L is generally less than that of NTU_S. When n_f is increased from 50 to 250, the overall Lewis number ($Le_{tot}=Nu_s/Nu_L$) varies from 0.60 to 0.57. The variations are relatively small.

When the fiber number is fixed, the larger the fiber diameter is, the less the tube-side pressure drop is, and the greater the shell-side pressure drop is. The larger the fiber diameter is, the larger the tube-side space is, and the less the shell-side space is. The packing fraction is increased from 0.14 to 0.56 when the fiber diameter is increased from 1.2 to 3 mm. For heat and moisture recovery, it is better to have a balanced pressure drop, so a 2.7 mm diameter is the optimum one. In such cases, the tube and shell sides have equal pressure drops.

The effectiveness and the outlet parameters can be calculated easily with the established correlations. They provide a power tool for system analysis and component design.

11.5 Liquid-to-air heat and mass exchanger with a hollow fiber membrane bundle

11.5.1 General information

In previous sections, ε-NTU correlations for air-to-air heat and mass exchangers were introduced. In this section, the ε-NTU correlations for liquid-to-air heat and mass exchangers will be described. Liquid desiccant air dehumidification is a typical liquid-to-air process which involves simultaneous heat and mass transfer. Correlations for these processes are complex because there are phase changes in the processes and the coupling between heat transfer and mass transfer is very considerable.

Compared to various air dehumidification technologies, liquid desiccants have some coherent advantages. (1) Compared to solid adsorbents, liquid desiccants have relatively higher heat mass transfer capabilities, which results in a compact structure. This is because the heat and mass transfer coefficients with liquid convection of desiccants are usually higher than those with solid conduction of the adsorption bed. (2) The regeneration temperature is low, which implies the possibility of using a renewable energy sources such as solar energy. (3) The absorption heat and evaporation heat can be utilized for desiccant regeneration through heat-recovery measures. Liquid desiccants can be used as energy-storage materials to store solar energy or off-peak electricity. Off-peak electricity can be used to regenerate desiccants. When there is no electricity, the stored concentrated desiccants can be used for cooling as well. For these reasons there have been many studies on liquid desiccant air dehumidification. Packed beds are the medium for liquid absorption and regeneration.

However, salt solutions such as LiCl solution are highly corrosive. When fresh air is dehumidified, it makes contact with the salt solution directly. The liquid droplets may be carried over to the air stream and they can be subsequently transported along the dehumidified air to the conditioned space. A common way to avoid this is to install a filter at the outlet of the packing column to capture most of the droplets from the air. However, droplets as small as several microns can still penetrate the filters, and would then be transported to the conditioned space. These small liquid droplets are detrimental to indoor equipment, furniture, and, most seriously, human health. Further, the dehydrated salts may form a "salt fog" with air, which is rather corrosive and harmful. These features have posed potential risks for indoor air quality and they have limited the application of liquid desiccant air dehumidification in many countries which have stringent hygienic laws. Therefore a non-contacting liquid desiccant air dehumidification technology is desired.

To solve this problem, in recent years hollow fiber membrane contactors have been adopted to replace the packed columns. The membranes can permit moisture permeation, but they can prevent the leaking of liquid droplets. In this way, liquid desiccant flows on the tube side, and the air stream flows on the shell side. The problem of liquid droplet crossover is prevented. This technology is called non-contacting liquid desiccant air dehumidification.

Heat and mass transfer in the hollow fiber membrane module is the key issue concerning performance analysis and system optimization. Detailed finite difference modeling is complex. Therefore simple ε-NTU correlations have been presented for engineering applications [6]. Correlations for a cross-flow hollow fiber membrane bundle were introduced. For a cross-flow arrangement, the governing equations have been presented, but the correlations are still not available due to the difficulties in analytically solving two-variable two-dimensional partial differential equations.

11.5.2 Module effectiveness and resistance

Consider a hollow fiber membrane module, as shown in Figure 11.4. It is now to be used for liquid desiccant air dehumidification. LiCl solution flows on the tube side and air flows on the shell side, in a counter-flow arrangement. The heat and mass transfer processes are schematically shown in Figure 11.7. The process takes place through a membrane. The coordinate system is also shown in the figure.

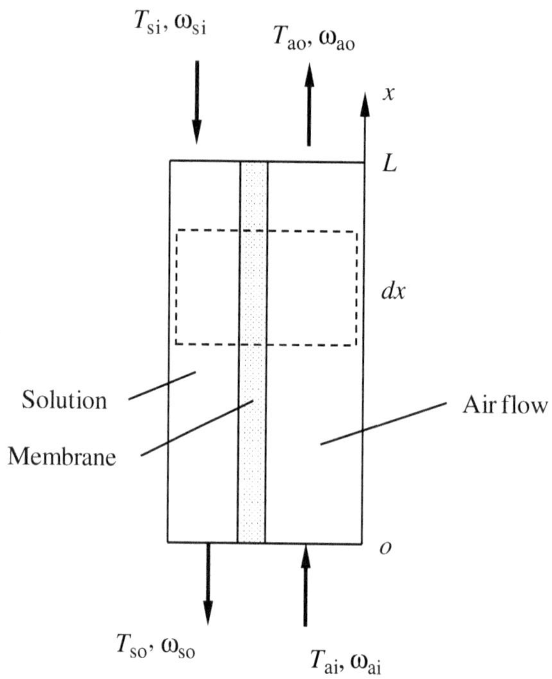

FIGURE 11.7

Schematics of a counter-flow liquid desiccant air dehumidification process through the membrane.

For air dehumidification applications, the sensible cooling and latent (moisture removal) effectiveness of the module are defined by

$$\varepsilon_s = \frac{T_{ao} - T_{ai}}{T_{si} - T_{ai}} \tag{11.46}$$

$$\varepsilon_L = \frac{\omega_{ao} - \omega_{ai}}{\omega_{si} - \omega_{ai}} \tag{11.47}$$

The latent effectiveness is also the air dehumidification effectiveness. ω_{si} is the equilibrium humidity of the LiCl solution in the inlet state. The sensible effectiveness reflects a temperature change in the dehumidification process. Differently from sensible-only heat exchangers, absorption heat is released in dehumidification. As a result, the sensible effectiveness may be negative in some cases due to the absorption heat generated. The transfers of sensible heat and moisture are in different directions now.

The enthalpy of the solution is a sum of the product of specific heat and temperature and the absorption heat

$$H_s = c_s T_s + (1 - X_s)\, h_{abs} \tag{11.48}$$

where c_s is the specific heat of solution (kJ kg^{-1} K^{-1}); h_{abs} is absorption heat (kJ/kg); the reference states where enthalpy is equal to zero are taken to be dry at 0 °C and liquid water at 0 °C.

The equivalent enthalpy of the solution is the enthalpy of the air that is in equilibrium with solution

$$H_s^e = c_{pa} T_s + \omega_s h_v \tag{11.49}$$

where h_v is the heat of evaporation of water at 0 °C (kJ/kg). Sometimes h_{abs} is also approximated by h_v, by ignoring the heat of mixing.

The enthalpy effectiveness of the module is defined by

$$\varepsilon_{en} = \frac{H_{ao} - H_{ai}}{H_{si}^e - H_{ai}} \tag{11.50}$$

Substituting Eqs. (11.46), (11.47) and (11.49) into Eq. (11.50), it becomes

$$\varepsilon_{en} = \frac{\varepsilon_s + H_2^* \varepsilon_L}{1 + H_2^*} \tag{11.51}$$

where the ratio of latent to sensible energy differences between the solution and the air is defined by

$$H_2^* = \frac{h_v(\omega_{si} - \omega_{ai})}{c_{pa}(T_{si} - T_{ai})} \approx 2501 \frac{\Delta\omega}{\Delta T} \tag{11.52}$$

Resistance from the solution through the membranes to air can be considered as a sum of resistances in series. The overall heat and mass transfer coefficients from solution to air on the basis of the outer surface of the fibers are calculated by [6]

$$\frac{1}{U_s} = \frac{1}{h_i}\left(\frac{d_0}{d_i}\right) + \frac{\delta_m}{\lambda_m}\left(\frac{d_0}{\bar{d}}\right) + \frac{1}{h_o} \tag{11.53}$$

$$\frac{1}{U_L} = \frac{1}{k_i^e}\left(\frac{d_0}{d_i}\right) + \frac{\delta_m}{D_{vm}}\left(\frac{d_0}{\bar{d}}\right) + \frac{1}{k_o} \tag{11.54}$$

The membranes here are different from those for heat and moisture recovery. The moisture diffusion mechanisms are also different. Water solution and diffusion is the mechanism for mass transfer in hydrophilic membranes, while gas Knudsen diffusion is the mechanism for mass transfer in hydrophobic membranes. So here the vapor diffusivity in membranes is directly used in the equation.

The solution-side convective mass transfer coefficient is calculated by the Sherwood numbers on the tube side (see Chapter 7), with water diffusivity in solution, D_{ws}. It is calculated on the basis of water concentration differences. Calculating on the basis of humidity differences, the equivalent moisture transfer coefficient on the solution side is defined by

$$\rho_a k_i^e \Delta\omega_s = \rho_s k_i \Delta\left(1\text{-}X_s\right) \tag{11.55}$$

or

$$k_{\mathrm{i}}^{\mathrm{e}} = \frac{\rho_{\mathrm{s}}}{\rho_{\mathrm{a}}} k_{\mathrm{i}} K_{\mathrm{ps}} \tag{11.56}$$

where ω_{s} is the equilibrium humidity with the solution at a given liquid temperature and concentration; K_{ps} is the partition coefficient of the solution. This is the slope of the water uptake in solution to equilibrium humidity.

$$K_{\mathrm{ps}} = \frac{\Delta(1 - X_{\mathrm{s}})}{\Delta\omega_{\mathrm{s}}} \tag{11.57}$$

11.5.3 Governing equations

Air flows along the x direction, and solution flows along the $-x$ direction. Normalized equations for heat and moisture conservation in the air stream are

$$\frac{dT_{\mathrm{a}}^{*}}{dx^{*}} = NTU_{\mathrm{s}}\left(T_{\mathrm{s}}^{*} - T_{\mathrm{a}}^{*}\right) \tag{11.58}$$

$$\frac{d\omega_{\mathrm{a}}^{*}}{dx^{*}} = NTU_{\mathrm{L}}\left(\omega_{\mathrm{s}}^{*} - \omega_{\mathrm{a}}^{*}\right) \tag{11.59}$$

where x is the air flow direction.

$$NTU_{\mathrm{s}} = \frac{U_{\mathrm{s}} A_{\mathrm{tot}}}{m_{\mathrm{a}} c_{\mathrm{pa}}} \tag{11.60}$$

$$NTU_{\mathrm{L}} = \frac{\rho_{\mathrm{a}} U_{\mathrm{L}} A_{\mathrm{tot}}}{m_{\mathrm{a}}} \tag{11.61}$$

$$Le_{\mathrm{tot}} = \frac{NTU_{\mathrm{s}}}{NTU_{\mathrm{L}}} = \frac{U_{\mathrm{s}}}{\rho_{\mathrm{a}} c_{\mathrm{pa}} U_{\mathrm{L}}} \tag{11.62}$$

where Le_{tot} is the overall Lewis number, which reflects the ratio of the overall heat transfer coefficient to the overall mass transfer coefficient. For membrane systems, mass transfer resistance is usually larger than heat transfer resistance, therefore the overall Lewis number is always greater than unity.

Normalized equations for heat and moisture conservation in liquid solution are

$$\frac{dT_{\mathrm{s}}^{*}}{dx^{*}} = R_3 NTU_{\mathrm{s}}\left(T_{\mathrm{s}}^{*} - T_{\mathrm{a}}^{*}\right) + R_4 \frac{d\omega_{\mathrm{a}}^{*}}{dx^{*}} \tag{11.63}$$

$$\frac{d\omega_{\mathrm{s}}^{*}}{dx^{*}} = \frac{E_{\mathrm{T}} h_{\mathrm{v}}}{H_2^{*} c_{\mathrm{pa}}} \frac{dT_{\mathrm{s}}^{*}}{dx^{*}} \tag{11.64}$$

where in the equations the two dimensionless heat capacity ratios are defined by

$$R_3 = \frac{m_a c_{pa}}{m_s c_s} \tag{11.65}$$

$$R_4 = \frac{m_a h_{abs} (\omega_{si} - \omega_{ai})}{m_s c_s (T_{si} - T_{ai})} \tag{11.66}$$

The slope of humidity to temperature for the solution

$$E_T = \left. \frac{\partial \omega_s}{\partial T_s} \right|_{\bar{X}_s} \tag{11.67}$$

Dimensionless variables

$$T^* = \frac{T - T_{ai}}{T_{si} - T_{ai}} \tag{11.68}$$

$$\omega^* = \frac{\omega - \omega_{ai}}{\omega_{si} - \omega_{ai}} \tag{11.69}$$

$$x^* = \frac{x}{L} \tag{11.70}$$

As seen, at the air stream outlet where $x^* = 1$

$$T^*_{ao} = \varepsilon_s \tag{11.71}$$

$$\omega^*_{ao} = \varepsilon_L \tag{11.72}$$

As can be seen, the first ratio defined by Eq. (11.65) is the traditional air-to-solution sensible heat capacity ratio, and the second one defined by Eq. (11.66) is a newly established latent-to-sensible heat ratio of the solution, which actually reflects the effect of absorption heat of moisture on the solution temperature.

Substituting Eq. (11.59) into Eq. (11.63), we have

$$\frac{dT^*_s}{dx^*} = R_3 NTU_s (T^*_s - T^*_a) + R_4 \frac{NTU_s}{Le_{tot}} (\omega^*_s - \omega^*_a) \tag{11.73}$$

Substituting Eq. (11.73) into Eq. (11.64), we have

$$\frac{d\omega^*_s}{dx^*} = \frac{E_T h_v}{H^*_2 c_{pa}} R_3 NTU_s (T^*_s - T^*_a) + \frac{E_T h_v}{H^*_2 c_{pa}} R_4 \frac{NTU_s}{Le_{tot}} (\omega^*_s - \omega^*_a) \tag{11.74}$$

11.5.4 Analytical solutions

Equations (11.58), (11.59) and (11.73), (11.74) are the governing differential equations for heat and moisture transfer in the module. They can be solved by finite difference iterations, plus the solution equilibrium state equations. However, analytical solutions can be obtained after transformations of the above equations. They are more convenient for engineering applications.

Subtracting Eq. (11.73) with (11.58), we have

$$\frac{d\left(T_s^* - T_a^*\right)}{dx^*} = (R_3 - 1)\, NTU_s \left(T_s^* - T_a^*\right) + R_4 \frac{NTU_s}{Le_{tot}} \left(\omega_s^* - \omega_a^*\right) \tag{11.75}$$

Subtracting Eq. (11.74) with (11.59), we have

$$\frac{d\left(\omega_s^* - \omega_a^*\right)}{dx^*} = \frac{E_T h_v}{H_2^* c_{pa}} R_3 NTU_s \left(T_s^* - T_a^*\right) + \left(\frac{E_T h_v}{H_2^* c_{pa}} R_4 - 1\right) \frac{NTU_s}{Le_{tot}} \left(\omega_s^* - \omega_a^*\right) \tag{11.76}$$

Let

$$\xi = T_s^* - T_a^* \tag{11.77}$$

$$\Psi = \omega_s^* - \omega_a^* \tag{11.78}$$

Equations (11.75) and (11.76) become a set of differential equations as

$$\frac{d\xi}{dx^*} = a_{11}\xi + a_{12}\Psi \tag{11.79}$$

$$\frac{d\Psi}{dx^*} = a_{21}\xi + a_{22}\Psi \tag{11.80}$$

where constants

$$a_{11} = (R_3 - 1)\, NTU_s \tag{11.81}$$

$$a_{12} = R_4 \frac{NTU_s}{Le_{tot}} \tag{11.82}$$

$$a_{21} = \frac{E_T h_v}{H_2^* c_{pa}} R_3 NTU_s \tag{11.83}$$

$$a_{22} = \left(\frac{E_T h_v}{H_2^* c_{pa}} R_4 - 1\right) \frac{NTU_s}{Le_{tot}} \tag{11.84}$$

The set of Eqs. (11.79), (11.80) have an analytical solution in the form of the following [6]:

$$\xi = C_1 e^{\lambda_1 x^*} + C_2 e^{\lambda_2 x^*} \tag{11.85}$$

$$\Psi = -K_1 C_1 e^{\lambda_1 x^*} + K_2 C_2 e^{\lambda_2 x^*} \tag{11.86}$$

The roots of the characteristic equations of (11.79) and (11.80) are given as

$$\lambda_{1,2} = \frac{1}{2}\left[(a_{11} + a_{22}) \pm \sqrt{(a_{11} + a_{22})^2 - 4(a_{11}a_{22} - a_{12}a_{21})}\right] \tag{11.87}$$

By satisfying the general set of Eqs. (11.85) and (11.86) to Eq. (11.79), the coefficients of K_1 and K_2 can be given as

$$K_1 = \frac{a_{11} - \lambda_1}{a_{12}} \tag{11.88}$$

$$K_2 = \frac{-a_{11} + \lambda_2}{a_{12}} \tag{11.89}$$

Similarly, by satisfying the general set of Eqs. (11.85) and (11.86) to Eq. (11.80), the coefficients of K_1 and K_2 can be given as

$$K_1 = \frac{a_{21}}{a_{22} - \lambda_1} \tag{11.90}$$

$$K_2 = \frac{a_{21}}{\lambda_2 - a_{22}} \tag{11.91}$$

Either (11.88) and (11.89) or (11.90) and (11.91) can be applied. In this study, Eqs. (11.88) and (11.89) are selected.

At $x^* = 0$,

$$T^*_{\text{ai}} = \omega^*_{\text{ai}} = 0 \tag{11.92}$$

Equations (11.85) and (11.86) become

$$\xi = T^*_{\text{so}} - T^*_{\text{ai}} = T^*_{\text{so}} = C_1 + C_2 \tag{11.93}$$

$$\Psi = \omega^*_{\text{so}} - \omega^*_{\text{ai}} = \omega^*_{\text{so}} = -K_1C_1 + K_2C_2 \tag{11.94}$$

At $x^* = 1$,

$$T^*_{\text{si}} = \omega^*_{\text{si}} = 1 \tag{11.95}$$

Equations (11.85) and (11.86) become

$$\xi = T_{\text{si}}^* - T_{\text{ao}}^* = 1 - T_{\text{ao}}^* = C_1 e^{\lambda_1} + C_2 e^{\lambda_2} \tag{11.96}$$

$$\Psi = \omega_{\text{si}}^* - \omega_{\text{ao}}^* = 1 - \omega_{\text{ao}}^* = -K_1 C_1 e^{\lambda_1} + K_2 C_2 e^{\lambda_2} \tag{11.97}$$

Further, integrating Eq. (11.76) from $x^* = 0$ to 1, we have

$$\omega_{\text{so}}^* - \omega_{\text{si}}^* = \frac{E_{\text{T}} h_{\text{v}}}{H_2^* c_{\text{pa}}} R_3 \left(T_{\text{so}}^* - T_{\text{si}}^*\right) \tag{11.98}$$

An overall heat balance in the hollow fiber membrane module will have the following equation

$$m_{\text{s}} c_{\text{s}} \left(T_{\text{si}} - T_{\text{so}}\right) = m_{\text{a}} c_{\text{pa}} \left(T_{\text{ao}} - T_{\text{ai}}\right) + m_{\text{a}} h_{\text{abs}} \left(\omega_{\text{ao}} - \omega_{\text{ai}}\right) \tag{11.99}$$

Normalizing the above equation to

$$T_{\text{si}}^* - T_{\text{so}}^* = R_3 \left(T_{\text{ao}}^* - T_{\text{ai}}^*\right) + R_4 \left(\omega_{\text{ao}}^* - \omega_{\text{ai}}^*\right) \tag{11.100}$$

Substituting Eqs. (11.92) to (11.97) into (11.98) and (11.100) respectively, we have

$$b_{11} C_1 + b_{12} C_2 + b_{13} = 0 \tag{11.101}$$

$$b_{21} C_1 + b_{22} C_2 + b_{23} = 0 \tag{11.102}$$

where in the equations the coefficients are

$$b_{11} = K_1 + \frac{E_{\text{T}} h_{\text{v}}}{H_2^* c_{\text{pa}}} R_3 \tag{11.103}$$

$$b_{12} = -K_2 + \frac{E_{\text{T}} h_{\text{v}}}{H_2^* c_{\text{pa}}} R_3 \tag{11.104}$$

$$b_{13} = 1 - \frac{E_{\text{T}} h_{\text{v}}}{H_2^* c_{\text{pa}}} R_3 \tag{11.105}$$

$$b_{21} = R_3 e^{\lambda_1} - R_4 K_1 e^{\lambda_1} - 1 \tag{11.106}$$

$$b_{22} = R_3 e^{\lambda_2} + R_4 K_2 e^{\lambda_2} - 1 \tag{11.107}$$

$$b_{23} = 1 - R_3 - R_4 \tag{11.108}$$

The algebraic solutions of the set of Eqs. (11.101) and (11.102) are

$$C_1 = \frac{b_{13}b_{22} - b_{12}b_{23}}{b_{12}b_{21} - b_{11}b_{22}} \tag{11.109}$$

$$C_2 = \frac{b_{11}b_{23} - b_{13}b_{21}}{b_{12}b_{21} - b_{11}b_{22}} \tag{11.110}$$

At this step, the analytical solutions of Eqs. (11.96) and (11.97) are obtained. With these coefficients, the sensible effectiveness, the latent effectiveness, as well as the outlet temperature and humidity can be calculated. It should be noted that

$$\varepsilon_S = T^*_{ao} = 1 - C_1 e^{\lambda_1} - C_2 e^{\lambda_2} \tag{11.111}$$

$$\varepsilon_L = \omega^*_{ao} = 1 + K_1 C_1 e^{\lambda_1} - K_2 C_2 e^{\lambda_2} \tag{11.112}$$

Water vapor partial pressure, temperature, and LiCl solute concentration are governed by a set of thermodynamic equations as described in Chapter 6. From these equations, the slope E_T and partition coefficient K_{ps} can be estimated.

11.5.5 Parametric analysis

A counter-flow hollow fiber membrane module is used for liquid desiccant air dehumidification. The geometric, physical and operating conditions are listed in Table 11.2. As can be seen, the geometric parameters are the same as for the module listed in Table 11.1. However the membranes used are different. They are porous membranes rather than hydrophilic membranes, so the diffusivity is higher. The ε-NTU correlations have been validated with a continuous LiCl solution desiccant cycle [6].

Setting the other parameters unchanged under the design operating conditions, the flow rates of the air stream are changed. Consequently all the dimensionless parameters are changed accordingly. The analytical solution is used to calculate the sensible and moisture effectiveness. The values are plotted in Figure 11.8. As can be seen, the higher the process air flow rates are, the less the sensible and moisture effectivenesses are. The sensible effectiveness is usually higherer than the latent effectiveness. This is reasonable considering a total Lewis number greater than unity. However, under high flow rates, the sensible and the latent effectiveness come closer. The two variables even intersect at flow rate of 20 kg/h. The reason will be explained below.

The variations of the dimensionless parameters defined in the study are plotted in Figure 11.9. As can be seen, all the parameters change with a change in air flow rates. The total Lewis number increases only a little. This is because both the convective heat and the mass transfer coefficients increase with flow rates. However, the membrane resistance doesn't change much. The membrane heat resistance is negligible, and the total heat transfer resistance decreases accordingly with convective resistance. However, the moisture transfer resistance of the membrane is higher and it is the dominant factor. The total mass transfer resistance changes little with convective coefficients. Therefore the total Lewis number increases only slightly with increasing air flow rates.

Table 11.2 Physical and transport properties of the hollow fiber membrane module for liquid air dehumidification under design operating conditions.

Name of Property	Symbol	Unit	Values
Shell diameter	D_o	mm	40
Module length	L	mm	300
Membrane type	Hydrophobic porous		
Number of fibers in a module	n_f		200
Fiber outer diameter	d_o	mm	1.5
Fiber inner diameter	d_i	mm	1.3
Packing fraction	ϕ		0.28
Packing density	A_v	m^2/m^3	750
Moisture diffusivity in air	D_{va}	m^2/s	2.82×10^{-5}
Water diffusivity in solution	D_{ws}	m^2/s	3.0×10^{-9}
Moisture diffusivity in membrane	D_{vm}	m^2/s	4.3×10^{-6}
Heat of absorption	h_{abs}	kJ/kg	2501
Thermal conductivity of membrane	λ_m	$Wm^{-1}K^{-1}$	0.47
Partition coefficient of solution	K_{ps}		14.6
Temperarure of air to module	T_{ai}	°C	35.0
Humidity of air to module	ω_{ai}	kg/kg	0.021
Mass flow rate of air	m_a	kg/h	7.55
Mass flow rate of solution	m_s	kg/h	9.89
Mass fraction of LiCl in solution to module	X_{si}	kg/kg	0.35
Temperature of solution to module	T_{si}	°C	25.0
Equilibrium humidity with solution to module	ω_{si}	kg/kg	0.0055
Specific heat of solution	c_s	$kJ\ kg^{-1}K^{-1}$	2.8
Thermal conductivity of solution	λ_s	$Wm^{-1}K^{-1}$	0.5
Viscosity of solution	ν_s	m^2/s	4.17×10^{-6}
Density of solution	ρ_s	kg/m^3	1215
Slope of equilibrium humidity to temperature of solution	E_T	$kg\ kg^{-1}K^{-1}$	5.39×10^{-4}
Number of transfer units for sensible heat	NTU_s		4.42
Total Lewis number	Le_{tot}		2.80

The parameter R_3 increases with increasing air flow rates. The total number of transfer units NTU_s becomes less, due to increased air flow rates. The sensible and the latent effectivenesses decrease with increased air flow rates and lower NTU_s values. Obviously, NTU_s is the dominant parameter determining

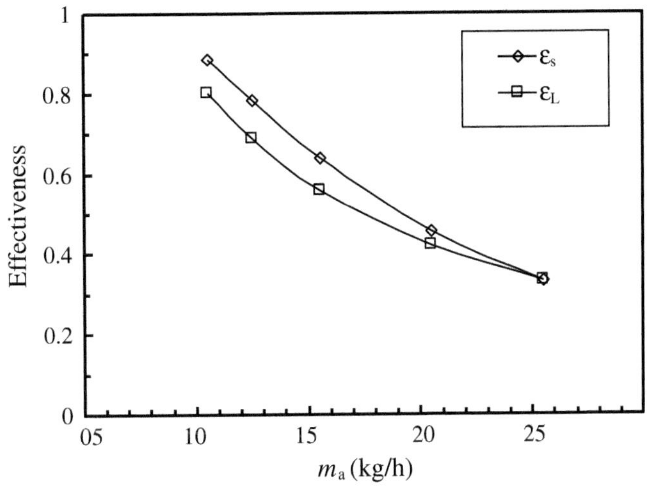

FIGURE 11.8

Sensible and latent effectiveness calculated with correlations for liquid desiccant air dehumidification with the hollow fiber membrane module.

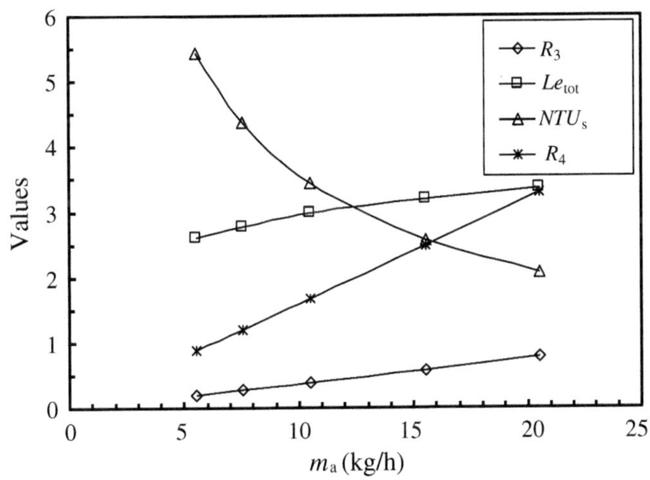

FIGURE 11.9

Variations of the dimensionless parameters in a counter-flow membrane module for liquid desiccant air dehumidification.

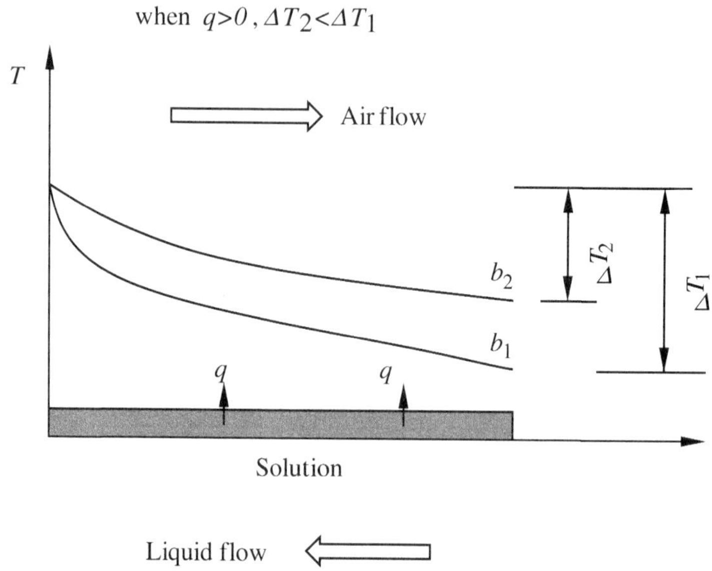

Figure 11.10

Temperature rises due to heat release via liquid desiccant air contacting.

heat and mass transfer. Therefore to increase the heat transfer area is an efficient way to increase heat and mass transfer. This is the reason why membrane contactors are so called. Compared to traditional packed-column liquid desiccant air dehumidification, the membrane system adds an additional mass transfer resistance from the membranes. The effective heat and mass transfer coefficients are sacrificed somewhat. However, since the hollow fiber membrane module is rather compact (the packing fraction can be as high as 0.6), the specific heat and mass transfer area can be as high as $2000\,m^2/m^3$. In contrast, a packed column only has a packing density around $300\,m^2/m^3$. In addition to its compactness, the mini-flows are well controlled in a membrane module. The liquid short-cuts that often occur in a packed column are prevented. As a result, the total number of transfer units for both sensible heat and moisture is rather high. Packed-column systems are usually very bulky. With the membrane technology, small-scale systems are possible for household use.

The parameter R_4 increases with air flow rates. The reason is that in this case more moisture is absorbed by the desiccant, more heat is released to the solution, which then leads to a higher temperature in the solution. The outlet temperature of the air stream becomes higher due to this solution temperature rise. The situation is shown in Figure 11.10. The absorption heat is like an internal heat source q. As can be seen, the figure depicts an air stream flowing through a fluid surface. If there is no heat source q, the air temperature change is ΔT_1. If there is a heat source q, the air temperature change is ΔT_2. Since $\Delta T_2 < \Delta T_1$, the sensible effectiveness becomes smaller. In extreme cases, negative sensible effectiveness can be obtained. This is the reason why the sensible and latent effectivenesses become closer in the figure. They are a result of the interactions between heat and mass transfer.

11.5.6 Cross-flow bundle

For a cross-flow bundle, if the desiccant flows on the tube side and in the x direction, while the air stream flows on the shell side and along the y direction, the normalized governing equations are [8]:

$$\frac{\partial T_a^*}{\partial y^*} = NTU_s \left(T_s^* - T_a^*\right) \tag{11.113}$$

$$\frac{\partial \omega_a^*}{\partial y^*} = NTU_L \left(\omega_s^* - \omega_a^*\right) \tag{11.114}$$

$$\frac{\partial T_s^*}{\partial x^*} = -R_3 NTU_s \left(T_s^* - T_a^*\right) - R_4 NTU_L \left(\omega_s^* - \omega_a^*\right) \tag{11.115}$$

$$\frac{\partial \omega_s^*}{\partial x^*} = \frac{E_T h_v}{H_2^* c_{pa}} \frac{\partial T_s^*}{\partial x^*} \tag{11.116}$$

The definitions of dimensionless parameters in the equations are the same as in the previous section for the counter-flow membrane module. For cross flow, the set of equations is two-dimensional. It is more complex to find analytical solutions for the above equations. At present, they can only be solved by finite difference schemes. Readers are encouraged to solve this problem.

11.6 Conclusion

Conjugate heat and mass transfer in ducts usually involve detailed numerical schemes to solve the temperature and concentration fields in the duct. The performance, in terms of sensible and latent effectiveness, can then be estimated with known outlet parameters. However, in engineering, it is usually too complicated and time-consuming to use these techniques. ε-NTU correlations provide a simple yet convenient tool to realize this goal. In fact, they are the exact analytical solutions to the governing equations for heat and mass transfer processes, with some extent of approximation. This chapter provides the correlations for air-to-air heat and mass exchangers, as well as for liquid-to-air heat and mass exchangers. The dimensionless parameters in the correlations are the factors influencing system performance.

References

[1] Kays WM, Crawford ME. Convective heat and mass transfer. New York: McGraw-Hil; 1993.

[2] Incropera FP, Dewitt DP. Fundamentals of Heat and Mass Transfer. 3rd ed.. New York: Wiley; 1990.

[3] Zhang LZ. Total heat recovery: heat and moisture recovery from ventilation air. New York: Nova Science; 2009.

[4] Zhang LZ, Niu JL. Effectiveness correlations for heat and moisture transfer processes in an enthalpy exchanger with membrane cores. ASME J Heat Transfer 2002;122:922929.

[5] Zhang LZ. An analytical solution for heat mass transfer in a hollow fiber membrane based air-to-air heat mass exchanger. J Membr Sci 2010;360:217–25.

[6] Zhang LZ. An analytical solution to heat and mass transfer in hollow fiber membrane contactors for liquid desiccant air dehumidification. ASME, J Heat Transfer 2011;133:092001-1–092001-8.

[7] Niu JL, Zhang LZ. Membrane-based enthalpy exchanger: material considerations and clarification of moisture resistance. J Membr Sci 2001;189:179–91.

[8] Zhang LZ. Coupled heat and mass transfer in an application-scale cross-flow hollow fiber membrane module for air humidification. Int J Heat Mass Transfer 2012;55:5861–9.

CHAPTER

12 Novel Materials for Heat and Mass Exchangers

12.1 Introduction

Fluid flow and conjugate heat and mass transfer in ducts are closely related to the properties of the duct wall materials. Modern heat and mass exchangers are made of functional materials. Every material has a specific function to fulfill an objective. For instance, membranes to make total heat exchangers are highly vapor-permeable. High thermal conductivity and high mass diffusivity in the materials are the prerequisite conditions for high efficiency heat and mass exchangers. In recent years, some novel duct wall materials have been invented. They have been successfully used in various industries. For example, several novel membranes are used. They include hydrophobic-hydrophilic dual-polar membranes, composite supported liquid membranes, membranes with finger-like macrovoids, and one-step-fabricated asymmetric membranes. In another application, to increase the thermal conductivity of duct walls, highly thermally conductive polymers are developed, with an inverse problem technique to optimize the filler shapes. To control indoor air quality, high selective adsorbents are fabricated with a core-shell technique. They can only adsorb moisture, while other pollutant gases cannot be adsorbed. These are the key materials for heat and mass exchangers such as total heat exchangers, adsorbent beds, desiccant wheels, and liquid-to-air membrane contactors. Needless to say, there will be no novel heat and mass exchangers without these new functional materials.

This chapter will give an overview of these new materials. Their design theories, fabrication techniques, and performances will be introduced and discussed. This knowledge is the output of cross-disciplinary research accumulated over the past 10 years.

12.2 Novel membranes

Control of indoor air humidity is a critical issue, especially in hot and humid regions. Membrane materials are the key components in total heat exchangers for fresh air ventilation. Hydrophilic polymer membranes that are permeable to vapor, but impermeable to air, have been considered for moisture permeation. Traditionally they are homogeneous membranes. Typical materials include [1]: cellulose triacetate, Nafion, regenerated cellulose, sulfonated poly(phenylene oxide), polyether-polyurethane, siloxane-amido copolymer, polystyrene-sulfonate, polyvinylidene fluoride and polyethersulfone, and cellophane. These materials have been used in air dehumidification applications. However, moisture diffusion coefficients in such polymer membranes are usually very low, of the order of 10^{-12}–$10^{-13}\,m^2/s$. They can be used successfully in compressed air dehumidification because this process uses compressed air to generate the trans-membrane partial pressure difference. Regrettably total heat exchangers only have limited trans-membrane vapor partial pressure difference (<2000 Pa). Consequently their performance is quite limited.

To address this problem, several novel membranes have been made. They are: hydrophobic-hydrophilic composite membranes [2,3], composite supported liquid membranes [4–7], membranes with finger-like macrovoids [8], and one-step fabricated asymmetric membranes [9]. These membranes, if porous, can also be used as membrane contactors for liquid desiccant air dehumidification.

12.2.1 Hydrophobic-hydrophilic composite membrane

Vapor diffusion in homogeneous membranes is rather low. To increase vapor permeation rates, a hydrophobic-hydrophilic dual-polar composite membrane has been made. According to this scheme, a thin active layer is cast onto a thick porous polyvinylidene fluoride (PVDF) support layer or a layer made from other materials. The porous support layer provides the necessary mechanical strength while the thin active layer provides the permselective separating effect. The permeation rates can be greatly improved due to the reduction in resistance. The structure is depicted in Figure 12.1. The support layer is thick and porous and the skin layer is thin and dense. The porous layer is hydrophobic and the skin layer is hydrophilic. The permeability with this structure is very high.

Method of membrane fabrication

The composite membrane is formed by coating cellulose triacetate (CTA) casting solution onto the PVDF support membrane. The fabrication process is composed of the following three steps [2,3]:

(1) Casting of PVDF porous support layer. A certain amount of PVDF powder is dissolved in dimethylformamide (DMF) solution. The casting solution is coated onto a clean glass plate. The plate is then dipped into a water bath controlling the temperature and time. Finally, the membrane is formed and dried. PVDF porous membranes are made.

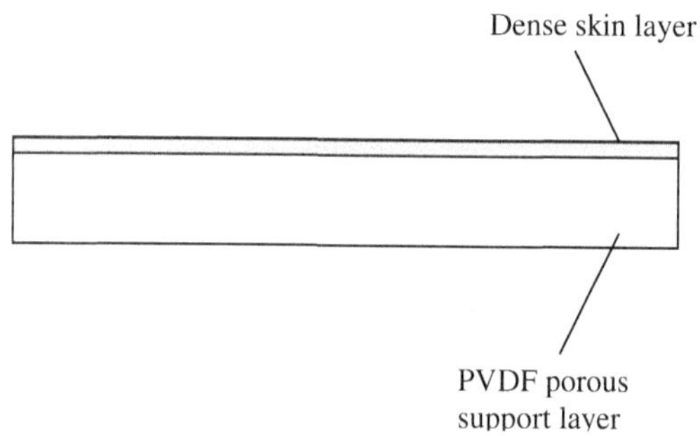

FIGURE 12.1

Schematic of a composite hydrophobic-hydrophilic dual-polar membrane.

(2) Formulation of casting solution. CTA powder is weighed and placed into a vessel with acetone solution at about 95 °C. The solution is heated and stirred until it is completely dissolved. This takes about 3 h.

(3) A certain amount of cross-linking agent (e.g. L-malic acid), catalyst (e.g. glacial acetic acid) and additive (LiCl) is added to the solution. The solution is continuously stirred at 70 °C until these different compositions are completely dissolved to form a homogeneous solution. The solution is cooled down and kept still for de-bubbling. This takes another 2 h.

The polymer solution is coated on the PVDF support membrane. The thickness of the active layer is controlled with a casting knife. The thickness of the final active layer can be from several microns to a dozen microns, depending on the gap between the knife edge and the support layer. Then the asymmetric membrane is placed into a vacuum drying oven for cross-linking at 90 °C for 1 h. The membrane is further dried at 60 °C for 3 h.

Mass diffusivity and heat conductivity in the membrane

The membrane is composed of two layers. Heat conduction and moisture diffusion through the membrane can be depicted by a model as shown in Figure 12.2.

Moisture flux through the composite membrane is expressed by

$$J = \rho_a D_e \frac{\omega_1 - \omega_2}{\delta_1 + \delta_2} \tag{12.1}$$

where D_e is the effective moisture diffusivity in the composite membrane (m^2/s), calculated by

$$D_e = \frac{\delta_1 + \delta_2}{\frac{\delta_1}{D_{e1}} + \frac{\delta_2}{D_{e2}}} \tag{12.2}$$

where D_{e1} and D_{e2} are the effective diffusivity in the first layer and second layer, respectively.

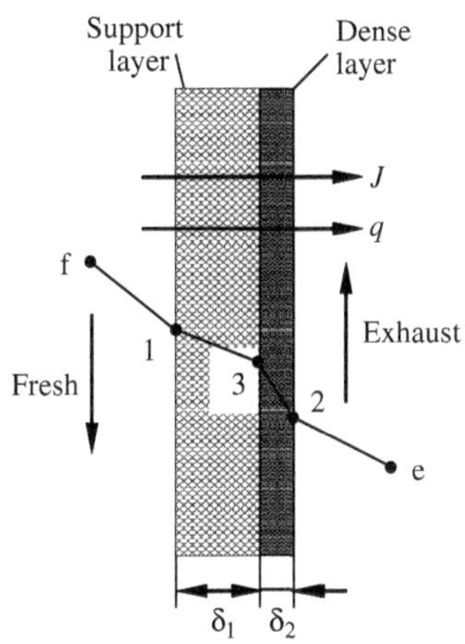

FIGURE 12.2

Transfer model through the composite membrane.

The effective diffusivity can be calculated according to their diffusion mechanisms. The established theory of gas diffusion in hydrophobic membranes considers three mechanisms: Poiseuille flow, ordinary molecular diffusion and Knudsen diffusion, or a combination of them. In the support layer, D_{e1} can be calculated as the combined Knudesn and ordinary diffusion, taking into account porosity and tortuosity.

The ordinary diffusion coefficient of water vapor molecule in air is expressed by [10]

$$D_{O} = \frac{1.86 \times 10^{-7} T^{3/2}}{P_{m} \sigma_{va}^{2} \Omega} \sqrt{\frac{1}{M_{v}} + \frac{1}{M_{a}}} \tag{12.3}$$

where D_O is in m^2/s; T is in K; M_v and M_a are molecule weight in g/mol for vapor and air respectively; P_m is the pressure of the system in atm; σ_{va} is the collision diameter in angstroms (10^{-10} m), which is the mean molecular diameter of vapor and air, $\sigma_{va}=0.5(\sigma_v+\sigma_a)$; for air $\sigma_a=3.711$ Å; for vapor $\sigma_v=2.641$ Å; Ω is a function of the interactions of molecules of vapor and air, which is around 0.45 for vapor-air mixture.

Knudsen diffusion coefficient

$$D_{K} = \frac{d_{P}}{3} \sqrt{\frac{8RT}{\pi M_{v}}} \tag{12.4}$$

where R is the gas constant, 8.314 J/(mol K).

Combined Knudsen and ordinary diffusivity is

$$D_{KO}^{-1} = \left(D_{K}^{-1} + D_{O}^{-1}\right) \tag{12.5}$$

Effective diffusivity in this layer

$$D_{e1} = \frac{\varepsilon_{t1}}{\tau_{s1}} D_{KO,1} \tag{12.6}$$

Rather than by gas diffusion as in hydrophobic layer, it is believed that moisture transfer in the second layer is as follows: The water vapor is first adsorbed to the membrane surface adjacent to the support layer, then it diffuses to the exhaust air side in liquid state, and finally it is desorbed from the surface of the hydrophilic layer adjacent to the exhaust air.

$$D_{e2} = \frac{\rho_{2} K_{p2} D_{w2}}{\rho_{a}} \tag{12.7}$$

D_{w2} is water diffusivity in the second layer (m^2/s), ρ_2 is the density of this layer (kg/m^3), and K_{p2} is the partition coefficient of moisture in layer 2, which reflects the sorption properties of moisture in membranes.

Heat flux through the composite membrane

$$q = \frac{\lambda_{e} (T_{1} - T_{2})}{\delta_{1} + \delta_{2}} \tag{12.8}$$

$$\lambda_{e} = \frac{\delta_{1} + \delta_{2}}{\frac{\delta_{1}}{\lambda_{1}} + \frac{\delta_{2}}{\lambda_{2}}} \tag{12.9}$$

The heat conductivity in porous media is [3]

$$\lambda_1 = \left(\lambda_a^{1/4}\varepsilon_t + \lambda_s^{1/4}\left(1-\varepsilon_t\right)\right)^4 \tag{12.10}$$

where subscripts a and s denote moist air and the solid material part, respectively; ε_t is the porosity of membrane. The estimation method of heat conductivity for layer 2 is the same as for layer 1.

A test rig has been set up to measure the vapor permeability through the fabricated membranes. The whole test set-up is shown in Figure 12.3. Two air streams, one humid and one dry, flow through a membrane exchanger to exchange moisture. For the humid side, ambient air is humidified and is driven to a heating/cooling coil in a hot/cool water bath. After the temperature and humidity reach test conditions, the air is then drawn through the exchanger for moisture exchange. For the dry stream, it is driven directly from ambient to the exchanger. The two inlet temperatures are set to the same values. The composite membrane is sandwiched by two stainless-steel half shells. Two parallel air passages on both sides of the membrane are formed, which is like a counter-flow one-plate plate-and-shell heat exchanger. A schematic of the membrane exchanger is shown in Figure 12.4. The flow channel height is 2 mm, and both the width and length are 10 cm. The effective membrane area is 100 cm^2. By measuring inlet and outlet temperature and humidity values, the moisture permeability is calculated.

After the inlet and outlet humidities are measured, the total mass transfer coefficients are calculated by:

$$k_{tot} = \frac{u_a A_c(\omega_{1i} - \omega_{1o})}{A_t \Delta\omega_{lm}} \tag{12.11}$$

where A_c is the cross-section area of air duct (m^2), A_t is the transfer area of membrane in the cell (m^2), $\Delta\omega_{lg}$ is the logarithmic mean humidity difference between the solution surface and air stream, and is calculated by

$$\Delta\omega_{lg} = \frac{\left(\omega_{1i} - \omega_{2o}\right) - \left(\omega_{1o} - \omega_{2i}\right)}{\ln\frac{(\omega_{1i}-\omega_{2o})}{(\omega_{1o}-\omega_{2i})}} \tag{12.12}$$

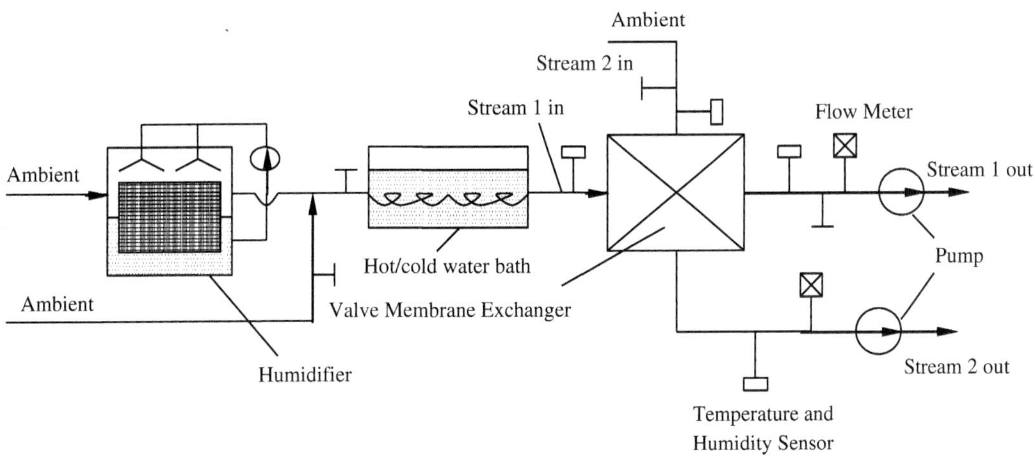

FIGURE 12.3

Experimental set-up of permeability measurement for new membranes.

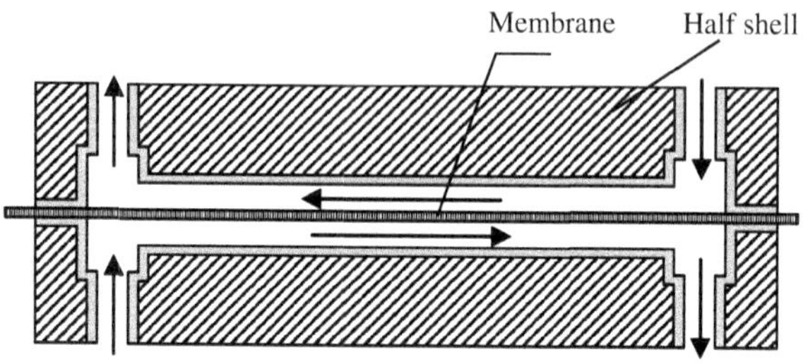

FIGURE 12.4

Structure of the membrane exchanger composed of two symmetric half shells.

where ω represents the temperature humidity ratio (kg/kg), subscripts 1 and 2 represent air stream 1 and air stream 2 respectively, and subscripts i and o represent inlet and outlet respectively.

On the other hand, the total mass transfer coefficient can be calculated by the following equation

$$\frac{1}{k_{\mathrm{tot}}} = \frac{1}{k_1} + \frac{\delta_{\mathrm{tot}}}{D_{\mathrm{e}}} + \frac{1}{k_1} \tag{12.13}$$

where k_1 and k_2 are convective mass transfer coefficients in flow 1 and 2 respectively. They can be calculated with the established Sherwood numbers in the fluids.

The measured vapor permeation rate J (kg/m^2 s) is calculated by the following equation

$$J = \frac{u_{\mathrm{a}}\rho_{\mathrm{a}}A_{\mathrm{c}}(\omega_{1\mathrm{i}} - \omega_{1_{\mathrm{o}}})}{A_{\mathrm{t}}} \tag{12.14}$$

where ρ_{a} is air density (kg/m^3). Under the same air flow rates, the higher the effective diffusivity of the membrane, the higher the permeability.

Permeability is defined by

$$Pe = \frac{J}{\Delta\omega_{\mathrm{lg}}}\delta_{\mathrm{tot}} \tag{12.15}$$

Theoretically, deleting air-side resistance, the permeability of the membrane [kg m^{-1} s^{-1}(kg/kg)$^{-1}$] itself is

$$Pe = \rho_{\mathrm{a}}D_{\mathrm{e}} \tag{12.16}$$

Performance

Various membranes are made with different LiCl additives in the skin layers. The obtained permeation rates of the dual-polar membranes are plotted in Figure 12.5. As can be seen, the higher the air flow rates are, the higher the moisture permeation rates are. The reason behind this is that the higher the air stream velocities are, the less the convective moisture transfer resistance in the two boundaries

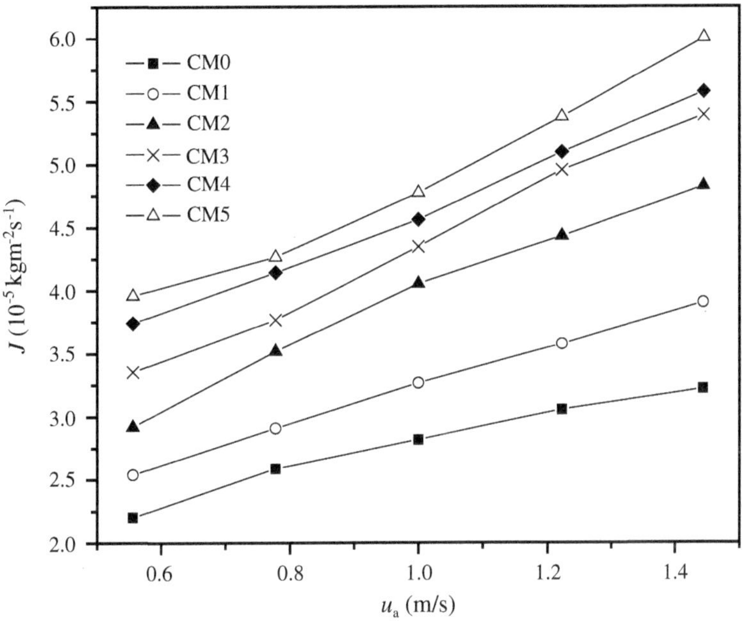

FIGURE 12.5

Effects of LiCl content on vapor permeation rates. CMO, 0%LiCl; CM1, 0.5% LiCl; CM2, 1% LiCl; CM3, 1.5% LiCl; CM4, 2% LiCl; CM5, 2.5% LiCl.

layers adjacent to the membrane is. Furthermore, both the moisture permeation rate and the total mass transfer coefficient increase with higher LiCl content in membranes. This proves that LiCl salts can indeed improve moisture permeation substantially. CM5, which has 2.3% LiCl, has the highest moisture permeation rate. Actually, under an air velocity of 1.0 m/s, its vapor permeation rate is 70% higher than CM0, which has no LiCl additives. Further increase of LiCl content in the casting solution is not suggested since the membrane will be difficult to dry.

The micro-structure of the composite membrane can be observed by scanning electron microscopy (SEM), as shown in Figure 12.6 for the cross section. The structure is very clear.

The sorption isotherms between the skin layer material of the membrane and water vapor are measured in a sorption analyzer. The adsorption and desorption isotherms of the skin layer material with 1% LiCl additive at three temperatures are shown in Figure 12.7. The relative humidity changes from 0 to 1.0. As can be seen, the higher the temperature is, the less the water adsorbed. Membranes can adsorb more water vapor with increased LiCl content. LiCl is a strong hygroscopic salt. Addition of LiCl in the CTA membrane can increase its potential for moisture adsorption. Considering the solution-diffusion mechanisms of vapor permeation in membranes, it is no wonder that vapor permeability was greatly improved. When *RH* approaches 1.0, the adsorbed moisture increases sharply due to multilayer chemical adsorptions. This on the other hand implies that the membranes are nonporous. This is in agreement with the SEM results.

FIGURE 12.6

SEM graph of the composite dual-polar membrane, showing the cross section.

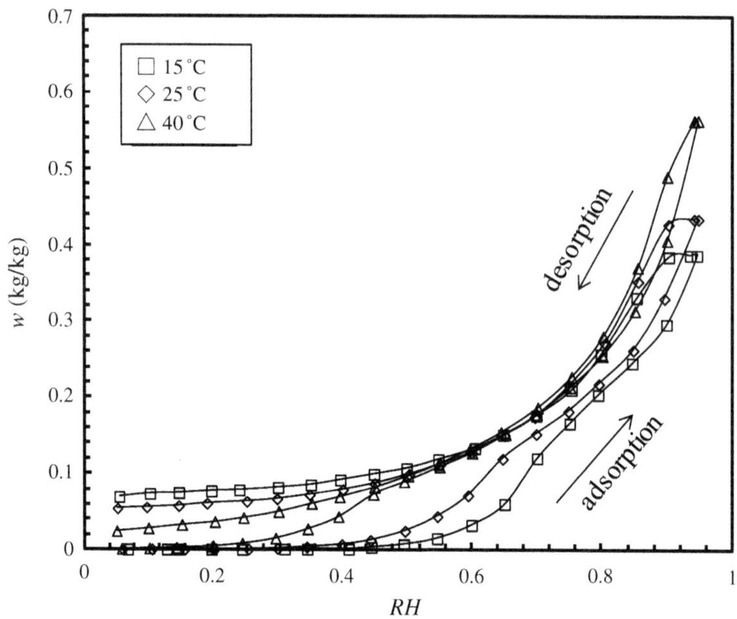

FIGURE 12.7

Adsorption and desorption curves of membrane materials at different temperatures, CTA skin layer with 1% LiCl additive.

It can also be observed from Figure 12.7 that the higher the relative humidity is, the steeper the sorption isotherms become. Considering that the driving force for vapor permeation is the moisture gradients across the membrane, therefore the higher the relative humidity, the greater the vapor permeation rates are.

12.2.2 Composite supported liquid membrane

Moisture diffusivity in solid polymer membranes is usually very low, of the order of 10^{-12}–10^{-13} m^2/s. In contrast, moisture diffusivity in liquid is several orders higher than in solid membranes. A new membrane has been proposed to enhance moisture transfer: the composite supported liquid membrane (SLM).

The concept is schematically depicted in Figure 12.8. It employs a hygroscopic liquid solution (such as LiCl, $CaCl_2$) immobilized in a porous support membrane to facilitate the transport of moisture. To protect the SLM, two hydrophobic PVDF layers are formed on the surfaces of the SLM. The permeability through the membrane can be improved [4–7].

Method of membrane fabrication

Under room temperature, well-stirred $CaCl_2$ ($CaCl_2$ is much cheaper than LiCl) solution with 35% mass fraction is first prepared in a closed glass container. Vacuum degassing is applied for 2 h for the three membranes, after which, a CA (cellulose acetate) membrane is dipped into the $CaCl_2$ solution. After 24 h, the CA membrane is moved from the solution and placed onto a clean glass plate which is cleaned with alcohol. Surplus $CaCl_2$ solution on the surfaces of the CA membrane is blotted off with paper tissue. To be sure that no $CaCl_2$ liquid is removed from the membrane pores, the cleaning procedure is very gentle. At this stage, PVC glue is brushed on one surface of the two PVDF membranes, and at the same time on both surfaces of the CA membrane. After a few seconds, the two PVDF membranes are glued to the CA membrane and are pressed together gently for a few seconds. The prepared composite membrane is placed in a constant-temperature constant-humidity chamber for another 24 h, before the experiment is performed.

Mass diffusivity and heat conductivity in the membrane

Heat and moisture transfer through the membrane can be depicted by a model as shown in Figure 12.9.

Moisture flux through the composite membrane is expressed by

$$J = \rho_a D_e \frac{\omega_1 - \omega_4}{\delta_1 + \delta_2 + \delta_3} \tag{12.17}$$

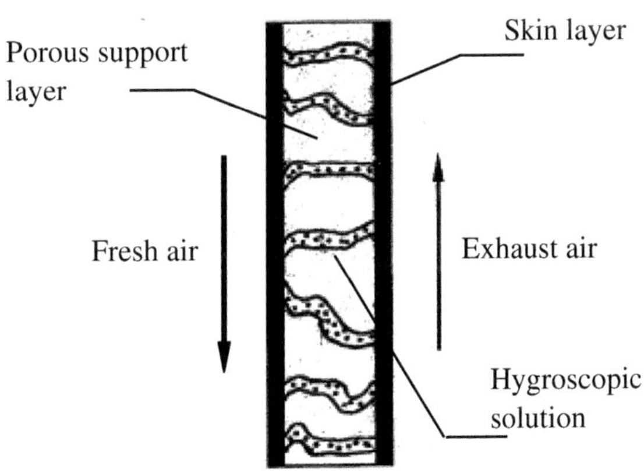

FIGURE 12.8

Concept of the composite supported liquid membrane (CSLM).

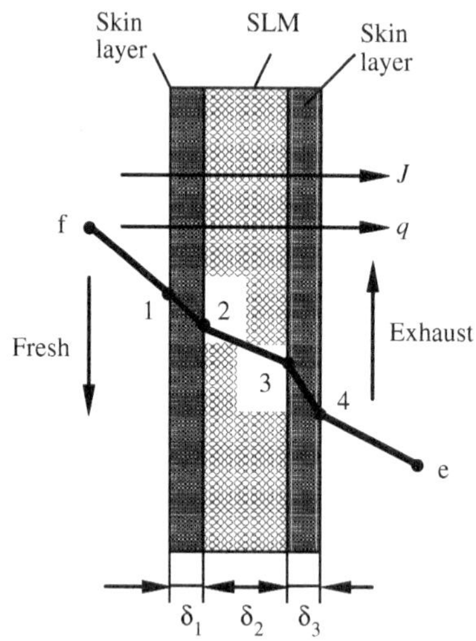

FIGURE 12.9

Heat and mass transfer model in the composite supported liquid membrane.

where D_e is the effective moisture diffusivity in the composite membrane (m^2/s), calculated by

$$D_e = \frac{\delta_1 + \delta_2 + \delta_3}{\frac{\delta_1}{D_{e1}} + \frac{\delta_2}{D_{e2}} + \frac{\delta_3}{D_{e3}}} \tag{12.18}$$

where D_{e1}, D_{e2}, D_{e3} are the effective diffusivities in the first, second, and third layer, respectively.

The estimation of moisture diffusivity in the first and the third layers is the same as for the first layer in the above section for the composite solid membrane. They can be estimated by Eqs. (12.13) to (12.16). The equivalent diffusion coefficient of vapor in liquid membrane is

$$D_{e2} = \frac{\rho_s K_{p,lq} \varepsilon_{t2}}{\rho_a \tau_{s2}} D_{w,lq} \tag{12.19}$$

where subscript "lq" means liquid solution, D_{ws} is the water diffusivity in liquid solution (m^2/s), τ_{s2} is tortuosity of layer 2, and $K_{p,lq}$ is the partition coefficient of the liquid solution. Water vapor partial pressure, temperature, and hygroscopic solution concentrations are strongly coupled and governed by thermodynamic equations.

The heat conductivity in the first and third layer can be analyzed based on Eq. (12.10). The heat conductivity in the second layer, where a liquid solution is stationed in the porous media, can be analyzed by

$$\lambda_2 = \lambda_{lq}\varepsilon_{t2} + \lambda_s \left(1 - \varepsilon_{t2}\right) \tag{12.20}$$

where subscript "s" denotes solid phase.

Performance

Membranes are made with the parameters of each layer as listed in Table 12.1. The SEM image is shown in Figure 12.10. The cross section can be clearly observed. The sandwich structure indicates a three-layer composition. Permeability is measured in the test rig built above. It can be observed that the skin layer thickness has a great influence on the overall permeation rates of the whole membrane. The effects of skin layer thickness on permeation rates are shown in Figure 12.11. The protective layers have a great influence on membrane performance. The permeation rate decreases with an increase in protective layer thickness, due to the increased moisture resistance. As for the permeability, it first decreases, and then increases with increasing thickness. The reason is that the permeability is co-determined by two contradictory factors: membrane resistance and membrane thickness. The final permeability is the result of the balance between these two factors. When the thickness is less than 20 μm, increased resistance with thickness is more influential, while when the protective thickness is greater than 20 μm, increased thickness is more influential. When the two protective thicknesses are reduced from 100 μm to 5 μm, the permeation rate can be improved by 20%.

Besides the skin layer thickness, the porosity of the middle support layer also has a great effect on performance. It is used to station the hygroscopic solution. Not only the barrier but also the active layer facilitates moisture transfer. Figure 12.12 shows the effects of middle layer porosity on performance. As can be seen, the permeation rate and the permeability increase with increasing porosity, due to the resistance decreasing. The porosity has more effect on performance than middle layer thickness does. An increase of porosity from 0.2 to 0.8 produces a 90% permeation rate improvement.

12.2.3 Membranes with finger-like macrovoids

For liquid-air processes involving membrane contactors, hydrophobic porous membranes are required. Such membranes should be barriers to liquid, but be selectively permeable to vapors. Porous and microporous membranes are the designed structures. The phase-inversion method has been the widely used technology for membrane preparation since it was successfully used by Loeb and Sourirajan to develop cellulose acetate membranes for seawater desalination in the 1960s [1]. According to this method, the formation of the membrane structure is controlled by both the thermodynamics of the casting solution and the kinetics of the transport process. Usually, depending on the rate of phase separation, two different structures, namely a symmetric sponge-like (from delayed-phase demixing) or an asymmetric finger-like structure (from instantaneous-phase demixing) can be expected. The finally formed membrane

Table 12.1 Parameters for the composite supported liquid membrane.

Symbol	Unit	Value	Symbol	Unit	Value
δ_1	μm	45	ε_{t1}		0.51
δ_2	μm	52	ε_{t2}		0.62
δ_3	μm	45	ε_{t3}		0.51
d_{p1}	μm	0.15	d_{p3}	μm	0.15
d_{p2}	μm	0.22	τ_{s1}		2.0
τ_{s2}		2.0	τ_{s3}		2.0

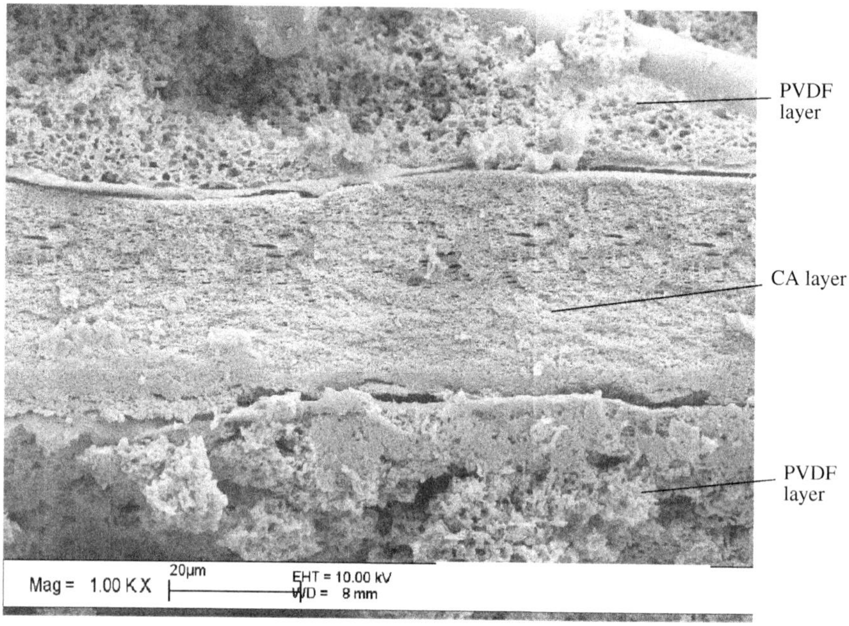

FIGURE 12.10

SEM graph of the composite supported liquid membrane, showing the cross section.

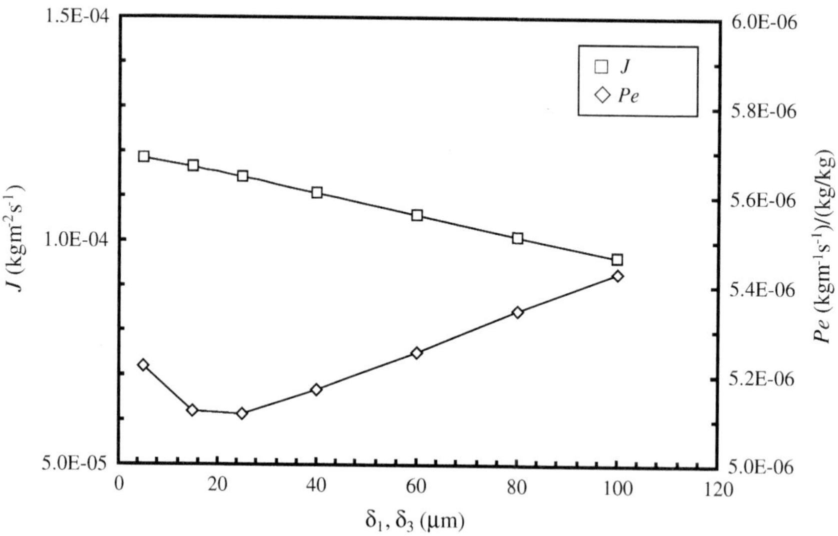

FIGURE 12.11

Effects of skin layer thickness on permeation rates and permeability of composite supported liquid membranes.

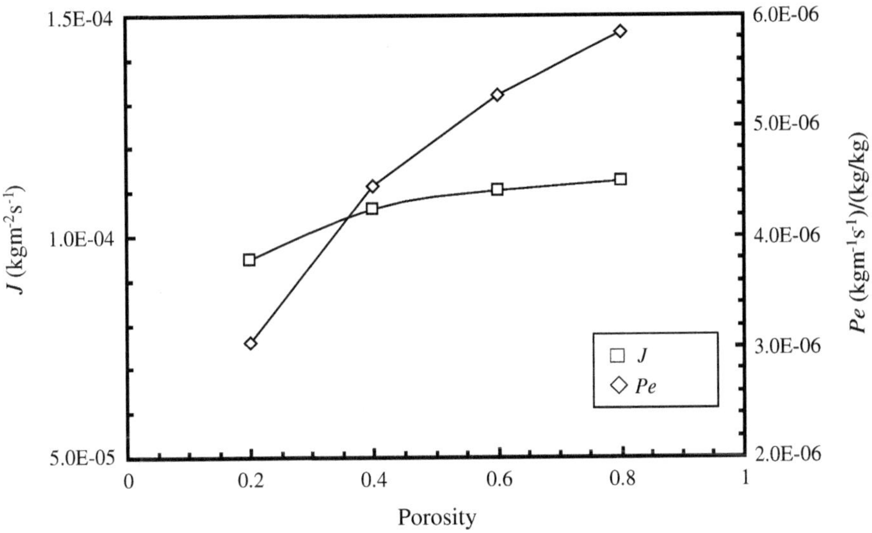

FIGURE 12.12

Effects of middle layer porosity on permeation rates and permeability of composite supported liquid membranes.

is interesting in structure. Figure 12.13 shows a typical asymmetric finger-like structure. It is composed of three parts from bottom to top: a sponge-like porous support, a porous media with finger-like and nearly parallel macrovoids, and finally a very thin skin layer with rather small pores. Due to the existence of the large finger-like macrovoids, it is qualitatively believed that the asymmetric membranes have less resistance than symmetric membranes, and thus are beneficial for permeation performance [8].

Method of membrane fabrication

Industrial-grade PVDF is used as the membrane raw material. The membrane is fabricated through the following steps: (1) 15–20 g PVDF powder is weighed and placed into a vessel with 100 g DMF (*N,N*-dimethylformamide) at about 90 °C. The solution is heated and stirred until it is completely dissolved. This takes about 2 h. (2) Three to 5 g polyethylene glycol (PEG) and polyvinyl pyrrolidone (PVP) is added to the solution and kept stirring for another 2 h. (3) The solution is degassed and cast onto a clean glass plate with appropriate thickness control by a knife edge. (4) The plate is kept in atmosphere for several seconds. (5) The glass with cast solution is dipped into a 25 °C water bath. Immediately, the solution solidifies and peels off the glass plate to form a membrane sheet. The membrane is soaked in a water bath for 24 h for solvent and non-solvent exchange. (6) The membrane is dried in a vacuum drying box for 10 h. (7) The membrane is ready for structural characterization.

The membrane structure is schematically depicted in Figure 12.14. It is in fact composed of three layers: a sponge-like porous layer, a sponge layer with finger-like macrovoids, and a skin layer. Their dimensions are plotted in the figure. Though the macrovoids may have differences, more or less, in length and dimension, in the model they are approximated as identical cylinders, to ease the mathematical analysis. To account for their zigzag nature, a tortuosity is used to compensate for this feature.

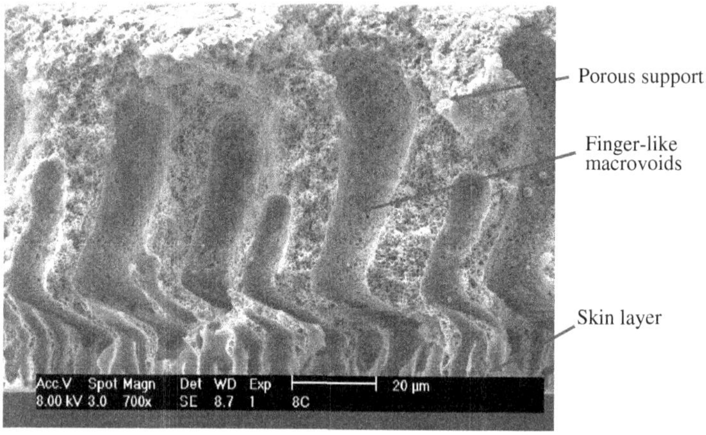

FIGURE 12.13

SEM graph of the cross section of an asymmetric porous membrane with finger-like macrovoids.

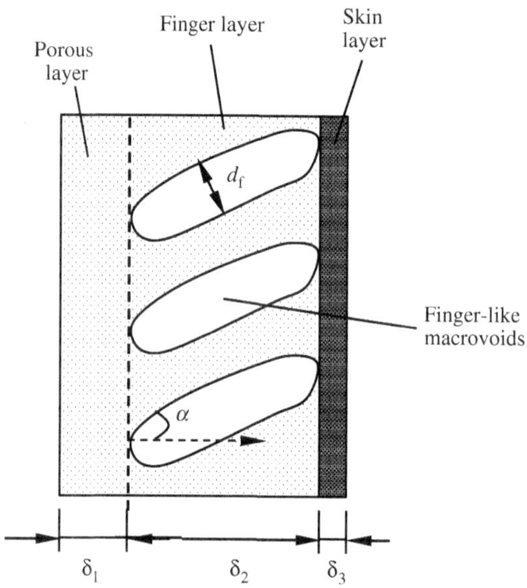

FIGURE 12.14

Simplified structure of a finger-like macrovoid asymmetric membrane.

Mass diffusivity and heat conductivity in the membrane

Generally, the resistance model is convenient in calculating heat mass transfer through a multi-component medium. Gas diffusion through the asymmetric membrane can be depicted by the

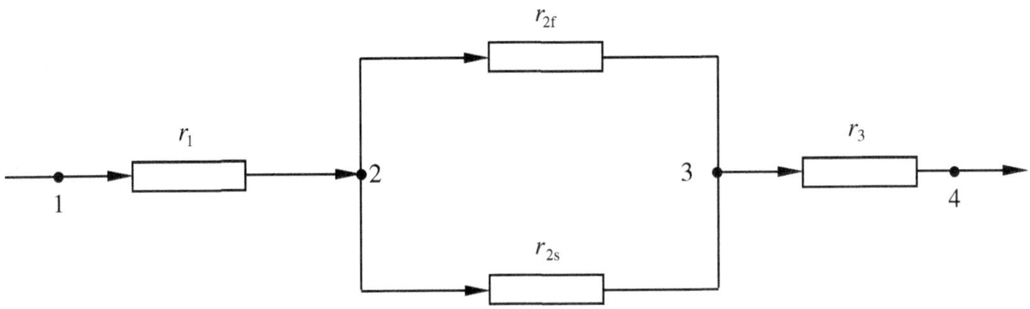

FIGURE 12.15

Resistance model in the asymmetric membrane.

resistance model as shown in Figure 12.15. The mechanism for gas diffusion is combined Knudsen and ordinary diffusion. The diffusivity can be estimated by Eqs. (12.3) to (12.6).

The gas diffusion resistance in the first layer is (s/m)

$$r_1 = \frac{\delta_1}{D_{e1}} \tag{12.21}$$

The resistance through the second layer consists of two parallel parts: the finger-like macrovoids and the sponge-like micropores. The resistance through the finger-like macrovoids

$$r_{2f} = \frac{\tau_{s,2f} l_f}{\varepsilon_{t,2f} D_{2f}} \tag{12.22}$$

where $\tau_{s,2f}$ is the tortuosity of the finger-like macrovoids in the second layer. For straight capillaries, the tortuosity value is 1.0; for rather zigzag channels, tortuosity is always given a value of 2.8–3.0. In the equation, l_f is the length of the macrovoids. D_{2f} is gas diffusivity through the macrovoids by substituting macrovoid mean diameter as the character pore size.

Besides gas diffusion through the macrovoids, there is still some diffusion through the residual sponge-like micropores in this layer. The resistance through this part is

$$r_{2s} = \frac{\tau_{s1} \delta_2}{\varepsilon_{t1}(1 - \varepsilon_{t,2f}) D_1} \tag{12.23}$$

where in the equation, the tortuosity, porosity, and diffusivity are considered as the same as for the first layer, since they are similar in structure from SEM observations. The finger-like macrovoid wall has the same structure as the porous substrate. The calculation of diffusivity in the third layer is the same as in the first layer.

According to Figure 12.15, the total resistance for membrane

$$r_{tot} = r_1 + \left(r_{2f}^{-1} + r_{2s}^{-1}\right)^{-1} + r_3 \tag{12.24}$$

Effective mass diffusivity through the whole membrane

$$D_e = \frac{\delta_1 + \delta_2 + \delta_3}{r_{tot}} \tag{12.25}$$

Heat conductivity in the first layer and the third layer can be calculated by Eq. (12.10). In the second layer, considering media having a porosity of $\varepsilon_{t,2f}$, and a solid-part heat conductivity of λ_1, a form similar to Eq. (12.10) can be used

$$\lambda_2 = \lambda_a \varepsilon_{t,2f} + \lambda_1 \left(1 - \varepsilon_{t,2f}\right) \tag{12.26}$$

Overall effective thermal conductivity

$$\lambda_e = \frac{\delta_1 + \delta_2 + \delta_3}{\delta_1/\lambda_1 + \delta_2/\lambda_2 + \delta_3/\lambda_3} \tag{12.27}$$

It should be noted that membrane thermal resistance is rather lower than boundary resistance on the two membrane surfaces, due to the small thickness of the membranes.

Resistance in the membrane is analyzed. It is found that the second layer has the highest gas diffusivity through a single pore or finger. This is because the larger the pore (finger) diameter is, the higher the gas diffusivity. Nevertheless, the diffusivity through a finger (D_{2f}) is of the same order as the diffusivity through a small pore (D_3). Due to the existence of straight finger voids, the total porosity in the second layer is greatly improved, and consequently the effective gas diffusivity through this layer (D_{2e}) is one order higher than that through the sponge support layer (D_{1e}). Overall, due to the increased porosity provided by straight finger voids, effective diffusivity can be greatly improved.

Performance

A membrane is made with the parameters listed in Table 12.2. The membrane is tested in a small-scale single-plate heat and mass exchanger, as described in Figure 12.3 [8]. The calculated heat mass transfer parameters for the membrane are also listed in Table 12.2. The second layer has the highest gas diffusivity through a single pore or finger. This is because the larger the pore (finger) diameter is, the higher the gas diffusivity. Nevertheless, the diffusivity through a finger (D_{2f}) is of the same order as the diffusivity through a small pore (D_3). The effects of pore diameter on diffusivity are not obvious. However, due to the existence of straight finger voids, the total porosity in the second layer is greatly improved, and consequently the effective gas diffusivity through this layer (D_{2e}) is one order higher than that through the sponge support layer (D_{1e}). Therefore, due to the increased porosity provided by straight finger voids, effective diffusivity can be greatly improved.

The calculated mass diffusion resistance is in agreement with the above analysis. The first sponge layer has the highest resistance. Though the skin layer has the lowest number of pores and the smallest pore diameters, due to the ultra-thinness it has the least resistance. This indicates that for asymmetric membranes, the sponge-like layer has the largest mass transfer resistance.

Besides higher gas diffusivity, the parallel finger voids provide higher mechanical support for the membrane due to the honeycomb networking structure. Therefore they are beneficial in industrial applications.

The higher the porosity is, the lower the thermal conductivity is. Table 12.2 shows that the calculated thermal conductivity is the lowest for the finger layer, which has the largest porosity. However,

Table 12.2 Characterization of parameters of the asymmetric membrane developed.

Symbol	Unit	Value	Symbol	Unit	Value
δ_1	μm	28	ε_{t1}		0.65
δ_2	μm	45	ε_{t3}		0.35
δ_3	μm	2.1	τ_{s1}		3.0
d_{p1}	μm	0.72	$\tau_{s,2f}$		1.0
d_{p3}	μm	0.17	τ_{s3}		3.0
n_f	1/μm	0.07	α	deg	75
d_f	μm	8.1	λ_s	$Wm^{-1}K^{-1}$	0.36
D_1	$10^{-5}m^2/s$	2.35	D_{e1}	$10^{-5}m^2/s$	0.51
D_{2f}	$10^{-5}m^2/s$	2.77	D_{e2}	$10^{-5}m^2/s$	1.76
D_3	$10^{-5}m^2/s$	1.53	D_{e3}	$10^{-5}m^2/s$	0.18
ε_{t1}		0.65	r_1	s/m	5.49
ε_{t2}		0.87	r_2	s/m	2.55
ε_{t3}		0.35	r_3	s/m	1.17
λ_1	$Wm^{-1}K^{-1}$	0.1431	λ_2	$Wm^{-1}K^{-1}$	0.0702
λ_3	$Wm^{-1}K^{-1}$	0.2432			

due to the small thickness of membrane, the final heat transfer resistance of the whole membrane is still negligible, compared to the convective resistance in the air boundary layers on the two air streams. Therefore, the bad effects on heat transfer are not obvious.

The mean heat flux and moisture permeation rate through the membrane are shown in Figure 12.16 under various air flow velocities. As can be seen, the higher the bulk air velocity, the higher are the heat flux and mass permeation rates. The reason behind this is that the higher the air velocity, the lower the boundary layer resistance is on the air sides. Heat flux can be as high as $200\,W/m^2$, and gas permeation can be as high as $1.8 \times 10^{-4}\,kg\,m^{-2}\,s^{-1}$, indicating high heat mass exchanger potentials with such porous membranes. They are the very crux for membrane contactors, either parallel-plate type or hollow fiber type.

12.2.4 One-step-fabricated asymmetric membrane

Various membranes have been sought in the last 20 years for total heat recovery; however, none of them is perfect. The problems are: (1) The homogeneous dense single-layer membranes have very low moisture permeation rates; (2) the composite membranes that have high vapor permeability, whether they are composite supported liquid membranes or composite solid membranes, require complex manufacturing processes. The different layers in the membrane need to be fabricated separately. To fabricate a membrane, it usually requires at least two to three steps: (a) casting the support layer; (b) drying; and then (c) making the skin layer. The processes are complex and costly. The finally prepared membranes are rather expensive. It's unrealistic to use these costly membranes in future large-scale engineering applications. (3) These processes rely on toxic solvents such as dimethylformamide (DMF) and dimethylacetamide (DMAC). Unfortunately these substances are prohibited for use because of environmental protection reasons.

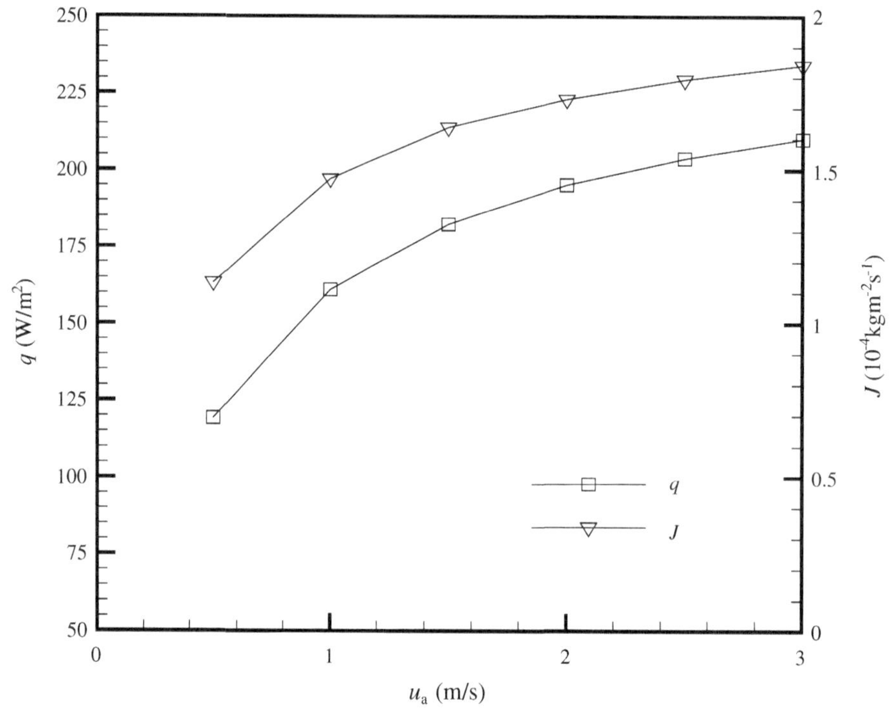

FIGURE 12.16

Heat flux and permeation rate of the asymmetric membrane with finger-like macrovoids.

There are several basic requirements for membranes used in total heat exchangers. They are: (1) The moisture permeability should be high. The reason is that the vapor partial pressure difference between the fresh air and the exhaust air is very small, of the order of 2000 Pa, so high vapor permeability is required to achieve an acceptable effectiveness. (2) The membrane should be impermeable to CO_2, to prevent the fresh air from being polluted by the exhaust air. A dense hydrophilic layer should be formed on the surface of the membranes. (3) The material should be cheap. This is necessary for large-scale engineering applications. (4) The manufacturing process should be simple and environmentally friendly.

Considering these requirements, this section proposes a new approach: to make cheap membranes with a one-step manufacturing process [9]. The materials constituting of the casting solution are CA (cellulose acetate), acetic acid and deionized water. CA is the chief membrane material. Acetic acid is the solvent and deionized water is the additive. Tap water is used as the coagulation bath. No other additives are required. All these materials are commonly used materials and they are cheap and easily available. They are environmentally friendly. During the membrane manufacturing process, the dense skin layer and the porous support layer are formed simultaneously in the water bath. In other words, the whole process is accomplished in one step. No toxic waste water is discharged. The membrane forming process is simple, economical, and green. It provides a solution for supplying high-performance yet cheap membranes for large-scale engineering applications of total heat recovery.

Method of membrane fabrication

The casting solution is prepared with CA (acetyl content 54.6%), acetic acid (reagent grade) and deionized water. Acetic acid is the solvent and deionized water is the additive. Water in the casting solution is double distilled and de-ionized before using. Tap water is used as the coagulant bath.

Membranes are prepared in a one-step approach [9]. This is a direct wet phase inversion method. Various solution compositions have been tested. By keeping the CA concentration unchanged, the acetic acid to deionized water (AW) weight ratios (χ) are varied. Five values of χ are selected: 100:0, 90:10, 80:20, 70:30, and 60:40. The compositions in the five casting solutions are listed in Table 12.3. Each solution is mixed and stirred at 60 °C for 3 h until it is completely dissolved. The homogeneous solution is cooled down and placed for another 12 h in order to be de-bubbled. Each solution is coated at room temperature on a glass plate with a knife. The thickness of the membrane is kept uniform at 150 μm. The glass plate is cleaned with alcohol before solution coating. Then the coated membrane is immersed into a coagulation bath containing 10 L tap water for at least 5 min. Instantly, the membrane is formed on the glass and is then peeled off. Subsequently, the sample is transferred to another water bath and is kept in the bath for 24 h to remove the excess reagent. Then the membranes are pressed on a flat plate and placed into a vacuum drying oven. It takes 1 h at 60 °C for drying. Finally, the membranes are placed in an air-conditioned room for 3 h.

Heat conductivity and moisture diffusivity

The structures of the one-step-fabricated membranes are similar to those of the composite hydrophobic-hydrophilic membranes, so the resistance models for heat conduction and moisture diffusion for the hydrophobic-hydrophilic dual-polar membranes are also applicable to the one-step asymmetric membranes. The detailed equations for estimation of heat conductivity and moisture diffusivity in Section 12.2 can be adopted.

Performance

SEM images of the asymmetric membranes are shown in Figure 12.17. As can be seen, the structures are rather asymmetric. They are in some degree similar to the composite hydrophobic-hydrophilic dual-polar membranes, but the manufacturing processes are largely simplified.

The fabricated membranes are tested in the small total heat exchanger test rig built before [9]. The permeation rates are measured. The results with the membranes in Table 12.3 are shown in

Table 12.3 Compositions of the casting solution for one-step made asymmetric membrane.

Synthesized Membrane	Compositions of Casting Solution (wt.%)			AW (χ)
	CA	Acetic acid	H_2O	
M1	10	90	0	100:0
M2	10	80	9	90:10
M3	10	72	18	80:20
M4	10	63	27	70:30
M5	10	54	36	60:40

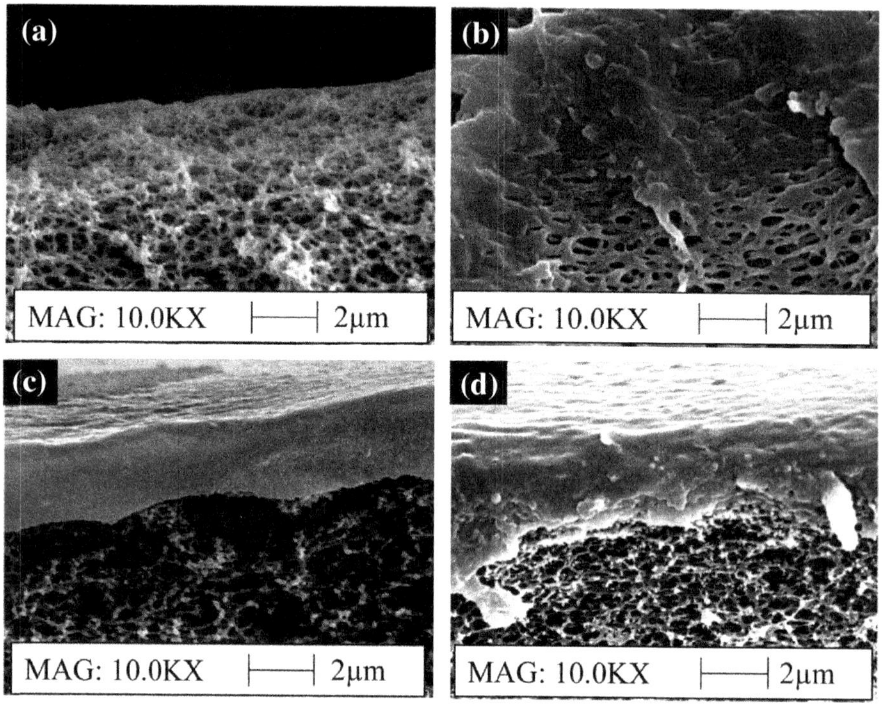

FIGURE 12.17

SEM graphs of the cross sections of the one-step-fabricated asymmetric membranes with different acetic acid to water ratios: (a) M2, (b) M3, (c) M4 and (d) M5.

Figure 12.18. For each membrane, the air flow rates are changed from 1 to 5 m/s. As can be seen, the permeation rates increase with air flow velocities due to decreased air-side resistance. Under the same air flow rates, the permeation rates decrease with increased additive (deionized water) content in the casting solution, from AW 100:0 to AW 60:40. This means the moisture transfer resistance rises with more additives (deionized water) in the solution, which is in agreement with the membrane structures.

12.2.5 Selectivity of membranes

It should be noted that fresh air ventilation is crucial for indoor environmental control. Total heat exchangers are the key equipment for energy conservation in ventilation. Membranes have been used for total heat exchangers for a dozen years. Much effort has been spent on the water vapor permeability of various membranes; however, relatively little attention has been paid to the selectivity of moisture over VOCs through such membranes. A membrane should be a barrier to other unwanted gases such as VOCs.

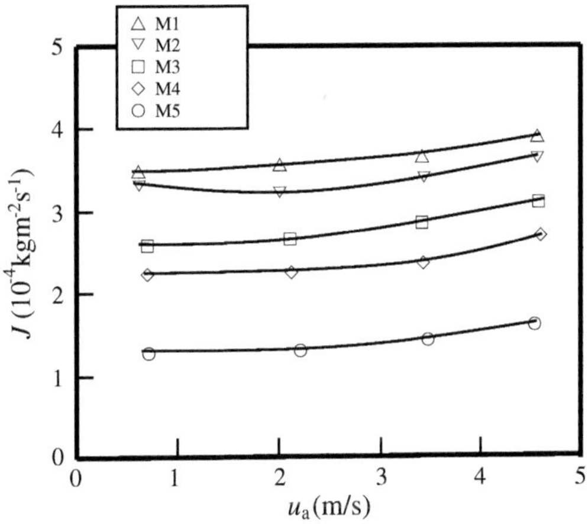

FIGURE 12.18

Permeation rates of the one-step-made asymmetric membranes.

The porous support layers are only used to support the active layers. The selectivity is accomplished by the active layers. So their selectivity has been investigated. The most commonly used membrane materials for the active layers, either hydrophilic ones (PVP, PVA, PAM, Na(Alg), CS, CA, EC) or hydrophobic ones (PP and PDMS), are tested for their permeability to moisture and five VOCs (acetic acid, formaldehyde, acetaldehyde, toluene and ethane) [11]. PVA-1 is PVA with LiCl additive. PVA-2 is PVA without LiCl. Various membranes are measured for the permeation of moisture and VOCs. The test rig is shown in Figure 12.19. According to the procedure, a composite membrane is sandwiched by two shells to form a mass exchanger. In the exchanger, two parallel air passages are formed. They are separated by the membrane. It is like a counter-flow one-plate plate-and-shell heat exchanger. Two streams, one humid and having no VOC (representing fresh air), the other dry and having VOC (representing exhaust air), exchange moisture and VOC through the membranes. The test rig is placed in a chamber at a constant temperature of 25 °C. The flow rates, temperature, humidity and VOC concentration values to and from the exchanger are recorded using flow meters, temperature and humidity sensors and GC (gas chromatograph), respectively. The flow channel length and width are 11 cm and 2.5 cm, respectively. The channel pitch is 2 mm. The effective exchange area of membrane is 27.5 cm^2. The uncertainties are: temperature ±0.1 °C; relative humidity ±1%; VOC concentration ±1%; volumetric flow rate ±1%. The final uncertainty is ±6.5% for moisture and VOC permeability. The permeability to moisture and VOCs is measured. Then the selectivity can be calculated by

$$\beta = \frac{Pe'_{\mathrm{v}}}{Pe'_{\mathrm{VOC}}} \tag{12.28}$$

$$Pe' = \frac{J_{\mathrm{i}}\delta}{\Delta p_{\mathrm{i}}} \tag{12.29}$$

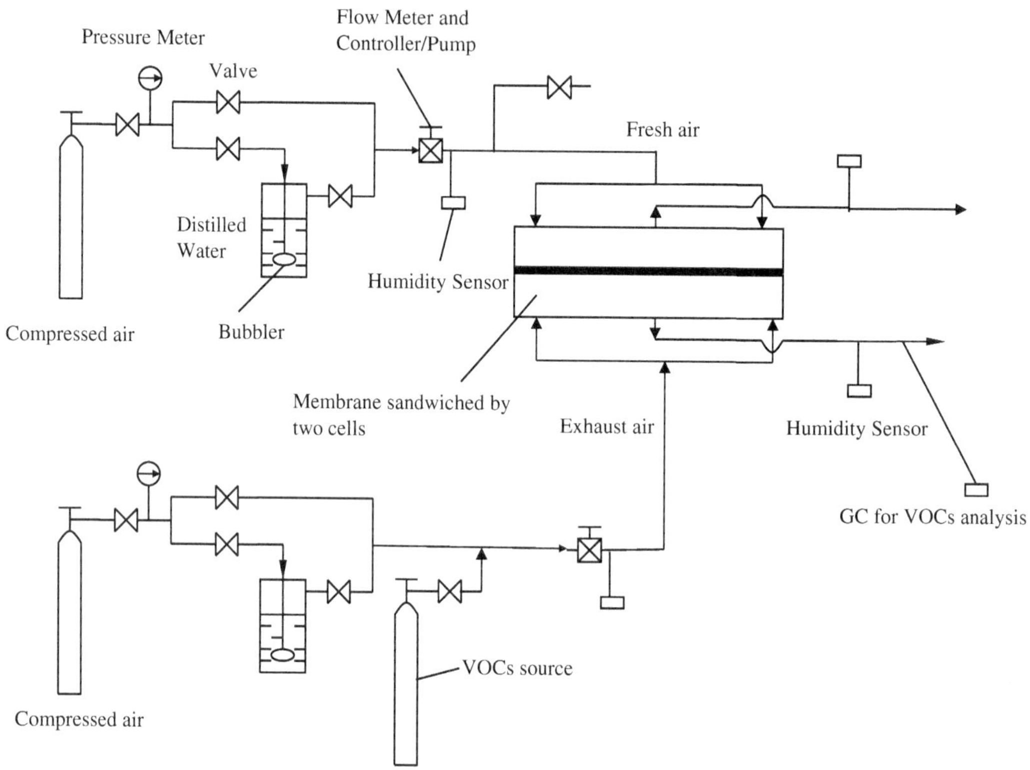

FIGURE 12.19

The test rig for selectivity analysis.

where permeability Pe' (g cm^{-1} cmHg^{-1}) is calculated by the transmembrane partial pressure difference of gas i. It should be noted that to compare the permeability of different gases, the permeability should be based on the same pressure differences.

With a solution-diffusion model, the selectivity of moisture to VOCs in these membranes is calculated. It provides some reference for future material selections [11]. Figure 12.20 shows the selectivity of vapor to formaldehyde for the ten active-layer membranes. As can be seen, the selectivity of vapor/formaldehyde is 200–300 for several membranes. The best selectivity is for the PVA-1 material. This indicates that VOCs cannot permeate the membranes while moisture can. Total heat exchangers are ideal to control indoor air quality, in addition to energy recovery. There are no cross-over problems with these materials.

12.3 Highly conductive polymers

Heat exchangers have been widely used in various industries. In many cases the operation conditions are rather harsh, and the corrosion of heat exchangers becomes a major problem. To solve this problem, plastics have been used as the materials to make heat exchangers. However, the thermal

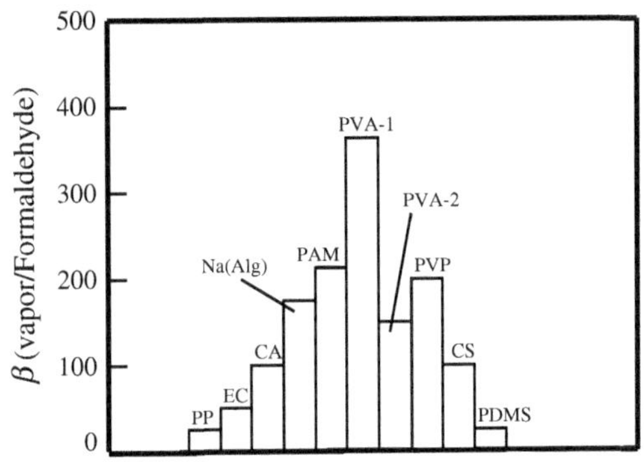

FIGURE 12.20

Selectivity of water vapor over formaldehyde for the ten membranes.

conductivity of these plastics is rather low, of the order of 0.1 W $m^{-1}K^{-1}$, which makes plastic heat exchangers poor in performance. In comparison, the thermal conductivity of stainless steel is around 20 W $m^{-1}K^{-1}$.

Adding fillers made of high conductivity materials to plastics to form new composite materials is a solution. In this way, the thermal conductivity of the mixed composite material can be improved, while its virtue of being corrosion-proof can be kept. The fillers investigated include graphite, carbon black, carbon fibers, ceramic or metal particles, and carbon nano-tubes. When the filler content is satisfactory and the fillers are properly distributed, the composite polymer materials could have a thermal conductivity as high as 10 W $m^{-1}K^{-1}$ [12].

It was discovered that filler shapes have a great influence on the final thermal conductivity of the filled composite materials. Recently filler shapes have been optimized to produce maximum thermal conductivity with specified filler content [12].

12.3.1 Direct problem

Heat conduction in the composite material is modeled. The material is assumed to be two-dimensional. A unit cell in the material is selected, which is composed of the base material and a filler in the center, as shown in Figure 12.21. Further, the fillers are assumed to be uniformly distributed in the base materials with no aggregates. To study fillers of various shapes, the filler in the unit cell itself is composed of three rectangular sub-fillers, F_1, F_2, and F_3. Their width and heights can be varied to form different overall shapes. The dimensionless axial positions of the unit cell are defined by

$$\xi = x'/L_0 \tag{12.30}$$

$$\eta = y'/H_0 \tag{12.31}$$

where L_0 and H_0 are the length and height of a unit cell. Clearly, the dimensions of the unit cell are $\xi_0 = \eta_0 = 1$. The volume content of the filler can be calculated by

$$V_F = x_1x_2 + x_3x_4 + x_5x_6 \tag{12.32}$$

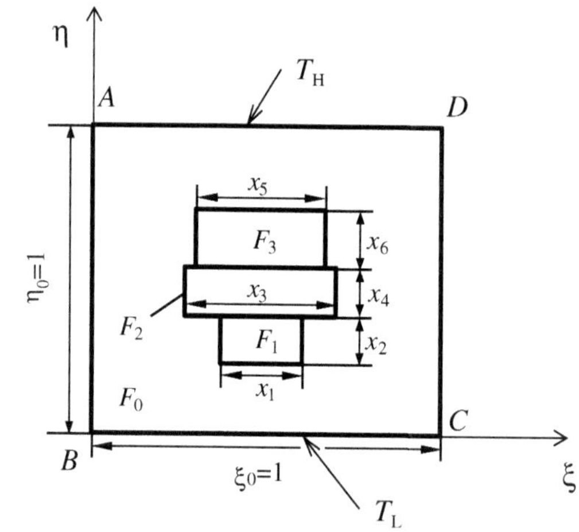

FIGURE 12.21

Schematic of a unit cell in the composite material. It is composed of a base material F_0, and a filler that has three rectangular sub-fillers F_1, F_2, and F_3.

where x_j ($j = 1{,}6$) is the dimensionless coordinates for the sub-fillers defined in the figure. As mentioned, the cell is assumed to be two-dimensional. Heat conduction in the base materials is satisfied by

$$\frac{\lambda_b}{(\rho c_p)_b}\left(\frac{\partial^2 T_b}{L_0^2 \partial \xi^2} + \frac{\partial^2 T_b}{H_0^2 \partial \eta^2}\right) = 0 \tag{12.33}$$

where T is temperature and subscript "b" means base plastic materials. The physical properties, heat conductivity λ_b, density ρ and specific heat c_p here are written out, to reflect the local properties of the grids.

Heat conduction in the filler

$$\frac{\lambda_F}{(\rho c_p)_F}\left(\frac{\partial^2 T_F}{L_0^2 \partial \xi^2} + \frac{\partial^2 T_F}{H_0^2 \partial \eta^2}\right) = 0 \tag{12.34}$$

Uniform and constant temperatures are assumed at the cell upper (upstream) and lower (downstream) surfaces. At the top surface of the cell, AD, the boundary condition is

$$T_b\big|_{\eta=1} = T_H \tag{12.35}$$

At the bottom surface of the cell, BC, the boundary condition is,

$$T_b\big|_{\eta=0} = T_L \tag{12.36}$$

The other two boundary conditions, AB, DC

$$\left.\frac{\partial T}{\partial \xi}\right|_{\xi=0,1} = 0 \tag{12.37}$$

They are assumed to be adiabatic.

At the boundaries between the base materials and the fillers, the temperatures in the base materials are equal to those in the fillers. They are conjugate boundary conditions. They are expressed as

$$T_{\text{b, interface}} = T_{\text{F, interface}} \tag{12.38}$$

where subscript "interface" means contact surfaces between the base materials and the fillers.

When the temperature fields in the cell are calculated, the effective thermal conductivity of the composite material in η direction for the cell is estimated by

$$\lambda_{\text{tot}} = \frac{q_m H_0}{T_H - T_L} \tag{12.39}$$

where q_m is the area mean heat flux at a cross section normal to the η direction (parallel to AD).

12.3.2 Reverse problem

For the inverse problem, the objective is to solve the filler shape parameters F_1, F_2, and F_3, at given filler contents to produce the highest conductivity. For this problem, the filler shape parameters are unknown, but the objective is known: to optimize λ_{tot}. They can be summarized mathematically in the following as an optimization problem.

$$\min f(\boldsymbol{X}) = -\lambda_{\text{tot}} \tag{12.40}$$

$$x_1x_2 + x_3x_4 + x_5x_6 = V_F \tag{12.41}$$

$$x_2 + x_4 + x_6 \le \eta_{\max} \tag{12.42}$$

$$x_j \ge 0, j = 1, \ldots, N \tag{12.43}$$

where f is the objective function, $\boldsymbol{X}$ is the vector of filler dimensions with elements x_j. For the three sub-fillers, here $N = 6$. In Eq. (12.42), $\eta_{\max}$ is the permitted maximum size of a filler. It can be estimated by

$$\eta_{\max} = \sqrt{V_{\text{F,max}}} \tag{12.44}$$

where the maximum filler content $V_{\text{F,max}}$ is assumed to be 0.64. Too high a filler content would make the composite material too brittle. Sometimes the maximum filler size is determined by the filler manufacturing processes.

The inverse problem is solved by the successive quadratic programming (SQP) optimization technique [12].

12.3.3 Optimized filler shapes

Using the methodology proposed above, the filler shapes are optimized for $V_F = 0.2$ and $\lambda_F/\lambda_b = 1363.0$. The base material is PP plastics (conductivity $0.11\,\mathrm{W\,m_{-1}\,K_{-1}}$), and the filler material is copper (conductivity $150\,\mathrm{W\,m^{-1}\,K^{-1}}$. The results are: $x_1 = 0.8$, $x_2 = 0.054$, $x_3 = 0.162$, $x_4 = 0.69$, $x_5 = 0.8$, $x_6 = 0.055$. The maximum final heat conductivity for the composite material with this filler shape is $\lambda_{tot}/\lambda_b = 4.63$. The filler shape is plotted in Figure 12.22. As can be seen, the filler is like a capital letter "I" [12].

Heat flux vectors in the unit cell are plotted in Figure 12.23. Two-dimensional heat conduction is observed. The effects of the filler can be seen from the figure. The filler acts as a bridge, transporting most of the heat fluxes through it. It absorbs heat at the upstream by F_3, releasing it at the downstream by F_1. The middle part of the filler, F_2, is capable of transporting this absorbed heat effectively, with little temperature gradients. The base material in the middle seldom takes part in heat transport. The filler is similar to a tree, absorbing water by the roots, supplying it via the stem and releasing the water to the leaves. The stem, though thin, is efficient in transporting most of the water. The contact area both at the upstream and at the downstream areas should be large enough to transport heat to most of the base material at upstream and downstream. The "I"-shaped filler is the best because it has the largest contact area with the base material at the two ends, like roots and branches. The filler should be extended at the two ends to minimize the overall thermal gradients in the composite material, to give a higher overall conductivity. Multi-branching is not possible at present, but may be possible in future when techniques have been developed.

12.3.4 Conductivity with various filler shapes

Filler shapes have a great effect on the conductivity of the composite materials. There have been many correlations for the estimation of the thermal conductivity of composites with fillers [13]. However,

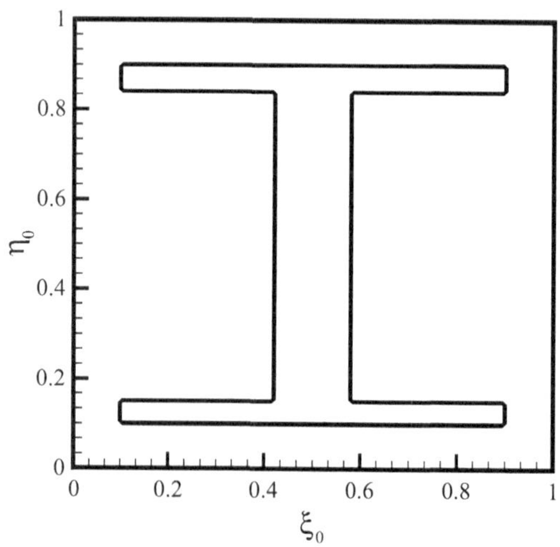

FIGURE 12.22

The optimized filler shape in the unit cell. $V_F = 0.2$, $\lambda_F/\lambda_b = 1363.0$, $\lambda_{tot}/\lambda_b = 4.63$.

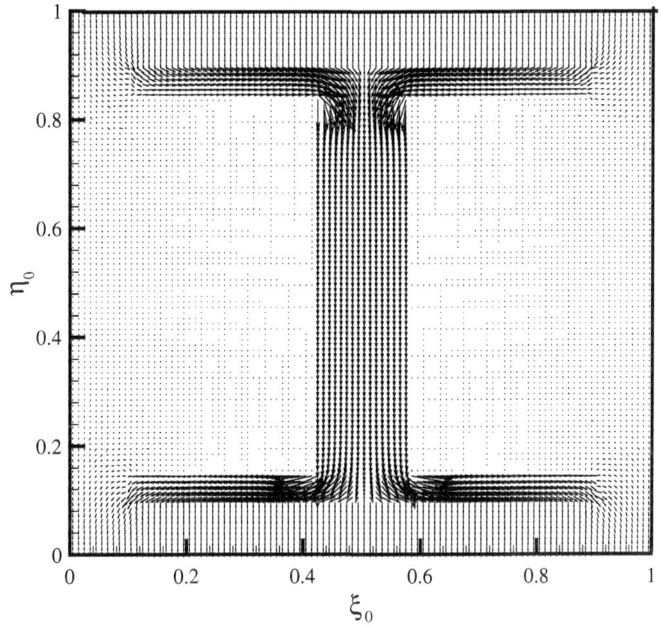

FIGURE 12.23

Heat flux vectors in the unit cell with the optimized filler shape.

they didn't consider the interactions between the base material and the fillers. To account for these effects, thermal conductivity with various filler shapes has been modeled. The filler shapes considered are listed in Table 12.4. Their orientation angles are 0°, meaning their main axis is parallel to the heat flow. The following equations are proposed for the estimation of the thermal conductivity with various fillers.

$$\lambda_{tot} = B_0\lambda_b + B_1\lambda_b V_F + B_2\lambda_F + B_3\lambda_b\psi \tag{12.45}$$

where

$$\begin{cases} \psi = \frac{\gamma}{\pi} - \frac{n-1}{2} \\ \gamma \in \left[\frac{n-1}{2}\pi, \frac{n}{2}\pi\right] \\ n \in (1, 3, 5 \ldots\ldots) \end{cases} \quad \text{or} \quad \begin{cases} \psi = \frac{n}{2} - \frac{\gamma}{\pi} \\ \gamma \in \left[\frac{n-1}{2}\pi, \frac{n}{2}\pi\right] \\ n \in (2, 4, 6 \ldots\ldots) \end{cases}$$

where γ is the orientation angle of the fillers (angle between main axis and heat flow), ψ is a function of the orientation angles. The uncertainty is within 5%.

The constants in the correlations for different filler shapes are listed in Table 12.5. The applicable ranges of filler content are $0.01 \le V_F \le 0.25$, and λ_F/λ_b less than 100. When λ_F/λ_b exceeds 100, λ_F must be replaced by 100 λ_b. The reason is that when the filler thermal conductivity is 100 times greater than that of the base material, there is no significant further improvement in the thermal conductivity of the

Table 12.4 The shapes of fillers with orientation angle 0°. The dashed lines are the main axes.

Filler name	Shapes	Parameters
Circular	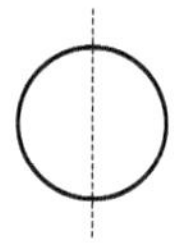	
Square	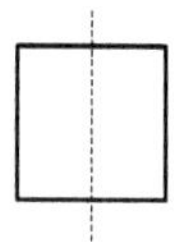	
Equilateral triangular	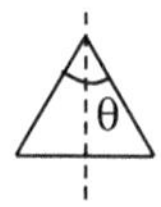	$\theta = 60°$
Rhombic	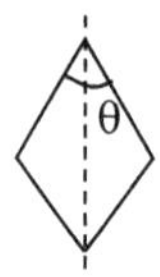	$\theta = 60°$
Elliptical	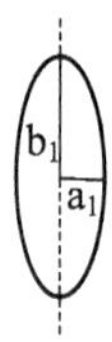	$\frac{a_1}{b_1} = \frac{1}{3}$
Rectangular	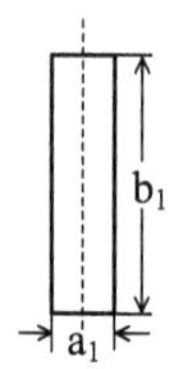	$\frac{a_1}{b_1} = \frac{1}{5}$
T-shaped	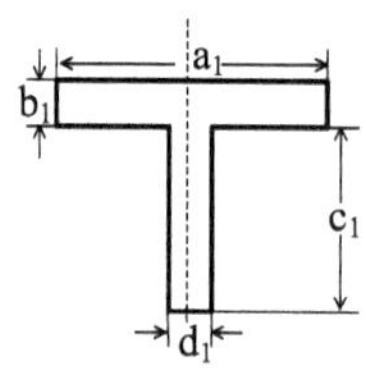	$a_1 = 5b_1 = \frac{5}{4}c_1 = 5d_1$
I-shaped	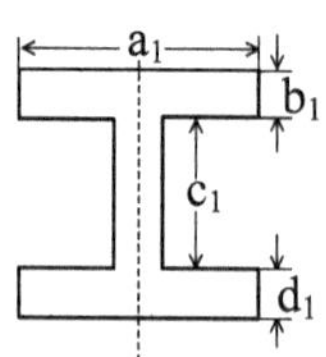	$a_1 = 5b_1 = \frac{5}{3}c_1 = 5d_1$

Table 12.5 Constants in the correlations for various filler shapes.

Shape of Filler	Constants of Correlations			
	B_0	B_1	B_2	B_3
I-shaped	0.537	6.685	0.005	−0.391
T-shaped	0.71	5.543	0.003	−0.338
Rectangular (1/5)	0.882	6.602	0.002	−1.402
Elliptical (1/3)	0.817	5.123	0.002	−0.679
Rhombic	0.874	3.944	0.001	−0.402
Triangular	0.911	2.857	0.001	0
Square	0.933	2.490	0.001	0
Circular	0.946	2.272	0.001	0

composite materials. In other words, there is little use in solely increasing the thermal conductivity of the fillers if the filler conductivity is already 100 times higher than that of the base materials. Further, for most composites, the filler content must be greater than 1% to have real improvements in conductivity. At higher filler contents (> 25%), filler aggregates would usually occur, and the correlations are no longer valid. With such high content, composite materials will become brittle and they will lose mechanical durability.

12.4 Highly selective adsorbents

Indoor air quality (IAQ) has drawn increasing attention because it can significantly affect people's health, comfort, satisfaction and productivity. Fresh air ventilation is an efficient way to improve IAQ. A ventilation rate of more than 25 L/s per person was recommended to minimize sick building syndrome (SBS) symptoms. Such a ventilation rate, if applied, will lead to a great increase in energy consumption in buildings. Therefore, heat and moisture recovery from ventilation air is necessary, for the sake of energy conservation.

Rotary wheels are one category of key equipment for heat and moisture recovery from ventilation air. In this process, the wheel coated with desiccant (such as silica gel) rotates between the outside fresh air and the inside exhaust air, and moisture can be transferred between the fresh air and the exhaust air, with cyclic moisture adsorption and desorption by the desiccants. However, it should be noted that besides moisture, volatile organic compounds (VOCs) could be adsorbed by the desiccant as well. The adsorbed VOCs would then be released to the fresh air (similar to the adsorption and desorption of moisture on the wheel). As a result, these harmful gases will enter the room with the dehumidified air, and consequently deteriorate the IAQ. As a consequence, a highly selective desiccant, which adsorbs more water vapor and less VOC, is highly desirable with the increasing expectations for the IAQ.

Although inorganic salts such as LiCl can absorb a large amount of water vapor from air, they are corrosive and deliquescent. A molecular sieve, a traditional desiccant having higher selectivity, suffers from low moisture adsorption capacity, high cost and high regeneration temperature. In comparison, silica gel and alumina are more popular because of their high water sorption capacity and affordable prices. However, little attention has been paid to their selectivity of water vapor over VOCs.

Recently, the composites "hygroscopic salt inside a porous matrix" (CSPM) have been developed for gas separation, drying and dehumidification [14]. The CSPM are composed of a kind of guest salt ($CaCl_2$, LiBr, $MgCl_2$, LiCl, etc.) which is confined to the pores in a host matrix (silica gel and mesoporous silica). The salt is connected to a sorbate molecule (water, methanol, or ethanol vapor and ammonia) to increase the sorption capacity. The benefit of these composites is that their adsorption properties can be nano-tailored by varying the chemical nature of the impregnated salt, the pore size of the host matrix, and the synthesis conditions. Recently, different hydrophilic polymers (cellulose, poly(acrylate-co-acrylamide), sulfonic styrene cation exchanger) have been applied in the preparation of selective adsorbents for water vapor by building special interactions between moisture molecules and the functional groups.

Recently, a new strategy to synthesize a selective desiccant has been adopted. By amine functionalization with a silane coupling agent, and subsequent self-assembly with malic acid on porous silica gel, a composite desiccant-like "organic salt" in a porous matrix was prepared (shown in Figure 12.24). The microstructure of the prepared desiccant was characterized by the N_2-adsorption method, and the adsorption performance for moisture and selectivity of moisture/VOCs was evaluated. The prepared desiccant can exhibit comparable adsorption capacity to silica gel but with a high selectivity, which is promising for future use in rotary wheels for heat and moisture recovery.

12.4.1 Method of material manufacture

(1) Activation of the silica gel
Activation of silica gel is necessary before amine functionalization in order to eliminate impurities and activate the silanol groups on the surface [14]. Thus, silica gel particles were immersed into 0.5 M HCl and treated for 30 min in an ultrasonic bath, then rinsed with pure water several times until neutrality and dried at 80 °C for 12 h.

(2) Amine functionalization of the silica gel surface
The activated silica gel (6 g) was added to the 100 mL triethoxyl 3-aminopropylsilane solution in toluene (4%, wt). After stirring at 80 °C for 12 h, the resultant mixture was filtered and then washed with ethanol to remove the residual organic compounds. The obtained white solid was dried to yield the amino-functionalized silica gel (NH_2-SiO_2).

(3) Self-assembly of malic acid
The above NH_2-SiO_2 was added to 100 mL aqueous solution of malic acid (4%, wt) at room temperature. After stirring for 2 h, the resultant mixture was filtrated and dried at 100 °C to yield the selective desiccant (MA-NH_2-SiO_2).

As described above, the selective desiccant was prepared via two steps. As shown in Figure 12.24, an amine group was grafted on the surface of the silica gel by hydrolyzation between the –$Si(EtO)_3$ of triethoxyl 3-aminopropylsilane and the hydroxyl group (–OH) on the silica gel surface. To keep the high hydrophilicity of the silica gel surface, malic acid (MA) containing two carboxyl groups and one hydroxyl group was chosen as cross-linker. MA was self-assembled on the silica gel surface via neutralization of the carboxyl group and the amine group on the NH_2-SiO_2 surface. Thus, an organic

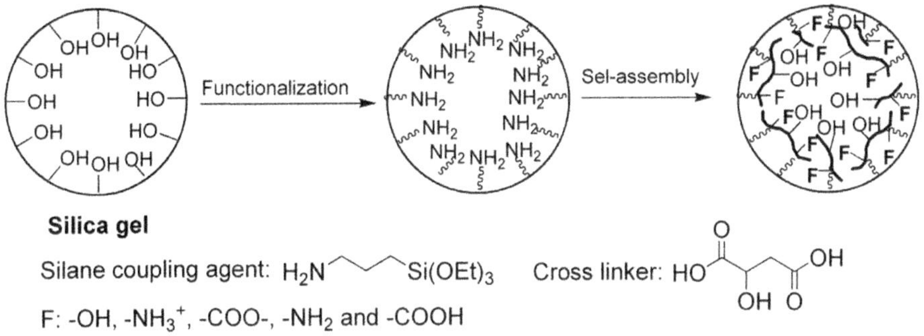

FIGURE 12.24

Preparation of selective desiccant.

network with abundant –OH, $–NH_3^+$, $–COO^-$, $–NH_2$ and –COOH was formed on the surface of the silica gel. The obtained product is porous material, similar to the composite "organic salt in the porous silica." As anticipated, there exists a stronger interaction of these hydrophilic groups in the organic network with H_2O. This strong interaction can endow the novel desiccant high selectivity for water vapor because of the preferential adsorption of H_2O over low-polar toluene and acetaldehyde.

12.4.2 Performance

Adsorption performance of MA-NH_2-SiO_2 for water vapor and VOCs was measured at 298 K in the breakthrough test rig (Figure 12.25) [14]. Toluene and acetaldehyde were used as model VOCs. The adsorption of moisture and VOCs was measured. The selectivity is defined as the ratio of adsorption capabilities of the two gases.

Considering the fact that comfortable RH is in the range from 0.4 to 0.6, the water vapor adsorption capacity of the desiccant was measured at RH from 0.7 to 0.9, under which a common desiccant wheel undertakes the moisture adsorption processes (as shown in Figure 12.26). MA-NH_2-SiO_2 and neat SiO_2 adsorb more water vapor with increasing humidity. In the measured RH range, MA-NH_2-SiO_2 has comparable adsorption capacity for water vapor to neat SiO_2 in spite of its reduced pore volume. This may be because abundant hydrophilic groups on MA-NH_2-SiO_2 surfaces offset the adverse effect of its reduced pore volume on moisture adsorption.

Aromatics and aldehydes are some of the most prevalent air pollutants. Toluene and acetaldehyde as model VOCs were thus chosen to investigate the VOC adsorption capacity of SiO_2 and MA-NH_2-SiO_2.

The toluene and acetaldehyde adsorption capacity of silica gel before and after modification was calculated. As can be observed in Figure 12.27, MA-NH_2-SiO_2 can adsorb much less toluene than neat SiO_2 at the same concentration. For example, the toluene adsorption capacity of MA-NH_2-SiO_2 is only 0.2 times that of SiO_2 at a toluene concentration of 0.02 mg/L. The toluene adsorption capacity of MA-NH_2-SiO_2 increased more slowly with increasing concentration than that of silica gel. As anticipated, the acetaldehyde adsorption capacity of MA-NH_2-SiO_2 shows a reducing trend. MA-NH_2-SiO_2 adsorbs 0.66 times as much acetaldehyde as silica gel does at an acetaldehyde concentration of 0.02 mg/L. The above results demonstrate that surface modification can reduce greatly VOC adsorption by silica gel. This is

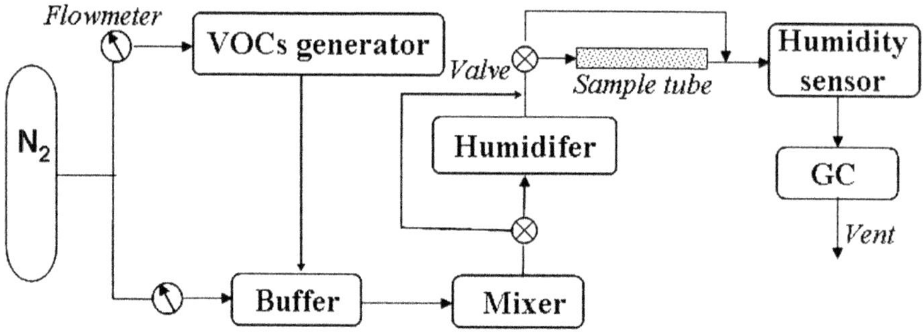

FIGURE 12.25

Adsorption performance test rig.

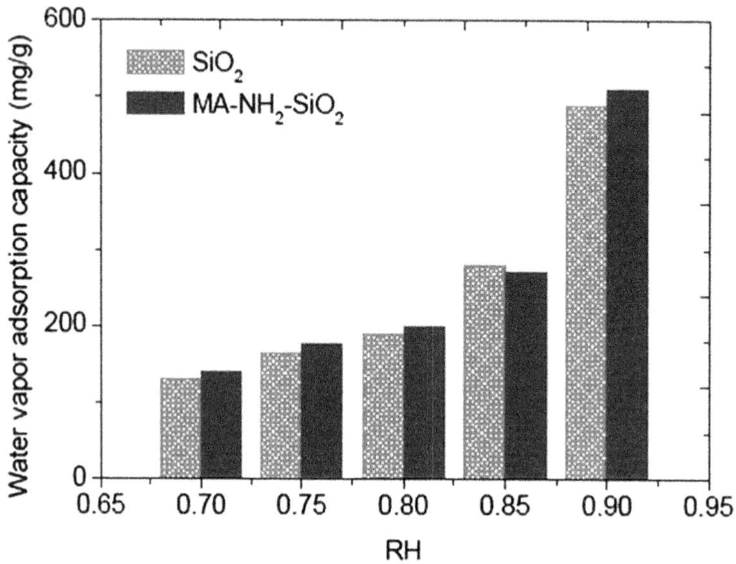

FIGURE 12.26

Moisture adsorption capacity of $MA-NH_2-SiO_2$ and SiO_2 at different *RH* values.

because the hydrophilic groups formed on the pore walls weaken the interaction of toluene/acetaldehyde with the $MA-NH_2-SiO_2$ surface.

The test data for selectivity are shown in Table 12.6. The moisture/toluene selectivity of $MA-NH_2-SiO_2$ increases significantly in comparison with silica gel. This is because H_2O is more strongly adsorbed on the $MA-NH_2-SiO_2$ surface than toluene. H_2O and toluene adsorb competitively, and as H_2O adsorption increases, fewer adsorption sites are available for the more weakly adsorbed toluene, so selectivity increases.

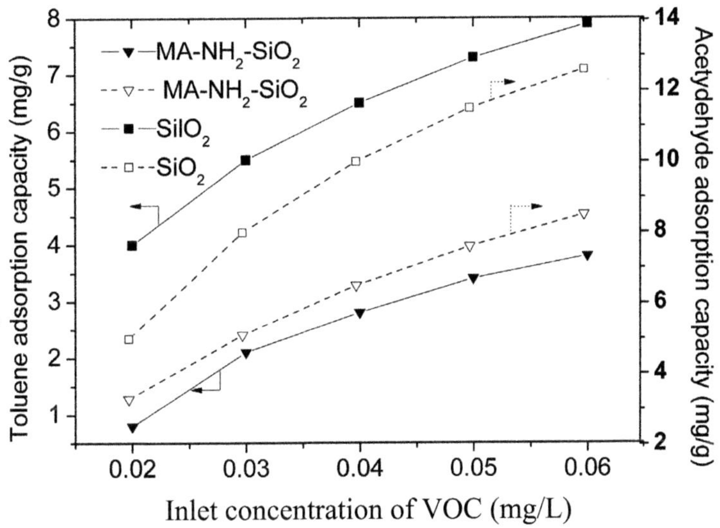

FIGURE 12.27

VOCs adsorption capacity of MA-NH$_2$-SiO$_2$ and SiO$_2$.

Table 12.6 Selectivity of moisture to VOCs at different RH.

RH	Moisture/Toluene		Moisture/Acetaldehyde	
	SiO$_2$	MA-NH$_2$-SiO$_2$	SiO$_2$	MA-NH$_2$-SiO$_2$
0.7	22.0	39.6	13.5	22.6
0.75	27.5	52.2	16.1	19.5
0.8	35.3	70.4	21.2	38.4
0.85	55.6	128.5	33.3	67.4
0.9	115.8	413.3	69.3	160.8

Under a relative humidity of 0.9, the moisture/toluene selectivity of MA- NH$_2$-SiO$_2$ is 413.3, which is 3.6 times that of silica gel. It can also be seen that the selectivity for moisture over acetaldehyde shows a similar tendency. The selectivity of MA-NH$_2$-SiO$_2$ is 2.3 times that of neat SiO$_2$ under a relative humidity of 0.9. Moreover, the selectivity of MA-NH$_2$-SiO$_2$ increases with increasing *RH* more quickly than that of silica gel. These results demonstrate that MA-NH$_2$-SiO$_2$ is a kind of adsorbent with excellent selectivity for water vapor over toluene/acetaldehyde, especially under high humidity. Also note that the selectivity for water vapor over acetaldehyde is lower than that for water vapor over toluene. Considering the fact that the polarity of the three molecules is H_2O >CH_3CHO >C_7H_8, there are stronger interactions between these hydrophilic groups and water molecules. It is safely concluded that the high selectivity of the MA-NH$_2$-SiO$_2$ arises from the preferential adsorption of H_2O over toluene and acetaldehyde on the MA-NH$_2$-SiO$_2$ surface with abundant hydrophilic groups.

12.5 Conclusion

Duct wall materials are crucial to the heat and mass transfer properties of ducts. Various novel functional materials have been developed for heat and mass exchangers. Four novel membranes have been developed: the composite hydrophobic-hydrophilic membrane, the composite supported liquid membrane, the asymmetric membrane with finger-like macrovoids, and the one-step-fabricated asymmetric membrane. All these membranes are highly porous. Some are for total heat exchangers, others are for membrane contactors. Engineering applications demand cheap, permselective, highly vapor-permeable membranes. Moisture diffusivity and heat conductivity have been evaluated. For sensible heat transfer, membrane resistance is very small and it can be ignored. However, for moisture transfer, membrane resistance is the major part of total resistance. Therefore it cannot be ignored. Novel membranes should be developed to reduce this part of the resistance. It has a determining effect on the latent effectiveness of a total heat exchanger.

To solve the problem of plastic heat exchangers, composite polymers have been developed. A reverse problem technique is used to optimize the filler shapes. The results are that the best fillers are in an "I" shape. Correlations are proposed for composite materials with various filler shapes. Finally, a desiccant material of high selectivity has been developed. It can be used in adsorbent beds or desiccant wheels. These kinds of heat and mass exchangers require adsorbents of high moisture adsorption capabilities, but lower VOC adsorption capabilities.

References

[1] Zhang LZ. Total heat recovery: heat and moisture recovery from ventilation air. New York: Nova Science; 2009.

[2] Zhang LZ, Wang YY, Wang CL, Xiang H. Synthesis and characterization of a PVA/LiCl blend membrane for air dehumidification. J Membr Sci 2008;308:198–206.

[3] Zhang LZ. Numerical study of heat mass transfer in an enthalpy exchanger with a hydrophobic-hydrophilic composite membrane core. Numer Heat Transfer A: Appl 2007;51:697–714.

[4] Zhang LZ. Fabrication of a lithium chloride solution based composite supported liquid membrane and its moisture permeation analysis. J Membr Sci 2006;276:91–100.

[5] Zhang LZ. Heat and mass transfer in a total heat exchanger: cross-corrugated triangular ducts with composite supported liquid membrane. Numer Heat Transfer A: Appl 2008;53:1195–210.

[6] Zhang LZ, Xiao F. Simultaneous heat and moisture transfer through a composite supported liquid membrane. Int J Heat Mass Transfer 2008;51:2179–89.

[7] Zhang LZ. Effects of membrane parameters on performance of vapor permeation through a composite supported liquid membrane. Sep Sci Technol 2006;41:3517–38.

[8] Zhang LZ. Coupled heat and mass transfer through asymmetric porous membranes with finger-like macrovoids structure. Int J Heat Mass Transfer 2009;52:751–9.

[9] Zhang XR, Zhang LZ, Liu HM, Pei LX. One-step fabrication and analysis of an asymmetric cellulose acetate membrane for heat and moisture recovery. J Membr Sci 2011;366:158–65.

[10] Cussler EL. Diffusion mass transfer in fluid systems. 2nd ed. Cambridge: Cambridge University Press; 2000.

[11] Zhang LZ, Zhang XR, Miao QZ, Pei LX. Selective permeation of moisture and VOCs through polymer membranes used in total heat exchangers for indoor air ventilation. Indoor air 2012;22:321–30.

[12] Zhang LZ, Wang XJ, Pei LX. Nonlinear programming optimization of filler shapes for composite materials with inverse problem technique to maximize heat conductivity. Int J Heat Mass Transfer 2012;55:7287–96.

[13] Wang JF, Carson JK, North MF. A new structural model of effective thermal conductivity for heterogeneous materials with co-continuous phases. Int J Heat Mass Transfer 2008;51:2389–97.

[14] Pei LX, Lv ZM, Zhang LZ. Selective adsorption of a novel high selective desiccant for prospective use in heat and moisture recovery for buildings. Build Environ 2012;49:124–8.

Research Papers from Our Group on Conjugate Heat and Mass Transfer

A.1 SCI papers

[1] Zhang LZ, Wang XJ, Quan YY, Pei LX. Conjugate heat conduction in filled composite materials considering interactions between the filler and base materials. Int J Heat Mass Transfer, in press.
[2] Zhang LZ, Huang SM, Zhang WB, Pei LX. Turbulent heat and mass transfer across a hollow fiber membrane bundle considering interactions between neighboring fibers. Int J Heat Mass Transfer, in press.
[3] Huang SM, Zhang LZ, Pei LX. Transport phenomena in a cross-flow hollow fiber membrane bundle used for liquid desiccant air dehumidification. Indoor Built Environ, in press.
[4] Chen ZY, Zhang LZ, Song H. Investigating the impacts of included angles on flow and heat transfer in cross-corrugated triangular ducts with field synergy principle. Therm Sci, in press.
[5] Zhang LZ, Li ZX, Zhong TS, Pei LX. Flow maldistribution and performance deteriorations in a cross flow hollow fiber membrane module for air humidification. J Membr Sci 2013;427(C):1–9.
[6] Zhang LZ, SM Huang, L.X. Pei. Conjugate heat and mass transfer in a cross-flow hollow fiber membrane contactor for liquid desiccant air dehumidification. Int J Heat Mass Transfer 2012;55(25-26):8065–76.
[7] Zhang LZ. Progress on heat and moisture recovery with membranes: from fundamentals to engineering applications. Energ Convers Manage 2012;63(C):173–95.
[8] Zhang LZ, Wang XJ, Pei LX. Nonlinear programming optimization of filler shapes for composite materials with inverse problem technique to maximize heat conductivity. Int J Heat Mass Transfer 2012;55(23-24):7287–96.
[9] Zhang LZ. Coupled heat and mass transfer in an application-scale cross-flow hollow fiber membrane module for air humidification. Int J Heat Mass Transfer 2012;55(21-22):5861–9.
[10] Zhang LZ, Zhang XR, Miao QZ, Pei LX. Selective permeation of moisture and VOCs through polymer membranes used in total heat exchangers for indoor air ventilation. Indoor Air 2012;22(4):321–30.
[11] Huang SM, Zhang LZ, Tang K, Pei LX. Turbulent heat and mass transfer across a hollow fiber membrane tube bank in liquid desiccant air dehumidification. ASME J Heat Transfer 2012;134(8):082001-1-10.

[12] Zhang LZ, Huang SM, Chi JH, Pei LX. Conjugate heat and mass transfer in a hollow fiber membrane module for liquid desiccant air dehumidification: a free surface model approach. Int J Heat Mass Transfer 2012;55(13-14):3789–99.
[13] Pei LX, Zhang LZ. Preparation and selective adsorption of core-shell desiccant for heat and moisture recovery. Colloids Surf A: Physicochem Eng Asp 2012;406C:68–74.
[14] Huang SM, Zhang LZ, Tang K, Pei LX. Fluid flow and heat mass transfer in membrane parallel-plates channels used for liquid desiccant air dehumidification. Int J Heat Mass Transfer 2012;55(3):2571–80.
[15] Pei LX, Lv ZM, Zhang LZ. Selective adsorption of a novel high selective desiccant for prospective use in heat and moisture recovery for buildings. Build Environ 2012;49(1):124–8.
[16] Zhang LZ, Zhong WC, Chen JM, Zhou JR. Fluid flow and heat transfer in plate-fin and tube heat exchangers in transitional flow regime. Numer Heat Transfer A: Appl 2011;60(9):766–84.
[17] Zhang LZ. An analytical solution to heat and mass transfer in hollow fiber membrane contactors for liquid desiccant air dehumidification. ASME J Heat Transfer 2011;133(9):092001-1-8.
[18] Zhang LZ. Heat and mass transfer in a randomly packed hollow fiber membrane module: a fractal model approach. Int J Heat Mass Transfer 2011;54(13-14):2921–31.
[19] Zhang LZ, Huang SM. Coupled heat and mass transfer in a counter flow hollow fiber membrane module for air humidification. Int J Heat Mass Transfer 2011;54(5-6):1055–63.
[20] Zhang XR, Zhang LZ, Liu HM, Pei LX. One-step fabrication and analysis of an asymmetric cellulose acetate membrane for heat and moisture recovery. J Membr Sci 2011;366(1-2):158–65.
[21] Zhang LZ, Chen ZY. Convective heat transfer in cross-corrugated triangular ducts under uniform heat flux boundary conditions. Int J Heat Mass Transfer 2011;54(1-3):597–605.
[22] Zhang LZ. Heat and mass transfer in a quasi-counter flow membrane-based total heat exchanger. Int J Heat and Mass Transfer 2010;53(23-24):5478–86.
[23] Pei LX, Zhao WJ, Zhang LZ. Preparation and characterization of porous PVDF membranes for dehumidification with PEG as additive. J Appl Polym Sci 2010;118(5):2696–703.
[24] Liang CH, Zhang LZ, Pei LX. Performance analysis of a direct expansion air dehumidification system combined with membrane-based total heat recovery. Energy 2010;35(9):3891–901.
[25] Zhang LZ. An analytical solution for heat mass transfer in a hollow fiber membrane based air-to-air heat mass exchanger. J Membr Sci 2010;360(1-2):217–25.
[26] Liang CH, Zhang LZ, Pei LX. Independent air dehumidification with membrane-based total heat recovery: modeling and experimental validation. Int J Refrig 2010;33(2):398–408.
[27] Zhang LZ, Liang Cai-hang, Pei Li-Xia. Conjugate heat and mass transfer in membrane-formed channels in all entry regions. Int J Heat Mass Transfer 2010;53(5-6):815–24.
[28] Zhang LZ. Performance deteriorations from flow maldistribution in air-to-air heat exchangers: a parallel-plates membrane core case. Numer Heat Transfer A: Appl 2009;56(9):746–63.
[29] Zhang LZ. Flow maldistribution and performance deteriorations in a membrane-based heat and mass exchanger. ASME J Heat Transfer 2009;131(11):111801-1-7.
[30] Zhang LZ. Flow maldistribution and thermal performance deterioration in a cross-flow air to air heat exchanger with plate-fin cores. Int J Heat Mass Transfer 2009;52(19-20):4500–9.
[31] Zhang LZ. Heat and mass transfer in plate-fin enthalpy exchangers with different plate and fin materials. Int J Heat Mass Transfer 2009;52(11-12):2704–13.
[32] Zhang LZ. Coupled heat and mass transfer through asymmetric porous membranes with finger-like macrovoids structure. Int J Heat Mass Transfer 2009;52(3-4):751–9.

[33] Zhang LZ, Liang CH, Pei LX. Heat and moisture transfer in application-scale parallel-plates enthalpy exchangers with novel membrane materials. J Membr Sci 2008;325(12):672–82.
[34] Zhang LZ. A fractal model for gas permeation through porous membranes. Int J Heat Mass Transfer 2008;51(21-22):5288–95.
[35] Zhang LZ, Xiao F. Simultaneous heat and moisture transfer through a composite supported liquid membrane. Int J Heat Mass Transfer 2008;51(9-10):2179–89.
[36] Zhang LZ. Heat and mass transfer in a total heat exchanger: cross-corrugated triangular ducts with composite supported liquid membrane. Numer Heat Transfer A: Appl 2008;53(11):1195–210.
[37] Zhang LZ. Heat and mass transfer in plate-fin sinusoidal passages with vapor-permeable wall materials. Int J Heat Mass Transfer 2008;51(3-4):618–29.
[38] Zhang LZ, Wang YY, Wang CL, Xiang H. Synthesis and characterization of a PVA/LiCl blend membrane for air dehumidification. J Membr Sci 2008;308(1-2):198–206.
[39] Zhang LZ. Thermally developing forced convection and heat transfer in rectangular plate-fin passages under uniform plate temperature. Numer Heat Transfer A: Appl 2007;52(6):549–64.
[40] Zhang LZ. Numerical study of heat mass transfer in an enthalpy exchanger with a hydrophobic-hydrophilic composite membrane core. Numer Heat Transfer A: Appl 2007;51(7):697–714.
[41] Zhang LZ. Laminar flow and heat transfer in plate-fin triangular ducts in thermally developing entry region. Int J Heat Mass Transfer 2007;50(7-8):1637–40.
[42] Zhang LZ. Heat and mass transfer in a cross flow membrane-based enthalpy exchanger under naturally formed boundary conditions. Int J Heat Mass Transfer 2007;50(1-2):151–62.
[43] Zhang LZ. Effects of membrane parameters on performance of vapor permeation through a composite supported liquid membrane. Separ Sci Technol 2006;41(16):3517–38.
[44] Zhang LZ. Mass diffusion in a hydrophobic membrane humidification/dehumidification process: the effects of membrane characteristics. Separ Sci Technol 2006;41(8):1565–82.
[45] Li F, Niu JL, Zhang LZ. A physically-based model for prediction of VOCs emissions from paint applied to an absorptive substrate. Build Environ 2006;41(10):1317–25.
[46] Zhang LZ. Fabrication of a lithium chloride solution based composite supported liquid membrane and its moisture permeation analysis. J Membr Sci 2006;276(1-2):91–100.
[47] Zhang LZ. Investigation of moisture transfer effectiveness through a hydrophilic polymer membrane with a field and laboratory emission cell. Int J Heat Mass Transfer 2006;49(5-6):1176–84.
[48] Zhang LZ. Energy performance of independent air dehumidification systems with energy recovery measures. Energy 2006;31(8-9):1228–42.
[49] Zhang LZ. Evaluation of moisture diffusivity in hydrophilic polymer membranes: a new approach. J Membr Sci 2006;269(1-2):75–83.
[50] Zhang LZ. Turbulent three-dimensional air flow and heat transfer in a cross-corrugated triangular duct. ASME J Heat Transfer 2005;127(10):1151–8.
[51] Zhang LZ. Convective mass transport in cross-corrugated membrane exchangers. J Membr Sci 2005;260(1-2):75–83.
[52] Zhang LZ. Numerical study of periodically fully developed flow and heat transfer in cross-corrugated triangular channels in transitional flow regime. Numer Heat Transfer A: Appl 2005;48(4):387–405.
[53] Zhang LZ, Zhu DS, Deng XH, Hua B. Thermodynamic modeling of a novel air dehumidification system. Energy Build 2005;37(3):279–86.

[54] Zhang LZ, Niu JL. Modeling VOCs emissions in a room with a single-zone multi-component multi-layer technique. Build Environ 2004;39(5):523–31.
[55] Zhang LZ, Niu JL. Effects of substrate parameters on the emissions of volatile organic compounds from wet coating materials. Build Environ 2003;38(7):939–46.
[56] Zhang LZ, Niu JL. Mass transfer of volatile organic compounds from painting material in a standard field and laboratory emission cell (FLEC). Int J Heat Mass Transfer 2003;46(13):2415–23.
[57] Zhang LZ, Niu JL. A pre-cooling munters environmental control cooling cycle in combination with chilled-ceiling panels. Energy 2003;28(3):275–92.
[58] Zhang LZ, Niu JL. Laminar fluid flow and mass transfer in a standard field and laboratory emission cell (FLEC). Int J Heat Mass Transfer 2003;46(1):91–100.
[59] Zhang LZ, Niu JL. Indoor humidity behaviors associated with decoupled cooling in hot and humid climates. Build Environ 2003;38(1):99–107.
[60] Zhang LZ, Niu JL. Effectiveness correlations for heat and moisture transfer processes in an enthalpy exchanger with membrane cores. ASME J Heat Transfer 2002;122(5):922–9.
[61] Zhang LZ, Niu JL. Performance comparisons of desiccant wheels for air dehumidification and enthalpy recovery. Appl Thermal Eng 2002;22(12):1347–67.
[62] Niu JL, Zhang LZ. Effects of wall thickness on the heat and moisture transfers in desiccant wheels for air dehumidification and enthalpy recovery. Int Commun Heat Mass Transfer 2002;29(2):255–68.
[63] Niu JL, Zhang LZ, Zuo HG. Energy savings potential of chilled-ceiling combined with desiccant cooling in hot and humid climates. Energy Buildings 2002;34(5):487–95.
[64] Niu JL, Zhang LZ. Heat transfer and friction coefficients in corrugated ducts confined by sinusoidal and arc curves. Int J Heat Mass Transfer 2002;45(3):571–8.
[65] Zhang LZ, Niu JL. A numerical study of laminar forced convection in sinusoidal ducts with arc lower boundaries under uniform wall temperature. Numer Heat Transfer A: Appl 2001;40(1):55–72.
[66] Niu JL, Zhang LZ. Membrane-based enthalpy exchanger: material considerations and clarification of moisture resistance. J Membr Sci 2001;189(2):179–91.
[67] Zhang LZ, Niu JL. Energy requirements for conditioning fresh air and the long-term savings with a membrane-based energy recovery ventilator in Hong Kong. Energy 2001;26(2):119–35.
[68] Zhang LZ. A three-dimensional non-equilibrium model for an intermittent adsorption cooling system. Solar Energy 2000;69(1):27–35.
[69] Zhang LZ, Jiang Y, Zhang YP. Membrane-based humidity pump: performance and limitations. J Membr Sci 2000;171(2):207–16.
[70] Zhang YP, Jiang Y, Zhang LZ, et al. Analysis of thermal performance and energy savings of membrane based heat recovery ventilator. Energy 2000;25(6):515–27.
[71] Zhang LZ. Design and testing of an automobile waste heat adsorption cooling system. Appl Therm Eng 2000;20(1):103–14.
[72] Zhang LZ, Wang L. Momentum and heat transfer in the adsorbent of a waste heat adsorption cooling system. Energy 1999;24(7):605–24.
[73] Zhang LZ, Jiang Y. Heat and mass transfer in a membrane-based Enthalpy Recovery Ventilator. J Membr Sci 1999;163(1):29–38.
[74] Zhang LZ, Wang L. Effects of coupled heat and mass transfers in adsorbent on the performance of a waste heat adsorption cooling unit. Appl Therm Eng 1999;19(2):195–215.
[75] Zhang LZ, Wang L. Performance estimation of an adsorption cooling system for automobile waste heat recovery. Appl Therm Eng 1997;17(12):1127–39.

A.2 EI papers

[1] Huang Si-Min, Zhang Li-Zhi. Advanced indoor humidity control: new impacts on conjugated heat and mass transfer. Comput Therm Sci 2012;4(6):539–47.

[2] Zhang Li-Zhi. Performance analysis of new air conditioning systems with novel total heat (energy) exchangers. In: sixth international conference on thermal engineering theory and applications (ICTEA) May 29–June 1, 2012, Istanbul, Turkey.

[3] Zhang Li-Zhi. Enhanced heat transfer and energy conservation: A review of the selected papers from ISHTEC2012. Int J Low Carbon Technol 2012;7(4):255–6.

[4] Zhang Xin-Ru, Zhang Li-Zhi, Pei Li-Xia. Sorption, permeation and selective transport of moisture/VOCs through a CA membrane for total heat recovery. Int J Low Carbon Technol 2012;7(2):122–6.

[5] Zhang Li-Zhi. A large eddy simulation (LES) of plate-fin and tube heat exchangers. In: 2010 14th international heat transfer conference, IHTC 14, August 8–13, 2010, Washington, DC, United States, 4:141–7.

[6] Pei Li-Xia, Zhang Li-Zhi. Improvement of hydrophilicity of porous PVDF membranes with LiCl additives. J Southeast Univ (English edition), 2010;26(2):201–4.

[7] Niu JL, Zhang LZ, Zuo HG. Analysis of energy and humidity performance of a system combining chilled-ceiling with desiccant cooling. ASHRAE Trans 2002;108(2):195–201.

[8] Niu JL, Zhang LZ. Potential energy savings for conditioning fresh air with a membrane-based energy recovery ventilator. ASHRAE Trans 2002;108(1):54–63.

[9] Niu JL, Zhang LZ, Zuo HG. Performance of a novel HVAC system in hot and humid weather conditions: chilled-ceiling combined with desiccant cooling. The Hong Kong Institution of Engineers (HKIE) Trans 2002;9(2):25–9.

[10] Niu JL, Zhang LZ. Membrane-based energy recovery ventilators: a solution to heat recovery for ventilation air in Hong Kong. The Hong Kong Institution of Engineers (HKIE) Trans 2001;8(2):58–63.

A.3 Books and proceedings

[1] Zhang Li-Zhi. Air dehumidification technologies. Beijing: China Chemical Industry Press; 2005.

[2] Zhang Yinping, Zhang Lizhi, Liu Xiaohua, Mo Jinhan. Mass transfer in built environment. Beijing: China Building and Construction Press; 2006.

[3] Zhang Li-Zhi. Total heat recovery: heat and moisture recovery from ventilation air. New York: Nova Science; 2009.

[4] Zhang Lizhi, Wang Shuangfeng, Fan Shuanshi, Zhang Zhengguo, Gao Xuenong. Advances in heat transfer and energy conservation. In: Proceedings of ISHTEC2012, Fourth international symposium on heat transfer and energy conservation, January 6–9, 2012, Guangzhou, China.

A.4 International keynote or invited speeches

[1] Zhang Li-Zhi. Research activities in South China University of Technology in indoor humidity control. Invited speech presented at Dept. of Engineering. University of Federal Fluminense, November 23, 2012, Rio de Janeiro, Brazil.

[2] Zhang Li-Zhi. Some key issues in active indoor humidity control technologies. ENCIT2012, 14th Brazilian Congress of Thermal Science and Engineering, November 18–22, 2012, Rio de Janeiro, Brazil.

[3] Zhang Li-Zhi. Critical issues in membrane-based liquid desiccant air dehumidification. In: SET2012, 11th symposium on sustainable energy technologies, September 2–5, 2012, Vancouver, Canada (invited talk).

[4] Zhang Li-Zhi. Advanced indoor humidity control: new impacts on conjugated heat and mass transfer. CHT-12, 12th symposium on advances in computational heat transfer, July 1–6, 2012, Bath, England.

[5] Zhang Li-Zhi. Performance analysis of new air conditioning systems with novel total heat (energy) exchangers. ICTEA2012, Sixth International Conference on Thermal Engineering: Theory and Application, May 29–June 1, 2012, Istanbul, Turkey.

[6] Zhang Li-Zhi. Heat and mass transfer problems in new moisture control technologies. In: Plenary keynote speech presented at ISHTEC2012, Fourth international symposium on heat transfer and energy conservation, January 6–9, 2012, Guangzhou, China.

[7] Zhang Li-Zhi. Progress on heat and moisture recovery from ventilation air: from fundamentals to engineering applications. In: Plenary keynote speech presented at VIII minsk international seminar heat pipes, Heat Pumps, Refrigerators, Power Source, September 12–15, 2011, Minsk, Belarus.

[8] Zhang Li-Zhi. Heat and moisture recovery with membranes: from fundamentals to engineering applications. In: Plenary keynote speech presented at SET2011, 10th symposium on sustainable energy technologies, September 4–7, 2011, Istanbul, Turkey.

[9] Zhang Li-Zhi. New moisture control technologies in subtropical regions. In: Sino-German forum on building energy conservation in South China, October 13, 2008, Guangzhou, China.

Glossary

Nomenclature

a	half channel height or contactor shell width (m)
A	area (m^2)
a	thermal diffusivity (m^2/s)
A_c	cross-section area (m^2)
A_{flow}	flow area (m^2)
A_t	transfer area (m^2)
A_v	specific area (m^2/m^3)
b	half channel width or contactor shell height (m)
Bi	Biot number
C	concentration (kg/m^3); constants; drag coefficient
COP	coefficient of performance
c_p	specific heat (kJ kg^{-1} K^{-1})
c_s	specific heat of solution (kJ kg^{-1} K^{-1})
C_s	shape factor for the adsorption isotherm
C_{w0}	constant (kg/m^3)
D	diffusivity (m^2/s)
d	diameter (m)
d_f	diameter of finger (m)
D_f	fractal dimension
D_h	hydrodynamic diameter (m)
d_p	pore diameter (m)
d_v	void diameter (m)
D_{va}	vapor diffusivity in air (m^2/s)
D_{vs}	equivalent vapor diffusivity in solid (m^2/s)
D_{wf}	water diffusivity in fin material (m^2/s)
D_0	module diameter (m)
E_T	slope of humidity to temperature (K^{-1})
f	friction factor
f	oscillation frequency (Hz)
$f(d_v)$	probability density function

F_p	fin pitch (m)
h	convective heat transfer coefficient (kW m^{-2} K^{-1}), enthalpy (kJ/kg)
H	specific enthalpy (kJ/kg)
H^*	ratio of latent to sensible energy differences
h_{abs}	absorption heat (kJ/kg)
h_v	evaporation heat of water (kJ/kg)
H_z	spacing (m)
H_0	height of unit cell (m)
J	Jacobian transformation operator; permeation rate (kg m^{-2} s^{-1})
k	convective mass transfer coefficient (m/s)
k	turbulent kinetic energy (m^2/s^2)
k_B	Boltzmann constant, 1.38×10^{-23} J/K
k_m	internal mass transfer coefficient of adsorbents (1/s)
k'_m	equivalent internal mass transfer coefficient of adsorbents (1/s)
Kn	Knudsen number
K_p	Henry coefficient (kg m^{-3} Pa^{-1}); partition coefficient [(kg water/kg material)/(kg vapor/kg air)]
L	length (m)
Le	Lewis number
l_f	length of finger (m)
L_f	length of fin (m)
L_H	overall height of the duct including wall (m)
L_W	overall width of the duct (m)
L_0	length of unit cell (m)
m	mass flow rate (kg/s); flux
M	molecule weight (g/mol); number of categories, measure
m_d	mass of the wheel (kg)
m_L^*	dimensionless latent heat ratio
m_s^*	dimensionless sensible heat capacity ratio
$\dot{m}$	mass flux (kg m^{-2} s^{-1})
n	number
N	number; rotary speed (rpm); row of fibers
n_{ch}	number of channels
n_f	number of fibers
NTU	number of transfer units
Nu	Nusselt number
p	partial pressure (Pa)
P_0	static pressure (Pa)
P	total pressure (Pa)
Pe	Peclet number; permeability (kg m^{-1} $s^{-1}$$(kg/kg)^{-1}$)
Pe'	permeability in (g cm^{-1} $cmHg^{-1}$)
P_f	perimeter (m)
P_L	longitudinal pitch (m)

P_m	mean pressure (atm); total pressure within the pores (atm)
Pr	Prandtl number
P_T	transverse pitch (m)
q	heat flux (kW/m^2) or mass flux (kg m^{-2} s^{-1})
Q	total heat transfer (kW)
q_{st}	adsorption heat (kJ/kg)
R	gas constant, 8.314 J/(mol K); ratio for heat/mass capacity
r	resistance (m^{2}K/kW for thermal and s/m for moisture); radius (m)
Re	Reynolds number
RH	air relative humidity
R_o	shell radius (m)
R_1	inlet temperature to humidity ratio
S	source term, strain rate (1/s)
s	tangent coordinate for fin (m)
Sc	Schmidt number
SDP	specific dehumidification power (g kg^{-1} s^{-1})
Sh	Sherwood number
Sh_{ω}	fully developed Sherwood number under uniform concentration condition
St	Stanton number
St_r	Strouhal number
T	temperature (K)
t	time (s)
u	velocity (m/s), instantaneous, streamwise
U	total heat or mass transfer coefficient (kW m^{-2} K^{-1} for heat and m/s for mass); velocity vectors matrix; dimensionless velocity coefficient
u_a	air velocity (m/s)
u_i	face velocity (m/s)
v	spanwise velocity (m/s)
V	volumetric flow rate (m^3/s); volume (m^3)
V_F	volumetric ratio of filler
w	normal velocity (m/s); water uptake in solid (kg water/kg dry solid)
W_i	fractional flow
W_{max}	maximum water uptake of solid (kg/kg)
x	axial or transverse coordinate (m)
X	mass fraction of solute
$\boldsymbol{X}$	vector
x, y, z	coordinates (m)
x', y'	coordinates (m)
x_1, y_1	tangent and normal coordinate to fin (m)
x_F	core or channel length (m)
x_j	*j*th dimensionless coordinates
X_s	mass fraction of solution (kg water/kg solution)
y_F	core or channel width (m)

z_1	inlet coordinate of duct, z for air and (L-z) for solution
z_F	core depth (m)
z_L	length of duct (m)

Greek letters

α	angle (rad); skew angle
α, β, γ	coefficients in coordinate transformation
α_f	aspect ratio
α_p	permeability in Darcy's Law (m_2)
α_r	ratio of diffusive to convective resistance for membrane
α_0	angle fraction for process stream; half apex angle
β	angle; flow nonuniformity; selectivity
β_f	ratio of wall thickness to duct width
β_r	ratio of total number of transfer units for moisture to that for sensible heat
Γ	diffusion coefficient
Γ_ϕ	generalized diffusivity
γ	orientation angle of fillers
δ	thickness (m)
ε	effectiveness; efficiency dissipation rate (m^2/s^3)
ε_t	porosity
η	efficiency; longitudinal coordinate on computational plane
η_{fin}	fin efficiency
θ	apex angle; dimensionless temperature; inlet header angle, or half apex angle for plate-fin duct (degree)
Θ	dimensionless mass fraction of solution
λ	mean free path (m); thermal conductivity (kW m^{-1} K^{-1})
μ	dynamic viscosity (kg m^{-1} s^{-1}) or (Pas)
ν	kinematic viscosity (m^2/s)
ξ	dimensionless humidity; transverse coordinate on computational plane; variable
ρ	density (kg/m^3)
σ	coefficient; molecule diameter (Å); pitch to diameter ratio
σ_T	slope of adsorption to temperature (K^{-1})
σ_ω	slope of adsorption to humidity
τ	deterioration factor; dimensionless time; membrane to air conductivity ratio
τ_s	tortuosity of pores
τ_1	shear stress (Pa)
ϕ	general variable, coordinates transformation operator; packing fraction
χ	weight ratio
ψ	correction factors; function of orientation angles; variable
ω	humidity ratio or moisture content (kg moisture/kg dry air)
Ω	dimensionless humidity; fin conductance parameter; function of collision

Subscripts

a	air; air side
ad	adsorption process
ave	average
b	base material; bulk
c	cool; cooling; convective; cross section
C	conjugate
cyc	cycle
d	desiccant, dehumidification
de	desorption process
e	effective, equilibrium; quivalent; exhaust air
en	enthalpy
eq	equilibrium
f	fiber; fin; finger; fluid; free surface; fresh air; friction
F	filler
fin	fin
G	geometric
gel	silica gel
h	hot; heating, humidification; hydrodynamic, heat transfer
H	uniform heat flux (for heat) or uniform mass flux (for mass) boundary conditions
i	inlet; inner; *i*th direction; node index
in	inlet
j	node index
k	node index
K	Knudsen
L	local; lower surface; latent heat
Lat	latent, moisture
lq	liquid
m	mass, mean, membrane
max	maximum
min	minimum
n	normal directions
o	outlet; outer
O	ordinary
out	outlet
p	plate
s	surface; sensible heat; solid; solution; subgrid
t	total; transfer; turbulence
T	under uniform temperature (concentration) conditions
tot	total
tube	tube
u	upper

v	vapor
w	wall, water
z	thickness
0	ambient; initial conditions; orderly array, static
1	fresh air side; membrane surface on air side
2	exhaust air side; membrane surface on solution side

Superscripts

*	dimensionless
′	exhaust air duct; fluctuation; solution duct

Index

I

J

K

L

M

N